Lecture Notes in Networks and Systems 790

The series "Lecture Notes in Networks and Systems" publishes the latest developments in Networks and Systems—quickly, informally and with high quality. Original research reported in proceedings and post-proceedings represents the core of LNNS.

Volumes published in LNNS embrace all aspects and subfields of, as well as new challenges in, Networks and Systems.

The series contains proceedings and edited volumes in systems and networks, spanning the areas of Cyber-Physical Systems, Autonomous Systems, Sensor Networks, Control Systems, Energy Systems, Automotive Systems, Biological Systems, Vehicular Networking and Connected Vehicles, Aerospace Systems, Automation, Manufacturing, Smart Grids, Nonlinear Systems, Power Systems, Robotics, Social Systems, Economic Systems and other. Of particular value to both the contributors and the readership are the short publication timeframe and the world-wide distribution and exposure which enable both a wide and rapid dissemination of research output.

The series covers the theory, applications, and perspectives on the state of the art and future developments relevant to systems and networks, decision making, control, complex processes and related areas, as embedded in the fields of interdisciplinary and applied sciences, engineering, computer science, physics, economics, social, and life sciences, as well as the paradigms and methodologies behind them.

Indexed by SCOPUS, INSPEC, WTI Frankfurt eG, zbMATH, SCImago.

All books published in the series are submitted for consideration in Web of Science.

For proposals from Asia please contact Aninda Bose (aninda.bose@springer.com).

Anna Burduk · Andre Batako · José Machado ·
Ryszard Wyczółkowski · Katarzyna Antosz ·
Arkadiusz Gola
Editors

Advances in Production

Intelligent Systems in Production Engineering and Maintenance

Springer

Editors
Anna Burduk
Wroclaw University of Science and Technology
Wrocław, Poland

José Machado
School of Engineering
University of Minho
Guimarães, Portugal

Katarzyna Antosz
Rzeszow University of Technology
Rzeszów, Poland

Andre Batako
Liverpool John Moores University
Liverpool, UK

Ryszard Wyczółkowski
Katedra Budowy Maszyn
Politechnika Śląska
Gliwice, Poland

Arkadiusz Gola
Lublin University of Technology
Lublin, Poland

ISSN 2367-3370 ISSN 2367-3389 (electronic)
Lecture Notes in Networks and Systems
ISBN 978-3-031-45020-4 ISBN 978-3-031-45021-1 (eBook)
https://doi.org/10.1007/978-3-031-45021-1

This Springer imprint is published by the registered company Springer Nature Switzerland AG
The registered company address is: Gewerbestrasse 11, 6330 Cham, Switzerland

Paper in this product is recyclable.

Preface

The Fourth International Conference on Intelligent Systems in Production Engineering and Maintenance ISPEM 2023 was organized by the Wrocław University of Science and Technology (Poland), Liverpool John Moores University (England) and the University of Minho (Portugal). The high prestige of the conference is confirmed by the honorary patronage of the rector of the Wroclaw University of Science and Technology, a pro-vice chancellor and the dean for Engineering and Technology at Liverpool John Moores University and Dean of School of Engineering University of Minho. The conference was supported by the Polish Academy of Sciences.

The conference was held in Wrocław (Poland) between September 13 and 15, 2023. The ISPEM conference was established as an interdisciplinary forum in the areas of intelligent systems, methods and techniques and their applications in production systems. The importance of these issues attracted both academic and industrial participants. The primary concerns of the conference attendees were new solutions for innovative factories, research results and case studies taking into account advances in production and maintenance from the point of view of Industry 4.0 and Industry 5.0.

Manuscripts submitted for ISPEM 2023 were accepted for publication in the Conference Proceedings based on two independent reviews by the members of the Program Committee appointed by the conference chairs. The reviewers were selected according to their competences and the principle of avoiding a conflict of interest—a personal relationship between a reviewer and an author, occupational subordination and direct scientific cooperation over the last two years prior to the review. Each review was provided in a written form and concluded with an explicit recommendation. Before the final decision, the authors' replies to reviewers' comments were required.

We would like to thank all the authors for inspiring presentations and the representatives of both industry and academia for participating in lively discussions. Special thanks go to the members of the Program Committee for the reliable process of reviewing the papers. We are also grateful to the Organizing Committee for the hard work at the conference preparatory stage and a good organization of all events.

September 2023

Anna Burduk
Andre D. L. Batako
José Machado
Katarzyna Antosz
Arkadiusz Gola
Ryszard Wyczółkowski

Organization

Executive Committee

Conference Chair

Anna Burduk	Wrocław University of Science and Technology, Poland

Conference Co-chairs

Andre D. L. Batako	Liverpool John Moores University, UK
José Machado	University of Minho, School of Engineering, Portugal

Patrons

Arkadiusz Wójs	Rector of Wrocław University of Science and Technology, Poland
Nduka Nnamdi Ekere	Pro Vice Chancellor and Dean for Engineering and Technology at Liverpool John Moores University, UK
Pedro Arezes	Dean of School of Engineering University of Minho, Portugal
Celina Pezowicz	Dean of Faculty of Mechanical Engineering, Wrocław University of Science and Technology, Poland
Edward Chlebus	Head of Center for Advanced Manufacturing Technologies, CAMT, Wrocław University of Science and Technology, Poland

Program Committee

Adjallah Kondo H.	Institut Nationale Polytechnique de Lorraine, France
Antonelli Dario	Politecnico di Torino, Italy
Antończyk Agnieszka	Wroclaw University of Environmental and Life Sciences, Poland

Antosz Katarzyna	Rzeszow University of Technology, Poland
Avram Camelia	Technical University of Cluj-Napoca, Romania
Banaszak Zbigniew	Koszalin University of Technology, Poland
Banaś Wacław	Silesian University of Technology, Poland
Batako Andre	Liverpool John Moores University, UK
Będza Tomasz	Wrocław University of Science and Technology, Poland
Błażej Ryszard	Wrocław University of Science and Technology, Poland
Bocewicz Grzegorz	Technical University of Koszalin, Poland
Borucka Anna	Military University of Technology, Poland
Burduk Anna	Wrocław University of Science and Technology, Poland
Butdee Suthep	Rajamangala University of Technology Krungthep, Thailand
Chaszczewska-Markowska Monika	Polish Academy of Sciences, Poland
Chudjuarjeen Saichol	Rajamangala University of Technology Krungthep, Thailand
Cyplik Piotr	Poznan University of Technology, Poland
Damasevicius Robertas	Kaunas University of Technology, Lithuania
Davidrajuh Reggie	University of Stavanger, Norway
Dąbrowska Magdalena	Collegium Witelona, Poland
Dereń Aldona Małgorzata	Wrocław University of Science and Technology, Poland
Diakun Jacek	Poznan University of Technology, Poland
Dostatni Ewa	Poznan University of Technology, Poland
Dulina Luboslav	University of Žilina, Slovakia
Gawlik Jozef	Cracow University of Technology, Poland
Gola Arkadiusz	Lublin University of Technology, Poland
Golińska-Dawson Paulina	Poznan University of Technology, Poland
Grabowik Cezary	Silesian University of Technology, Poland
Grabowska Sandra	Silesian University of Technology, Poland
Grznár Patrik	University of Žilina, Slovakia
Grzybowska Katarzyna	Poznan University of Technology, Poland
Gwiazda Aleksander	Silesian University of Technology, Poland
Helman Joanna	Wrocław University of Science and Technology, Poland
Herbuś Krzysztof	Silesian University of Technology, Poland
Husár Jozef	Technical University of Kosice, Slovakia
Ivanov Vitalii	Sumy State University, Ukraine
Jacyna-Gołda Ilona	Warsaw University of Technology, Poland

Jagielski Dariusz	Centre for Heart Diseases 4th Military Hospital, Poland
Jasiulewicz-Kaczmarek Małgorzata	Poznań University of Technology, Poland
Jurdziak Leszek	Wrocław University of Science and Technology, Poland
Karoluk Michał	Wrocław University of Science and Technology, Poland
Kasprzyk Rafał	Military University of Technology, Poland
Kirianów-Błażej Agata	Wrocław University of Science and Technology, Poland
Kłos Sławomir	University of Zielona Góra, Poland
Knapčíková Lucia	Technical University of Kosice, Slovakia
Kochańska Joanna	Wrocław University of Science and Technology, Poland
Koliński Adam	Poznan School of Logistics, Poland
Kopczewski Marian	General Tadeusz Kościuszko Military University Land Force, Poland
Krajewska-Spiewak Joanna	Cracow University of Technology, Poland
Krenczyk Damian	Silesian University of Technology, Poland
Kulińska Ewa	Opole University of Technology, Poland
Kuric Ivan	University of Žilina, Slovakia
Kurzynowski Tomasz	Wrocław University of Science and Technology, Poland
Lewandowski Tadeusz	Wrocław University of Science and Technology, Poland
Li Nan	Beijing University of Technology, China
Li Peng	Beijing University of Technology, China
Liu Xiucheng	Beijing University of Technology, China
Ładysz Rafał	George Mason University, USA
Łapczyńska Dagmara	Wrocław University of Science and Technology, Poland
Machado José	University of Minho, Portugal
Maletič Damjan	University of Maribor, Slovenia
Markowska-Chaszczewska Monika	Polish Academy of Sciences, Poland
Medyński Daniel	Collegium Witelona, Poland
Mikołajewski Dariusz	Kazimierz Wielki University in Bydgoszcz, Poland
Molasy Mateusz	Wrocław University of Science and Technology, Poland
Monica Zbigniew	Silesian University of Technology, Poland
Nowacki Krzysztof	Silesian University of Technology, Poland

Olender-Skóra Małgorzata	Silesian University of Technology, Poland
Pakšiová Renáta	University of Economics in Bratislava, Slovakia
Patalas-Maliszewska Justyna	University of Zielona Góra, Poland
Pereira Alejandro	University of Vigo, Spain
Pihnastyi Oleh	National Technical University, Kharkiv Polytechnic Institute, Ukraine
Plinta Dariusz	University of Bielsko-Biala, Poland
Prządka Przemysław	Wroclaw University of Environmental and Life Sciences, Poland
Rehm Mathias	Technical University of Chemnitz, Germany
Reiner Jacek	Wrocław University of Science and Technology, Poland
Rojek Izabela	Kazimierz Wielki University, Poland
Rosienkiewicz Maria	Wrocław University of Science and Technology, Poland
Rusińska Małgorzata	Wrocław University of Science and Technology, Poland
Rutkowska Małgorzata	Wrocław University of Science and Technology, Poland
Saniuk Sebastian	University of Zielona Góra, Poland
Skonieczny Jan	Wrocław University of Science and Technology, Poland
Smolina Iryna	Wrocław University of Science and Technology, Poland
Srikhumsuk Phatchani	Rajamangala University of Technology Krungthep, Thailand
Stadnicka Dorota	Rzeszow University of Technology, Poland
Stylios Chrysostomos	University of Ioannina, Greece
Szymańska Elżbieta	Warsaw University of Life Sciences, Poland
Świderski Andrzej	Motor Transport Institute, Poland
Tarapata Zbigniew	Military University of Technology, Poland
Tomé Eduardo	Henrique University, Portugal
Trojanowska Justyna	Poznan University of Technology, Poland
Tubis Agnieszka	Wrocław University of Science and Technology, Poland
Vahonova Oleksandra	National Technical University, Ukraine
Varela Leonilde	University of Minho, Portugal
Wang Yujue	Beijing University of Technology, China
Waszkowski Robert	Military University of Technology, Poland
Więcek Dorota	University of Bielsko-Biala, Poland
Wojciechowski Szymon	Poznan University of Technology, Poland
Woźna Anna	Wrocław University of Science and Technology, Poland

Wyczółkowski Ryszard	Silesian University of Technology, Poland
Xavior Anthony M.	Vellore Institute of Technology, VIT, India

Organizing Committee

Balashov Artem	Wrocław University of Science and Technology, Poland
Będza Tomasz	Wrocław University of Science and Technology, Poland
Helman Joanna	Wrocław University of Science and Technology, Poland
Kochańska Joanna	Wrocław University of Science and Technology, Poland
Krowicki Paweł	Wrocław University of Science and Technology, Poland
Łapczyńska Dagmara	Wrocław University of Science and Technology, Poland
Molasy Mateusz	Wrocław University of Science and Technology, Poland
Musiał Kamil	Wrocław University of Science and Technology, Poland
Parmar Parthkumar	Wrocław University of Science and Technology, Poland
Rosienkiewicz Maria	Wrocław University of Science and Technology, Poland
Rusińska Małgorzata	Wrocław University of Science and Technology, Poland
Wojciechowski Szymon	Poznan University of Technology, Poland
Woźna Anna	Wrocław University of Science and Technology, Poland
Wyczółkowski Ryszard	Silesian University of Technology, Poland

Contents

The Mediating Role of Auditing in Created Shared Value Optimization Through Big Data Analytics: A Conceptual Review 1
Salem Udoh and Khaled Mohamed

NIRF Imaging with Indocyanine Green (ICG) in a Veterinary Minimally Invasive Surgery 15
Przemysław Prządka, Zdzisław Kiełbowicz, and Joanna Tunikowska

Development of Cloud ERP Model and Its Application in Smart Mining 28
Vidosav Majstorovic, Vladimir Simeunovic, Radivoje Mitrovic, Dragan Stosic, Sonja Dimitrijevic, and Zarko Miskovic

Location Suitability for the Implementation of Unmanned Aerial Vehicles in the Vaccine Supply Chain of Sri Lanka 43
Kalani Gunaratne, Harsha Chamara Hewage, Izabela Iwa Nielsen, Grzegorz Bocewicz, Amila Thibbotuwawa, and Zbigniew Banaszak

Modern Trends in Project Management – Selected Issues – Case Study 60
Anna Bluszcz and Marcin Chabior

Cyber Resilience as a New Strategy to Reduce the Impact of Cyber Threats 75
Jerzy Stanik and Jaroslaw Napiórkowski

Simulation, Sensorization and Testing of a Hybrid Inspection Robot 93
Erika Ottaviano, Pierluigi Rea, and Lorenzo Miele

Two Kinds of New Flexible Eddy Current Sensors for Curved Surface Defect Detection 106
Xinhua Zhou, Bin Wu, Xiucheng Liu, Yujue Wang, and Cunfu He

Determining the Range of Image Base of ARM Firmware 120
Jinmiao Wang, Ruijin Zhu, Xiaofei Cao, and Yuejin Zhang

Supporting the Selection of Quality Tools Using Neural Networks 133
Beata Starzyńska and Izabela Rojek

Information Design Management of Machining Parts on Metal Cutting Machines 146
Vira Shendryk, Petro Pavlenko, and Justyna Trojanowska

Virtual Training at ELLA Platform – The Case Study 159
Ľuboslav Dulina, Beáta Furmannová, Mariana Máchová, Michaela Bodingerová, and Tomáš Michulek

Web Application for the Learning of Emotions in Children with Autism Spectrum Disorder 174
Christophe Magalhães, João Sena Esteves, Ana Paula Pereira, Vinícius Silva, and Filomena Soares

Lean Manufacturing Knowledge Correlation Model 188
Iwo Podloch and Krzysztof Nowacki

Assessment of the Effectiveness of Six Sigma Methodology Implementation - A Literature Review 200
Paula Kolbusz and Katarzyna Antosz

Analysis of Basic Characteristics of Textile Yarn Using Image Processing Techniques 215
Marta Rodrigues, Filipe Pereira, and José Machado

Modeling the Impact of Leadership on Change Management Decisions Using System Dynamics Simulation 228
Manuka Lakdinu, M. Mavin De Silva, Kavith Balachandra, Amila Thibbotuwawa, H. Niles Perera, Tadachika Nakayama, and Peter Nielsen

Using an ISO 9001 Based Framework as a Benchmark to Identify Best Practices Used by Sri Lankan Practitioners when Selecting Suppliers 243
Nikila Dharmadasa, M. Mavin De Silva, Amila Thibbotuwawa, Tadachika Nakayama, and Izabela Iwa Nielsen

Requirements for Flow Control Systems for Discrete Production of Mechanical Products 255
Jaroslaw Chrobot

An Augmented System Based on Machine Learning for Boccia Assisted Gameplay 268
João Cruz, Vinícius Silva, João Sena Esteves, and Filomena Soares

Ultrasound Tissue Differentiation (UTC) an Objective Diagnostic Tool in Prevention and Healing Monitoring Tendon Problems in Race Horses 281
Andrzej Golachowski and Barbara Golachowska

Location of Obstacles on Transport Routes in a Centralized Multi-agent System ... 292
Grzegorz Iskierka, Bartosz Poskart, and Kamil Krot

Converging Data Mesh and Microservice Principles into a Unified Logical Architecture ... 300
Francisco Morais, Nuno Soares, João Bessa, José Vicente, Pedro Ribeiro, José Machado, and Ricardo J. Machado

Development and Comparative Analysis of IT Tools Supporting Data Processing in Custom Manufacturing of Orthoses ... 315
Anna Dudkowiak, Ewa Dostatni, Filip Górski, Aleksander Tomczak, and Karolina Krzyżanek

Design of Experiments: An Overview and Future Paths ... 328
Hugo Silva, André S. Santos, Leonilde R. Varela, Justyna Trojanowska, and Kristina Berladir

SOC Evaluation of Lithium-ion Battery Based on MFC Sensors ... 342
Lyu Yan, Geng Liang, and Gao Jie

Indicators of Hierarchical Structure Model of Supporting the Production Management Process – A Framework ... 356
Artur Meller, Mariusz Piechowski, and Arkadiusz Gola

Gamification-Based Crowdsourcing as a Tool for New Product Development in Manufacturing Companies ... 368
Mateusz Molasy, Maria Rosienkiewicz, Joanna Helman, and Mariusz Cholewa

Analysis Safety and Quality of Products for Internal Supply Chains ... 380
Julia Giera, Ewa Kulińska, and Małgorzata Dendera – Gruszka

Development of Acoustic Beam Deflection Electromagnetic Acoustic Transducer for Surface Crack Detection of Wind Turbine Main Shaft ... 393
Gao Jie, Dong Baoyu, Lyu Yan, and He Cunfu

Analysis of Human Errors in the Traditional and Automated Order-Picking System ... 406
Agnieszka A. Tubis, Juni Rohman, Anna Smok, and Dominika Dopart

Generating Complex Dynamic Forms in Low-Code Development Platforms ... 420
Robert Waszkowski and Maciej Dybowski

Optimizing the Form Data Presentation in Low-Code Platforms with the Use of Artificial Intelligence ... 434
Robert Waszkowski and Marcin Sirant

Algorithm for the Assignment of Courses and Instructors in Class Planning ... 445
Tadeusz Nowicki and Patryk Wicherek

Application of Clustering and SOM Analysis for Identification of Conveyor Belt Damage Based on Data from the Diagbelt + Magnetic System ... 461
Aleksandra Rzeszowska, Leszek Jurdziak, Ryszard Błażej, and Agata Kirjanów-Błażej

Performance Marketing in the Management of Supply Chains Sustainable Development ... 476
Shulhina Liudmyla

The Use of Multi-criteria Decision-Making Methods in the Outsourcing of Logistics Services ... 490
Xenie Lukoszová, Elżbieta Jadwiga Szymańska, and Ewa Kulińska

Strategic Activities in the Area of Intellectual Property Management in the Enterprise ... 506
Aldona Małgorzata Dereń and Jan Skonieczny

A Systematic Literature Review of Management and Technological Approaches Integration with Industry 4.0 ... 515
Danielle Barreto Blanco, Leonilde Varela, José Vicente, and José Machado

Impact Analysis and Evaluation of the COVID-19 Pandemic on Decision-Making Processes in Enterprises Based on Just-In-Time Methodology ... 533
Anna Woźna, Małgorzata Rusińska, and Mateusz Bryłkowski

Digital Twins of the Manufacturing System Based on a Simulation Model with Bi-directional Process Data Exchange ... 548
Damian Krenczyk

Determination of the Scope and Parameters of Testing the Total Resistance of Military Equipment in Class N.11 to the Impact of Critical Environmental Conditions ... 560
Kamil Dworak, Cezary Grabowik, Oleh Pihnastyi, and M. Anthony Xavior

Implementation of Industry 4.0 Solutions in the Agricultural Field of Ukraine 575
Olga Tsesliv and Igor Sikorsky

Optimization of Control Activities Using Poka Yoke Tool on the Example of a Selected Production Company 585
Paweł Jokiel, Magdalena Dąbrowska, Phatchani Srikhumsuk, Jozef Huzar, Piotr Motyka, Daniel Medyński, and Krzysztof Kolbusz

Author Index 605

The Mediating Role of Auditing in Created Shared Value Optimization Through Big Data Analytics: A Conceptual Review

Salem Udoh(✉) and Khaled Mohamed

Faculty of Management and Economics, Tomas Bata University, Mostní 5139, 760 01 Zlín, Czech Republic
{udoh,mohamed}@utb.cz

Abstract. The essence of big data and big data analytics as technological breakthroughs is to create and share value among stakeholders. Numerous studies focus on the impact of big data analytics upon firm's performance, big data analytics and auditing and big data with created shared value. There is very little if any study on the mediating role of auditing in the relationship between big data analytics and shared value optimization. As a methodology, we used 'Publish or Perish' which is an academic "search engine" to search all the articles on the big data analytics and firm performance; big data analytics and auditing; big data analytics and shared value optimization topics available on Crossref without time limit. Initial search generated 1000 studies as articles, books and other publications. Based on our criteria, we included only 32 articles focusing on big data analytics and auditing (n = 18); big data analytics and created shared value (n = 14); and, without big data analytics and firm performance (n = 18) already substantially covered in extant literature. The main finding is that there is no study to our knowledge and study's limitation on the mediating role of auditing in the tripartite relationship following our conceptual model based on optimization theory of engineering and given the importance of information integrity for stakeholders.

Keywords: big data analytics · created shared value optimization · auditing

1 Introduction

In this paper we ask the question, 'to what extent has the mediating role of auditing in the relationship between big data analytics and created shared value optimization been investigated in extant literature?' Our literature review shows growing research on the link between big data analytics and firm's performance [3, 27, 31] but little, if any, research on created shared value optimization neither the mediating role of auditing in the reporting process of the enhanced firm performance from BDA insights. Created shared value is communicated to stakeholders as operating performance information which credibility the audit process is designed to enhance [4, 24, 41]. This prompts us to ask the question, 'to what extent has the mediating role of auditing in the relationship between big data analytics and created shared value optimization been investigated in extant

A. Burduk et al. (Eds.): ISPEM 2023, LNNS 790, pp. 1–14, 2023.
https://doi.org/10.1007/978-3-031-45021-1_1

literature?' The context is that value of information or valence from data mining process still requires external validation process [7, 44, 45] to bridge the credibility gap and is where auditing comes in. Moreover, information becomes valuable when transformed into knowledge which guides internal decision making and/or external decision making and evaluation purposes [3]. This is particularly necessary in the context of 'created shared value' principle proposed by Porter and Kramer [35].

Porter and Kramer hold that corporations and society exist on a mutual-dependence relationship [35]. This review therefore aims to map extant literature on the extent to which the mediating role of auditing in created shared value optimization through big data analytics has been investigated. For clarity, we provide definition for created shared value optimization here. According to Porter and Kramer [35], the shared value principle is characterized as policies and operating practices that improve a company's competitiveness while also promoting economic and social progress in the communities where it operates. The underlying concept suggests that the optimization of created shared value occurs when the value report undergoes auditing [4, 41].

Theoretically, big data analytics has proven to impact firm performance through various mediators, and moderators in various contexts, but none on the mediating role of auditing in the value creation process. Considering that ultimately the output of the process is value often communicated as information with credibility gap tendency it becomes an imperative to integrate into the value creation chain the audit process. This paper therefore seeks to elevate the firm performance discussion to shared value optimization on one hand and on the other hand the need to pay attention to credibility of the performance information as the means for communicating that value to stakeholders. This background informs our review methodology.

We conducted literature search in 'Publish or Perish' academic search engine in Crossref and Google Scholar. We find that eighteen articles focused on big data analytics and auditing; fourteen articles on created shared value; and eighteen articles on big data analytics and firm performance. To our knowledge and the limit of this review no study focused on the mediating role of auditing in a conceptual model based on optimization theory in big data analytics- shared value relationship.

In the remaining sections we will examine the theoretical and conceptual review; methods which delineate the search stratégy for the selected 32 included journals articles; results and discussion; and, conclusion.

2 Theoretical Review and Conceptual Review

The review is organized by conceptual issues with their supporting theories related to the topic, and empirical reviews of scholars' works on big data analytics and external auditing as well as created shared value. The structure aims at identifying gaps in literature, theories used in earlier studies, and gaps discovered in extant empirical works. Moreover, the conceptual framework in Fig. 1 below is based on the theory of optimization and built around three constructs/variables: big data analytics (BDA), audit process and shared value optimization.

2.1 Theoretical Framework

This study is built on the theories of optimization, stakeholders' theory and the principle of 'created shared value' propounded by Porter and Kramer [35]. In their breakthrough seminal paper, 'The big idea: Creating shared value', Porter and Kramer [34] conceptualized how to promote profits while solving societal problems. The theory of optimization as borrowed from engineering holds that every (engineering) process or system is built with a specific purpose, which might be a desired performance of the operation of the product being created or of the manufacturing process itself [12]. Furthermore, del Castillo [12] continues that '*in many cases, engineering design activities involve tests or experimentation since the product or process is not well understood, and the desired performance cannot be guaranteed*'.

Essentially, process optimization aims at minimizing the cost of operation of a production process or maximizing its yield or minimizing the variability of a quality characteristics. In the same vein, we extend this logic to the role of auditing in reducing noises in the shared value creation process including inadvertent or deliberate distortions in the reported value or performance of the system or its agents. Auditing thus provides guarantees/assurances on the credibility of the communicated information which can be relied upon for performance evaluation and/or decision needs of the stakeholders. Besides, auditing also improves operating decisions through its policing function on behavior and constructive feedback on observed systemic shortcomings.

2.2 Conceptual Framework

The research model below depicts the tri-partite relationship among Big Data Analytics (BDA), audit process (AP) and shared value optimization (SVOp).

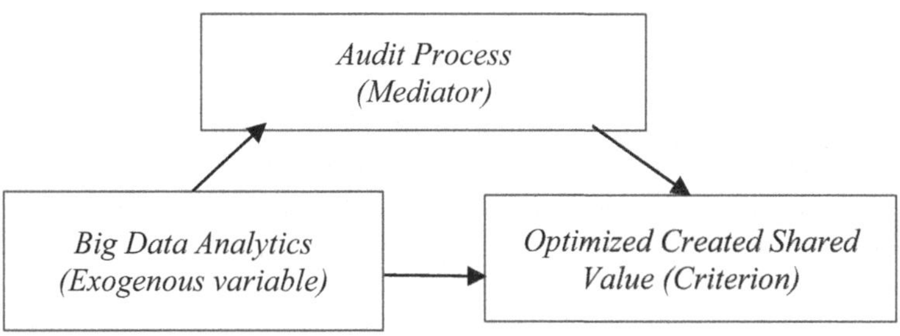

Fig. 1. Conceptual Framework Theory of optimization [12]

Big Data Analytics

Big data analytics is still evolving and has generated much interest in practice and academia. It is used as a term to describe data mining and statistical analysis using business intelligence and analytics (BI&A) technologies [6]. The analytics include text analytics, web analytics, network analytics and mobile analytics [6] of the volume, velocity and variety of data upon which data mining can be conducted. Big data analytics has been linked with various industries, processes and operations including accounting, auditing among others [16, 49, 50].

Big data analytics is information extraction process or mining from big data. The term "big data" pertains to the vast amount of data present in numerous organizations, largely made possible by recent advancements in information technology [17]. Big data sources include email messages, social media postings, phone calls, online purchase transactions, website traffic and video streams. Big data is categorized under digitalization with artificial intelligence and internet of things [17]. Artificial intelligence simulates a limited range of human intelligence under what Drury [17] describes as machine-based learning using the application Siri or Alexa as illustration: 'when customers use internet sites to search for products or services, a search engine on the company's site learns from the results of successful purchases and adapts its future responses accordingly.

Internet of Things (IoT) on the other hand describes '*the interconnection via the internet of computing devices embedded in everyday objects enabling them to send and receive data over a network without requiring any human interaction*' [17]. The raw unstructured data estimated at about 90% from the IoT have been identified to hold so much fortune however with uncertainties hitherto unknown but need to be analyzed. According to Warren et al. [47] '*it is not the volume of information that matters but what is done with the available information and its usefulness.*' This quantum of volume poses both opportunities and challenges. The opportunities are in the 'market' demand for this intelligence and how best organizations can analyze and produce actual organizational value from this data through better decisions. Warren et al. [48] hold that the volume of data makes it difficult for any reasonable analysis or sorting to be conducted using the established database management system or traditional software programs. Other issues include determining what data is likely to be valuable; determining its worth; and exploring and transforming this raw data into relevant information for decision making.

Optimized Created Shared Value

The next concept or principle in the model is created shared value. Created shared value is distinct from Porter and Kramer's [35] "Creating Shared Value" principle. Porter and Kramer's [35] principle explains the interdependency between organizations and society. As distinct from value-based management which seeks to maximize shareholder wealth [35], created shared value maximizes wealth for stakeholders. Porter and Kramer [35] argue that a healthy society requires successful companies creating jobs, wealth and improved living standards as objectives beyond pure economic objectives' pursuit. Porter and Kramer [35] admonish that to avoid any short-lived illusory success from pursuit of merely economic objectives companies should follow the principle of shared value. According to Porter and Kramer [35], shared value is described as "policies and operating practices that improve a company's competitiveness while also advancing economic and social conditions in the communities where it operates."

Below are few results by Drury [17] on how Porter and Kramer's [19] created shared value thinking is transforming the value chain:

- *Energy use and logistics*: £ 175 million annual savings by Marks and Spencer from 75% reduced carbon emissions due to stoppage of supplies from one hemisphere and transporting to another.

- *Resource usage*: Technological advances are enabling companies to reduce usage of water, raw materials and packaging. Coca Cola for example reduced its worldwide water consumption by nearly 20% between 2004 and 2012.
- *Distribution*: Kindle, iTunes and the print media have established alternative electronic distribution methods that reduce paper and transportation usage.
- *Procurement*: Nestle created shared value by improving lives of local farmers/suppliers while securing a stable supply of high-quality specialized coffee. It achieved this by intense collaboration with local farmers, providing consultancy on farming practices, guaranteeing bank loans and brokering pesticides and fertilizers supplies.

Created shared value coined for this study is the product/outcome of Porter and Kramer's [35] "Creating shared value" as more of a process. We argue that the value created as performance is what is shared with stakeholders but first as stewardship information which needs to be validated by the external audit process to earn legitimacy.

The Audit Process as Mediator

Auditing processes both internal and external have critical role in guaranteeing the integrity of information provided to stakeholders [4, 41]. The stakeholders' concept is the thrust of this review and secondly that created shared value is communicated as reports and is where external auditors come in.

International Federation of Accountants (IFAC) in International Standards of Auditing 100 defines auditing as '*an assurance engagement in which a practitioner expresses a conclusion designed to enhance the degree of confidence of intended users other than the responsible party about the outcome of evaluation or measurement of a subject matter against criteria.*' [24]. This definition reveals the underlying principle in every auditing exercise. In the context of shared value, the endpoint of innovation or technological breakthrough such as big data is the creation of value from the mass of unstructured data. The value is the net gain from the differential between increased revenues and reduced cost of operations, satisfied customers and improved environment. But this must be communicated through operations reports which often questioned credibility is validated by the auditors' opinion. The usual assumption is that a priori organization's information might have been compromised deliberately or inadvertently and therefore users would need some independent validation. Auditing thus provides needed guarantees on the credibility of the communicated information which can then be relied upon for performance evaluation and/or decision needs of the stakeholders [11, 41]. Besides, auditing also improves operating decisions through its policing function on agents' behavior and constructive feedback on observed systemic shortcomings.

3 Methods

Given that audit work is gradually turning to information audit in the era of 'internet +' [25], coupled with prior systematic reviews on big data analytics [27] and created shared value [28], we focus our review more on studies that link the mediating role of auditing with big data analytics and shared value optimization. In doing so, we performed a scoping review following the six main steps delineated by Durach et al. [18]: 1) clarifying the research question, 2) specifying the characteristics of focal studies, 3) gathering the potentially related literature, 4) choosing the studies that clearly fit criteria, 5) integrating the literature, and 6) reporting or describing the findings. Two guiding questions were therefore formulated for the search: 'what are the various research themes used for describing the interactions among BDA, shared value optimization and mediating role of auditing? How many studies have investigated the mediating role of auditing in the relationship between BDA and created shared value optimization?'.

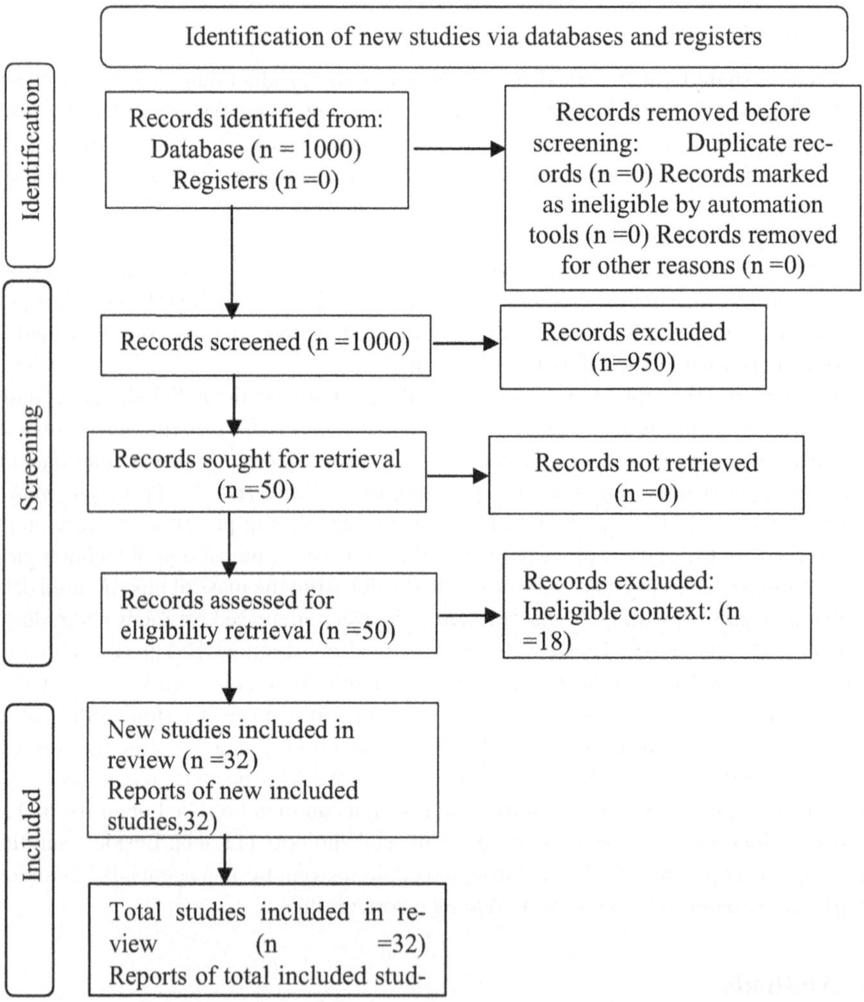

Fig. 2. Search results and source selection and inclusion process Source: Authors

Initial search without time limit in "Publish or Perish" academic search engine generated 1000 results and their key words via Crossref: 'big data analytics without specific mention of auditing and firm performance or value' (950 articles); 'big data analytics, auditing, firm performance' (68 articles). The 68 articles were further reviewed, 5 duplicates and 13 articles with ineligible contexts were identified and removed. Finally, full text of the included studies was reviewed in Google scholar for relevance check using their URLs. Based on our criteria, we included only 32 articles focusing on big data analytics and auditing (n = 18); big data analytics and created shared value (n = 14); and, without big data analytics and firm performance (n = 18) already substantially covered in extant literature. The results are given as Fig. 2 below:

4 Results and Discussion

Four categories were identified for the search results: 18 articles linked BDA with auditing; 18 articles discussed improving firm performance with big data analytics; and 14 articles discussed created shared value optimization. Despite various discussions linking BDA and auditing, gaps exist over the mediating role of external auditing in the BDA-shared value relationship based on optimization theory. This paper fills this gap by adding the mediating role of auditing to the relationship between BDA and shared value optimization.

Moreover, given that significant research has been conducted on big data analytics and firm performance we focus only on big data analytics and auditing as well as created shared value optimization.

4.1 Review of Findings

The results of this scoping review are discussed under the follow sections: i) big data analytics and auditing; and ii) big data analytics and created shared value.

4.2 Big Data Analytics and Auditing

18 sources in our review linked BDA with auditing. 3 sources proposed models for internal auditing and external auditing process respectively [16, 49, 50]. Yuna [49] for instance constructs big data analysis model based on decision tree algorithm and process optimization theory [12]. He found that his Fuzzy ID3 model outperformed standard Min-Ambiguity fuzzy decision tree algorithms. Ding [16] explores the use of artificial intelligence in internal audit process to optimize the firm's financial performance based on deep learning theory. He observed an improvement in performance of the deep learning neural network model optimized by the genetic algorithm. Zhang and Wang [50] discuss the use of financial audit model based on artificial neural network technology to assist in audit of large number of loan verification calculations. The model achieved an accuracy rate of above 94% with an error rate less than 5%. This demonstrates how this approach can assist external auditors in audit of liabilities. Wang [47] adds that in the field of finance, "Internet + accounting" and big data auditing make it possible to obtain direct activity information of enterprises through computers.

Two studies focused on motivations and challenges for the use of big data and big data analytics in external auditing. Dagilienė and Klovienė [11] discuss organizational intentions to use Big Data and Big Data Analytics (BDA) in external auditing. They combined in one methodology contingency theory approach, a literature review and 21 in-depth interviews with three different types of respondents. They identify two categories of motivating factors (company-related and institutional) and the circumstances that will produce intended audit outcomes when BDA is used in the audit process. They further find that external auditors are likely to focus on the BDA procedures to satisfy regulatory requirements and create more value for business clients. Perera, and Abeygunasekera [34] also examine the motivations and challenges of using BD and BDA in the Sri Lankan audit industry. Their survey of Big 4 audit firms in Sri Lanka, audit client companies, regulators, and professionals for data analytics found that usage of innovative tools is mainly driven by both internal and external pressures. Similarly, De Santis and D'Onza [15] find that the Big Four lead the BDA-driven audit innovation process and BDA is used to complement traditional audit procedures. However, slow adoption of BDA is observed outside the audit community due to issues such as digital maturity of audit clients, the lack of audit standards and the audit oversight authority's negative view of BDA. Iguma and Riccio [23] corroborate De Santis and D'Onza [15] result in their finding with Brazilian internal auditors' behavioural intention to adopt big data analytics. The results showed performance expectancy and social influence as the two most important factors in increasing the likelihood of BDA adoption, whereas effort expectancy and enabling circumstances had little effect.

Furthermore, Thogmartin [40] contributes to the discussion in the area of internal auditing and risk assessment. He observes that data analytic techniques are becoming a key tool in internal auditing for providing assurance on risk, control, and governance processes consistent with its definition by the Institute of Internal Auditors as "an independent, objective assurance and consulting activity designed to add value and improve an organization's operations.' Sayedahmed et al., [38] also contribute to internal audit and Big Data Analytics literature in two ways. First, they identify big data sources to include accounting software, mobile phone signals, GPS devices, online searches, barcodes, cameras, video clips, transactional databases, and submissions to financial, governmental, and other agencies. These have the potential to produce more insights, improve decision making, and find trends that improve planning and lower risks. Next, they introduce internal auditors' roles in the three major stages of Big Data production, which include data mining, analysis, and utilization.

Other perspectives include audit approach for shared data in cloud storage facility [7, 25]. Chi-Wei Liu et al., [7] explain that audit of shared data in cloud storage domains require different approaches for signature, public audits, dynamic data and user revocation. The approaches will include analysis function, security, and performance requirements. Lin [25] recommends pedagogical shift in universities to reflect the realities of "Internet +" era. Warren et al. [48] support this view. According to Warren et al. [48], the availability of video, audio, and textual information through Big Data can lead to improvements in managerial accounting, financial accounting, and financial reporting practices. For managerial accounting, Big Data can contribute to the development and improvement of management control systems and budgeting processes. In terms of financial accounting, Big Data has the potential to enhance the quality and relevance of accounting information, thereby increasing transparency and aiding stakeholders in decision-making. Additionally, in the field of financial reporting, Big Data can assist in the creation and refinement of accounting standards, ensuring that the accounting profession continues to provide valuable information as the dynamic, real-time, global economy evolves.

Martins et al. [28] argue for accountants and auditors to fill the current capacity gap in business fundamentals lacking among mainstream data and computer scientists. This review is an answer to this call. Wadhawan and Mahajan [44] explain auditors' role as data analysts in providing credible information for decision needs of investors, companies and government. They present auditing concepts and techniques that could be employed in auditing in BDA era to detect and manage operational, transactional and auditing risk. Wagner et al. [45] discuss the risk of basing public regulatory decisions on unverified data and suggested the concept of auditing intermediates to address data access issues, potential intermediaries misuse, and the general lack of standards for data provision by large online platforms.

4.3 Big Data and Created Shared Value

We identify fourteen articles in this category. The concept of creating shared value is the brainchild of Porter and Kramer [35] in which they advocated for organizations to share with society value created from operations in a win-win situation. Corazza et al. [9] compares created shared value (CSV) to Plato and Baudrillard's concept of 'simulacrum', in which companies adopting CSV create an interpretation through their sustainability disclosure. They find that 'CSV is not viewed as something unrelated to corporate social responsibility (CSR), not just philanthropy, but is a strategically oriented shift.' This view is extended by Vitari and Raguseo [1, 8, 21, 36] who argue on the impact of environmental characteristics of created shared value. However, Crane et al. [10] critique Porter and Kramer's [35] concept of creating shared value; holding that despite its popularity among practitioner and academic audiences, its shortcomings include a shallow conception of the corporation's role in society. de los Reyes et al. [13] in their critique claim that 'While Porter and Kramer's Creating Shared Value (CSV) works well as a management framework to address "win-win" business and society issues, it leaves managers ill-equipped to legitimately manage issues where they the prospect of "win-lose" or "lose-win" social engagements.' They suggest a framework that combines CSV with ethical frameworks. Our study introduces auditing dimension in response. Moreover, Pavlovich and Corner [33] propose conscious awareness through spiritual practices as one of the ethical frameworks. They claim this can shape entrepreneurial intentions and venture characteristics focused on shared value.

Arora [2] discusses the opportunities and challenges of big data for value creation including high data complexity and uncertainty, capacity for storage and analysis as well as data integrity. He presents the capabilities of Hadoop and Apache Spark technologies in handling big data challenges at advanced level of data mining, data optimization and predictive forecasting techniques. Florin and Schmidt [20] develop a strategy process model for hybrid ventures that occupy the middle ground between nonprofits and for-profits. Elke den Ouden [14] presents an approach to designing shared value for businesses, non-profit organizations, end-users and society by providing solutions to challenges such as the aging population, energy scarcity and environmental issues. Edu et al. [19] propose a framework using the resource-based view and the capability view to explore the integration of digital capabilities to support value creation in an organization. Mandeep et al. [26] present how organizations can truly create value for firm and customer in diverse sectors such as agriculture, healthcare, cyber security, and smart cities. Sáenz et al. [37] use information value chain approach to theorize about how quality data and IT-enabled data analytics sensing capability in the marketing unit relate differently to the unit performance and to firm-level performance. This was confirmed with results from a survey of 346 firms showing partially and fully mediated effects for quality data, and direct and partially mediated effects for sensing capability.

In their study, Michelini et al. [30] aimed to distinguish the characteristics of social and inclusive business models, and to explore the benefits and risks associated with each model. Through an analysis of ten case studies, Michelini et al. [30] identified the distinctive features of inclusive business and social business models, as well as the advantages and risks they entail. The findings of their study led the authors to conclude that social and inclusive business models share commonalities in terms of partner networks, knowledge utilization, and value chain, particularly in the development of innovative distribution strategies. However, it should be noted that this similarity may not hold when the market being considered is not in an emerging country. Importantly, both types of models align in their capacity to generate social benefits.

Verhoef et al. [42] hold that only intentional plans and strategies by organizations can create value from information processing of the available vast and complex data. They suggest that big data is not a revolution in and of itself, but rather an evolution of the recent increase in data availability.

Grover et al. [22] address an existing gap in the literature by examining the impact of big data and analytics (BDA) on organizational value. They enhance the understanding of BDA value by extending established frameworks for assessing the value of information technology. The authors offer practical illustrations of BDA applications to demonstrate the efficacy of the framework. Furthermore, they discuss how this framework facilitates the exploration of different constructs and relationships that are relevant to the creation and realization of BDA value.

Menghwar and Daood [29] undertook a systematic review encompassing 242 articles published between 2010 and 2020, focusing on the concept of creating shared value (CSV) introduced by Porter and Kramer [35]. The authors defined CSV as a strategic process that enables corporations to transform social challenges into business prospects. Through their comprehensive review and analysis, several key findings emerged. Firstly, the study highlights that CSV represents a significant and valuable addition to the existing literature, rather than being merely a revolutionary concept or a passing trend. Secondly, the authors argue that a firm's decision to adopt a CSV strategy is contingent upon rational considerations, particularly concerning opportunity costs and transaction costs. Lastly, it is revealed that there is no universally applicable approach to creating shared value. Instead, a multitude of external and internal factors exert influence on a firm's ability to effectively pursue a CSV strategy.

The review includes several sources that examine the relationship between big data analytics (BDA) and auditing. Some studies focus on the application of BDA in internal and external auditing procedures, such as the use of decision tree algorithms and process optimization theory to build BDA models for improved performance. Other studies explore the motives and challenges of utilizing BDA in auditing, highlighting variables such as regulatory requirements, creating value for clients, and internal/external pressures. Factors such as performance expectation and social impact influence BDA adoption and acceptance in auditing. The review additionally examines how BDA can help with risk assessment, internal audits, and the utilization of shared data in cloud storage facilities. Furthermore, it underlines the potential benefits of BDA in managerial accounting, financial accounting, and financial reporting, as well as the necessity for accountants and auditors to bridge the data analytics knowledge gap. The review also expresses concerns regarding data verification, standards, and auditors' roles as data analysts in providing credible information.

5 Conclusions

Value created from technological innovations like big data analytics are communicated as information and optimized when believed by the users [26, 39]. Walker [46] describes big data analytics as "Big Data to Big Profits" but Wadhawan and Mahajan [44] caution also of corresponding big risks. Auditing mitigates those risks by giving credibility to created shared value information [22]. This review therefore highlights the role of auditing in providing credibility to information and optimizing value created through technological innovations and big data analytics.

Moreover, evidence abound on big data analytics and firm performance [1, 5, 43], big data analytics and auditing [22, 31, 32] as well as created shared value introduced by Porter and Kramer [35]; however, there are gaps regarding the mediating role of auditing in optimizing shared value through big data analytics. These gaps include limited focus on auditing as a mediator, lack of empirical research, the need for a comprehensive conceptual model, and limited consideration of contextual factors and practical implications.

The dominance of authors with non-accounting and auditing backgrounds contributes to these gaps. Conducting empirical studies and developing a comprehensive conceptual model in this area can advance knowledge, bridge gaps, and enhance auditing practices related to big data analytics. It can also support decision-making processes and promote responsible and ethical data use, leading to positive operational and social impacts. Auditors need to adapt to technological advancements and invest in acquiring the necessary skills to stay relevant. The importance of dependable information and the role of auditing in providing it are emphasized to avoid historical failures and address challenges in the field. The main limitation of this research is that it is a review however together with the identified gaps present a rich area for future research.

References

1. Akter, S., Wamba, S., Gunasekaran, A., Dubey, R., Childe, S.: How to improve firm performance using big data analytics capability and business strategy alignment? Int. J. Prod. Econ. **182**, 113–131 (2016). https://doi.org/10.1016/j.ijpe.2016.08.018
2. Arora, B.: Big Data analytics: the underlying technologies used by organizations for value generation. In: Chahal, H., Jyoti, J., Wirtz, J. (eds.) Understanding the Role of Business Analytics. Springer, Singapore (2019). https://doi.org/10.1007/978-981-13-1334-9_2

3. Bogdan, M., Borza, A.: Big data analytics and firm performance: a text mining approach. In: Proceedings of the 14th International Management Conference "Managing Sustainable Organizations" 5th–6th Nov. 2020, Bucharest, Romania. https://doi.org/10.24818/IMC/2020/03.08
4. Causholli, M., Knechel, R.: An examination of the credence attributes of an audit. Account. Horiz. **26**(4), 631–656 (2012)
5. Chatterjee, S., Chaudhuri, R., Vrontis, D., Thrassou, A.: Impacts of big data analytics adoption on firm sustainability performance. Qual. Res. Financ. Mark. 1755–4179 (2022). https://doi.org/10.1108/qrfm-01-2022-0005
6. Chen, H., Chiang, R.H., Storey, V.C.: Business intelligence and analytics: from Big Data to big impact. MIS **36**, 1165–1188 (2012)
7. Liu, C.-W., Hsein, W.-F., Yang, C.-C., Hwang, M.-S.: A survey of public auditing for shared data storage with user revocation in cloud computing. Int. J. Netw. Secur. **18**(4), 650–666 (2016)
8. Claudio, V., Raguseo, E.: Big data analytics business value and firm performance: linking with environmental context. Int. J. Prod. Res. **58**(18), 5456–5476 (2020). https://doi.org/10.1080/00207543.2019.1660822
9. Corazza, L., Scagnelli, S.D., Mio, C.: Simulacra and sustainability disclosure: analysis of the interpretative models of creating shared value. Corp. Soc. Responsib. Environ. Manag. **24**(5), 414–434 (2017). https://doi.org/10.1002/csr.1417
10. Crane, A., Palazzo, G., Spence, L.J., Matten, D.: Contesting the value of "creating shared value." Calif. Manage. Rev. **56**(2), 130–153 (2014). https://doi.org/10.1525/cmr.2014.56.2.130
11. Dagilienė, L., Klovienė, L.: Motivation to use big data and big data analytics in external auditing. Manag. Audit. J. **34**(7), 750–782 (2019). https://doi.org/10.1108/maj-01-2018-1773
12. De del Castillo, E.: Process Optimization. A Statistical Approach. International Series in Operation Research and Management Science. Springer (2007)
13. de los Reyes, G., Scholz, M., Smith, N.C.: Beyond the "win-win": creating shared value requires ethical frameworks. California Manag. Rev. **59**(2), 142–167 (2017). https://doi.org/10.1177/0008125617695286
14. den Ouden, E.: Innovation Design Creating Value for People, Organizations and Society. Springer Verlag, London (2012). https://doi.org/10.1007/978-1-4471-2268-5
15. De Santis, F., D'Onza, G.: Big data and data analytics in auditing: in search of legitimacy. Medit. Account. Res. **29**(5), 1088–1112 (2021). https://doi.org/10.1108/medar-03-2020-0838
16. Ding, R.: Enterprise intelligent audit model by using deep learning approach. Comput. Econ. **59**, 1335–1354 (2022). https://doi.org/10.1007/s10614-021-10192-9
17. Drury, C.: Cost and Management Accounting 11. CENTAGE, London (2021)
18. Durach, C., Wieland, A., Machuca, J.A.D.: Antecedents and dimensions of supply chain robustness: a systematic literature review. Int. J. Phys. Distrib. Logist. Manag. **45**(1/2), 118–137 (2015). https://doi.org/10.1108/IJPDLM-05-2013-0133
19. Edu, S.A., Agoyi, M., Agozie, D.Q.: Integrating digital innovation capabilities towards value creation: a conceptual view. Int. J. Intell. Inf. Technol. **16**(4), 37–50 (2020). https://doi.org/10.4018/IJIIT.2020100103
20. Florin, J., Schmidt, E.: Creating shared value in the hybrid venture arena: a business model innovation perspective. J. Soc. Entrep. **2**(2), 165–197 (2011). https://doi.org/10.1080/19420676.2011.614631
21. Wamba, S.F., Akter, S., de Bourmont, M.: Quality dominant logic in big data analytics and firm performance. Bus. Process. Manag. J. **25**(3), 512–532 (2019). https://doi.org/10.1108/bpmj-08-2017-0218

22. Grover, V., Chiang, R., Liang, T.-P., Zhang, D.: Creating strategic business value from big data analytics: a research framework. J. Manag. Inf. Syst. **35**(2), 388–423 (2018). https://doi.org/10.1080/07421222.2018.1451951
23. Iguma, M., Riccio, E. : Factors influencing Brazilian internal auditors' behavioral intention to adopt big data analytics. Int. J. Audit. Technol. 217–239 (2021). https://doi.org/10.1504/IJAUDIT.2020.115929
24. International Federation of Accountants®: Handbook of International Quality Control, Auditing, Review, Other Assurance, and Related Services Pronouncements, 2014 edn. I (2014)
25. Lin, R.: Construction of information auditing curriculum for auditing major in big data era. In: Sugumaran, V., Xu, Z., Zhou, H. (eds.) Application of Intelligent Systems in Multi-modal Information Analytics, MMIA 2020. Advances in Intelligent Systems and Computing, vol. 1233. Springer, Cham (2021). https://doi.org/10.1007/978-3-030-51431-0_103
26. Saggi, M.K., Jain, S.: A survey towards an integration of big data analytics to big insights for value-creation. Inf. Process. Manage. **54**(5), 758–790 (2018). https://doi.org/10.1016/j.ipm.2018.01.010
27. Maroufkhani, P., et al.: Big data analytics and firm performance: a systematic review. Information **10**, 226 (2019). https://doi.org/10.3390/info10070226
28. Martins, A., Paula, A., Silva, P., Fontes, A.: A paradigm shift in accounting and auditing in the era of big data. In: Martins, A., Paula, A., Silva. A.P., Fontes, A. (eds.) Business Intelligence and Analytics in Small and Medium Enterprises, 1st edn., p. 16. CRC Press (2019). https://doi.org/10.1201/9780429056482-4
29. Menghwar, P.S., Daood, A.: Creating shared value: a systematic review, synthesis and integrative perspective. Int. J. Manag. Rev. **23**, 466–485 (2019). https://doi.org/10.1111/ijmr.12252
30. Michelini, L., Fiorentino, D.: New business models for creating shared value. Soc. Responsib. J. **8**(4), 561–577 (2012). https://doi.org/10.1108/17471111211272129
31. Mikalef, P., Boura, M., Lekakos, G., Krogstie, J.: Big data analytics and firm performance: findings from a mixed-method approach. J. Bus. Res. **98**, 261–276 (2019). https://doi.org/10.1016/j.jbusres.2019.01.044
32. Müller, O., Fay, M., vom Brocke, J.: The effect of big data and analytics on firm performance: an econometric analysis considering industry characteristics. J. Manag. Inf. Syst. **35**(2), 488–509 (2018). https://doi.org/10.1080/07421222.2018.1451955
33. Pavlovich, K., Corner, P.D.: Conscious enterprise emergence: shared value creation through expanded conscious awareness. J. Bus. Ethics **121**, 341–351 (2014). https://doi.org/10.1007/s10551-013-1726-y
34. Perera, M.C.D., Abeygunasekera, A.W.J.C.: Big data and big data analytics in external auditing: motivations and challenges. Int. J. Account. Bus. Finance **7**(Special Issue), 1–16 (2021). https://doi.org/10.4038/ijabf.v7i0.105
35. Porter, M., Kramer, N.: Creating shared value: competitive advantage through social impact. Shared Value Initiative (2011). https://www.sharedvalue.org/about/what-is-shared-value/
36. Raguseo, E., Vitari, C.: Investments in big data analytics and firm performance: an empirical investigation of direct and mediating effects. Int. J. Prod. Res. **56**(15), 5206–5221 (2018). https://doi.org/10.1080/00207543.2018.1427900
37. Sáenz, J., Ortiz de Guinea, A., Peñalba-Aguirrezabalaga, C.: Value creation through marketing data analytics: the distinct contribution of data analytics assets and capabilities to unit and firm performance. Inf. Manag. **59**(8), 103724 (2022). https://doi.org/10.1016/j.im.2022.103724
38. Sayedahmed, N., Anwar, S., Shukla, V.K.: Big data analytics and internal auditing: a review. In: 3rd International Proceedings on Computation, Automation and Knowledge Management (ICCAKM), pp. 1–5, Dubai (2022). https://doi.org/10.1109/iccakm54721.2022.9990045

39. Shabbir, M.Q., Gardezi, S.B.W.: Application of big data analytics and organizational performance: the mediating role of knowledge management practices. J. Big Data **7**, 47 (2020). https://doi.org/10.1186/s40537-020-00317-6
40. Thogmartin, J.R.: Internal auditing seeking action from top management to mitigate risk. In: Thogmartin, J.R. (ed.) Data Analytics, 1st edn., p. 12. Imprint Auerbach Publications (2019)
41. Udoh, S.: Auditing 2 Lecture Notes [Class Handout]. University of The Gambia, Kanifing Campus (2013)
42. Verhoef, P.C., Kooge, E., Walk, N.: Creating Value with Big Data Analytics: Making Smarter Marketing Decisions, 1st edn. Routledge (2015). https://doi.org/10.4324/9781315734750
43. Whitelock, V.: Business analytics and firm performance: role of structured financial statement data. J. Bus. Anal. **1**(2), 81–92 (2018). https://doi.org/10.1080/2573234X.2018.1557020
44. Wadhawan, A., Mahajan, R.: Detecting and managing operational, transactional and auditing risk using data analytics. In: 5th International Conference on Reliability, Infocom Technologies and Optimization (Trends and Future Directions) (ICRITO), pp. 171–174 (2016). https://doi.org/10.1109/ICRITO.2016.7784947
45. Wagner, B., Kuklis, L.: Establishing auditing intermediaries to verify platform data. In: Martin, M., Damian, T. (eds.) Regulating Big Tech: Policy Responses to Digital Dominance 2021 online edn. Oxford Academic, New York (2021). https://doi.org/10.1093/oso/9780197616093.003.0010
46. Walker, R.: From Big Data to Big Profits: Success with Data and Analytics, vol. 352. Oxford University Press (2015). https://doi.org/10.1093/acprof:oso/9780199378326.001.0001
47. Wang, X.: Reflections on the application of AI in auditing practice in the context of big data. In: Xu, Z., Choo, K.K., Dehghantanha, A., Parizi, R., Hammoudeh, M. (eds.) Cyber Security Intelligence and Analytics CSIA 2019, Advances in Intelligent Systems and Computing, vol. 928. Springer, Cham (2020). https://doi.org/10.1007/978-3-030-15235-2_170
48. Warren, D.J., Moffitt, K.C., Byrnes, P.: How big data will change accounting. Account. Horiz. **29**(2), 397–407 (2015). https://doi.org/10.2308/acch-51069
49. Yuna, S.: Construction and application of enterprise internal audit data analysis model based on decision tree algorithm. Discrete Dyn. Nat. Soc. 1–13 (2022). https://doi.org/10.1155/2022/4892046
50. Zhang, Z., Wang, Z.: Design of financial big data audit model based on artificial neural network. Int. J. Syst. Assur. Eng. Manag. 1–10 (2021). https://doi.org/10.1007/s13198-021-01258

NIRF Imaging with Indocyanine Green (ICG) in a Veterinary Minimally Invasive Surgery

Przemysław Prządka(✉), Zdzisław Kiełbowicz, and Joanna Tunikowska

Department and Clinic of Surgery, Faculty of Veterinary Medicine, Wroclaw University of Environmental and Life Sciences, Wroclaw, Poland
przemyslaw.przadka@upwr.edu.pl

Abstract. The use of fluorescence technology with indocyanine green has been shown to provide real-time visualization of lymphatic flow, identification of sentinel lymph nodes, assessment of intestinal perfusion, and detection of the extrahepatic bile duct. These capabilities could potentially reduce the mortality and complication rate following surgery. However, the lack of standardization in dosage and time of ICG administration requires further clinical and experimental studies. This applies in particular to veterinary medicine, which results from the lack of general access to the latest technology. Near-infrared indocyanine green (ICG) fluorescence is a great hope for the further development of minimally invasive surgery in humans and animals. The article presents the possibilities of using near-infrared indocyanine green in thoracic and general surgery and compares it with the current development of selected surgical procedures in veterinary medicine. The rapid progress of veterinary surgery, including minimally invasive surgery, makes the wide use of near-infrared indocyanine green a matter of the near future.

Keywords: surgery · animals · ICG · dog · cat

1 Introduction

Despite the rapid advancement of veterinary Minimally Invasive Surgery (MIS), including laparoscopy, around the world, it is believed that this field is approximately two decades behind that of human medicine [19]. The laparoscopic and thoracoscopic techniques provide many advantages, with decreased trauma being the most commonly cited. This reduces recovery time, allowing for a quicker return to normal activities. An example of this is ovariectomy in bitches, which is seen to be superior to open surgery [12, 57]. There is a constant need for new laparoscopic and thoracoscopy procedures, which can be developed from existing procedures used in human medicine or created entirely from new. The advancement of veterinary laparoscopy and thoracoscopy is driven by the clear benefits of endoscopic surgery, which both veterinarians and pet owners find appealing.

Indocyanine green (ICG) is a water-soluble tricarbocyanine dye that is used for various medical purposes, such as measuring cardiac output, assessing hepatic function,

A. Burduk et al. (Eds.): ISPEM 2023, LNNS 790, pp. 15–27, 2023.
https://doi.org/10.1007/978-3-031-45021-1_2

evaluating liver blood flow, and performing ophthalmic angiography [10]. Near-Infrared (NIR) fluorescent light, which is invisible to the human eye, can be used to produce images due to its low autofluorescence background and ability to deeply penetrate tissue. This light has a wavelength of 700–900 nm [18].

Surgery can benefit from the use of Invisible Near-Infrared light fluorescence imaging to provide real-time visualization of certain anatomical structures. This NIR fluorescence imaging system, when combined with fluorescent contrast agents, can help to more clearly identify specific structures [10]. NIR fluorescence imaging with ICG has been utilized in a variety of surgical procedures, such as sentinel lymph node mapping [10, 26, 73], intraoperative solid tumor identification [10, 31, 45], organ perfusion assessment in gastrointestinal anastomosis and flap reconstruction surgery [13, 42, 46]. This paper presents current and possible applications of this technique in veterinary surgical procedures.

1.1 Chylothorax and Thoracic Duct Imaging

Chylothorax is a medical condition in which Chyle leaks into and accumulates in the pleural space, leading to respiratory impairment, weight loss, and exercise intolerance. To treat this condition, thoracic duct ligation, pleuroperitoneal shunting, thoracic omentalisation, cisterna chyli ablation, and pericardiectomy have been suggested [5, 17, 22, 34, 59]. Thoracic duct ligation is the most common surgical intervention for the treatment of chylothorax [37]. Unfortunately, recurrence is often seen which is believed to be due to blockages in the various divisions of the thoracic duct [4]. It is very important to correctly identify the lymphatic branches during surgery to ensure a successful thoracic duct ligation and to avoid missing any ducts. To determine the optimal ligation site for each individual chylothorax case, it is necessary to precisely assess the thoracic duct's tributaries [28]. Treating chylothorax due to thoracic duct injury is complex, as it can be hard to detect and repair. Lymphangiography can be used pre-operatively to detect Chyle leakage, but this has limited usefulness for detecting leaks during surgery [10]. It can be challenging to differentiate thoracic ducts from the surrounding connective tissue and they are often embedded in a large amount of mediastinal fat. To identify these ducts, an intraoperative lymphatic contrast agent, such as methylene blue, can be used. This has been done by administering the dye into popliteal or mesenteric lymph nodes [33]. Due to the depth limitations of visible light in penetrating fat, as well as peri-aortic fat deposition in the mediastinum, methylene blue is often not visible. To ensure safe and effective identification of thoracic duct branches in veterinary chylothorax patients, a high-contrast method of intraoperative thoracic duct identification is recommended [20, 61].

Light in the visible spectrum can only penetrate a few hundred microns into blood and tissue, making the deeper structures invisible [20, 61]. However, NIR fluorescence imaging systems can utilize light spectra with longer wavelengths than the human eye can detect, which allows them to penetrate several millimeters deeper into tissues [20]. The use of Indocyanine green as a near-infrared fluorophore for operative imaging of the lymphatic system is supported by its high binding to plasma lipoproteins [55]. The longer wavelengths of photons emitted by ICG allow for the visualization of NIR-highlighted structures deeper in mediastinal fat than is possible with white light viewing.

The use of near-infrared fluorescence imaging with ICG during liver procedures has been studied for the purpose of detecting liver tumors in both humans and animals [36, 53]. In addition, ICG fluorescence has been tested to detect kidney tumors in humans [65], and has been used to identify sentinel lymph nodes in breast cancer [32], malignant melanoma [64] and gastro-intestinal cancers [40, 44]. Veterinary oncological surgery of the liver has also investigated the application of near-infrared fluorescence imaging with indocyanine green for the surgical treatment of hepatic masses in dogs [53]. Intravenous ICG is administered 12 to 48 h prior to surgery. It glows on the liver surface when exposed to a near-infrared source. After injection, ICG is taken up by both tumoral and non-tumoral hepatocytes. With normal or subnormal hepatic function and in the absence of biliary obstruction, it is excreted in the bile and is cleared from the healthy liver parenchyma after a few hours [36]. The fluorescent light emitted by ICG, due to its particular features, allows the detection of both hepatocellular and non-hepatocellular tumors, with ICG remaining fixed in tumoral hepatocytes and in pathological areas of the liver, particularly around non-hepatocellular tumors [36]. The use of infrared light to cause excitation and emission of ICG in human medicine has been shown to be effective for intraoperative identification of hepatic masses and assessing complete resection [49, 53].

Sakurai et al. [53] showed that ICG fluorescence imaging could be of clinical use for detecting and assessing the margins of canine hepatic tumors. However, they noted that the fluorescence imaging was not able to provide a differential diagnosis of the tumor. Nonetheless, the cited authors noted that fluorescence imaging could be a useful tool for assessing complete resection during surgical operations on canine hepatic tumors exhibiting increased fluorescence [53].

3 Colorectal Surgery

Surgery is the primary form of treatment for non-metastatic colon and rectal cancer in humans. According to multiple research studies, the extent of lymphadenectomy (removal of lymph nodes) is a critical factor, as it gives an indication of prognosis as well as potentially improving survival rates if more lymph nodes are removed [51]. The prognostic value of lymph node status in patients with colorectal cancer is widely acknowledged and is a key factor in deciding whether adjuvant chemotherapy is necessary. Unfortunately, it is believed that up to 30% of those with negative lymph node status may still suffer from recurrence within five years following surgery. This potential issue of under-staging can be addressed through immunohistochemistry and multi-level sectioning of lymph nodes; though these techniques can be costly and time-consuming. An alternative approach is sentinel lymph node mapping, which identifies the node most likely to have tumoral infiltrates [43]. If the sentinel lymph node within a given lymphatic drainage area is tumor-free, according to Hackethal et al. [21], radical lymphadenectomy is not necessary as the sentinel node is considered to be representative of the other lymph nodes in the area.

The near-infrared indocyanine green fluorescence can be a useful tool for making informed decisions during minimally invasive surgeries [9]. Peritumoral or interstitial injection of ICG or similar substances can help transport them through lymphatic pathways to the sentinel lymph nodes [21]. Real-time visualization of lymph flow (lymphatic mapping) has been shown to be a viable option, which can potentially assist surgeons in more precise staging [9]. The use of ICG fluorescence for sentinel lymph node detection has already been demonstrated to be promising in humans for a variety of carcinomas, such as breast, endometrial and cervical, gastric, colorectal and pancreatic [3, 9, 41, 48]. In colo-rectal and gastric surgery, the value of ICG-fluorescence guided surgery has been established [9]. Moreover, the real-time visualization of lymphatic flow using ICG has been found to be feasible, potentially aiding the surgeon in identifying sentinel lymph nodes and performing a better lymphadenectomy [21]. Furthermore, ICG is becoming the preferred modality for lymphatic mapping in colon and rectal cancer in humans, as it migrates in lymphatics and lymph nodes after direct tissue injection, providing an intraoperative map of the tumor-specific draining area [51].

Ribero et al. [51] found that lymphadenectomy was changed in 50% of cases, with 41.4% of the 70 patients having lymph nodes outside the traditional resection field. This confirms the results of other studies that reported a similar rate of modification of lymphadenectomy, mostly in patients undergoing right colectomies [47]. The biggest challenge remains determining the optimal time for preoperative submucosal tattooing, as there are no standard protocols regarding the dose, number of injections, and time of administration of ICG. It has been suggested that the ideal time for injection is between 3 h and 3 days preoperatively [47, 51, 67].

Few studies have been conducted to examine the use of near-infrared indocyanine green fluorescence in oncological surgeries on animals, primarily focusing on skin and mammary gland tumors [15, 72]. In our clinic, we have recently been able to use near-infrared indocyanine green fluorescence for intestinal surgery, including mapping sentinel lymph nodes and assessing the blood supply to the performed anastomosis – Fig. 1.

In our opinion, the most important aspect in the administration of ICG to a large intestine tumor in operated dogs allowed for the assessment of lymph drainage and the determination of the sentinel lymph node. At the same time, in each operated case, ICG was administered intravenously in order to assess the blood supply to the anastomotic fragments of the intestines – Fig. 2.

Despite advances in postoperative care and surgical techniques, the rate of anastomotic leakage after rectal surgery remains at approximately 10% worldwide [69]. Many factors are associated with the formation of anastomotic leakage, but adequate blood supply is essential for anastomotic healing [69]. Therefore, an accurate evaluation of the intestinal blood perfusion is essential for avoiding anastomotic leakage. Various methods have been proposed to measure this intraoperatively, but most of them are not used in clinical practice due to costly equipment and lack of reliability [69].

Recent studies have indicated that intraoperative indocyanine green fluorescence angiography can help prevent anastomotic leakage in colorectal surgery by providing visual evidence of the perfusion of the anastomotic region [58]. Similarly, fluorescence imaging for tissue perfusion assessment has been shown to be beneficial in improving

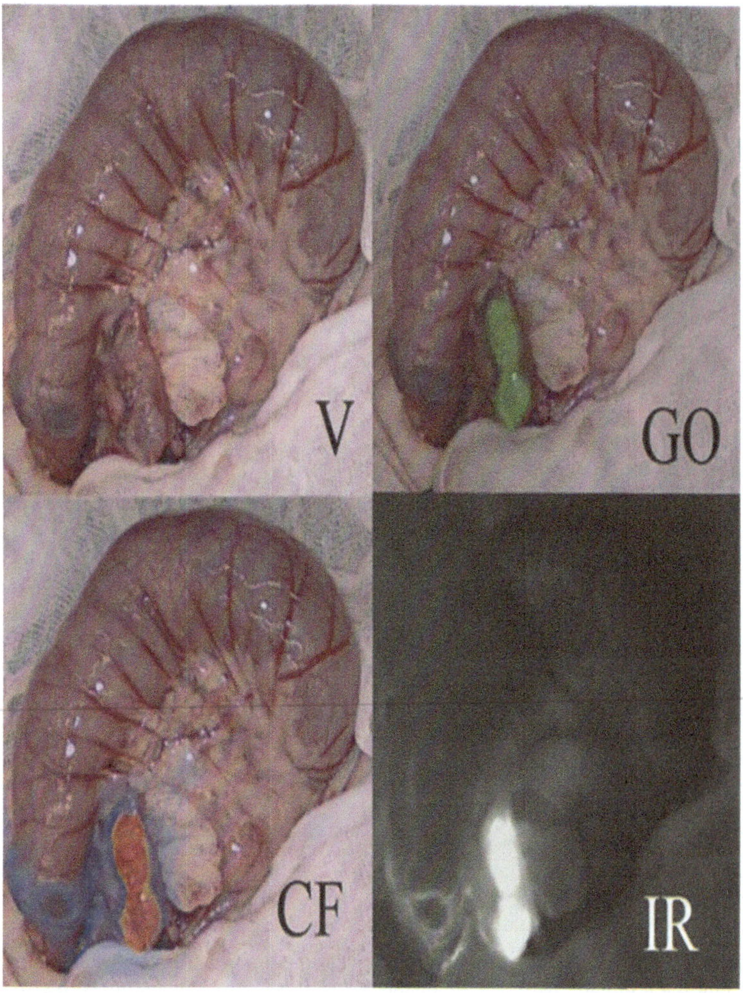

Fig. 1. ICG Fluorescence in Sentinel Lymph Nodes of the Area Affected by Cancer. V: Visible Light, GO: Green Overlay, CF: Visible Light Image with Overlaid Infrared Image, IR: Infrared Fluorescent Image.

surgical outcomes in plastic, cardiothoracic, hepatobiliary, and foregut surgeries [6, 58]. In colorectal surgery specifically, ICG fluorescence imaging has been found to be useful for assessing anastomotic perfusion [68].

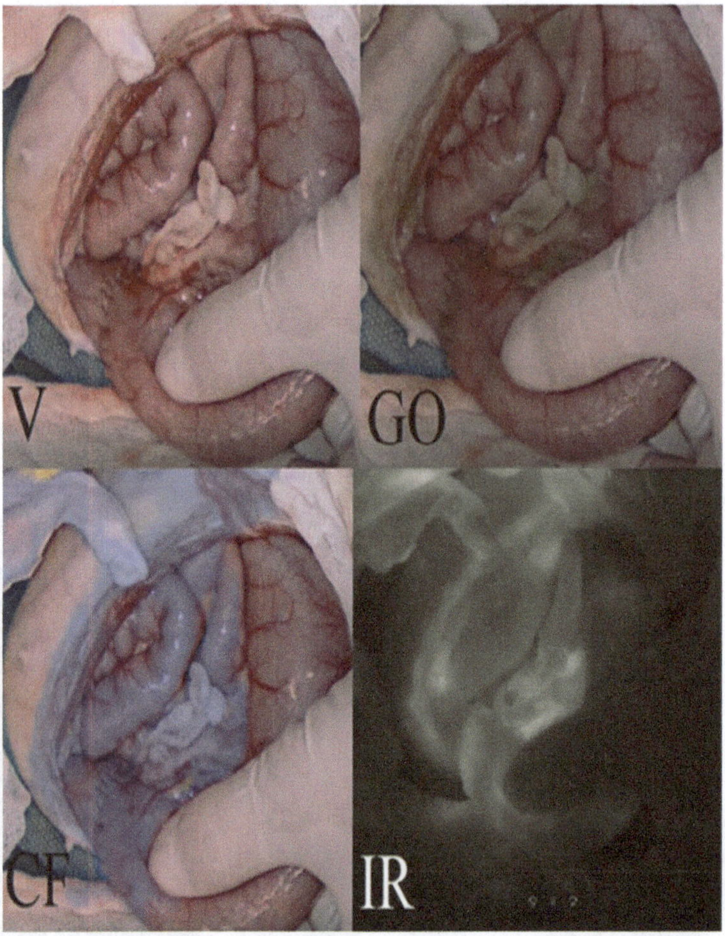

Fig. 2. Intravenous ICG Fluorescence visible in the area after resection and anastomosis. V: Visible Light, GO: Green Overlay, CF: Visible Light Image with Overlaid Infrared Image, IR: Infrared Fluorescent Image

4 Conclusions

The use of near-infrared indocyanine green fluorescence technology has the potential to revolutionize minimally invasive surgery in humans and animals. It can be used to visualize lymphatic flow in real time, identify sentinel lymph nodes, measure intestinal perfusion, and identify the extrahepatic bile duct, among other applications. These advancements could lead to a decrease in mortality rates and postoperative complications. However, due to the lack of standardization in dosage and timing of ICG administration, further clinical and experimental studies are needed, particularly in veterinary medicine where access to the latest technology is limited.

References

1. Aoki, T., et al.: Image guided liver mapping using fluorescence navigation system with indocyanine green for anatomical hepatic resection. World J. Surg. **32**, 1763–1767 (2008). https://doi.org/10.1007/s00268-008-9620-y
2. Ashitate, Y., Tanaka, E., Stockdale, A., Choi, H.S., Frangioni, J.V.: Near-infrared fluorescence imaging of thoracic duct anatomy and function in open surgery and video-assisted thoracic surgery. J. Thorac. Cardiovasc. Surg. **142**, 31–38 (2011). https://doi.org/10.1016/j.jtcvs.2011.03.004. Epub 2011 Apr 7
3. Baiocchi, G.L., Diana, M., Boni, L.: Indocyanine green-based fluorescence imaging in visceral and hepatobiliary and pancreatic surgery: state of the art and future directions. World J. Gastroenterol. **24**(27), 921–930 (2018). https://doi.org/10.3748/wjg.v24.i27.2921
4. Bilbrey, S.A., Birchard, S.J.: Pulmonary lymphatics in dogs with experimentally induced chylothorax. JAAHA **30**, 86–91 (1994)
5. Birchard, S.J., Smeak, D.D., McLoughlin, M.A.: Treatment of idiopathic chylothorax in dogs and cats. J. Am. Vet. Med. Assoc. **212**, 653–657 (1998)
6. Boni, L., David, G., Dionigi, G., Rausei, S., Cassinotti, E., Fingerhut, A.: Indocyanine green-enhanced fluorescence to assess bowel perfusion during laparoscopic colorectal resection. Surg. Endosc. **30**, 2736–2742 (2016). https://doi.org/10.1007/s00464-015-4540-z
7. Buchs, N.C.: Intraoperative fluorescent cholangiography using indocyanin green during robotic single site cholecystectomy. Int. J. Med. Robot. **8**, 436–440 (2012). https://doi.org/10.1002/rcs.1437
8. Buddingh, K.T., Nieuwenhuijs, V.B.: The critical view of safety and routine intraoperative cholangiography complement each other as safety measures during cholecystectomy. J. Gastrointest. Surg. **15**, 1069–1070 (2011). https://doi.org/10.1007/s11605-011-1413-1
9. Casaccia, M., Mora, M., Santori, G., Ghiggi, C., Angelucci, E.: Laparoscopic lymph node biopsy for lymphoma with a novel use of indocyanine green fluorescence in a 66- year-old male patient. Int. J. Surg. Case. Rep. **90**, 106692 (2022). https://doi.org/10.1016/j.ijscr.2021.106692
10. Chiu, C.-H., et al.: Clinical use of near-infrared fluorescence imaging with indocyanine green in thoracic surgery: a literature review. J. Thorac. Dis. **8**(Suppl 9), 744–748 (2016). https://doi.org/10.21037/jtd.2016.09.70
11. Chiu, H.C.: Successful usage of intra metatarsal pad injection of ICG/near-infrared fluorescent lymphography for VATS thoracic duct ligation in two Shiba Inu dogs with recurrent chylothorax. Vet. Surg. **48**, O151 (2019). https://doi.org/10.1111/vsu.13213
12. Culp, W.T.N., Mayhew, P.D., Brown, D.C.: The effect of laparoscopic versus open ovariectomy on postsurgical activity in small dogs. Vet. Surg. **38**, 811–817 (2009). https://doi.org/10.1111/j.1532-950X.2009.00572.x
13. Degett, T.H., Andersen, H.S., Gögenur, I.: Indocyanine green fluorescence angiography for intraoperative assessment of gastrointestinal anastomotic perfusion: a systematic review of clinical trials. Langenbecks Arch. Surg. **401**, 767–775 (2016). https://doi.org/10.1007/s00423-016-1400-9
14. Enwiller, T.M., Radlinsky, M.G., Mason, D.E., Roush, J.K.: Popliteal and mesenteric lymph node injection with methylene blue for coloration of the thoracic duct in dogs. Vet. Surg. **32**, 359–364 (2003). https://doi.org/10.1053/jvet.2003.50044
15. Favril, S., et al.: Fluorescence-guided surgery using indocyanine green in dogs with superficial solid tumours. Vet. Rec. **187**(7), 273 (2020). https://doi.org/10.1136/vr.105554
16. Flum, D.R., Dellinger, E.P., Cheadle, A., Chan, L., Koepsell, T.: Intraoperative cholangiography and risk of common bile duct injury during cholecystectomy. JAMA **289**, 1639–1644 (2003). https://doi.org/10.1001/jama.289.13.1639

17. Fossum, T.W., et al.: Thoracic duct ligation and pericardectomy for treatment of idiopathic chylothorax. J. Vet. Intern. Med. **18**, 307–310 (2004). 10.1892/0891-6640(2004)18<307:tdlapf>2.0.co;2
18. Frangioni, J.V.: New technologies for human cancer imaging. J. Clin. Oncol. **26**, 4012–4021 (2008). https://doi.org/10.1200/JCO.2007.14.3065
19. Fransson, B.A., Mayhew, P.D.: Small Animal Laparoscopy and Thoracoscopy, pp. 236–242. Wiley-Blackwell, Iowa (2015)
20. Gioux, S., Choi, H.S., Frangioni, J.V.: Image-guided surgery using invisible near-infrared light: fundamentals of clinical translation. Mol. Imag. **9**, 237–255 (2010)
21. Hackethal, A., Hirschburger, M., Eicker, S.O., Mücke, T., Lindner, C., Buchweitz, O.: Role of indocyanine green in fluorescence imaging with near-infrared light to identify sentinel lymph nodes, lymphatic vessels and pathways prior to surgery – a critical evaluation of options. Geburtshilfe Frauenheilkd. **78**(1), 54–62 (2018). https://doi.org/10.1055/s-0043-123937
22. Hayashi, K., Sicard, G., Gellasch, K., Frank, J.D., Hardie, R.J., McAnulty, J.F.: Cisterna chyli ablation with thoracic duct ligation for chylothorax: results in eight dogs. Vet. Surg. **34**, 519–523 (2005). https://doi.org/10.1111/j.1532-950X.2005.00078.x
23. Holzinger, F., Krahenbuhl, L., Schteingart, C.D., Ton-Nu, H.T., Hofmann, A.F.: Use of a fluorescent bile acid to enhance visualization of the biliary tract and bile leaks during laparoscopic surgery in rabbits. Surg. Endosc. **15**, 209–212 (2001). https://doi.org/10.1007/s00464 0000265
24. Iida, G., et al.: Intraoperative identification of canine hepatocellular carcinoma with indocyanine green fluorescent imaging. J. Small Anim. Pract. **54**(11), 594–600 (2013). https://doi.org/10.1111/jsap.12148
25. Ishizawa, T., Bandai, Y., Kokudo, N.: Fluorescent cholangiography using indocyanine green for laparoscopic cholecystectomy: an initial experience. Arch. Surg. **144**, 381–382 (2009). https://doi.org/10.1001/archsurg.2009.9
26. Ito, N., Fukuta, M., Tokushima, T., Nakai, K., Ohgi, S.: Sentinel node navigation surgery using indocyanine green in patients with lung cancer. Surg. Today **34**, 581–585 (2004). https://doi.org/10.1007/s00595-004-2780-y
27. Kaburagi, T., et al.: Intraoperative fluorescence lymphography using indocyanine green in a patient with chylothorax after esophagectomy—report of a case. Surg Today. **43**, 306–310 (2013). https://doi.org/10.1007/s00595-012-0391-6
28. Kamijo, K., Kanai, E., Oishi, M., Ichihara, N., Asari, M., Yamada, K.: Perirectal injection of imaging materials for computed tomographic lymphography and near infrared fluorescent thoracoscopy in cats. Vet. Med. **64**, 342–347 (2019). https://doi.org/10.17221/32/2019-VETMED
29. Kawaguchi, Y., et al.: Portal uptake function in veno-occlusive regions evaluated by real-time fluorescent imaging using indocyanine green. J. Hepatol. **58**, 247–253 (2013). https://doi.org/10.1016/j.jhep.2012.09.028
30. Khan, O.A., Balaji, S., Branagan, G., Bennett, D.H., Davies, N.: Randomized clinical trial of routine ontable cholangiography during laparoscopic cholecystectomy. Br. J. Surg. **98**, 362–367 (2011). https://doi.org/10.1002/bjs.7423
31. Kim, H.K., et al.: Intraoperative pulmonary neoplasm identification using near-infrared fluorescence imaging. Eur. J. Cardiothorac. Surg. **49**, 1497–1502 (2016). https://doi.org/10.1093/ejcts/ezv367
32. Kitai, T., Inomoto, T., Miwa, M., Shikayama, T.: Fluorescence navigation withindocyanine green for detecting sentinel lymph nodes in breastcancer. Breast Cancer **12**(3), 211–215 (2005). https://doi.org/10.2325/jbcs.12.211
33. Korpita, M.F., et al.: Thoracoscopic detection of thoracic ducts after ultrasoundguided intrahepatic injection of indocyanine green detected by near-infrared fluorescence and methylene

blue in dogs. Vet. Surg. **51**(Suppl 1), O118–O127 (2022). https://doi.org/10.1111/vsu.13682. Epub 2021 Jul 23
34. LaFond, E., Weirich, W.E., Salisbury, S.K.: Omentalization of the thorax for treatment of idiopathic chylothorax with constrictive pleuritis in a cat. J. Am. Anim. Hosp. Assoc. **38**, 74–78 (2002). https://doi.org/10.5326/0380074
35. Larose, P.C., Brisson, B., Sanchez-Lazaro, A., Monteith, G., Singh, A., Zhang, M.: Near-infrared fluorescence cholangiography in dogs: a pilot study. In: ACVS (2021). https://www.eventscribe.net/2021/ACVS/fsPopup.asp?Mode=presInfo&PresentationID=915337
36. Lim, C., et al.: Indocyanine green fluorescence imaging in the surgical management of liver cancers: current facts and future implications. J. Visc. Surg. **151**(2), 117–124 (2014). https://doi.org/10.1016/j.jviscsurg.2013.11.003
37. MacDonald, N.J., Noble, P.J.M., Burrow, R.D.: Efficacy of en bloc ligation of the thoracic duct: descriptive study in 14 dogs. Vet. Surg. **37**(7), 696–701 (2008). https://doi.org/10.1111/j.1532-950X.2008.00437.x
38. Matsui, A., et al.: Real-time intra-operative near-infrared fluorescence identification of the extrahepatic bile ducts using clinically available contrast agents. Surgery **148**, 87–95 (2010). https://doi.org/10.1016/j.surg.2009.12.004
39. Mayhew, P.D., Mehler, S.J., Radhakrishnan, A.: Laparoscopic cholecystectomy for management of uncomplicated gall bladder mucocele in six dogs. Vet. Surg. **37**, 625–630 (2008). https://doi.org/10.1111/j.1532-950X.2008.00428.x
40. Miyashiro, I., et al.: Intraoperative diag-nosis using sentinel node biopsy with indocyanine green dyein gastric cancer surgery: an institutional trial by experiencedsurgeons. Ann. Surg. Oncol. **20**(2), 542–546 (2012). https://doi.org/10.1245/s10434-012-2608-8
41. Moukarzel, L.A., Feinberg, J., Levy, E.J., Leitao, M.M.: Current and novel mapping substances in gynecologic cancer care. Int. J. Gynecol. Cancer **30**(3), 387–393 (2020). https://doi.org/10.1136/ijgc-2019-001078
42. Muntean, M.V., Muntean, V., Ardelean, F., Georgescu, A.: Dynamic perfusion assessment during perforator flap surgery: an upto-date. Clujul. Med. **88**, 293–297 (2015). https://doi.org/10.15386/cjmed-484. Epub 2015 Jul 1
43. Noguera, J., Castro, L., Garcia, L., Mosquera, C., Gomez, A.: Lymphadenectomy guided by indocyanine-green (ICG) in colorectal cancer: a pilot study. J. Surg. Tech. Proced. **3**(1), 1023 (2019)
44. Noura, S., et al.: Feasibility of a lateral regionsentinel node biopsy of lower rectal cancer guided by indo-cyanine green using a near-infrared camera system. Ann. Surg. Oncol. **17**(1), 144–151 (2009). https://doi.org/10.1245/s10434-009-0711-2
45. Okusanya, O.T., et al.: Intraoperative near-infrared imaging can identify pulmonary nodules. Ann. Thorac. Surg. **98**, 1223–1230 (2014). https://doi.org/10.1016/j.athoracsur.2014.05.026
46. Pacheco, P.E., Hill, S.M., Henriques, S.M., Paulsen, J.K., Anderson, R.C.: The novel use of intraoperative laser-induced fluorescence of indocyanine green tissue angiography for evaluation of the gastric conduit in esophageal reconstructive surgery. Am. J. Surg. **205**, 349–352 (2013). https://doi.org/10.1016/j.amjsurg.2012.11.005
47. Park, S.Y., Park, J.S., Kim, H.Y., Woo, I.T., Park, I.K., Choi, G.-S.: Indocyanine green fluorescence imaging-guided laparoscopic surgery could achieve radical d3 dissection in patients with advanced right-sided Colon Cancer. Dis. Colon. Rectum. **63**, 441–449 (2020). https://doi.org/10.1097/DCR.0000000000001597
48. Peltrini, R., et al.: Intraoperative use of indocyanine green fluorescence imaging in rectal cancer surgery: the state of the art. World J. Gastroenterol. **27**(38), 6374–6386 (2021). https://doi.org/10.3748/wjg.v27.i38.6374
49. Purich, K., et al.: Intraoperative fluorescence imaging with indocyanine green in hepatic resection for malignancy: a systematic review and meta-analysis of diagnostic test accuracy studies. Surg. Endosc. **34**(7), 2891–2903 (2020). https://doi.org/10.1007/s00464-020-07543-2

50. Quinlan, A.S.F., Wainberg, S.H., Phillips, E., Oblak, M.L.: The use of near infrared fluorescence imaging with indocyanine green for vascular visualization in caudal auricular flaps in two cats. Vet. Surg. **50**(3), 677–686 (2021). https://doi.org/10.1111/vsu.13577
51. Ribero, D., Mento, F., Sega, V., Lo Conte, D., Mellano, A., Spinoglio, G.: ICG-guided lymphadenectomy during surgery for colon and rectal cancer—interim analysis of the GREENLIGHT trial. Biomedicines. **10**, 541 (2022). https://doi.org/10.3390/biomedicines 10030541
52. Rossanese, M., Williams, P., Tomlinson, A., Cinti, F.: Long-term outcome after cholecystectomy without common bile duct catheterization and flushing in dogs. Animals **12**, 2112 (2022). https://doi.org/10.3390/ani12162112
53. Sakurai, N.: Clinical impact of near-infrared fluorescence imaging with indocyanine green on surgical treatment for hepatic masses in dogs. BMC Vet. Res. **18**(1), 374 (2022). https://doi.org/10.1186/s12917-022-03467-2
54. Sanford, D.E.: An update on technical aspect of cholecystectomy. Surg. Clin. N. Am. **99**, 245–258 (2018). https://doi.org/10.1016/j.suc.2018.11.005
55. Schaafsma, B.E., et al.: The clinical use of indocyanine green as a near infrared fluorescent contrast agent for image-guided oncologic surgery. J. Surg. Oncol. **104**, 323–332 (2011). https://doi.org/10.1002/jso.21943
56. Scott, J., et al.: Perioperative complications and outcome of laparoscopic cholecystectomy in 20 dogs. Vet. Surg. **45**(S1), O49–O59 (2016). https://doi.org/10.1111/vsu.12534
57. Shariati, E., Bakhtiari, J., Khalaj, A., Niasari-Naslaji, A.: Comparison between two portal laparoscopy and open surgery for ovariectomy in dogs. Vet. Res. Forum. **5**, 219–223 (2014)
58. Shen, R., Zhang, Y., Wang, T.: Indocyanine green fluorescence angiography and the incidence of anastomotic leak after colorectal resection for colorectal cancer: a meta-analysis. Dis. Colon Rectum. **61**(10), 1228–1234 (2018). https://doi.org/10.1097/DCR.0000000000001123
59. Smeak, D.D., et al.: Treatment of chronic pleural effusion with pleuroperitoneal shunts in dogs: 14 cases (1985–1999). J. Am. Vet. Med. Assoc. **219**, 1590–1597 (2001). https://doi.org/10.2460/javma.2001.219.1590
60. Spinoglio, G., et al.: Real-time near-infrared (NIR) fluorescent cholangiography in single-site robotic cholecystectomy (SSRC): a single-institutional prospective study. Surg. Endosc. **27**, 2156–2162 (2013). https://doi.org/10.1007/s00464-012-2733-2
61. Steffey, M.A., Mayhew, P.D.: Use of direct near-infrared fluorescent lymphography for thoracoscopic thoracic duct identification in 15 dogs with chylothorax. Vet. Surg. **47**, 267–276 (2018). https://doi.org/10.1111/vsu.12740. Epub 2017 Nov 6
62. Suh, Y.J., et al.: Indocyanine green as a near-infrared fluorescent agent for identifying parathyroid glands during thyroid surgery in dogs. Surg. Endosc. **29**(9), 2811–2817 (2015). https://doi.org/10.1007/s00464-014-3971-2
63. Tan, I.C., et al.: Investigational lymphatic imaging at the bedside in a pediatric postoperativechylothorax patient. Pediatr. Cardiol. **35**, 1295–1300 (2014). https://doi.org/10.1007/s00 246-014-0946-y
64. Tanaka, R., Nakashima, K., Fujimoto, W.: Sentinel lymph nodedetection in skin cancer using fluorescence navigation withindocyanine green. J. Dermatol. **36**(8), 468–470 (2009). https://doi.org/10.1111/j.1346-8138.2009.00679.x
65. Tobis, S., et al.: Near infrared fluorescenceimaging after intravenous indocyanine green: initial clinicalexperience with open partial nephrectomy for renal corticaltumors. Urology **79**(4), 958–964 (2012). https://doi.org/10.1016/j.urology.2011.10.016
66. Townsend, K.L., Milovancev, M., Bracha, S.: Feasibility of near-infrared fluorescence imaging for sentinel lymph node evaluation of the oral cavity in healthy dogs. Am. J. Vet. Res. **79**(9), 995–1000 (2018). https://doi.org/10.2460/ajvr.79.9.995
67. Ushijima, H., et al.: Visualization of lymphatic flow in laparoscopic colon cancer surgery using indocyanine green fluorescence imaging. Sci. Rep. **10**, 14274 (2020)

68. Wada, T., et al.: ICG fluorescence imaging for quantitative evaluation of colonic perfusion in laparoscopic colorectal surgery. Surg. Endosc. **31**, 4184–4193 (2017). https://doi.org/10.1007/s00464-017-5475-3
69. Wada, T., Kawada, K., Hanada, K., Obama, K.: Quantitative analysis of colonic perfusion using ICG fluorescence angiography and its consequences for anastomotic healing in a rat model. Cancers **14**, 4024 (2022). https://doi.org/10.3390/cancers14164024
70. Wakaiki, S., et al.: Indocyanine green angiography for examining the normal ocular fundus in dogs. J. Vet. Med. Sci. **69**(5), 465–470 (2007). https://doi.org/10.1292/jvms.69.465
71. Wan, J., Oblak, M.L., Ram, A., Singh, A., Nykamp, S.: Determining agreement between preoperative computed tomography lymphography and indocyanine green near infrared fluorescence intraoperative imaging for sentinel lymph node mapping in dogs with oral tumours. Vet. Comp. Oncol. **19**, 295–303 (2021). https://doi.org/10.1111/vco.12675
72. Wang, C., et al.: Application of near-infrared fluorescent cholangiography using indocyanine green in laparoscopic cholecystectomy. J. Int. Med. Res. **48**(12), 300060520979224 (2020). https://doi.org/10.1177/0300060520979224
73. Yamashita, S., et al.: Video-assisted thoracoscopic indocyanine green fluorescence imaging system shows sentinel lymph nodes in non-small-cell lung cancer. J. Thorac. Cardiovasc. Surg. **141**, 141–144 (2011). https://doi.org/10.1016/j.jtcvs.2010.01.028
74. Yang, F., Zhou, J., Hao, L.: Near-infrared fluorescence-guided thoracoscopic surgical intervention for post-operative chylothorax. Interact Cardiovasc. Thorac. Surg. **26**, 171–175 (2018). https://doi.org/10.1093/icvts/ivx3

Development of Cloud ERP Model and Its Application in Smart Mining

Vidosav Majstorovic[1], Vladimir Simeunovic[2], Radivoje Mitrovic[1], Dragan Stosic[2], Sonja Dimitrijevic[2], and Zarko Miskovic[1](✉)

[1] Faculty of Mechanical Engineering, University of Belgrade, Kraljice Marije 16, 11120 Belgrade 35, Serbia
zmiskovic@mas.bg.ac.rs
[2] Institute Mihajlo Pupin, Volgina 15, 11060 Belgrade, Serbia

Abstract. The concept of smart mining allows for more efficient and sustainable power production, as well as improved safety and worker conditions. It also enables companies to be more responsive to market demands and changes in resource availability. The use of digital technologies and smart systems in mining and power production is therefore crucial for the development of Industry 4.0 and the transition towards a more sustainable and interconnected economy. Based on these approaches, the paper presents a developed and applied private cloud ERP model, which has three modules: virtual part – based on cloud computing model (SaaS model) or private cloud, business part – uses IoT to connect business processes (procurement, sales, management, finance, warehouse, downtime tracking), production part – tracking production (at the open-pit coal mine) and technological processes (tracking and maintenance of auxiliary equipment). Most of these elements have already been applied, as shown in detail in the paper.

Keywords: Industry 4.0 · Mining · ERP · Cloud

1 Introduction

Smart mines are able to optimize their production processes with the help of real-time data analysis, automation and remote control of mining operations, and predictive maintenance. This translates to significant improvements in production efficiency and output, as well as reduction in downtime and repair costs. Moreover, smart mines are able to improve safety in mining operations by providing real-time tracking of workers, equipment, and vehicles. This allows for better monitoring of safety risks and reducing the likelihood of accidents. Additionally, through the use of advanced analytics, smart mines are able to identify trends and patterns in safety incidents and address them proactively, thereby improving safety performance [1].

In the mining industry, ERP systems are used to manage production planning and control, inventory management, and resource allocation. This includes scheduling and tracking of equipment, labor, and materials, as well as monitoring of production processes and performance metrics. The development of cloud-based ERP systems has

A. Burduk et al. (Eds.): ISPEM 2023, LNNS 790, pp. 28–42, 2023.
https://doi.org/10.1007/978-3-031-45021-1_3

further improved the accessibility and flexibility of production management. These systems can be accessed from anywhere with an internet connection and are often more cost-effective than traditional on-premise software installations. With the emergence of Industry 4.0, the focus has shifted to smart, connected systems that use real-time data and advanced analytics to optimize production processes and improve efficiency. This includes the use of automation, remote monitoring, and predictive maintenance to reduce downtime and improve safety. An overview of the development of this model is given in Table 1 [1–4], with specifically added application characteristics for mining. Additionally, each subsequent level contains all the characteristics of the previous level.

Table 1. History of ERP model development (adopted according [1–4])

Year/model/Level	Characteristics of MRP/ERP model (in manufacturing)	Aspects of function (in manufacturing)	Application in mining
1960s/IC/I level	Inventory management and control	Warehouse control	Spare parts control for maintenance
1970s/MRP I/II level	Material Requirements Planning	Bill of product	Bill of equipment maintenance component
1980s/MRP II/III level	Manufacturing Resources Planning	Bill of manufacturing process	Bill of technological component of equipment maintenance
1990s/ERP/ IV level	Enterprise Resource Planning	Integrate business activities across organization units	Enterprise Resource Planning for mining
2000s/ERP II/V level	Enterprise Resource Planning by Internet	Services Oriented Architecture (SOA)	Enterprise Resource Planning by Internet for mining
2010s/Cloud based ERP/ VI level	Cloud based ERP	ERP as software a service (SaaS) model	Cloud based ERP for mining
2020s/Industry 4.0 ERP/VII level	ERP of Industry 4.0 model	Industry 4.0 concept introducing	ERP of Industry 4.0 model for mining

The ERP model has evolved through different levels of development, each building on the previous one. The fifth level, ERP II, uses internet web browsers with a service-oriented architecture (SOA) and is supported by mobile devices. The sixth level, cloud ERP, is a business application delivered as a service model (SaaS) and is designed for small and medium-sized enterprises. The seventh and final level is the ERP model for Industry 4.0, which is based on a networking concept that leverages the internet, cloud computing, and AI/ML tools. The development of the first four levels of the ERP model is based on the development of information systems, with IC technology being included in the development of the ERP model. On the other hand, the latest model is based on a networking concept that leverages the internet, cloud computing, and AI/ML tools in the Industry 4.0 model. The research presented in this paper focuses on the last two levels of the ERP model. Specifically, it explores the use of cloud ERP and Industry 4.0 concepts in the context of open-pit coal mine maintenance resource planning and management, highlighting the benefits and challenges associated with these technologies.

The aim of this research is to develop a model for mining based on analogies with the ERP model for production [1] in the context of the Industry 4.0 model, as a private cloud model. Also, this can be called a hybrid model, because it was created on the existing information system, to which the elements of Industry 4.0 are added.

This paper is divided into five main sections. The first section is the introduction, which gives an overview of the development of the ERP model in the context of production and its application in smart mining. The introduction also highlights three areas where the ERP model can be applied in smart mining, as stated in the abstract. The second section of the paper analyses the basic concept of Industry 4.0 in mining. It discusses the importance of data analytics, automation, and digital transformation in creating smart and efficient mining operations. The third and four section of the paper presents the ERP model for smart mining as a private cloud solution. This section provides examples of how this model can be applied in practice, particularly in maintenance resource planning and management at open-pit coal mines. The five as final section of the paper looks at directions for future research. It highlights the need for big data analytics (BDA) supported by artificial intelligence/machine learning (AI/ML) to further enhance the efficiency of mining operations.

2 The Key Characteristics of Implementing Industry 4.0 Elements in Mining

The first characteristic relates to better management of operational costs [5–8]. Wireless monitoring provides connectivity among entities in the mine, today using the Internet of Things (IoT). Smart mines are essentially those whose key machines and devices (cyber-physical systems – CPS) are digitized through embedded sensors, which transmit data to a central system (server or private cloud) via a wireless network. As such, mines reduce costs not only of human labor – as people no longer have to physically perform, for example, manual data readings on site – but also because these systems move to predictive maintenance (condition monitoring). Mines not only save money by reducing these unnecessary costs, but by using intelligent operational management solutions as part of their wireless smart network, they are able to predict hazardous situations, reducing costs by avoiding sudden breakdowns that can completely shut down the mine for days or weeks. Smart mines, therefore, have a higher return on investment (ROI) and are more competitive than traditional mines. With wireless monitoring, mines will digitize their work machines, turning them into CPS and track their performance and condition monitoring (for predictive maintenance) in real-time, using embedded sensors, IoT and GPS systems.

The second dimension of smart mines is the automatization of mining operations [8, 9]. Smart mines with already integrated technologies can more easily automate their operations than those that lag behind in the application of these technologies. Mines can build on their existing wireless network of embedded sensors, for example, by installing automatization software that not only allows them to remotely control stationary and mobile assets but actually program automated interactions between them. Self-driving trucks are an excellent example of this, relying on both the physical network of sensors and remotely programmed and controlled software for autonomous mining operations transport, which previously required a fleet of truck drivers. Digitizing a mine in one way – for example, installing a wireless network of sensors for remote reading – can make further digitalization easier, such as automatizing key mining processes. Autonomous vehicles are revolutionizing mining, allowing humans to communicate with machines and control them remotely.

The third dimension of advanced mine models is workforce management [10]. Progress in IoT technologies makes them an ideal tool for monitoring the workforce in mines. They are particularly useful for field maintenance and real-time machine check-in instructions, as well as for improving care and safety for people – especially for underground mines. These IoT devices allow for RFID tracking of personnel and real-time visibility of the entire workforce, enabling instant location of key workers. They can also be used to identify areas where productivity can be increased, and to track work hours and employee engagement. Regarding workforce safety, personal devices and their connectivity to an IoT-based tracking network significantly improve emergency response time, allowing both operators and personnel to be immediately notified when an incident occurs. They can also minimize safety issues and prevent future accidents by restricting access to potentially dangerous areas in the mine. Workforce tracking also enables greater mining productivity, as workers are managed more efficiently, and potential incidents that would reduce working time are reduced or even eliminated.

The next dimension of smart mines is intelligent operational management [10]. Wireless monitoring and predictive maintenance of mining equipment and machines, enabled through operational management solutions, allow mines to remotely and continuously monitor working machines. Real-time digital data generation replaces expensive manual readings of critical operational data. Intelligent operational management is a key technology for creating digital twins of mines. It allows for the digitalization of mining processes and systems to get real-time tracking of everything, enabling data-driven decision-making in the mine. This means the digitalization of dynamic entities such as working machines, processes, and critical assets (slopes, dams, mining fields, etc.) to visualize their behavior, collect this information, and generate recommendations through AI/ML-based software, which are based on historical facts and real-time data. These solutions learn from past events through AI/ML, improving decision-making processes. Operational intelligence generates an overview of all mining operations in real-time and allows operators to make data-driven, informed decisions, both current and past (BDA).

Finally, the last dimension of smart mines is data safety and digital twins [4, 5]. Mining companies heavily protect their data, and they must especially have control over this information in the event of an incident or accident. One common way to protect data is by using local data collection systems, which help to avoid intrusion as their wireless network of geophysical sensors is concentrated in the monitored location. Instead of data "floating" in the cloud located on the internet, these systems collect and process their data on servers or a private cloud, which is only connected to the mine network. Mines traditionally struggle to manage engineering and asset information throughout the lifecycle of machines and equipment. Digital twins can help with this: they are virtual replicas of physical assets, processes, and systems, which provide a representation of the basic elements and dynamics of IoT devices used in the displayed space or system. Digital twins use AI (artificial intelligence), machine learning (a part of AI), and big data software analytics (BDA) to display digital simulation models in real-time (the digital world), which are continuously updated as their real, physical models (the real world).

3 ERP Model for Mining – A Literature Review

If the open-pit coal mining production is analyzed, smart production on it is based on the application of four dimensions of basic technologies for Industry 4.0 [1, 5, 11, 12]: (i) cloud manufacturing, (ii) Internet of Things (for mining), (iii) big data – BDA (for mining), and (iv) AI/ML analytics (for mining). This concept defines a business model framework for organizations in the circular economy – eco-sustainable smart production [11]. In this concept, the ERP model is a key element of the vertical integration model, but is also part of the smart chains of basic technologies for Industry 4.0 (simulation, big data, cloud computing, integration, AI and ML), for the organization that applies it, which in this case is the mine.

The Industry 4.0 model includes three integration concepts that are applied in various industries [1, 11, 13]. The first is vertical integration, which involves connecting processes and sensors to the organization's corporate planning systems, such as ERP. The integration center at this level is the cyber-physical system (CPS), which is used to realize production. An example of this in the mining industry would be the use of sensors on an excavator to collect data on its performance, which is then fed into the organization's ERP system to inform production planning and scheduling. The second integration concept is horizontal, which connects the demand for the product with its design, production, and delivery. The integration center here is an intelligent product with added value. This approach is commonly used in the mechanical engineering industry, where smart products are designed to meet customer needs and are integrated across the production process. The third integration concept is focused on supply chains and sustainable production, with a center for managing customer relationships (CRM) integration. This approach is also commonly used in the production industry, but is not directly applicable to the mining sector.

In the observed case, the open-pit coal mine is building a new business model for its Industry 4.0 concept, with a center on a large database and its reporting, where the ERP model plays a key role in planning and managing maintenance of machinery systems. On the other hand, a detailed analysis of the application of the ERP model in mining shows the same facts about their characteristics as explained for our model in the following text [14–17].

The above analysis did not include the various Platforms for Industry 4.0 which can also be used in mining (IoT, BDA, ML), because our approach was different.

4 ERP Model for Smart Mining

The planning and management of maintenance resources at the open-pit coal mine are carried out using a cloud private ERP model [1]. This model is an upgrade to the existing information and communication system (ICS) applied at the mine. The model has been developed and presented in a paper, and is designed to improve the efficiency of maintenance resource planning and management. The cloud private ERP model provides a centralized system for managing maintenance resources and planning maintenance activities. It allows for real-time monitoring of equipment performance and condition, enabling predictive maintenance and reducing downtime. The ERP model is integrated with the mine's existing information and communication system, as shown in Fig. 1 [18, 19]. This allows for seamless communication between different departments and stakeholders involved in maintenance planning and management.

The developed model consists of three parts [1, 13–15]: (i) a virtual part based on the private cloud computing (SaaS) model. It contains a virtual model of the ERP system that connects business processes (procurement, sales, management, finance, warehouse, downtime), production (at the mines), and technological processes (monitoring and maintenance of machinery) using IoT.

The ERP generates a large database (BDA), which, through analysis and synthesis, along with machine learning and AI tools, optimizes maintenance decisions based on condition, plans preventive maintenance, and manages spare parts inventory, (ii) interface (suppliers, services, and maintenance management). Their function is to allow maintenance managers to monitor the entire system online, and (iii) core ERP/MES model, in addition to the above function, also includes other business-technical and management functions of the open-pit coal mines, usually in online mode.

Considering the complexity and specificity of basic and auxiliary mining operations, as well as comprehensive business processes, the development and implementation of the private cloud ERP model as a digital platform is carried out in phases [1, 18, 19].

The first module discussed in this section focuses on procurement of fuel, technical fluids, and lubricants for mining operations. The planning process is based on consumption data from the previous period and an analysis of the scope and type of planned work for the upcoming period. Digital data and reports available in the MES (Mining Execution System) provide essential information for the procurement planning process. The MES is a web-based information system that is specifically developed for the auxiliary mining operations sector. It covers end-to-end business processes, including operations management and maintenance management. By connecting the Operations and Maintenance departments through the digital platform, business processes are simplified, information is obtained quickly, and work can be done without relying on paper-based systems. One of the many online reports available on the MES is an overview of planned maintenance operations on a daily basis, as shown in Fig. 2. This report allows for effective planning and efficient allocation of resources for maintenance operations.

The module discussed in this section provides dynamic reports that display key performance indicators (KPIs) related to daily machine work between two specified dates. The report provides information on the average and normative fuel consumption of the machine (represented as a percentage), as well as the ratio of idle time to machine movement time (also represented as a percentage). The report also includes detailed data

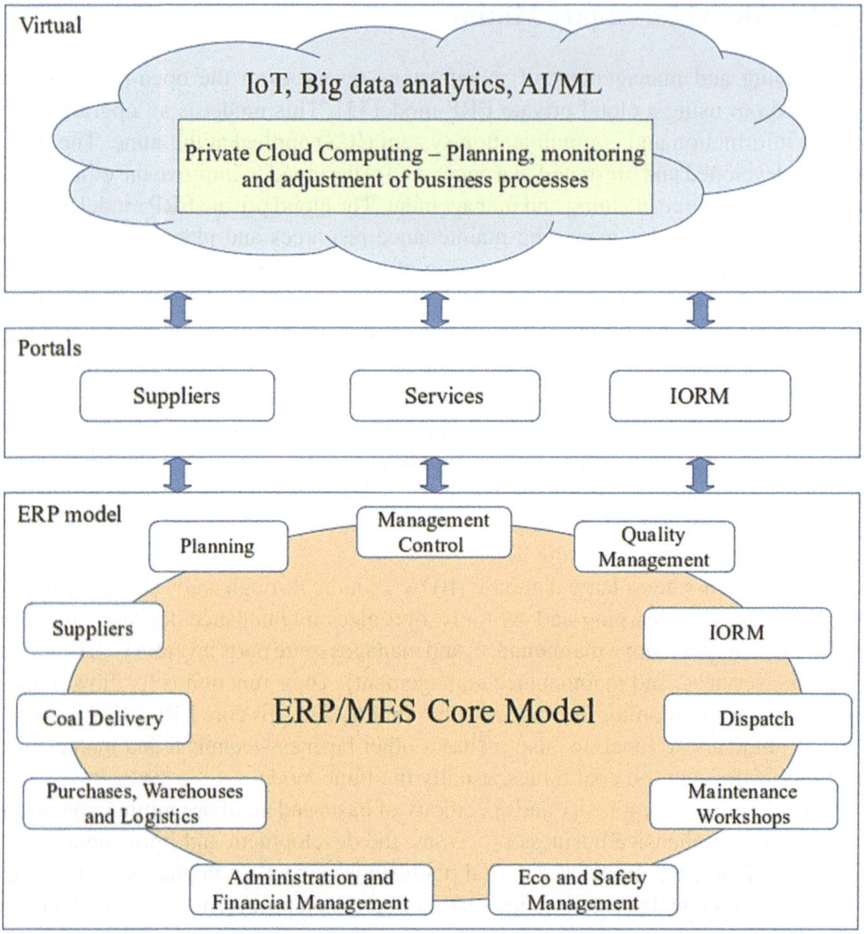

Fig. 1. Cloud ERP model for smart mining (adopted according [1, 13–15, 18, 19]).

relevant to each KPI. This report is dynamic and can be adjusted based on the report parameters, allowing users to control the data displayed and adjust the presentation of the report. Similar reports are generated for other KPIs relevant to production and maintenance operations, as shown in Fig. 3.

The second module discussed in this section is the fleet tracking system used in an open-pit coal mine. This system utilizes a web-based geo-tool and integrates GPS/GPRS and GIS technologies to support dispatching of excavated coal, fleet management, and maintenance management. The system is fully integrated with the MES and provides real-time tracking of machine positions on the mine map, including key vehicle parameters such as GPS position, engine running time, device running time, fuel volume in liters and percentages, and time of last message and panic alarm. The fleet tracking system also generates machine/vehicle movement history reports, which provide detailed analysis and visualization of route movements, fuel consumption, working hours, and

1 2 3 4 5 6 7 8 9 10 11-20 Next Last Records 1 to 15 of 24690 15

	Number	Int. no.	Work place (workshop/on site)	Maintenance type	Object of maintenance	Diagnostics	Regular service	Date	Verification	Remark
1.	1050/22	**PT6**	Rad_PM MAŠINE	Preventivno	Mašina	**Potreban redovan servis**		06.05.2022 07:34	**Da**	
2.	1047/22	**IVEKO 9**	Rad_PM VOZILA	Korektivno	Sistem za kočenje	**potrebna popravka kočnica**		05.05.2022 14:19	**Da**	
3.	1046/22	**TO-27**	Rad_PM VOZILA	Korektivno	Mehanizam za upravljanje	**potrebna popravka servo upravljača**		05.05.2022 11:32	**Da**	Vozilo ide na popravku u Autoturing
4.	1045/22	**Lada Niva 24**	Rad_PM VOZILA	Korektivno	Elektrika i elektronika	**potrebna zamena kaisa alternatora**		05.05.2022 12:09	**Da**	
5.	1045/22	**Lada Niva 24**	Rad_PM VOZILA	Korektivno	Elektrika i elektronika	**potrebna zamena pk kaisa**		05.05.2022 12:10	**Da**	
6.	1044/22	**C38**	Rad_PM MAŠINE	Preventivno	Mašina	**Potreban redovan servis**		05.05.2022 12:44	**Da**	
7.	1043/22	**UAZ112**	Rad_PM VOZILA	Korektivno	Kardanski prenosnik	**potrebno nameštanje kardana**		05.05.2022 11:25	**Da**	
8.	1042/22	**RENAULT 2**	Rad_PM VOZILA	Korektivno	Vozilo	**potrebna popravka blatobrana**		05.05.2022 13:14	**Da**	
9.	1037/22	**PT2**	Rad_PM MAŠINE	Korektivno	Elektrika i elektronika	**Potrebna zamena akumulatora**		05.05.2022 07:46	**Da**	
10.	1035/22	**RK21**	Rad_PM MAŠINE	Preventivno	Mašina	**Redovni servis**	**ms1**	05.05.2022 09:00	**Da**	
11.	1034/22	**K2**	Rad_PM MAŠINE	Preventivno	Mašina	**Redovni servis**	**ms1**	05.05.2022 07:52	**Da**	
12.	1033/22	**B26**	Rad_PM MAŠINE	Preventivno	Mašina	**Redovni servis**	**ms1**	06.05.2022 09:48	**Da**	
13.	1032/22	**Lada Niva 14**	Rad_PM VOZILA	Korektivno	Motor	**potrebno saniranje curenja goriva**		04.05.2022 13:33	**Da**	
14.	886/22	**UAZ103**	Rad_PM VOZILA	Korektivno	Sistem za kočenje	**potrebna popravka kočnica**		04.05.2022 13:40	**Da**	
15.	1031/22	**TO-2**	Rad_PM VOZILA	Korektivno	Prednji most	**potrebna popravka prednjeg trapa**		04.05.2022 13:23	**Da**	

1 2 3 4 5 6 7 8 9 10 11-20 Next Last Records 1 to 15 of 24690 15

Fig. 2. Report on machinery maintenance operations for a selected period by the day.

mileage, among other parameters. These reports are critical for decision-making related to fleet management, including the allocation of resources, identification of maintenance requirements, and identification of potential hazards.

The third module relates to order management (MES) for production, maintenance, fuel, lubricants, and technical fluids. On the other hand, online data for this module are obtained based on data collected from machine/vehicle sensors in the fleet tracking system. Comparative reviews of these reports from online MES (e.g. on working hours, fuel consumption, etc.) and the previous offline manual data source have contributed to rapid improvement in accuracy, speed, and quality of input data for operations and maintenance management, and therefore, an improvement in the procurement process.

Any significant deviation in data from these two sources is an alarm for a detailed check. Such careful checks have revealed some causes of data inaccuracy, such as human error in delayed data entry from paper/fuel order to MES or negligence of some employees in filling out paper forms. The paper forms were initially used when filling machines/vehicles at the workplace or in workshops. Examples of two maintenance reports from MES for a machine, as well as a summary of work orders, are provided in Figs. 4 and 5.

Also, part of this module is the implementation of industrial PDAs (Personal Digital Assistants) for digital record keeping of fuel dispensing in machines and vehicles. The PDAs are used by maintenance workers in workshops, operators in the technical fluids and lubricants warehouse (maintenance department), as well as fuel tank operators (operations department). Lubricants and technical fluids are used in maintenance for breakdowns (on site) as well as preventive maintenance (in workshops). Thanks to the fleet tracking system, the dispatcher can suggest the optimal route for reaching the machines and vehicles on site. This application is integrated with the MES so that data obtained in this way is linked to other relevant data using the operator's

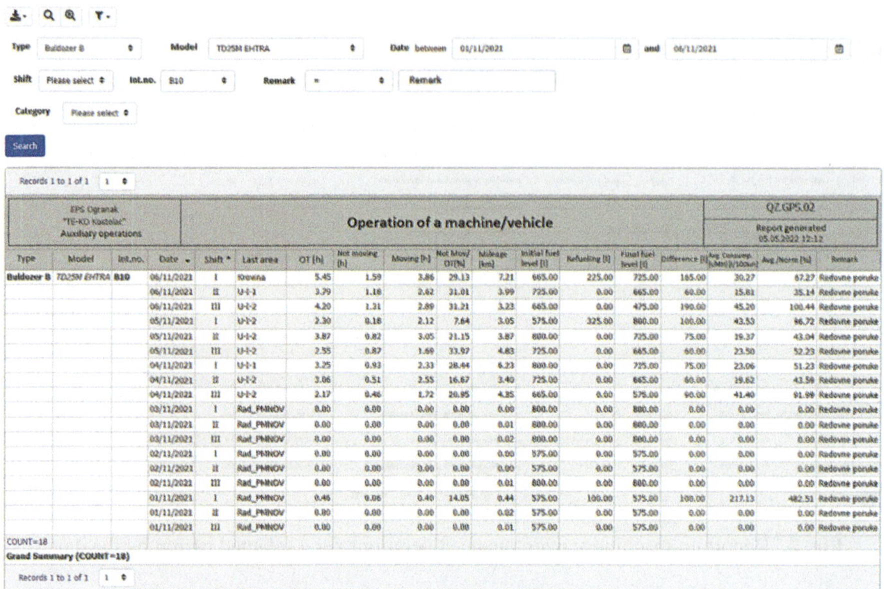

EPS Ogranak "TE-KO Kostolac" Auxiliary operations			Operation of a machine/vehicle													QZ.GPS.02 Report generated 05.05.2022 12:12	
Type	Model	Int.no.	Date	Shift	Last area	OT [h]	Not moving [h]	Moving [h]	Not Mov/OT[%]	Mileage [km]	Initial fuel level [l]	Refueling [l]	Final fuel level [l]	Difference [l]	Avg. Consump. [l/MH] [l/100km]	Avg./Norm [%]	Remark
Buldozer B	TD25M EHTRA	B10	06/11/2021	I	Krevina	5.45	1.59	3.86	29.13	7.21	665.00	225.00	725.00	165.00	30.27	67.27	Redovne poruke
			06/11/2021	II	U-I-1	3.79	1.18	2.62	31.01	3.99	725.00	0.00	665.00	60.00	15.81	35.14	Redovne poruke
			06/11/2021	III	U-I-2	4.20	1.31	2.89	31.21	3.23	665.00	0.00	475.00	190.00	45.20	100.44	Redovne poruke
			05/11/2021	I	U-I-2	2.30	0.18	2.12	7.64	3.05	575.00	325.00	800.00	100.00	43.53	96.72	Redovne poruke
			05/11/2021	II	U-I-2	3.87	0.82	3.05	21.15	3.87	800.00	0.00	725.00	75.00	19.37	43.04	Redovne poruke
			05/11/2021	III	U-I-2	2.55	0.87	1.69	33.97	4.83	725.00	0.00	665.00	60.00	23.50	52.23	Redovne poruke
			04/11/2021	I	U-I-1	3.25	0.93	2.33	28.44	6.23	800.00	0.00	725.00	75.00	23.06	51.23	Redovne poruke
			04/11/2021	II	U-I-2	3.06	0.51	2.55	16.67	3.40	725.00	0.00	665.00	60.00	19.62	43.59	Redovne poruke
			04/11/2021	III	U-I-2	2.17	0.46	1.72	20.95	4.35	665.00	0.00	575.00	90.00	41.40	91.99	Redovne poruke
			03/11/2021	I	Rad_PMNOV	0.00	0.00	0.00	0.00	0.00	800.00	0.00	800.00	0.00	0.00	0.00	Redovne poruke
			03/11/2021	II	Rad_PMNOV	0.00	0.00	0.00	0.00	0.01	800.00	0.00	800.00	0.00	0.00	0.00	Redovne poruke
			03/11/2021	III	Rad_PMNOV	0.00	0.00	0.00	0.00	0.02	800.00	0.00	800.00	0.00	0.00	0.00	Redovne poruke
			02/11/2021	I	Rad_PMNOV	0.00	0.00	0.00	0.00	0.00	575.00	0.00	575.00	0.00	0.00	0.00	Redovne poruke
			02/11/2021	II	Rad_PMNOV	0.00	0.00	0.00	0.00	0.00	575.00	0.00	575.00	0.00	0.00	0.00	Redovne poruke
			02/11/2021	III	Rad_PMNOV	0.00	0.00	0.00	0.00	0.01	800.00	0.00	800.00	0.00	0.00	0.00	Redovne poruke
			01/11/2021	I	Rad_PMNOV	0.46	0.06	0.40	14.05	0.44	575.00	100.00	575.00	100.00	217.13	482.51	Redovne poruke
			01/11/2021	II	Rad_PMNOV	0.00	0.00	0.00	0.00	0.02	575.00	0.00	575.00	0.00	0.00	0.00	Redovne poruke
			01/11/2021	III	Rad_PMNOV	0.00	0.00	0.00	0.00	0.01	575.00	0.00	575.00	0.00	0.00	0.00	Redovne poruke
COUNT=18																	
Grand Summary (COUNT=18)																	

Fig. 3. Report with KPIs on the daily operation of the machine.

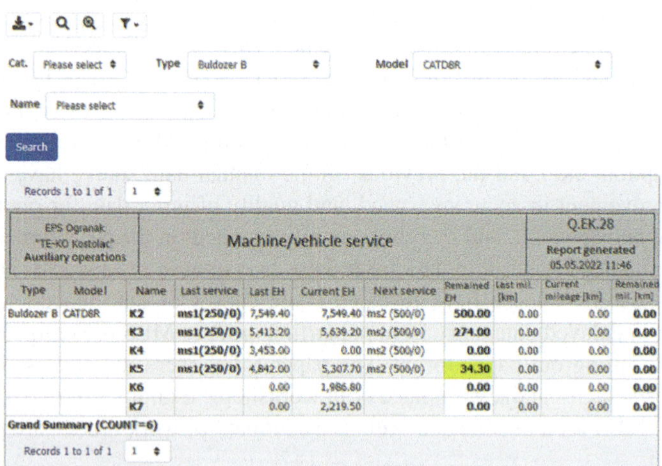

EPS Ogranak "TE-KO Kostolac" Auxiliary operations			Machine/vehicle service							Q.EK.28 Report generated 05.05.2022 11:46	
Type	Model	Name	Last service	Last EH	Current EH	Next service	Remained EH	Last mil. [km]	Current mileage [km]	Remained mil. [km]
Buldozer B	CATD8R	K2	ms1(250/0)	7,549.40	7,549.40	ms2 (500/0)	500.00	0.00	0.00	0.00
		K3	ms1(250/0)	5,413.20	5,639.20	ms2 (500/0)	274.00	0.00	0.00	0.00
		K4	ms1(250/0)	3,453.00	0.00	ms2 (500/0)	0.00	0.00	0.00	0.00
		K5	ms1(250/0)	4,842.00	5,307.70	ms2 (500/0)	34.30	0.00	0.00	0.00
		K6		0.00	1,986.80		0.00	0.00	0.00	0.00
		K7		0.00	2,219.50		0.00	0.00	0.00	0.00
Grand Summary (COUNT=6)										

Fig. 4. Report on regular services for a selected machine type (EH – engine hours; mil. – mileage) – MES.

work/maintenance order number. Several reports/KPIs are provided based on this data, including shift activity overview, average consumption, percentage of average versus normative consumption, fuel/lubricant/technical fluid consumption over time period, by machine/vehicle, working hours or mileage, etc. Therefore, timely and more accurate tracking is enabled, leading to better control and planning of fuel/lubricant/technical

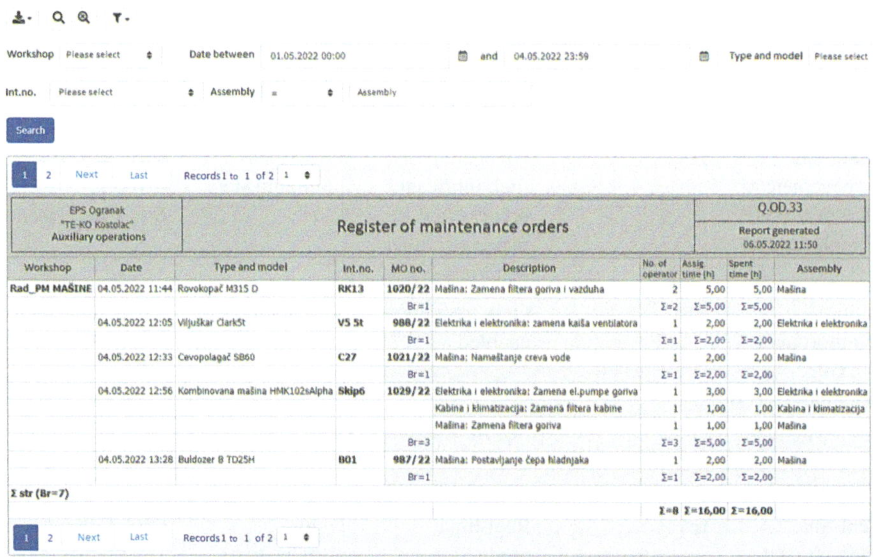

Workshop	Date	Type and model	Int.no.	MO no.	Description	No. of operator	Assig time [h]	Spent time [h]	Assembly
Rad_PM MAŠINE	04.05.2022 11:44	Rovokopač M315 D	**RK13**	**1020/22**	Mašina: Zamena filtera goriva i vazduha	2	5,00	5,00	Mašina
				Br=1		Σ=2	Σ=5,00	Σ=5,00	
	04.05.2022 12:05	Viljuškar Clark5t	**V5 5t**	**988/22**	Elektrika i elektronika: zamena kaiša ventilatora	1	2,00	2,00	Elektrika i elektronika
				Br=1		Σ=1	Σ=2,00	Σ=2,00	
	04.05.2022 12:33	Cevopolagač SB60	**C27**	**1021/22**	Mašina: Nameštanje creva vode	1	2,00	2,00	Mašina
				Br=1		Σ=1	Σ=2,00	Σ=2,00	
	04.05.2022 12:56	Kombinovana mašina HMK102sAlpha	**Skip6**	**1029/22**	Elektrika i elektronika: Zamena el.pumpe goriva	1	3,00	3,00	Elektrika i elektronika
					Kabina i klimatizacija: Zamena filtera kabine	1	1,00	1,00	Kabina i klimatizacija
					Mašina: Zamena filtera goriva	1	1,00	1,00	Mašina
				Br=3		Σ=3	Σ=5,00	Σ=5,00	
	04.05.2022 13:28	Buldozer B TD25H	**B01**	**987/22**	Mašina: Postavljanje čepa hladnjaka	1	2,00	2,00	Mašina
				Br=1		Σ=1	Σ=2,00	Σ=2,00	
Σ str (Br=7)									
						Σ=8	**Σ=16,00**	**Σ=16,00**	

Fig. 5. Register of maintenance orders – MES.

fluid consumption on a daily, monthly, and annual basis, better procurement planning, and finally reduced business costs.

Finally, the fourth module covers inventory management, administration, and financial management as an approach to traditional branch-level ERP. Therefore, MES and traditional ERP should share inventory data such as fuel and spare parts to support inventory planning, inventory issuance, and other inventory management sub-processes. Additionally, these information systems should share data for procurement management, as well as cost management. The both information systems use the same tagging system to track fixed assets and consumables inventory. Integrating MES with ERP in our case allowed for sharing data on relevant inventory items (e.g., item type, subtype, technical details), warehouse input/output logistics, warehouse inventory status, and inventory cost.

An overview of the basic characteristics of the elements of Industry 4.0 for the "Drmno" open-pit coal mine is shown in Table 2 [18–20].

Table 2. Industries 4.0 elements on the open-pit coal mine

Activity/State of development	Work in real time	Function/area of application	Characteristic	Software support
Automatic management of all orders (WIP) – implemented	Yes	All spare parts, fuel, lubricant, process raw materials, other materials	Prescribed minimum stock level	ERP, MES – WIP, Stock movement
Automatic monitoring of working machines (BTO) in the open-pit coal mine – implemented	Yes	Only tracking is enabled	Virtual GPS	All vehicles have GPS (MMS)
Real-time open-pit coal mine status overview control panel – implemented	Yes	Real-time production monitoring	Field visualization	GPS, GIS
Real-time maintenance – implemented	Yes	Working machines (BTO), mining equipment and plants	Integration of condition monitoring sensors	SCADA, ERP – MT, MES – WIP
Big data predictive analytics – under development	Yes	At the open-pit coal mine level, fuel, lubricants and spare parts, optimization of maintenance of working machines (BTO) and other facilities. The entire value chain in the open-pit coal mining	Big data analysis	ERP, BDA

(*continued*)

Table 2. (*continued*)

Activity/State of development	Work in real time	Function/area of application	Characteristic	Software support
On-line quality management (QMS/IMS) (production of coal, spare parts – tires and conveyor belts/BTO) – implemented	Yes	Processing plant, mining production, tire and conveyor system (BTO)	Big data, vertical integration	ERP, MES, QMS (IMS)
Greater safety (ergonomic conditions) at work through extended automation based on Industry 4.0 – under development	Yes	All operational and production systems at the open-pit coal mine (processing, geology and open-pit coal mine planning)	All systems and equipment on open-pit coal mine are integrated (BTO)	ERP, MES, OH&S (IMS)
Improved teamwork in the production environment through transparency and availability of data – implemented	Yes	Mining Production (BTO), processing plant, geology and mine planning. All drive and production systems	Vertical integration	ERP, MES, GPS, IMS
Improved environment through optimized use of resources, especially energy – under development	Yes	Mining production (BTO), processing plant, geology and the open-pit coal mine planning	Integration of condition monitoring sensors. Smart readers	MES, EMS

(*continued*)

Table 2. (*continued*)

Activity/State of development	Work in real time	Function/area of application	Characteristic	Software support
Expanded innovation capabilities through new technological capabilities in production – under development	Yes	Mining production (BTO), processing plant, geology and the open-pit coal mine planning	Vertical, horizontal and end-to-end integration layers	ERP, MES, IMS, MMS, GPS, GIS
Inventory Reporting – implemented	Yes	All spare parts, fuel, lubricant, process raw materials, other materials	Real-time reporting, BDA, vertical integration	ERP, MES

5 Conclusions and Future Research

Although there are many obstacles for mines which want to implement new technologies (elements of Industry 4.0), they must be overcome if mining companies want to remain competitive and successful in the digital era. Smart mines are the future and bring multiple benefits, from improving workforce safety and data protection, to increasing productivity and efficiency throughout the mining production value chain, to reducing costs of monitoring and safety, maintenance, etc.

The cloud-based ERP/MES model described and implemented above has enabled significantly faster decision-making and facilitated management activities at the operational, sector, and company levels. Covered business processes have improved in terms of KPIs, and costs have been significantly reduced. In addition, redesigning some business processes has eliminated repetitive activities, such as data entry from paper forms, and led to better working conditions for all employees. The foundation has been laid for further steps towards developing and implementing a full smart mining model.

Regarding to this, the next step is the development of a smart IoT network that will encompass all the most important elements of mining production in this organization, especially for the areas from Table 2, which are under development.

References

1. Majstorovic, V., Simeunovic, V., Miskovic, Z., Mitrovic, R., Stosic, D., Dimitrijevic, S.: Smart manufacturing as a framework for smart mining. Procedia CIRP **104**, 188–193 (2021). https://doi.org/10.1016/j.procir.2021.11.032
2. N., N., A Brief History of ERP – since 1960 and the future of ERP, available at: https://www.erp-information.com/history-of-erp.html. Accessed Feb 2023

3. Hochmuth, C.A., Bartodziej, C., Schwägler, C.: Industry 4.0 Is your ERP system ready for the digital era? https://www2.deloitte.com/content/dam/Deloitte/de/Documents/technology/Deloitte_ERP_Industrie-4-0_Whitepaper.pdf. Accessed Feb 2023
4. Tavana, M., Hajipour, V., Oveisi, S.: IoT-based enterprise resource planning: challenges, open issues, applications, architecture, and future research directions. Internet of Things **11**, 100262 (2020). https://doi.org/10.1016/j.iot.2020.100262
5. Barrie, G., Whyte, A., Bell, J.: IoT security: challenges and solutions for mining. In: Proceedings of the Second International Conference on Internet of things and Cloud Computing – ICC'17 – IoT security, pp. 1–9 (2017). https://doi.org/10.1145/3018896.3018933
6. Sethi, P., Sarangi, S.R.: Internet of things: architectures, protocols, and applications. J. Electr. Comput. Eng. **2017**, 1–25 (2017). https://doi.org/10.1155/2017/9324035
7. Molaei, F., Rahimi, E., Siavoshi, H., Afrouz, S.G., Tenorio, V.: A comprehensive review on internet of things (IoT) and its implications in the mining industry. Am. J. Eng. Appl. Sci. **13**(3), 499–515 (2020). https://doi.org/10.3844/ajeassp.2020.499.515
8. Atkins, A., Zhang, L., Yu, H.: Applications of RFID and mobile technology in tracking of equipment for maintenance in the mining industry. In: Aziz, N., Kininmonth, B. (eds.) Proceedings of the 2010 Coal Operators' Conference, Mining Engineering, University of Wollongong, 18–20 Feb 2019. https://ro.uow.edu.au/coal/326
9. Patel, P., Ali, M.I., Sheth, A.: On Using the Intelligent Edge for IoT Analytics. IEEE Intelligent Systems **32**(5), 64–69 (2017). https://doi.org/10.1109/MIS.2017.3711653
10. Battulwar, R., et al.: A practical methodology for generating high-resolution 3D models of open-pit slopes using UAVs: flight path planning and optimization. Remote Sens. **12**(14), 2283 (2020). https://doi.org/10.3390/rs12142283
11. Chen, B., Wan, J., Shu, L., Li, P., Mukherjee, M., Yin, B.: Smart factory of industry 4.0: key technologies, application case, and challenges. IEEE Access **6**, 6505–6519 (2018). https://doi.org/10.1109/ACCESS.2017.2783682
12. Paolanti, M., Romeo, L., Felicetti, A., Mancini, A., Frontoni, E., Loncarski, J.: Machine learning approach for predictive maintenance in industry 4.0. In: 2018 14th IEEE/ASME International Conference on Mechatronic and Embedded Systems and Applications (MESA), pp. 1–6 (2018). https://doi.org/10.1109/MESA.2018.8449150
13. Tongsuksai, S., Mathrani, S.: Integrating cloud ERP systems with new technologies based on industry 4.0: a systematic literature review. In: 2020 IEEE Asia-Pacific Conference on Computer Science and Data Engineering (CSDE) (2020). https://doi.org/10.1109/csde50874.2020.9411570
14. Navaneethakrishnan, C.: A comparative study of cloud based ERP systems with traditional ERP and analysis of cloud ERP implementation. Int. J. Eng. Comput. Sci. **2**(09) (2018). http://www.ijecs.in/index.php/ijecs/article/view/1921
15. Haddara, M., Elragal, A.: The readiness of ERP systems for the factory of the future. In: Conference on ENTERprise Information Systems/International Conference on Project MANagement/Conference on Health and Social Care Information Systems and Technologies, CENTERIS/ProjMAN/HCist 7–9 Oct 2015. https://doi.org/10.1016/J.PROCS.2015.08.598
16. Peng, G.C., Gala, C.J.: Cloud ERP: a new dilemma to modern organisations? J. Comput. Inform. Syst. **54**(4), 22–30 (2014). https://doi.org/10.1080/08874417.2014.11645719
17. Abd Elmonem, M.A., Nasr, E.S., Geith, M.H.: Benefits and challenges of cloud ERP systems – a systematic literature review. Future Comput. Inform. J. **1**(1–2), 1–9 (2016). https://doi.org/10.1016/j.fcij.2017.03.003
18. Simeunović, V., et al.: Development of industry 4.0 model for open pit coal mine (in Serbian). In: 26th YU INFO Conference, Procedia YU INFO 26 (2020), pp. 279–284. http://www.yuinfo.org/ZBORNIK_YU_INFO_2020.pdf. Accessed: Feb 2023

19. Stošić, D., et al.: Monitoring feeding fuel, lubricant and technical fluid in open-pit coal mine supported by modern ICT (in Serbian). In: 25th Conference, Procedia YU INFO 25, pp. 170–175 (2019). http://www.yuinfo.org/zbornici/2019/YUINFO2019.pdf. Accessed Feb 2023
20. Telukdarie, A., Sishi, M.: Implementation of Industry 4.0 technologies in the mining industry – a case study. Int. J. Mining Mineral Eng. **11**(1), 1 (2020). https://doi.org/10.1504/IJMME.2020.10027477

Location Suitability for the Implementation of Unmanned Aerial Vehicles in the Vaccine Supply Chain of Sri Lanka

Kalani Gunaratne[1], Harsha Chamara Hewage[1], Izabela Iwa Nielsen[2], Grzegorz Bocewicz[3], Amila Thibbotuwawa[1(✉)], and Zbigniew Banaszak[3]

[1] Center for Supply Chain, Operations and Logistics Optimization, University of Moratuwa, Katubedda 10400, Sri Lanka
amilat@uom.lk

[2] Department of Materials and Production, Aalborg University, 9220 Aalborg, Denmark

[3] Koszalin University of Technology, 75-453 Koszalin, Poland

Abstract. Considering the faster speed and the low response time of unmanned aerial vehicles (UAV), it is suggested as a solution to overcome the system inefficiencies in the current vaccine cold chain of Sri Lanka. Implementing the unmanned aerial system (UAS) at the district level is recommended to maintain the end-to-end effectiveness of the program. Under the suggested distribution network, vaccines are transported directly from the Regional Medical Supplies Division (RMSD) warehouse to the respective health facilities, surpassing the Medical Officer of Health (MOH) unit warehouse. However, the high fixed cost of a UAS doesn't support the implementation of UAVs at every RMSD. Therefore, to determine an appropriate division, a suitability analysis was conducted by intersecting eight factors with their weights of importance. Suitable factors were determined using previous literature and weights of importance were calculated by an expert survey. From the analysis, it was determined that the Kurunegala division is the most appropriate for UAV implementation. Therefore, Kurunegala is recommended as a starting point for implementing the proposed UAV-inclusive delivery system in Sri Lanka to realize potential benefits and practical implications.

Keywords: Drone · UAV · Vaccine delivery · Suitability analysis · Healthcare supply chain

1 Introduction

The Expanded Program on Immunization (EPI) has been the main component of the National Immunization Program since its implementation in 1978 [1]. The program ensures the availability of routinely recommended vaccines to mothers and infants to prevent infectious diseases. Currently, Sri Lanka has more than 90% coverage in all vaccines under the program [2]. But the vaccine supply chain is suffering from issues in the distribution system causing major cost accumulations, soaring wastage, and heavy time consumption [3]. The complete reliance on the road network and the persisting issues in the cold chain are the main sources of such inefficiencies [4].

A. Burduk et al. (Eds.): ISPEM 2023, LNNS 790, pp. 43–59, 2023.
https://doi.org/10.1007/978-3-031-45021-1_4

The continuous expansion of the vaccination program and the above-mentioned issues in the delivery network are threatening the sustainability of the vaccination program. It not only burdens the government with extreme costs but also endangers the quality of life of citizens. According to expert interviews, amidst the noticeable problem areas in the EPI program, the country has continued to execute the traditional delivery system with little to no technological advancements. Therefore, it is timely to identify technological improvements to help diminish the inefficiencies.

In the ever-growing journey of development to find solutions and reduce inadequacies UAVs, are effective in being a way out of the physical and economical impediments in delivery networks [5, 6]. Previous studies have assured the potential of UAVs in improving healthcare availability and delivering healthcare products to otherwise unreachable locations such as islands, swamps, or deserts [5–7]. Additionally, UAVs have proven their capability of successfully delivering small weights, whilst ensuring reduced financial, environmental, and social impact [8, 9]. Therefore, a replacement delivery network inclusive of UAVs is proposed for Sri Lanka to overcome the issues in vaccine distribution.

However, considering UAV's high fixed cost and technical limitations, implementing a nationwide UAV delivery network is challenging. Therefore, it is crucial to determine locations that are most benefited from the suggested UAV distribution system for the initial execution. Hence, our study analyzes the appropriateness of the local RMSDs for UAV implementation to determine and suggest a location for initiation. Through the results we expect to act as a decision support tool aiding in further studies on the utilization of drones in healthcare logistics in Sri Lanka. Moreover, will address the limitations to such an application along with future research trajectories that can aid in promoting the technology use in healthcare logistics.

Accordingly, we've initially identified a suitable model of drone execution in the Sri Lankan vaccine cold chain through expert interviews and literature. Then using a factor analysis, we have determined the most ideal location of implementation for the said model in Sri Lanka.

2 Background

2.1 Sri Lanka Vaccine Cold Chain

The very origin of immunization in Sri Lanka is recorded from the 19th century when the compulsory vaccination against smallpox, known as the Vaccination Ordinance of 1886 was introduced [1]. Since established National Immunization Program (NIP), continued to expand including more vaccine categories leading to the officially implement the Expanded Program on Immunization (EPI) in 1978 [1].

The EPI program is the fundamental component of the National Immunization Program [4]. The primary purpose of the program is to reduce mortality associated with vaccine-preventable diseases [4]. Accordingly, the program targets children (below 12 years of age), pregnant mothers, and females in the childbearing age group as the population categories for vaccine administration [4].

Vaccine importation and island-wide distribution is a complex and intricate task because of the high monetary value and sensitivity of vaccine vials. Therefore, many government entities and individuals are involved in the distribution network ensuring end-to-end reliability. The implementation and monitoring of vaccine cold chain within Sri Lanka are responsibilities of the Epidemiology Unit [4]. Accordingly to facilitate the countrywide regular supply, the vaccine distribution system adheres to an established healthcare administrative system comprised of four tiers [3, 4]. The system has successfully ensured the reachability of primary healthcare services to the smallest social units of the country. The administration of the network is decentralized [3]. Thus, each tier is responsible of storing vaccines at the best condition and distributing them to the following tier so that vaccines can ultimately reach the requisite population on time via hospitals or Maternal and Child Health (MCH) clinics. Figure 1 shows the entities and delivery arcs under the said four levels when the EPI cold chain is concerned [4, 10].

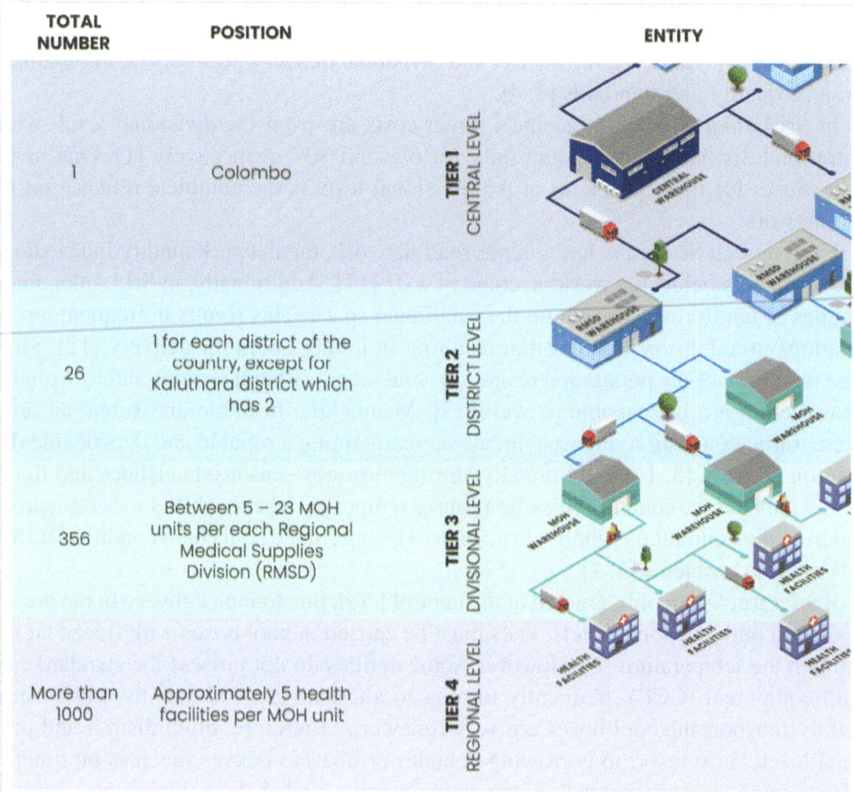

Fig. 1. Administrative Levels in the Sri Lankan Cold chain

Central Vaccine Store is replenished annually. Every 2 months, respective amounts of vaccine are distributed to the twenty-six RMSDs from the central vaccine store. Then, each RMSD ensures the monthly distribution of vaccines to their respective MOH units. Finally, every MOH unit; distributes vaccine vials to the MCH clinics and hospitals under their authority. It is evident that, through the four-tier distribution network, healthcare industry has established a push network rather than a pull network for vaccine administration. It might be the reason for achieving more than 90% of population coverage in all vaccine types in the program [2]. However, even though the program is effective, the efficiency of the system is in question due to the persistent problems in the distribution network.

2.2 Issues in the Cold Chain of Sri Lanka

At the central level the cold room complex is equipped with modern temperature controlling facilities along with space for the expanding vaccine schedule. Additionally, reefer trucks are available for the delivery of vaccines to the twenty-six RMSD warehouses, ensuring the right quality of the environment. Additionally, since the central delivery network caters from Colombo to the main towns in each district, they use the class A roads or the main roads for delivery. Hence delivery disruptions are minimum in the central level [4]. However, the district and divisional delivery networks do not maintain the same quality and standards [3, 4].

In Sri Lanka 64% of a vaccine's direct costs are from the divisional level, while central and district levels account only for 6% and 30% respectively [1]. One of the main causes for these problems at the divisional level is the complete reliance on the road network.

Even though Sri Lanka has a dense road network, the network quality index stands at 3.9, which is below the world average of 4.07 [11]. Additionally, in Sri Lanka, many varieties of landforms exist within the small land space. This results in frequent terrain variations in a delivery round, often resulting in longer hours for delivery [12]. Since more than 78% of the population resides in rural areas, difficult terrains and low-quality infrastructure are inescapable as well [13]. Meanwhile, in urban and suburban areas severe traffic congestion imposes threats on maintaining a reliable and dependable distribution network [3, 14]. Additionally, during monsoon seasons, landslides and floods are inevitable in the country. This often causes temporary inaccessibility to certain areas, hindering the continuous supply of vaccines. The unreliable cold chain condition further adds to these problems [3, 4].

Reefer trucks are only granted at the central level, due to bulk delivery in the arc. At the district and divisional levels, vials must be carried in cool boxes with freeze tags to maintain the temperature [4]. However, some entities do not possess the standard cold chain equipment (CCE), frequently turning to alternatives. Additionally, most trucks used to transport the cool boxes are very timeworn. Therefore, often district and divisional levels must resort to borrowing vehicles or bikes to deliver vaccines on time[3]. Since vaccine vials are highly sensitive to temperature and shaking these issues severely threaten the potency of vials.

The ever-growing technology industry has continued to impress the world with new and innovative technological findings for many problems. Transportation and delivery are such areas that the technology industry has continued to invent novel and impressive solutions in. Among the many inventions, Unmanned Aerial Vehicle (UAVs) or drone in delivery is one of the most resounding findings. It has been piloted and experimented in the healthcare field and has successfully borne solutions [7]. Hence is suggested as a possible solution to the existing inefficiencies in the Sri Lankan cold chain.

2.3 Delivery Using UAV in Healthcare

Usage of unmanned aerial vehicle (UAV) or dynamic remotely operated navigation equipment (Drone), is one of the foremost developments in industry 4.0. By definition, a UAV is an aircraft that is controlled distantly via radio waves or autonomously without any aircrew on board [7]. The continuous growth in technology has resulted in expanding levels of autonomy in UAVs. This has enabled automated take-off, landing, obstacle evasion, and course planning as well [15]. Additionally, its low maintenance cost, fast response time, driverless mobility, and absence of road network dependency suggests an alternative mode of delivery for road distribution networks [16].

Among the many public use cases of UAVs in delivery, its placement in the healthcare field is one of the most promising applications. Universal access to proper healthcare is a basic human right, emphasizing its importance to a quality life. But many physical and economical impediments such as expensive operational and fixed costs, improper infrastructure, and natural barriers impose threats and difficulties in the widespread availability of quality healthcare [6, 17]. In the ever-growing journey of technology to find solutions and reduce inefficiencies, drones have proven to be effective in being a way out to the above-mentioned challenges The reason is faster and more efficient healthcare delivery will not only improve the quality of life of people but also aid healthcare suppliers to further expand their services [6].

Evidently, vaccines, emergency kits, organs, blood, and laboratory samples are the possible healthcare products for drone delivery within its limited functional specifications. It's ability of fast response, constant high speed and no obligation to physical infrastructure has made UAVs immensely popular in emergency situations [18]. UAV applications in disaster relief campaigns in Haiti and Taiwan after 2010 and 2016 earthquakes, emergency blood or anti venom dose delivery are few such examples [19]. However, over the years its application in routine delivery of medical items has become popularized. Because it not only defrays the initial set-up cost through the large number of flights but also improves health coverage providing a bigger service in the medical field [7].

Previous studies have assured the potential of drones in improving healthcare coverage by delivering healthcare products to otherwise unreachable locations such as islands, swamps, or deserts [6]. According to HERMES generated simulation model by [5], usage of UAVs could increase the vaccine availability from 94% to 96% in Mozambique, Gaza province. Furthermore, a pilot project in Madagascar successfully delivered drugs and laboratory samples to remote areas by avoiding difficult terrains [20]. In 2019, UNICEF was successful in conducting a pilot project, yielding positive results from using UAVs to transport vaccines and medicine to islands in Vanuatu in replacement of boats [6]. Moreover, according to a pilot project conducted by Zipline, they were able to improve accessibility to vaccines in Rwanda, by avoiding poor road infrastructure [21]. However, it is worth noticing that the main objective of thus studies or projects was to improve healthcare availability and reach better healthcare coverage within the country. In fact, countries with low transport index or low district level vaccine coverage are deemed to be approaches that can be used to identify countries that are potentially suitable for drone use to increase accessibility [6]. Considering the specialties and capabilities of drones, improving accessibility to health is the most apparent application in healthcare field. However, studies also have identified that UAVs are capable of reducing delivery inefficiencies [8].

Therefore, despite the reasonable infrastructure, reachability of the destination and considerable healthcare coverage drones can further aid the healthcare field by increasing efficiency and reducing operational cost [22]. [8] identified that, use of drones in the routine supply of pharmaceutical products to customers in Rotterdam, replacing one of three delivery trucks with a UAV could realize a reduction in annual logistic cost, time consumption, and carbon dioxide emission by 12.5%, 12%, and 9% respectively. Furthermore, UAVs have been successful in delivering small weights, whilst ensuring reduced financial, environmental, and social impact [9]. Additionally, according to [5] cost savings can be derived from drones being able to delivery vaccines more quickly and cheaply than land-based methods. Moreover, the ability to transverse difficult terrains, reduced labor cost, diminished human error and replaced large fleets of vehicles assist in such cost reductions [23, 24]. Despite such possibilities, most studies have shed light towards improving healthcare coverage using drones, with very few studies focusing on the ability of drones in reducing the system inadequacies in the healthcare field.

However, within the healthcare system UAVs might offer advantages over existing delivery networks in selected scenarios only. Such as just-in-time delivery of emergency medicine, collection of time-sensitive lab samples, routine resupply of high-cost items, and vaccine delivery for better cold chain control [7]. The reasons for this are that the success of UAVs in delivery is dependent on the number of flights, product parameters (such as weight and volume), and the importance of UAV value addition [7]. Therefore, identifying the right applications is important. But when properly applied, major benefits can be realized.

3 Proposed Network and Research Objectives

The Sri Lankan cold chain is an ideal use case for UAV implementation considering the existing issues hindering the routine supply of important and costly vaccines [7]. However, implementing UAVs at the central level won't yield many benefits because of the considerable availability of CCE and the use of main roads. In contrast, district and regional levels are more suitable considering the severity of logistic issues.

As a rule of thumb, it is crucial to identify applications that will yield benefits over the end-to-end delivery rather than a portion of the network [7]. Therefore, it is recommended that the UAS be implemented at the district level to transfer vaccines directly from the RMSD warehouse to the appropriate health facilities, bypassing the warehouses at MOH units. This will remove the delivery arc at the divisional level, reducing the requirement and management of CCE, human resources, and delivery vehicles in the system. Further, vaccine wastage will reduce due to the elimination of storage and transportation under poor conditions. Additionally, it will maintain an end-to-end effective program whilst utilizing the UAVs at their maximum capacity. Figure 2 shows an overview of the suggested novel delivery network. Even though the MOH warehouse is avoided in the proposed system, the divisional level will continue to exist for administration.

However, considering the high fixed cost and technical limitations of a UAS, implementing UAV delivery networks in all twenty-six RMSDs is challenging. Therefore, it

Fig. 2. Suggested delivery network

is fitting to identify the RMSD most benefitted from the suggested UAV distribution network as an initiation. Thus, it is intended to rank the twenty-six RMSDs from most to least benefited, to realize the division most fitting for the application.

4 Methodology

A suitability analysis was conducted to achieve the objective of ranking the divisions based on the value addition. Suitability analysis is a decision support tool that is used to evaluate the appropriateness of a location for a certain use and is conducted by intersecting different aspects of the terrain [25]. In our research, the division's appropriateness for UAV implementation is realized by intersecting factors that can determine the value addition UAVs can attract under the suggested application.

To obtain a solution, a plethora of models could have been utilized depending on the decision environment and whether ordinal or nominal information is provided along with the level of expertise. That is, methods such as outranking models (ELECTRE, PROMETHEE) [26, 27], closest to ideal (TOPSIS, VIKROR) [28], linear combination methods (WSA) [22], hierarchical approaches (AHP, ANP) [29, 30], along with a large set of extensions concerning the integration of fuzzy or uncertain environment. From those methods, a linear combination method was used in our research due to the decision environment of the suitability analysis. The analysis was conducted in four steps, as mentioned below [22].

4.1 Factor Identification

The factors considered should determine the value addition from UAVs whilst being comparable among locations. Therefore, factors that are common and would not be influenced by the UAV application were neglected. Accordingly, using previous literature, the below factors were identified and finalized.

Demand Factors: Characteristics of the cargo and demand patterns influence the UAV value addition [7]. Nevertheless, since the type of product is common to all divisions, product factors such as financial value, health value, and shelf life cannot be used to compare the division's suitability. Instead, the quantity of demand was considered a factor. The quantity of demand has a positive impact on UAV implementation because more demand requires faster supply [22].

Accessibility Factors: The extent of accessibility to healthcare centers also reflects the suitability and value addition UAVs can attract to a location [7]. Therefore, to represent accessibility, road density and average elevation were considered as factors. Road density represents the total kilometers of road network available within one square kilometer of land area. Hence, suggests that the lower the road density is, the more difficult it is to access the health centers, resulting in better value addition from UAV implementation. Meanwhile, the average elevation of the location represents the topographical condition of the area. Thus, higher average elevation indicates that the area is more mountainous causing transportation difficulties when roads are used. Therefore, higher average elevation suggests better value addition from a UAS because UAVs can cruise faster and easier in difficult terrains.

Road Network Quality Factors: UAVs' lack of reliance on the road infrastructure brings benefits over a delivery system using a low-quality road network. To represent the quality of the road network within each division, class D road density and the number of damaged infrastructures were used as factors. Class D Road density (gravel roads that are usually motorable during dry weather only) denotes the number of kilometers of class D roads available in one square kilometer of the land area [13]. Further, the number of damaged infrastructures shows the damages to critical infrastructure within the division due to environmental disasters. Higher values in the above parameters suggest better benefit addition from UAVs due to lack of road dependency.

Supply Chain Factors: Health facility density, vehicle density, and area of service were considered as supply chain factors. Health facility density represents the number of health facilities available in one square kilometer of area. It suggests that the lower the health facility density is, far apart the facilities are located increasing delivery problems. Therefore, can harness more benefits from UAVs [7]. Since all rural and urban districts were considered in our research, vehicle density must be included as a factor highlighting the traffic conditions. Finally, the area of service also influences the benefits that can be gained from UAV implementation. Because higher the service area is higher the fuel cost and time consumption are when trucks are used, suggesting more appropriateness for UAV usage.

4.2 Factor Normalization

When the above-mentioned factors are considered together, values need to be normalized on a common scale. That is to neglect the bias effect occurring from the different values ranges in each factor [22]. To further explain the statement, if the average elevation values are in the hundreds while the quantity of demand values are in the thousands, the influence of the quantity of demand factor will be high if not normalized. This yields incorrect results. Therefore, the values of each factor were normalized to a scale of 1 (least suitable) to 10 (most suitable) using the min-max method. The min-max method was used as it returns better results compared to other methods when explaining variances [31].

Among the eight factors used in this research, health facility density and road density factors have an inversely proportional influence on suitability. Therefore, before normalizing, those two factors were converted to their negative form to reverse their effect on the scale.

4.3 Factor Weight Calculation

The above-mentioned factors influence the suitability of a location for UAV implementation, to different extents. Therefore, to identify the weight of influence of each factor, the weighting by ranking method was used. Accordingly, factors first need to be ranked assigning 1 to the most important factor, 2 to the second most important factor, etc. [25]. To determine accurate ranks, individuals with a clear rationale on the subject are required [22]. Therefore an expert survey was conducted allowing professionals from healthcare and UAV industries, to rank the factors based on their knowledge [22]. Additionally, equal contribution between the two fields was maintained to avoid any biased results. After collecting responses, the weights of influence were calculated in three steps as mentioned below [25].

First, a weight (w_i^a) for every factor was calculated, from each separate expert response using the reciprocal ranking method, as in Eq. (1) [25]. Where r is the rank assigned by the expert for the particular factor. i is the respective expert index. And a is the factor to which the weight is calculated.

$$w_i^a = 1/r_i^a \tag{1}$$

Every expert opinion is given equal significance. Therefore, to combine the opinions of all experts, a composite factor weight (w^a) was calculated for each factor. To compute the composite factor weight, Eq. (2) was used. Where n is the total number of expert responses collected.

$$w^a = \sum_{i=1}^{n} w_i^a \tag{2}$$

Finally, to bring the composite factor weights to a comparable state, Eq. (3) was used [25]. Where Wa is the comparable weight of the factor in concern. And m is the total number of factors used.

$$W_a = \frac{w^a}{\sum_{a=1}^{m} w^a} \tag{3}$$

The comparable weight calculated for each factor was used as their corresponding weight of influence in the analysis [25].

4.4 Factor Overlay

The suitability score was derived for each division using the sum-product of normalized factor scores and their respective weights of influence. The most suitable to least suitable divisions for UAV implementation are determined by arranging the suitability scores from highest to least.

5 Data Collection

5.1 Factor Overlay

Table 1 depicts the sources used for the data collection of the eight factors realized. Data was collected using online sources for 2019, to avoid any discrepancies caused by the influence of COVID-19.

The RMSD boundaries are equivalent to the district administrative boundaries in all cases except the Kalutara district. Unlike other districts, Kalutara is divided into two RMSDs. However, considering the ability of a UAV to serve both divisions and the data availability, the two RMSD divisions at Kalutara, were considered as one. Therefore, for the data collection and analysis, twenty-five divisions were considered, identical to the 25 administrative districts in the country.

Table 1. Data Sources

Aspect	Factor	Measuring parameter	Data source
Demand factors	Quantity of demand	Number of childbirths	Annual Health Statistics 2019 [10]
Accessibility factors	Road density	Total road kilometers (km)/Total land area[a] (km^2)	Economic and Social Statistics of Sri Lanka 2019 [13]
	Average elevation	Average elevation (m)	Topographic map.com [32]
Road network quality factors	Class D road[b] density	Total class D road[b] kilometers (km)/Total land area[a] (km^2)	Economic and Social Statistics of Sri Lanka 2019 [13]
	Number of damaged infrastructure	Damages to critical infrastructure due to natural disasters	Natural Disaster Relief Services Center, situation summary report 2019 [33]
Supply chain factors	Health facility density	Number of health facilities/Total area (km^2)	Annual Health Statistics 2019 [10] Economic and Social Statistics of Sri Lanka 2019 [13]
	Vehicle density	Number of vehicle registrations/Total land area[a] (km^2)	Economic and Social Statistics of Sri Lanka 2019 [13]
	Area of service	Total area (km^2)	Economic and Social Statistics of Sri Lanka 2019 [13]

[a]Total land area excluding inland waters
[b]Gravel roads that are usually motorable during dry weather only

5.2 Expert Survey

To collect the expert responses, quota sampling technique was used. Quota sampling is a non-probability sampling technique that is used when certain specific characteristics are expected from the sample [34]. It was applied to our research because our sample focus was individuals with expert knowledge of the healthcare or UAV industry.

In non-probability sampling, sample size and fields of expertise depending on the research requirement [35]. Because of the low heterogeneity in the areas of expertise required, a smaller sample is sufficient [36]. Additionally, maintaining an equal contribution between the two fields was crucial to diminish result bias. Considering the above-mentioned requirements, six expert responses were collected and used to derive factor weights. In the form, the experts were asked to rank the criteria assigning a value from one to eight, where one is the most important and eight is the least important. The figure below displays the ranks received for each criterion through the six expert responses. The names of the experts are not revealed to maintain confidentiality (Fig. 3).

Fig. 3. Expert Survey Results

6 Results

6.1 Normalize Factor Score

Figure 4 shows the normalized factor scores across the divisions. According to the reclassification, the quantity of demand, road density, and health facility density show frequent variations among divisions. Meanwhile, the number of damaged infrastructure and vehicle density shows little variation with most values in the lower spectrum.

Factors / Divisions	Demand factors	Accessibility factors		Road network quality factors		Supply chain factors		
	Quantity of demand	Road density	Average elevation	Class D road density	Number of damaged infrastructure	Health facility density	Vehicle density	Service area
Ampara	5	10	3	2	1	9	1	6
Anuradhapura	5	9	2	4	1	10	1	10
Badulla	4	7	5	4	1	8	1	4
Batticaloa	4	10	1	1	1	9	1	4
Colombo	10	4	1	7	3	1	10	1
Galle	6	7	1	2	1	7	3	2
Gampaha	9	4	1	10	2	5	5	2
Hambantota	4	9	2	1	2	9	2	4
Jaffna	3	1	1	2	1	6	2	1
Kaluthara	5	7	1	4	2	9	3	2
Kandy	3	5	9	5	4	6	1	3
Kegalle	4	5	5	9	1	8	2	2
Kilinochchi	2	3	1	2	1	9	1	2
Kurunegala	8	8	3	4	10	8	2	7
Mannar	1	10	1	1	1	9	1	3
Matale	7	9	4	3	1	9	2	3
Matara	4	7	2	1	1	8	3	2
Monaragala	3	10	4	2	1	10	1	8
Mullativu	1	10	1	1	1	10	1	4
Nuwaraeliya	4	8	10	2	1	8	2	2
Polonnaruwa	3	10	2	1	1	10	1	5
Puttalam	5	8	1	4	2	9	2	4
Ratnapura	6	7	4	7	1	8	2	5
Trincomalee	3	10	1	2	1	9	1	4
Vavuniya	2	9	1	1	1	10	1	3

Legend
10
9
8
7
6
5
4
3
2
1

Fig. 4. Normalized factor scores across divisions

6.2 Factor Weight

Table 2 represents the factor weights calculated for each parameter using experts' opinions. Class D Road density had the highest weight of importance at 21% while vehicle density and quantity of demand had the lowest at 7%. Interestingly, 65% of the weights were distributed between road network quality and accessibility factors, highlighting that those aspects were influential to the value addition of UAVs compared to demand and supply chain factors.

Table 2. Weight of Importance of Factors

Factor	Weight of importance
Quantity of demand	6.75%
Road density	11.96%
Average elevation	16.86%
Class D road density	20.81%
No. of damaged infrastructure	15.02%
Health facility density	14.06%
Vehicle density	6.56%
Area of service	7.98%

6.3 Suitability Score

Table 3 displays the suitability scores of each division arranged from largest to smallest. According to the results, the Kurunegala division (Score = 6.0) is the most suitable for the UAV system implementation, whilst Jaffna (Score = 2.3) is the least suitable. Further, the score gap between Kurunegala and Kegalle is the highest compared to other adjacent values in the table. It highlights that the value addition UAVs can attract to Kurunegala is significant compared to other divisions.

Table 3. Suitability Score

Rank	Division	Score	Rank	Division	Score
1	Kurunegala	6.0	14	Colombo	4.1
2	Kegalle	5.4	15	Hambantota	4.0
3	Kandy	5.1	16	Polonnaruwa	3.9
4	Rathnapura	5.1	17	Trincomalee	3.9

(continued)

Table 3. *(continued)*

Rank	Division	Score	Rank	Division	Score
5	Gampaha	4.9	18	Batticaloa	3.8
6	Anuradhapura	4.9	19	Mullaitivu	3.6
7	Monaragala	4.8	20	Vavuniya	3.5
8	Nuwaraeliya	4.7	21	Mannar	3.5
9	Badulla	4.6	22	Galle	3.4
10	Matale	4.5	23	Matara	3.2
11	Ampara	4.4	24	Kilinochchi	2.8
12	Puttalam	4.3	25	Jaffna	2.3
13	Kalutara	4.1			

7 Limitations to the Application

UAS is a promising technology to uplift the cold chain network in Sri Lanka. However, the lack of proper regulations surrounding commercial UAV applications might demotivate suitable stakeholders and investors on the implementation [37]. Lack of proper jurisdiction results in privacy concerns for the public, security and theft threats for drones, and lack of insurance coverage [6, 24]. Furthermore, standard operating procedures need to be developed and established for the usage of UAVs in the healthcare field, setting guidelines on temperature control, altitude usage, and prohibited airspace [38]. The development of a timely regulation framework is highly suggested to attract and implement more UAV applications beneficial to Sri Lanka.

8 Conclusion

The high vaccination coverage in Sri Lanka has been shadowing the issues in the EPI cold chain for many years. This has not only caused high-cost accumulation but also threatened the longevity of the program. The main pain points causing those issues are high reliance on the road network and lack of quality CCE. Previous studies and pilot projects suggest UAVs as an alternative delivery mode to reduce such system inadequacies [7, 8, 21, 25]. Therefore, considering the cruciality and routine supply of EPI vaccines, a novel drone delivery network was suggested.

In the proposed delivery system, UAVs transport vaccines directly from the RMSD warehouse to the respective health facilities avoiding the vaccine storage at MOH units. The faster delivery, low turnaround time, and better use of the UAV, support such applications [7]. The proposed system reduces the requirement and management of CCE, human resources, and vehicles in the system. Further, the end-to-end effective vaccine distribution lessens damage to vials reducing vaccine stockouts and wastage. Additionally, the healthcare system can realize environmental benefits due to the use of electricity instead of fuel [6]. Therefore, the suggested solution will not only reduce the financial burden but also support the all-around sustainability of the EPI program.

However, the high fixed cost challenges the implementation of the suggested UAV distribution system at every RMSD. Therefore, the division that is most in need of the proposed system was identified as an initiation.

Our study identified eight factors under demand, accessibility, road network quality, and supply chain aspects to determine the value addition from the suggested system. According to the experts' review, among those aspects, accessibility, and road network quality factors were the most influential, accounting for 65% of the weights. Nevertheless, from the suitability analysis, it was determined that Kurunegala is the most positively influenced division from the proposed system. Therefore, as a starting point for the implementation of the suggested novel delivery plan, Kurunegala RMSD is proposed. Initially, conducting a pilot project will aid in realizing benefits and practical implications.

However, the lack of attention towards developing regulations surrounding large-scale drone implementations in Sri Lanka might demotivate potential stakeholders and investors. Therefore, developing a regulatory framework supporting similar technological applications is highly recommended.

Through this research, it is expected to open doors for more studies on technological implementations in Sri Lankan healthcare logistics, especially in UAV applications due to its proven capabilities.

Moreover performance factors such as CCE condition, wastage volume, and holding cost are also influential factors that were not considered in this research due to the lack of data sources [7, 22]. Therefore, an extended suitability analysis including those factors is suggested. Additionally, to identify the extent of value addition, it is suggested to quantitatively determine the potential benefits to the Sri Lankan cold chain from the drone implementation. Moreover, there are many more drone applications in the healthcare field, that can add value to the Sri Lankan context. Such as lab sample delivery within hospital premises and medicine delivery to patients. Therefore, studies on such applications are also encouraged to aid the technological upliftment in the Sri Lankan healthcare system.

Acknowledgment. The authors would like to acknowledge the financial support given by the Norwegian Program for Capacity Development in Higher Education and Research for Development (NORHED II - Project number 68085), the "Politics and Economic Governance" sub-theme, the project "Enhancing Lean Practices in Supply Chains: Digitalization", which is a collaboration between the University of Stavanger (Norway), ITB (Indonesia), and the University of Moratuwa (Sri Lanka).

References

1. Jayasekara, H., De Silva, A., Amarasinghe, A.: Costing of immunization service provision in Kalutara district, Sri Lanka: a cross-sectional study. WHO South-East Asia J. Public Heal. **5**(2), 149 (2016). https://doi.org/10.4103/2224-3151.206252
2. W. H. Organization. Expanded Program On Immunization Fact Sheet (2021)
3. A. D. Bank: Sri Lanka National Immunization Program – Financial Assessment. Asian Development Bank (2001)

4. Paba Palihawadana, T.S.R.P., Amarasinghe, A., Ginige, P.S.: "Comprehensive multi-year plan for immunization 2012–2016. Sri Lanka
5. Haidari, L.A., et al.: The economic and operational value of using drones to transport vaccines. Vaccine **34**(34), 4062–4067 (2016). https://doi.org/10.1016/j.vaccine.2016.06.022
6. Wurbel, H.: Framework for the evaluation of cost-effectiveness of drone use for the last mile delivery of vaccines (2017)
7. Wright, C., Rupani, S., Nichols, K., Chandani, Y., Machagge, M.: What should you deliver by unmanned aerial systems? (2018)
8. Zubin, I., Van Arem, B., Wiegmans, B., Van Duin, R.: Using drones in the last-mile logistics processes of medical product delivery : a feasibility case study in Rotterdam. In: Proceedings of the 99th Annual Meeting TRB, no. January, pp. 1–17 (2020)
9. Bocewicz, G., Nielsen, P., Banaszak, Z., Thibbotuwawa, A.: A declarative modelling framework for routing of multiple UAVs in a system with mobile battery swapping stations. In: Burduk, A., Chlebus, E., Nowakowski, T., Tubis, A. (eds.) ISPEM 2018. AISC, vol. 835, pp. 429–441. Springer, Cham (2019). https://doi.org/10.1007/978-3-319-97490-3_42
10. M. of H. Medical Statistics Unit, "Annual Health Statistics 2019 (2019)
11. W. E. Forum: Sri Lanka Road Quality. The Global Economy.com
12. de Silva, S.S.: Reservoirs of Sri Lanka and their fisheries. In: FAO FISHERIES TECHNICAL PAPER 298
13. C. B. of S. Lanka, Economic and Social statistics of Sri Lanka, vol. XLI (2019)
14. Weisbrod, G., Fitzroy, S.: Traffic congestion effects on supply chains: accounting for behavioral elements in planning and economic impact models. In: Renko, S. (ed.) Supply Chain Management – New Perspectives. InTech (2011). https://doi.org/10.5772/23057
15. Thibbotuwawa, A., Nielsen, P., Bocewicz, G., Banaszak, Z.: UAVs Fleet Mission planning subject to weather fore-cast and energy consumption constraints. In: Szewczyk, R., Zieliński, C., Kaliczyńska, M. (eds.) Automation 2019: Progress in Automation, Robotics and Measurement Techniques, pp. 104–114. Springer International Publishing, Cham (2020). https://doi.org/10.1007/978-3-030-13273-6_11
16. Thibbotuwawa, A.: Unmanned Aerial Vehicle Fleet Mission Planning Subject to Changing Weather Conditions, Aalborg Universitetsforlag. Ph.d.-serien for Det Ingeniør-og Naturvidenskabelige Fakultet, Aalborg Universitet (2019)
17. Rabta, B., Wankmüller, C., Reiner, G.: A drone fleet model for last-mile distribution in disaster relief operations. Int. J. Disaster Risk Reduct. **28**, 107–112 (2018). https://doi.org/10.1016/j.ijdrr.2018.02.020
18. Moshref-Javadi, M., Winkenbach, M.: Applications and Research avenues for drone-based models in logistics: a classification and review. Expert Syst. Appl. **177**, 114854 (2021). https://doi.org/10.1016/j.eswa.2021.114854
19. Johnson, A.M., Cunningham, C.J., Arnold, E., Rosamond, W.D., Zègre-Hemsey, J.K.: Impact of using drones in emergency medicine: what does the future hold? Open Access Emerg. Med. **13**, 487–498 (2021). https://doi.org/10.2147/OAEM.S247020
20. Hoerder, D.: Revolutionary technology for lifesaving vaccine deliveries. In: UNICEF Pacific (2017)
21. Ezell, S.: Zipline enables real-time delivery of essential medical supplies in Rwanda. In: Information Technology and Innovation Foundation (2017)
22. Berninzon, A.L., Vongasemjit, O.: Potential Benefits of Drones for Vaccine Last-Mile Delivery in Nepal (2021)
23. Gunaratne, K., Thibbotuwawa, A., Vasegaard, A.E., Nielsen, P., Perera, H.N.: Unmanned aerial vehicle adaptation to facilitate healthcare supply chains in low-income countries. Drones **6**(11), 321 (2022). https://doi.org/10.3390/drones6110321
24. Shyamsunder Chitta, R.J.: Last mile delivery using drones. In: Technology Convergence, innovation & Decision SciencesAt: Seoul, South Korea, p. 13 (2017)

25. Helmut Flitter, S.H., Laube, P., luscher, P., Rogers, S.: Suitability Analysis (2013)
26. Joerin, F., Thériault, M., Musy, A.: Using GIS and outranking multicriteria analysis for land-use suitability assessment. Int. J. Geogr. Inf. Sci. **15**(2), 153–174 (2001). https://doi.org/10.1080/13658810051030487
27. Nasiri, H., Boloorani, A.D., Sabokbar, H.A.F., Jafari, H.R., Hamzeh, M., Rafii, Y.: Determining the most suitable areas for artificial groundwater recharge via an integrated PROMETHEE II-AHP method in GIS environment (case study: Garabaygan Basin, Iran). Environ. Monit. Assess. **185**(1), 707–718 (2013). https://doi.org/10.1007/s10661-012-2586-0
28. Sarı, F., Ceylan, D.A., Özcan, M.M., Özcan, M.M.: A comparison of multicriteria decision analysis techniques for determining beekeeping suitability. Apidologie **51**(4), 481–498 (2020). https://doi.org/10.1007/s13592-020-00736-7
29. Parry, J.A., Ganaie, S.A., Sultan Bhat, M.: GIS based land suitability analysis using AHP model for urban services planning in Srinagar and Jammu urban centers of J&K, India. J. Urban Manag. **7**(2), 46–56 (2018). https://doi.org/10.1016/j.jum.2018.05.002
30. Aghmashhadi, A.H., Azizi, A., Zahedi, S., Hoseinkhani, M., Cirella, G.T.: Land suitability mapping using GIS-based ANP for residential zoning: case research from central Iran. Trans. GIS **26**, 1017–1039 (2022). https://doi.org/10.1111/tgis.12896
31. Aksu, G., Güzeller, C.O., Eser, M.T.: The effect of the normalization method used in different sample sizes on the success of artificial neural network model. Int. J. Assess. Tools Educ. **6**(2), 170–192 (2019). https://doi.org/10.21449/ijate.479404
32. Maps, T.: Sri Lanka Topography Map. https://en-in.topographic-map.com/places/gla/Sri-Lanka/
33. Muthukuda, M.A.D. : Situation Summary Report by National Disaster Relief Services Center (2019)
34. Taherdoost, H.: Sampling methods in research methodology; how to choose a sampling technique for research. SSRN Electron. J. (2016). https://doi.org/10.2139/ssrn.3205035
35. O. of the C. of the Currency. Sampling Methodologies, Version 1 (2020)
36. W. F. Programme: Sampling. In: Emergency Food Security Assessment Handbook (EFSA), p. 297 (2009)
37. Madushi Dissanayake, R.K.: Drone technology in sri lanka: urgency for an effective legal framework. In: 12th International Research Conference 2019, KDU, p. 6 (2019)
38. Jeyabalan, V., Nouvet, E., Meier, P., Donelle, L.: Context-specific challenges, opportunities, and ethics of drones for healthcare delivery in the eyes of program managers and field staff: a multi-site qualitative study. Drones **4**(3), 44 (2020). https://doi.org/10.3390/drones4030044

Modern Trends in Project Management – Selected Issues – Case Study

Anna Bluszcz(✉) and Marcin Chabior

Silesian University of Technology, Akademicka, 2A 44-100 Gliwice, Poland
Anna.bluszcz@polsl.pl

Abstract. Today's enterprises operate in a global market characterized by the following conditions: volatility, uncertainty, complexity and ambiguity. These conditions require the implementation of adequate management methods, ensuring sustainable foundations for business activity in the long term. In enterprises, currently about 30% of activities are carried out in the form of projects. The article presents an innovative approach to project management, which evolves from the traditional approach to a new sustainable approach. In particular, the stage of initiating a new project, in which it is required to perform a project impact analysis under the P5 Standard, was discussed. Based on the conclusions of this analysis, it is possible to plan the steps that will be taken by the project team in order not only to reduce the negative impact of the project on sustainability, but also to increase the benefits of the project. The article presents Projects integrating Sustainable Methods (PRiSM), which means a comprehensive methodology for managing sustainable projects. PRiSM goes beyond the typical project lifecycle model to maximize value by focusing on the entire asset lifecycle. This five-step approach includes pre-project planning, product/service adoption and integration, and benefit realization. The publication contains practical examples of project implementation based on sustainable project management, which is a valuable source of information for both theoreticians and business practitioners.

Keywords: Project Management · Sustainable Development · Business

1 Introduction

The world's population has just surpassed 8 billion, which means that global industry must supply an increasing amount of goods and services to the market in order to meet the needs of a growing human population. This process generates global ecological threats in the form of environmental pollution and the crisis of ecosystems. This issue has been popularized in the literature for years since the early 1970s. It is referred to as the "limits to growth" [1–4], i.e. barriers that, if exceeded, may seriously threaten even the very survival of the human species on Earth. In response to these dangers, the idea of sustainable development was formulated [5, 6], which, in the face of the global risk of an ecological catastrophe, introduces many rules that allow business practice to adjust the level of economic activity to environmental opportunities, securing not only the present, but

A. Burduk et al. (Eds.): ISPEM 2023, LNNS 790, pp. 60–74, 2023.
https://doi.org/10.1007/978-3-031-45021-1_5

also future needs of humanity. Popularizing the principles of sustainable development is, on the one hand, a challenge, and on the other, a necessity for business, as the growing environmental awareness of consumers has a very strong impact on purchasing decisions. Hence, the changing megatrends and business conditions, where there are strong influences of various stakeholder groups on projects, mean that economic profitability is no longer the only factor taken into account. We are currently observing a strong increase in the importance of social relations and ecological conditions in all areas of society's life, in particular in the business zone, where the implementation of social and ecological goals ensures running a business in the long term [7]. Eco-friendly business models are receiving increasing attention in the literature [8]. In addition, it is worth quoting the results of research confirming that the implementation of individual sustainable development goals, i.e. ensuring inclusive and equitable quality education and promoting lifelong learning opportunities and strengthening the means of implementation and reinvigorating the global partnership for sustainable development clearly facilitate the implementation of environmental policies. Additionally, other SDGs have an indirect impact on environmental goals through their reinforcing interactions with the variables of eradicating poverty, ensuring healthy lives and promoting prosperity, promoting sustainable, inclusive and sustainable economic growth and productive employment, and protecting, restoring and promoting sustainable land use ecosystems, sustainable forest management and halting biodiversity loss [9].

A modern organization operates in business conditions called VUCA for short. VUCA is an acronym introduced by US military strategists from the U.S. Army War College to describe the situation that arose after the end of the Cold War: Volatility Uncertainty Complexity and Ambiguity. Volatility manifests itself in rapid changes in a chaotic manner, making it impossible to identify any trends or find patterns of behavior. The changes may concern, for example, the instability of the legal or geopolitical system or changing climate conditions, including the consequences of climate change. As a result, long-term detailed plans quickly become obsolete. Uncertainty is a challenge because it is difficult to draw conclusions about the future based on past experiences. Under conditions of uncertainty, managers have little influence on current events in business markets and the resulting problems. As a result, it is difficult to predict what will be the effect of the actions taken, which may lead to the extension of the planning process (to reduce uncertainty). Reducing uncertainty can lead to the fulfillment of the managerial saying: if you want to be 100% sure of something, you will be 100% late.

Complexity means the multidimensionality of the company's areas of activity and mutual correlations of factors influencing the given areas, which causes difficulties in identifying the cause-and-effect chain of actions taken and emerging problems. An example is international activity, which means operating in markets that differ in legal regulations, procedures, customs tariffs, languages and cultures. As a result of this complexity, managers can predict only some events, and the amount and nature of the data analyzed affects the decision-making process. Ambiguity means a high probability of misreading signals coming from the environment, a multitude of meanings and no previous experience in a given area. This means operating in what is known as "unknown unknowns" terrain. Managers go beyond their comfort zone and take actions with increased risk, or refrain from making decisions out of fear of risk.

In response to the changing conditions of the functioning of enterprises on the global market, the approach to project management in organizations is also changing. There is a strong focus on compliance with environmental standards and taking into account social conditions during the implementation of projects. Currently, new project management frameworks are being developed that indicate the directions of change in enterprises, and generally this approach is called green management or sustainable project management.

The aim of the article is to present a comprehensive methodology for project management in a sustainable approach and to present practical examples of projects in which the principles of a sustainable approach are taken into account. An innovative aspect of the publication is the popularization of a new approach to project management, which is currently being implemented in Poland, but is increasingly widely used globally.

2 Project and Project Management

2.1 Traditional Approach to Project Management

Generally speaking, a project means a temporary activity undertaken by a company in order to create a unique product or service. Temporary activity means that each project has a definite beginning and a definite end. The term unique means that the product or service differs in some way from all other products or services. For many organizations, projects are a way of doing things when a situation cannot be resolved as part of the organization's normal operations. Projects are interdisciplinary in nature, can be implemented by diverse consortia and require management functions. According to the basic definition of the Project Management Institute, project management means the use of available knowledge, skills, tools and techniques to meet the requirements and expectations of project principals [10]. In more practical terms, it can be assumed that project management includes a set of managerial activities, which include planning, organizing, coordinating, motivating and controlling the implementation of the assumed activities in the project, according to the established rules and methods of the project, the scope of which is predetermined and the financial resources for its implementation are limited.

All aspects of a global and turbulent environment affect project management conditions. The evolution of project management systems shows how markets change and how the approach to project management changes. Starting from the 1960s, operational management based on streamlining and optimizing the process flow (operational approach) was sufficient. Then, from the 1980s, as a result of the increase in requirements for the quality of project results, the study and improvement of project implementation in the full life cycle (process approach) takes place. Since 2000, organizations have been facing the problem of increased project risk, the implementation of a portfolio of various projects, and the increased role of stakeholders in projects. These conditions encourage organizations to research and improve the organizational system of project implementation (organizational approach). Since 2010, we can distinguish a significant increase in the aspects of sustainable development and social responsibility in business practice. These aspects force managers and investors to implement projects taking into account the requirements and limitations of stakeholders, the implementation of the postulates of social responsibility, which in practice means an increase in complexity and volatility and conditions during project implementation (green project management approach).

2.2 Modern Approach – Green Project Management

The globalization of the world's economic markets is constantly changing the conditions for business operations. The previous traditional approach to project management is undergoing an evolutionary process. The main aspects of project management such as time, cost, scope and quality still remain the main business aspects, but project management is currently being done based on the implementation of sustainable development principles. In practice, this means expanding project evaluation criteria to include social, economic and environmental areas, as well as project and process aspects. Changes in the approach to project management are presented in Fig. 1.

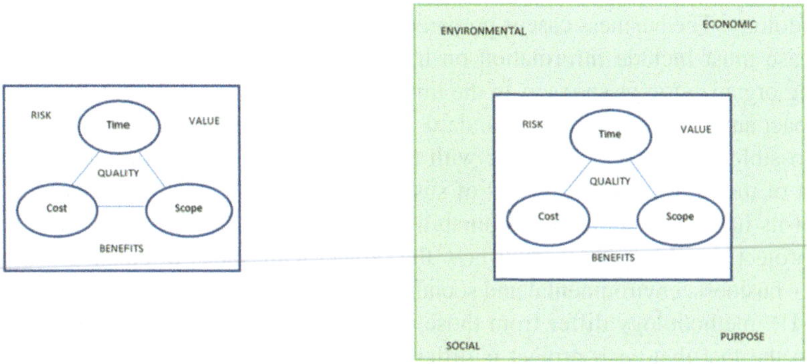

Fig. 1. Traditional project management vs green project management [11].

Green Project Management should be understood as an integrated approach to project management taking into account sustainable perspectives in business practice (social, environmental, economic). Popularization and implementation of activities based on sustainable principles ensures organizations' long-term development and success in a global and competitive market, where caring for the environment is currently gaining key importance. Sustainability was considered a new trend until recently. However, in the face of global social and environmental changes, it has become a requirement in today's world.

2.3 Project Integrating Sustainable Method – PRiSM Framework

The literature currently emphasizes the need to pay attention to the social sphere of industry, in particular in the era of business digitization [12, 13]. The main determinant of the development of Industry 5.0 is the emphasis on the welfare of the employee in the industry, including the growing ecological awareness of societies [14, 15].

Projects integrating Sustainable Methods (PRiSM) is a world-renowned comprehensive methodology for Managing Sustainable Projects. In other words The

PRiSM™ framework (Projects integrating Sustainable Methods) provides a comprehensive methodology who integrates sustainability in project management, by an effective fusion of best practices from the Standards Families for Quality ISO 9000 for the Environment ISO 14000, from the Standards for Energy ISO 50001, for Project Management ISO 21500, for Social Responsibility ISO 26000 and from the International Competence Baseline of the International Project Management Association. By adopting this framework and its proposed tools and methods, organizations can achieve balance between scarce resources, Social Responsibility and sustainable results in their projects. It is a structured methodology for sustainable – Green Project Management, which is based in a series of standards and incorporates their use in the standard ISO 21500:2012 Guidance on Project Management [16].

PRiSM goes beyond the typical project lifecycle model to maximize value by focusing on the entire asset lifecycle. The five-step approach includes pre-project planning, product/service adoption and integration, and benefit realization. The key factors differentiating PRiSM from the traditional approach are the features specific only to this methodology: The business case is prepared at the pre-project planning stage. The business case must include information on how the project supports the strategy of the funding organization or sponsor. In the initiation phase, it is also necessary to perform an impact analysis under the P5 Standard. Based on the conclusions of this document, it is possible to plan steps together with the project team that will limit the negative impact of the project in the context of sustainable development and increase the benefits of its implementation. A Sustainability Management Plan (SMP) is prepared for each project. It takes into account how the project team plans to balance conflicting needs – business, environmental and social responsibility. Project success criteria in the PRiSM™ methodology differ from those in traditional approaches. Much attention is paid to the fact that each project is different and success means something different for each stakeholder. Instead of asking, "Was this project successful?", the Sustainable Project Manager will ask, "How successful was this project in all respects?". One of the conditions for success is the achievement of the SDGs during the project. According to the authors of the PRiSM™ methodology, sustainable development begins with a value system and a rule-based approach to doing business. Institutions should operate in such a way as to fulfill at least basic responsibilities in the areas of human rights, labor, environmental protection and anti-corruption [17].

Sustainable project management is implemented in organizations using the so-called 5P aspects: **products** (technical aspect), **processes** (governance aspect), **profit** (financial aspect), **planet** (environmental aspect) and **people** (social aspect), matters into the project [18]. The 5P standard combines the sustainable development goals set by the United Nations and project management. This is done first at a basic level, such as identifying and implementing values such as healthy labor practices, human rights, local procurement to reduce CO2 emissions, or taking care of the local economic impact of projects. But the most complex part is implementing these values at all levels of management to transform the entire project and its product into a sustainable outcome. The model for such management is based on ISO standards, which indicate the normative framework and at the same time support its credibility. According to the ISO 21500 Guidance on Project Management, "a project consists of a unique set of processes involving coordinated and controlled activities with start and end dates, performed to achieve project objectives. The achievement of project objectives requires the provisions of deliverables in accordance with specified requirements. The implementation of the Standard P5 document takes the form of an analysis of the impact of the project assumptions on the environment, society and economy, as well as the convergence with the organization's strategy. It is performed at the start of the project in accordance with the PRiSM methodology. The purpose of this observation is to identify both threats and opportunities related to sustainable development. Enterprises are constantly looking for ways to gain a competitive advantage. Increasing efficiency in project implementation can significantly contribute to the market success of the entire organization. Improving the company's activities in the field of project management becomes an important factor influencing its development. The enterprise should, as far as possible, take care of sustainable development in each of the areas of its activity in order to develop continuously. The business case is prepared during the pre-project planning phase. The business case must include information on how the project supports the strategy of the funding organization/sponsor. The information contained in this document should therefore be sufficient to make a positive decision regarding project financing. During the initiation phase, it is also necessary to perform an impact analysis under Standard P5. Based on the conclusions of this document, it is possible to plan the steps that will be taken by the project team not only to reduce the negative impact of the project in terms of sustainability, but also to increase the benefits of the project.

A Sustainability Management Plan (SMP) is prepared for each project. It takes into account, among other things, how the project team plans to balance conflicting needs – business, environmental and social responsibilities. Sustainability Management Plan (SMP) describes how sustainability will be addressed during a project. P5 has an integral role in developing an SMP as P5 identifies the subjects to be addressed [11].

The project success criteria in the PRiSM™ methodology differ from those in traditional approaches. Much attention is paid to the fact that every project is different, and success means something different to each stakeholder. Instead of asking: "Was this project a success?" a sustainable project manager will ask: "How successful was this project in every aspect considered?". One of the conditions for success is the achievement of sustainability goals during the project.

According to the developers of the PRiSM methodology, sustainability starts with a value system and a principles-based approach to doing business. Institutions should operate in a way that meets at least basic human rights, labor and environmental protection and anti-corruption obligations.

Another element that differentiates PRiSM from other approaches is the six principles of delivering sustainable change [17].

- Commitment and Responsibility Recognize the fundamental rights of all people to a healthy, clean and safe environment, equal opportunity, fair pay, ethical procurement and adherence to the rule of law.
- Ethics and decision-making Support organizational ethics and decision-making on universal principles by identifying, reducing and preventing negative long- and short-term impacts affecting society and the environment.

- Integration and transparency fostering interdependence between economic development, social integrity and environmental protection in all aspects of management, practice and reporting.
- Grounded in principles and values Protecting and strengthening our natural resource base by improving the ways in which we develop and use technology and resources.
- Social and environmental justice Assess human vulnerability in environmentally sensitive areas and population centers through demographic dynamics.
- Economic prosperity Adhering to strategies, goals and fiscal targets that balance the needs of stakeholders and the needs of future generations.

2.4 Applying a Sustainable Standards in a Project Management Practice

Standard The GPM P5 Standard for Sustainability in Project Management was published in 2022 and is a relatively new standard that should not yet be assessed in terms of popularization in practice. The aim of the article, however, is to indicate examples of various projects in Poland that significantly emphasize the P5 issues and indicate that the trend of sustainable project management will be more and more clearly reflected in Polish business practice.

Based on other studies, it has been shown that new approaches in project management are very quickly disseminated and implemented in practice, an example of which is the use of social media in project management. The research was to indicate whether they constitute an opportunity or a threat to the projects. The research results clearly indicated that social media usually supports project managers and their members in such areas as communication, cooperation, involvement, knowledge management, work productivity and project promotion as well as employee development. And because effective project management requires creating and maintaining relationships between all project stakeholders, social media are a very good communication tool [19]. In conclusion, it can be said that project management is constantly changing to adapt the methods and tools necessary for the effective work of project teams. As indicated in the following examples, the issue of sustainable project management is already popularized in business practice.

Case study 1 – Green Project Management (GPM) – research Initiative.

Green Project Management (GPM) is a global organization that sets trends and methods in the field of sustainable projects. As a social organization, it presented a comprehensive study that provides current information on the implementation of sustainable development methods in project management. The survey was conducted in 2020 and involved 33,348 people from 94 countries. The largest group of respondents, i.e. 51%, came from 8 countries such as: USA, UK, China, Costa Rica, France Australia and others. The remaining 49% came from the 86 other countries. One of the survey questions was "Have extreme weather events such as flash floods, wildfires, rising sea levels affected your work project?" The answers of the respondents in the positions of project managers answered affirmatively and constituted 38%. This is an increase from just 4% in 2019. The response was even higher among the Program and Portfolio Managers who noticed this affected 42%. Among Management, 28% said yes while 72% were unsure or said no. 97% of the executives – respondents believe that projects and project management are an integral part of sustainable development. Only 13% of them believe that they are doing enough to combat climate change. Even though 91% of executives see a tangible effect as a result of the increase in sustainability performance of implementing the P5 standard, there are concerns that mapping the project goals into the organization's sustainability strategies and reporting is not enough. When we assume that the estimated global GDP in 2021 amounted to USD 94 trillion, and approximately 28% of project managers declared that their projects are affected by the effects of extreme weather events, this means that 30% of global GDP* (in some countries more), is USD 7.9 trillion may be at risk [20].

Case study 2. Analysis of involvement in the implementation of 5P on the basis of a selected mining company.

Access to information on the involvement of enterprises in the implementation of sustainable development at the level of projects implemented in the organization is quite a difficult task, therefore the article focuses on the organization's involvement in sustainable development at the strategic level of the entire organization. This commitment can be assessed based on the analysis of available information published in social responsibility reports. Selected reports of a mining company whose operations have a key impact on environmental issues were selected for analysis. The aim of the study was to analyze the level of the organization's commitment to the aspects of long-term sustainability for People, Planet, Prosperity, Product and Process (P5 Standards). The company transparently publishes goals and creates lasting, trust-based relationships with stakeholders – both internal and external as well as external ones. The enterprise is an important part of the local area community and has had a positive impact on it for years and solves a lot of regional problems. The company leads constructive dialogue, information exchange and compromise, which is a tool for building further cooperation, good relations and reputation. Social responsibility reports prepared according to the Global Reporting Initiative standard were analyzed. These are very detailed studies containing several dozen indicators characterizing the examined areas of the company's functioning. Indicator areas include: Profile of the organization, strategy, Ethics and integrity, Corporate governance, Stakeholder engagement, Reporting practice, Performance and indirect economic impacts, market presence, Procurement practices and social and environmental assessment of suppliers, Compliance and anti- corruption, Energy and emissions, Water and effluents, Biodiversity, Emissions, waste, Employment and labor relations, Health and safety, Training and education, Diversity, equal opportunity and anti-discrimination. The above-mentioned indicators are also included in the 5P standard, so the audited company reporting concentration on all the researched areas at the strategic level has the potential to implement these issues also at the level of implemented projects. The results of the GRI report analysis of a selected mining company are presented in the Table 1:

Case study 3. Examples of social and environmental aspects taken into account during the assessment of projects in the mining industry by financing institutions.

The specificity of mining projects is that they are closely related to the exploitation of natural resources and thus have a negative impact on the environment. The only example of mining projects that we can evaluate positively are projects related to the revitalization of post-industrial areas, including post-mining areas. A typical mining example may be the construction of a mine and surface and transport infrastructure necessary for the operation of a mining plant. Projects of this type are largely carried out in the past, currently revitalization of post-mining spaces is taking place. Nevertheless, such types of potential

Table 1. Analysis of involvement in the implementation of 5P on the basis of a selected mining company.

Aspects:	Example
People	The surveyed mining company has a large impact on society. It employs nearly six thousand people. The daily work of employees, their competence, experience and commitment enabled the operation of this one of the most modern mines in Poland and Europe. The vast majority of employees are employed full-time and for an indefinite period. Salaries in the company is one of the highest in the hard coal industry. The company provides economic security to thousands of employees and their families. Thanks to the company, it is possible to develop other small and medium-sized enterprises in the region, which has one of the highest unemployment rates in the country. Social activity of a company was carried out in the areas such as: supporting charities, aid organizations, sports, social, cultural sponsorships. The company actively influences the community by supporting schools (including vocational schools). Schools), educational projects, equal opportunities for students, supporting entrepreneurial attitudes, creativity and innovation, and education in this field sustainable development. The company established the Foundation to organize support for employees and the local community in a difficult life situation
Planet	The company identifies and assesses environmental aspects related to mining activities and monitors and develops environmental programs. The company has defined priorities in the field of environmental protection and conducts active activities in this respect, which relate to: prevention and removal of mining damage, maintenance of recovery methods mining waste, increasing efficiency use of electricity. The management board of the coal company declares to maintain the highest quality standards for all products. The company conducts activities in the field of environmental protection and limiting the undesirable impact on the environment through, among others, such activities as: sustainable use of spoil tips, development of new technologies for utilization of mine waste, action in the local territory through transformation of mining areas. The company has taken steps towards estimating the carbon footprint for the base year by which the process of optimizing this level can be started. The operation of each mine is related to emissions dust and gases into the atmosphere. Is not possible with the current technological and available knowledge equipment, such organization of mine work, to eliminate harmful emissions. The company monitors the level of emissions and reports their level. The company reports in detail many tasks completed in the field of biodiversity in which it actively participates
Prosperity	Value creation in a company is based on the analysis of six capitals and their correlation with strategic goals companies. These capitals include: financial, productive, intellectual, human, social and natural capital. Objectives and capital are correlated, moreover, the implementation of the assumed objectives requires constant dialogue with stakeholders. The company transparently publishes goals and creates lasting, trust-based relationships with stakeholders – both internal and external as well as external ones. The enterprise is an important part of the local area community and has had a positive impact on it for years and solves a lot of regional problems. The company leads constructive dialogue, information exchange and compromise, which is a tool for building further cooperation, good relations and reputation

projects require an administrative decision, which is made after a detailed assessment of the investment's impact on the environment and public consultations. This stage of the decision-making process depends on the ability to objectively weigh the environmental and social benefits or losses included in the analysis. The problem of assessing the impact of mining projects on the natural environment is well known due to their strong impact on nature and the local economy. A part of this assessment is the analysis of the area occupied for the development of the angle of the so-called ecosystem services, i.e. services provided by ecosystems located in the area of impact of the planned investment. The purpose of this analysis is to rationalize the placement of various elements of investment in such a way that the environmental cost is eliminated or minimized. Elaboration should include, among others mapping of ecosystems, identification and characterization of functions ecosystems in the project impact area, characteristics of ecosystem services provided by ecosystems, including their description, categorization, quantitative assessment, valuation services and analysis of trends in the flow of services in a selected period of time. The investor provides the Report on the impact of the project on the environment via public administration body for public inspection. The assessment documentation will be an appropriate, accurate and objective assessment and presentation of environmental and social hazards and impacts and can be prepared by the client, consultants and external experts. IFC's Environmental and Social Performance Standards define IFC clients' responsibilities for managing their environmental and social risks. The 2012 edition of IFC's Sustainability Framework, which includes the Performance Standards, applies to all investment and advisory clients whose projects go through IFC's initial credit review process after January 1, 2012 [21].

The Equator Principles Financial Institutions (EPFIs), have adopted the Equator Principles [22] in order to ensure that the Projects we finance and advise on are developed in a manner that is socially responsible and reflects sound environmental management practices. For Projects located in Designated Countries, the Assessment process evaluates compliance with relevant host country laws, regulations and permits that pertain to environmental and social issues. As part of the consultation requirement, the investor is required to prepare and publish a Stakeholder Engagement Plan (SEP). The above document contains detailed information on the actions taken to publicly inform about the investment and on the mechanism for examining complaints and submitting inquiries by interested parties. In addition, the investor should prove that the opinions expressed by the local community have been considered. This is done by indicating which comments and requests of the public have been taken into account in the project variant prepared for implementation. Providing information about the project allows interested parties to assess the impact of the investment project on the environment and the local community, as well as to learn about the benefits of the project itself. The implementation of the required activities is reported to the financing institution at least once a year and may be subject to an audit or other assessment throughout the duration of the Project. A part of the environmental impact assessment is the analysis of the area occupied for the investment in terms of services provided by ecosystems located in the area of impact of the planned investment. The purpose of this analysis is valuation economic costs related to the replacement of ecosystem services in the event that they are negatively affected and rationalization of the location of various elements of the investment in such a way that the environmental cost is eliminated or minimized. Early estimation of environmental and social costs reduces the risk of Project failure. Due to the globalization of environmental aspects and the growing social conflicts, large mining projects must meet international requirements in this area[23].

Case study 4. Evaluation of the implementation of the principle of sustainable development under program PO IR 2014–2020.

This report contains an assessment of the ongoing implementation of the sustainable development principle under PO IR 2014–2020. The main objective of the study was to assess and possibly verify the method of implementing the sustainable development principle at the implementation stage of PO IR 2014–2020. In addition, the project was aimed at developing a guide in the form of "good practices" in the field of compliance with the principle of sustainable development. This catalog refers to activities undertaken as part of projects co-financed from PO IR 2014–2020. The study was conducted in the period July – September 2017. The methodology used by the research contractor included both the analysis of the existing data and field research (quantitative and qualitative). The starting point for the evaluation was the understanding of the fact that PO IR 2014–2020 is aimed at supporting the innovativeness and competitiveness of the Polish economy. The principle of sustainable development, on the one hand, sets the framework for actions that can be taken under the Program, and on the other hand, it is a means that can be used to achieve its objective. In other words, while the evaluation part of the report focuses on assessing the implementation of the analyzed principle as one of the horizontal principles, the part related to good practices also includes examples of projects that contribute to the creation of competitive products or services through innovative pro-environmental activities. In the first case, the compliance of the PO IR 2014–2020 intervention with the principle of sustainable development was assessed in terms of the Program's neutral or positive impact on the environment and climate. Although the Forecast of the environmental impact of the draft Smart Growth Operational Program 2014–2020 contained a provision that the analysis of priority axes and investment priorities, as well as the environmental impact assessment, showed that a significant part of projects undertaken under PO IR 2014–2020 will not have a direct impact on the environment, however, the efforts of the Program implementation system were focused on such construction of the selection system and project selection criteria that would lead to the possible maximization of the positive impact (by requiring it at the stage of application evaluation). However, difficulties during the recruitment process and experience from the implementation of valuable projects that meet the objectives of the Program – albeit environmentally neutral – forced us to meet the applicants' expectations. The restrictiveness of the approach to environmental issues has been reduced, which translates into the modification of project selection criteria, understood as a departure from the absolute requirement of a positive impact of the project on the principle of sustainable development in favor of the required neutrality and rewarding positive impact [24]. The results of the report indicate numerous projects implemented in accordance with the principles of sustainable project management. These are projects such as:

1. Drive system for electric bicycles using a remotely controlled three-speed front bicycle gear. Project description: Adaptation of a mechanical front bicycle gear for use

in an electric bicycle and its integration with a drive system with a motor mounted in the rear bicycle wheel. Implementation of the principle of sustainable development: Implementation of the project does not involve any negative impact on the environment, the entire production process is relatively energy-saving and does not release negative substances into the environment. The result of the project as a green means of transport is in line with the implementation of the principle of sustainable development.

2. Obtaining innovative polymer surfactants and plasticizers Project description: Development of surfactants (SPC) that will meet the increasingly stringent requirements in the field of environmental protection standards. The method of implementation of the principle of sustainable development: Within this project, an important behavior constituting good practice in the implementation of the principle of sustainable development is the use of natural production inputs. They are not only more ecological, but also reproducible. This translates into a reduction in the impact of the production process on the natural environment, including minimizing the negative impact on the health of employees involved in the production of plasticizers. The emission of harmful substances into the atmosphere is limited. The fact that the raw materials used for production are recyclable translates into maintaining a greater diversity of raw materials in the environment, as the project does not affect the depletion of non-renewable resources.
3. Development and implementation of the technology of an innovative combined biofilter, which will be used in the industry in the future. Project description: The subject of the project is the development of an innovative combined biofilter using a filler made of waste materials, adapted to the purification and deodorization of specific process gases. These gases are to come from the mechanical and biological processing of municipal waste, from wastewater treatment and processing of sewage sludge, and from the food industry. The concept of a combined biofilter is an innovative solution, not used on an industrial scale, allowing for the optimization of operating parameters and the effects of reducing emissions, and at the same time contributes to a better use of waste materials in accordance with the circular economy model. The combined biofilter is intended to enable: conducting a two-stage process of process gas purification, technical and operational modification of biofiltration on a bed of waste materials, aimed at ensuring optimal parameters of the process and its effects. a significant reduction in emissions with relatively small investment.

Case study 5. A new horizontal principle "Do No Significant Harm" (DNSH).

Detailed implementation and implementation of the principles of sustainable development is required in projects implemented under the European Union funds. In the new financial perspective 2021–2027, a new horizontal principle "Do No Significant Harm" (DNSH) was introduced. In practice, this means that it is to be applied in projects universally, cross-sectionally, to the widest possible extent. In accordance with the legal norms of the Cohesion Policy, the European Regional Development Fund and the Cohesion Fund should support actions that comply with the Union's climate and environmental standards and priorities and do not cause serious damage. The environmental goals considered are: climate change mitigation, adaptation to climate change, sustainable use and protection of water and marine resources, circular economy, including waste prevention and recycling, prevention and control of air, water or soil pollution, protection and restoration of biodiversity and ecosystems. Member States were required to assess compliance with the DNSH principle of each of the programs implementing the Partnership Agreement. This is to ensure that co-financing from the European Union will be granted only to projects that comply with the DNSH principle.

2.5 Conclusion

The article is a review, describing new trends in project management. The content describes the characteristics of the traditional approach to project management, in which the most important issues were related to the scope of the project, implementation time, budget and quality of the final product or service. The article presents how the approach to managing the implementation of business projects is changing as economic conditions change and global megatrends change.

The constant change of geopolitical conditions on the markets requires flexibility in operation, which increases the need for the ability to implement corrections in current operations. Environmental awareness of societies, including customers, leads to changes in the assessment of key aspects of projects. Currently, more and more emphasis in business practice is not only economic issues, but the importance of social and environmental issues is growing, which is the basis for a new approach to project management called green management. The article presents the PRiSM methodology, which mainly concerns the analysis of the entire project life cycle in which the 5P standard is used. The last part of the article indicates the increase in popularity of the sustainable approach to project management in practice, based on surveys conducted by the Green Management organization and the practical implementation of sustainable environmentally friendly projects, the popularity of which will increase in particular through the implementation of the new horizontal principle in European funds.

The numerous examples presented in the article confirm that the use of good practices and frameworks of sustainable development standards in project management in business practice is necessary in order to gain both the acceptance of financing institutions and the acceptance of significant stakeholder groups.

As shown in the article, aspects of sustainable project management will be increasingly implemented in practice. Voluntary implementation of sustainable development principles in practice could be too slow to notice and measure the effects of their implementation. Thus, we can see the activity of financing institutions that determine the financing of projects that will implement the principles of sustainable development in practice and demonstrate the achievement of the assumed goals and effects.

References

1. Bluszcz, A.: Ecological growth boundaries, management systems in production engineering. P.A. NOVA S.A. **25**(1), 55–59 (2017)
2. Bluszcz, A.: What if not gross domestic product per capita? In: Kapounek, S., Kocis Krutilova, V. (eds.) 21st Annual International Conference 2018, Enterprise and Competitive Environment, pp. 83–91. Mendel University, Brno, Czech Republic (2018). ISBN 978-80-7509-561-9

3. Bluszcz, A.: Analysis of decoupling economic growth from material and energy use for several EU countries. In: 18th International Multidisciplinary Scientific Geoconference SGEM 2018, STEF92 Technology, pp. 157–166. Vienna, Austria (2018). ISBN 978-619-7408-70-6
4. Bluszcz, A., Manowska, A.: Panel analysis to investigate the relationship between economic growth, import, consumption of materials and energy. In: World Multidisciplinary Earth Sciences Symposium WMESS 2019, Earth and Environmental Science, nr 362, 2019art. no. 012153 1–9. Prague, Czech Republic (2019)
5. Bluszcz, A.: Conditions for maintaining the sustainable development level of EU member states. Soc. Indicators Res. **139**(2), 679–693 (2018)
6. Bluszcz, A.: Multidimensional comparative analysis as a tool for assessing the level of development of energy markets in selected European countries. In: Yilmaz, I., Marschalko, M., Drusa, M. (eds.) 6th World Multidisciplinary Earth Sciences Symposium WMESS 2020, Earth and Environmental Science, 609, art. no. 012020, pp. 1–10. Prague, Czech Republic (2020)
7. Wyrwa, J., Barska, A.: Disparities in social development of rural areas in the context of sustainable development of polish voivodeships. Probl. Agric. Econ. **4**(369), 45–77 (2021)
8. Domaracká, L., Seňová, A., Kowal, D.: Evaluation of eco-innovation and green economy in EU countries. Energies **16**, 962 (2023). https://doi.org/10.3390/en16020962
9. Kluza, K., Zioło, M., Bąk, I., Spoz, A.: Achieving environmental policy objectives through the implementation of sustainable development goals. The case for European union countries. Energies **14**(8), 2129 (2021). https://doi.org/10.3390/en14082129
10. Project Management Institute: A Guide to the Project Management Body of Knowledge, Newtown Square, Pennsylvania USA (2000)
11. The GPM5 Homepage: http://www.greenprojectmanagement.org. Last accessed 1 Feb 2023
12. Grabowska, S., Saniuk, S., Gajdzik, B.: Industry 5.0: improving humanization and sustainability of Industry 4.0. Scientometrics **127**(6), 3117–3144 (2022). https://doi.org/10.1007/s11192-022-04370-1
13. Kowal, B., Włodarz, D., Brzychczy, E., Klepka, A.: Analysis of employees' competencies in the context of industry 4.0. Energies **15**(19), 7142 (2022). https://doi.org/10.3390/en15197142
14. Grabowska, S.: Key components of the business model in an Industry 5.0 environment, Scientific Papers of Silesian University of Technology – Organization & Management Series, vol. 158, pp. 191–199 (2022)
15. Saniuk, S., Grabowska, S., Straka, M.: Identification of social and economic expectations: contextual reasons for the transformation process of Industry 4.0 into the Industry 5.0 concept. Sustainability **14**(3), 1391 (2022). https://doi.org/10.3390/su14031391
16. Katsarelis, T., Adamopoulou, I.: Evolving Project Management: The PRiSM™ Framework – Projects Integrating Sustainable Methods. Conference: 5th ENEPROT. Thessaloniki, Greece (2014)
17. PRiSM Homepage: https://gpm-emea.org/gpm/prism-methodology/. Last accessed 21 Jan 2023
18. Dobrovolskiiene, N., Tamošiunienie, R.: An Index to measure sustainability of a business project in the construction industry: lithuanian case. Sustainability **8**(14), 1–14 (2015)
19. Hysa, B., Spałek, S.: Opportunities and threats presented by social media in project management. Heliyon **5**(4), 1–28 (2019)
20. Insight into sustainable project management. A GPM research Initiative. Report (2022)
21. IFC Homepage: https://www.ifc.org/wps/wcm/connect/Topics_Ext_Content/IFC_External_Corporate_Site/Sustainability-At-IFC/Policies-Standards/Performance-Standards. Last accessed 20 Jan 2023
22. EP Homepage: https://www.equator-principles.com/. Last accessed 10 Jan 2023

23. Szamałek, K., Wierchowiec, J.: The importance and role of the JORC standard as the basis for a bank feasibility study of mining projects for the assessment of project profitability. Miner. Res. Manag. **31**(3), 25–44 (2015)
24. ECORYS Homepage: https://ecorys.pl. Last accessed 8 Feb 2023

Cyber Resilience as a New Strategy to Reduce the Impact of Cyber Threats

Jerzy Stanik and Jaroslaw Napiórkowski(✉)

Faculty of Cybernetics, Military University of Technology, Warsaw, Poland
jaroslaw.napiorkowski@wat.edu.pl

Abstract. This article explores the relationship between today's threats and cyber resilience by defining key concepts, highlighting the role of cyber resilience, and examining how cyber resilience can play a significant role in stopping cyberattacks and keeping digital data secure. The essence of cyber resilience is the use of such a level of cyber protection that the cost of carrying out an attack on the subject of action exceeds the cost of possible damage. The aim of the article is also to encourage further discussion and progress towards the integration of cybersecurity and cyber resilience by proposing new concepts and models of cyber resilience systems, as well as assessing how to implement these concepts in different actors.

Keywords: cybersecurity · cyber resilience · resilience system · cyber · threats

1 Introduction

In the 21st century, in the face of an unprecedented and exponentially growing global cyber threat matrix, the security of data, applications and processes is of paramount importance. Therefore, all actors – large, medium, and small – should be called upon to build cyber resilience. Cyber resilience is a concept that is rapidly gaining ground. It is a broad umbrella that covers information security, IT infrastructure, business processes and organizational continuity. Cyber resilience requires a holistic, adaptive, and global approach that can withstand evolving threats to the essential services and infrastructure of any actor, including:

- Real-time response capabilities to proactively detect threats and vulnerabilities.
- Protection against known attacks and proactive actions against new and predicted attack vectors, including the ability to self-heal.
- Automatic reconfiguration, recovery, and redundancy in case of disruption.
- Core Cyber Resilience Culture.

Even if companies do the right thing, they will still be at a huge disadvantage because unlike many other cyber threats or operational threats, cyber risk is primarily a borderless criminal activity. While the concept is now widely used among information security practitioners and political and business leaders in many countries, cyber resilience as an academic research topic is still in its infancy. For example, only 482 articles in the

A. Burduk et al. (Eds.): ISPEM 2023, LNNS 790, pp. 75–92, 2023.
https://doi.org/10.1007/978-3-031-45021-1_6

Google Researcher Index have "cyber resilience" at all, of which only 35 articles include it in the title [8].

Each entity of action – individual, company, organization, or state – is subject to tensions resulting from events, changes, cyber threats, and incidents taking place in their environment. These stressful situations are new challenges, the solution of which will affect the functioning of the entity until the situation can be brought under control through automation. Paradoxically, most companies that have fallen victim to a cyberattack would probably say that they were well protected at the time of the attacks. Cyber resilience has been extensively analyzed as part of the risk analysis, particularly in relation to critical infrastructure [2, 4, 5]. Knowledge in mature research fields is becoming more grounded in practices, laws and regulations, practitioner education, etc. Cyber resilience is currently undergoing many changes [1]. Our understanding of cyber resilience factors is growing but must adapt to rapid changes. Deepening our understanding of cyber resilience becomes imperative. It is essential to map the relevant factors in the cyber defense domain [24].

The article assesses the state of knowledge on cyber resilience based on a review of numerous literature items and thematic websites. This review aimed to identify the essential elements of cyber resilience to facilitate development beyond the state of the art. Common definitions of cyber resilience are presented and then clarified and refined. A new concept of cyber resilience was defined, and then a model of a cyber resilience system was built on its framework. The basic elements of the proposed model were characterized, which consists of: the subject of action, object of action, purpose of action, cyber resilience strategy, cyber resilience measurement indicators as well as technologies and tools supporting the measurement of the effectiveness of the cyber resilience system.

Section 2 reviews the literature. Section 3 describes the research process. Section 4 defines cyber resilience and the cyber resilience system model. In addition, Sect. 4 characterizes cyber resilience resources, competencies, technology, tools and strategies. Section 5 ends.

2 Literature Review

2.1 History and Cyber Resilience Review

Hult and Sivanesan [13] provide a historical context for the evolution of cybersecurity and outline what. They suggest that relying on traditional protection and preventive controls is not enough and recommend a more agile approach. The latest Panda Security report, presented at #PASS2018, defines cyber resilience as "the key to business security". Kott and Linkov [17] review and assess the cyber resilience of systems and networks, distinguishing between security, risk and resilience. ("Cyber resilience in firms, organizations and societies") They describe current and possible future cyber resilience practices and techniques, including technical considerations. Mattson [11] review critical national infrastructure and cyber security measures in the UK and the US. They assess the surrogate nature of well-planned and executed cyberattacks, illustrated by well-known historical examples, and find key steps in detection, deterrence, and disruption. Kaufmann [18] – cyber resilience is an organization's ability to protect against, detect, respond to, and recover from cyberattacks. The latest Panda Security report, presented at #PASS2018, defines cyber resilience as "the key to business security".

2.2 Policy

Gisladottir, Ganin, Keisler, Kepner and Linkov [7] consider the resilience of cybernetic systems with over-regulation and under-regulation. They recognize a common approach to implementing new regulations against new threats to harden the system, resulting in the stress of spending more time on training and policy implementation. Bostick, Connelly, Lambert and Linkov [3] consider how risk-based policy and investment decision-making affects resilience in damaged communities, and how economic motivations for managing resilience play a role in large-scale complex systems. Linkov et al. [19] note that cyber-attacks can cause damage disproportionate to the severity and cost of carrying out the attack. They identify and apply a variety of quantitative and qualitative metrics from the literature to develop resilience metrics for cyber systems in the physical, informational, cognitive, and social domains. Their general approach is to integrate real data, technical judgment, and literature-based measures. They link national policy objectives with specific systemic measures so that the allocation of resources translates into actionable interventions and investments.

2.3 Management

Linkov, Roslicky and Trump [22] present multiple perspectives on managing hybrid threats involving digital systems, which has an impact on various fields such as medicine, social media, and homeland security. Different protection and recovery strategies have been proposed under different disruption scenarios, focusing on risk and resilience in the information domain. ("Cyber resilience in firms, organizations and societies") Linkov and Palma-Oliveira [21] advocate a systemic view of resilience and risk related to the environment, cyberspace, and other domains. To address emerging threats, they propose resilience-based management in infrastructure, cybersecurity, and socialization methodologies and tools. Flamingos [6] considers paradigms and techniques to move from risk modeling to threat mitigation to ensure the resilience of cyber-physical systems. Both exposure to new threats and the potential of CPS to counter threats by managing and mitigating them are considered.

2.4 Internet of Things, Artificial Intelligence, Machine Learning and Cloud Computing

Almomani, Gupta, Wan, Altaher, and Manickam [1] present a neural fuzzy structure that applies hybrid learning and adapts an evolving connectionist system to dynamically detect online zero-day phishing emails. ("Internet of Things") Psannis, Kim Stergiou, and Gupta [27] test whether the integration of cloud computing and the Internet of Things (IoT) is secure. They found that cloud computing technology improves IoT function. Jiang et al [14] use deep learning, such as convolutional neural networks and iterative neural networks, to multi-channel attack detection to ensure information security. They generate classifiers based on training the neural network and introduce a voting algorithm to decide whether the input is an attack or not. Plageras, Psannis, Stergiou, Wang, and Gupta [25] assess the role of cloud computing, IoT and monitoring in collecting and managing sensor data in a smart building, which can enable energy efficiency. Gupta, Agrawal, and Wang [9] consider principles, algorithms, applications, and perspectives in computer security management and engineering. Gupta, Agrawal and Yamaguchi [18] identify research, perspectives, techniques and best practices in cryptology and cyber threat prevention. Memos, Psannis, Ishibashi, Kim and Gupta [23] present a media-based surveillance algorithm that combines information and communication technology and IoT, which can benefit so-called "smart city" users and companies in terms of user privacy, media security and requirements regarding the memory of sensor nodes.

3 Methodology

For cyber resilience to gain momentum as a research topic as well, it is important to define the term as well as assess how to implement these concepts across different actors. When there is a common understanding of what cyber resilience is and how to implement it, research and education will be more efficient and effective. Individuals, businesses, and societies need efficient and effective cyber resilience, and to achieve this we need, among other things, a common language. The research presented in this article is academic research, understood as a systematic effort aimed at gaining new knowledge in the domain of cyber security and boils down to a systematic search for relevant information on cyber resilience. This research is an academic activity and as such all terms and key concepts related to the concept of cyber resilience are used in a technical sense. The research presented in the article covers all the models, methods and techniques that have been undertaken to answer the question of how cyber resilience can play a significant role in deterring and responding to cyberattacks and affect a company's ability to continue its mission, even if will be attacked.

The conclusion from the analysis of the research is that cyber resilience requires a holistic, adaptive, and global approach that can withstand changing threats to essential services and infrastructure and is business oriented in the sense that it aims to continuously deliver the intended business results despite adverse cyber events. Achieving this goal boils down to building an effective cyber resilience system based on a previously developed cyber resilience model, strategy, or plan.

The basic research process led to the definition of the immune system model, described in detail in Sect. 4, which is also an essential part of this paper. In the research process, when formulating the individual components of this model, the following were considered:

- Research commissioned by industry and academia.
- Research results from a variety of sources, including academia, to keep you aware of cybersecurity and resilience thinking.
- Forecasts of future cyber threats and the possibilities of technological mitigation of their effects.

4 Results and Discussion

4.1 Defining Cyber Resilience

Simply put, cyber resilience is a measure of how well an actor/company can manage a cyber attack or data breach while continuing to operate effectively. Let's analyze the most important definitions and terms of cyber resilience, and thus highlight the differences between cyber resilience and a cyber resilience system. It is important to remember that each approach to a cyber resilience system may include different elements and characteristics related to cyber resilience.

Below, the concepts or definitions of cyber resilience are specified, based on which the author's definition of cyber resilience was constructed:

- Cyber resilience is a new approach to security that recognizes the threat of constant, evolving attacks. Organizations need a resilient cybersecurity strategy that enables them to respond to unavoidable breaches rather than hoping their data will never be compromised.
- Cyber resilience is an organization's ability to sustainably maintain, build and deliver intended business results despite adverse cyber events. Organizational practices to achieve and sustain cyber resilience must be comprehensive and aligned throughout the organization (i.e., including the supply chain). They must include a formal and well-equipped information security program, team and management that is effectively integrated with the organization's risk, crisis, business continuity and education programs [26].
- Cyber resilience is the result of bringing business, security, and IT together to develop an integrated strategy and action plan that aligns cyber security and business continuity. Its goal is to transform business expectations and guarantee the company a less than significant impact of a cyberattack. To achieve this, organizations must invest in developing and maturing a remediation program that can be reliably used to restore operations in case of an attack [28].
- Cyber resilience is an essential part of any organization's security strategy. It involves detecting and responding to cyberattacks quickly and taking the necessary steps to ensure your organization is resilient against future attacks [17].

Our definition of cyber resilience builds on the above definitions and other research by incorporating the entity of action defined [in definition 2 given in the next section], the object of the action, the purpose of the action, and by limiting the scope to cyber threats and cyber incidents.

Definition 1. Cyber resilience is the ability of an actor to resist, respond to and recover from cyber incidents to ensure the operational continuity of the actor.

In addition, according to a common definition in the literature, cyber threats are aimed at actors in cyberspace, causing cyber incidents, which are events with a certain probability that express the cyber risk for these events. Cyber resilience requires the existence of an effective cyber resilience system in the operating entity, the primary task of which is to quickly detect and respond to cyberattacks, as well as take the necessary steps/decisions to ensure the entity's resilience against future attacks. In addition:

- Cyber resilience is a form of business continuity management that mitigates security breaches and keeps your business running successfully during a cyberattack. ("Cyber Resilience: What It Is and Why It's Important").
- Cyber resilience affects the digital data and information processes required by the entity or the IT systems that the entity uses to conduct its business.
- Cyber resilience must be pursued at all levels of the actor/company and is a risk-based, proactive approach to threats, vulnerabilities, and consequences for critical data and supporting assets.
- Cyber resilience complements cybersecurity by creating a comprehensive risk management strategy.

A set of key issues affecting cyber resilience include:

- insecure configuration of active directory,
- no or limited use of endpoint detection and response,
- skills gaps in security operations,
- vulnerabilities in security monitoring and integration,
- ineffective soc processes and operating model,
- no patch and vulnerability management,
- no zero trust in the adoption of the security framework,
- insecure design and setup on cloud platforms,
- lack of good sdl practices in devops.

Cyber resilience, i.e. the ability to counteract and resist cyber threats, requires taking into account many factors. Here are some key factors that can guarantee cyber resilience:

1. Awareness and Training: Educating staff on cybersecurity is key. All employees should be aware of security risks and practices, such as strong passwords, identifying phishing, and not opening suspicious attachments,
2. Strong network security: Network security such as firewalls, antivirus, intrusion detection and prevention systems (IDS/IPS) and Internet traffic filtering are essential to protect against external attacks,
3. Updates and patches: Regularly updating operating systems, applications and firmware is important to prevent security vulnerabilities that can be exploited by cybercriminals,
4. Monitoring and responding to threats: Effective monitoring tools and security incident management (SIEM) systems allow you to detect, track and respond to potential cyber incidents in real time,
5. Contingency and recovery plan: Having a contingency plan that includes the recovery process and actions in the event of an attack is essential. Regular backups and keeping them in a safe place are also important,
6. Security Policy and Compliance: Implementing a clear IT security policy that includes policies on passwords, data access, user monitoring, etc. is crucial. Compliance with industry regulations and standards is also important,
7. Penetration testing and risk assessments: Regular penetration testing and risk assessments help identify potential vulnerabilities and make necessary fixes,
8. Partnerships with service providers: If the company uses third party providers, it is important to choose reliable partners who also apply high security standards,

9. Threat Monitoring and Machine Learning: Leveraging machine learning and AI solutions can help monitor and detecting new, evolving threats,
10. Security culture: Creating a security culture where all employees are responsible for cybersecurity and willing to report suspicious activity is key to ensuring cyber resilience.

It is worth remembering that cybersecurity is a dynamic and constantly changing environment, so it is necessary to constantly update strategies and practices to maintain cyber resilience.

4.2 A Model of a Cyber Resilience System

Today, no serious cybersecurity expert will give us a 100% guarantee that we are safe. With the current level of development of information technology, anything can be hacked – the only question is how much work the hackers are able to put in. Therefore, the only way to avoid an attack is to make it economically unviable for potential attackers. Of course, this is not an easy task. The biggest barrier here is that most of today's information systems have not been designed with cybersecurity in mind and therefore have vulnerabilities that are often covered by solutions such as suboptimal cyber resilience systems. This problem is particularly common in industrial cybersecurity. We are not in favor of eliminating everything that is older and designing new systems from scratch, but we believe that new systems, both information and cyber resilience, should be created in accordance with a concept that is guided by security at the design stage. Following the above and definition 1 [see Sect. 4.1.], we adopt an ordered six as a model for a cyber resilience system:

$$CRM = \langle SoA, OoA, CRS, PoA, TWT, CRI \rangle, \qquad (1)$$

where:

CRM – cyber resilience system model

SoA – The subject of action, which is a set of functionaries who are part of the organization's security service,
OoA– Object of action, which are SIO objects in relation to which the required level of security should be maintained,
CRS – Cyber Resilience Strategy
PoA – Purpose of action specified on subject activities,

TWT – Technology and work tools.
CRI – indicators of cyber resilience.
Above – mentioned elements are discussed in the following subsections of this article.

4.2.1 The Subject of Action

Definition 2. An actor is any individual or collective entity that is normally assumed to maintain its own cyber resilience. Definition 2 uses the term "usually" because there are exceptions for cybercriminals, such as suicide bombers or masochistic groups that may compromise their own cyber resilience. A non-threat entity is an entity that seeks or is indifferent to the cyber resilience of another entity.

$$SoA = \left\{ s_k^m, k \in K, m \in M \right\}, \tag{2}$$

where:

s_k^m – k-th entity of action of m-type,
$M = \{M_i(i = 1, 2, \ldots, m)\}$ – a set of types of entities or actors,
$K = \{K_i(i = 1, 2, \ldots, k)\}$ – a set of names of entities of action.

Several types of entities or actors can be distinguished. Some actors and their involvement in cyber resilience are shown in Fig. 1 to identify key actors relevant to maintaining and breaching cyber resilience.

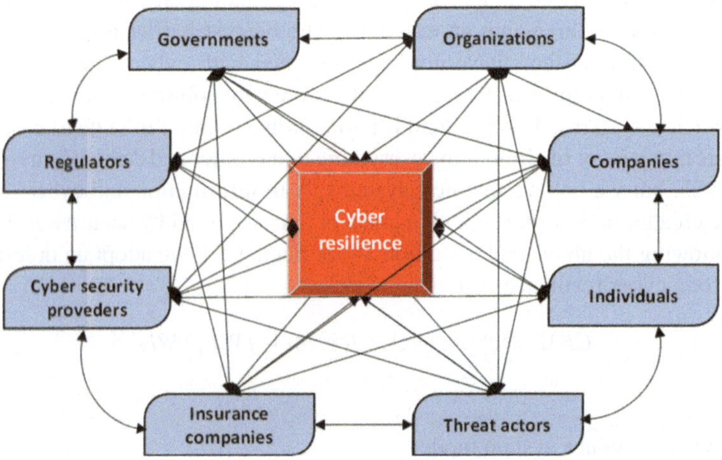

Fig. 1. Examples of actors involved in cyber resilience (Source: own elaboration).

Two-way arrows in Fig. 1 define any type of relationship or interaction. We generally assume double arrows, although one-way arrows may be applicable in special cases.

Non-threat actors can include governments, regulators, incident responders, insurers, organizations, and individuals. Cybercriminals can be hackers and criminals. Hybrid entities can be enterprises that sometimes, intentionally, or unintentionally, may compromise the cyber resilience of other entities. Actors work at various levels, ranging from the individual, through a group, organization, region, etc., to the global level. Each entity chooses strategies based on beliefs and preferences that influence cyber resilience.

Actors differ in their preferences regarding the resilience of other actors. For mutually exclusive entities, which are therefore not embedded with each other, we can distinguish how entities prefer cyber resilience to other entities. Some actors influence cyber resilience more than they are affected by it, and vice versa for others. Eight types of actors [Fig. 1] play different cyber resilience roles in many senses.

4.2.2 Object of Action

From the point of view of quickly detecting and responding to cyberattacks, as well as taking the necessary decisions/steps to ensure the organization's resistance to future attacks, the subject of the action is a set of such elements $e_j \in E^{SoA}$ of the operating entity (SoA), the desired state of which can be determined by the decision-making entity, which is a key element the subject of action. The elements of the collection E^{SoA} may be:

- cyber attacks, threats [external and internal] related to cyber resilience, evolving cyber threats,
- key processes of cyber resilience,
- solutions securing the digital environment of the entity,
- vulnerabilities in technical and organizational security,
- the risk of exploiting vulnerabilities.

"Cyberattacks are inevitable, so being able to repel them and then recover from them is critical. Nowhere is this more important than in critical infrastructure industries where any loss or reduction of services can be both socially and economically catastrophic, so it's worrying, that only just over a third of suppliers is what we consider highly resilient," said Dirk Schrader, cyber resilience architect at Greenbone Networks.

Understanding and effectively managing cyber resilience threats across the enterprise is a challenge, even for the most mature companies.

Given the rapid evolution of threats to cyber systems, new management approaches are needed that address risk across all interdependent domains (i.e., physical, informational, cognitive, and social) of cyber systems. Moreover, the traditional approach of securing cyber systems against identified threats has proven impossible. Here are the typical security-related cyber-resilience risks that businesses can experience [12]:

- Cybercrime: Cybercrime is a digital criminal activity with the goal of making money, accessing personal information, or disabling a device.
- Cyber espionage: Cyber espionage involves accessing sensitive material such as government or military information using proxy servers.
- Hacktivism: Hacktivism is a type of cybercrime that involves war against information or data that is waged for political, social, religious, or anarchist reasons.

Currently, every business entity has vulnerabilities in its IT infrastructure, as attack vectors and methods are growing day by day. However, organizations can enforce continuous vulnerability management and penetration exercises to establish a sound security posture. Vulnerabilities can be described as vulnerabilities in any IT asset, whether they are software defects or hardware part defects. These vulnerabilities or entry points allow another actor, such as a hacker, to break into an organization's IT infrastructure, website, operating systems, or network. In addition to the IT component (software or hardware) having vulnerabilities in the IT system, vulnerabilities can also be introduced by human error, misconfigurations or simply due to the lack of implemented security controls.

4.2.3 Purpose of a Cyber Resilience System

The operation of a cyber resilience system can be defined as:

1. Regarding ensuring greater resilience to cyber threats as an ordered pair:

$$A^{CRS} = \langle \alpha^{CRS}, Z^{CRS} \rangle, \tag{3}$$

   where:
2. Regarding the control of the operational properties of the cybersecurity team, established under the cyber resilience system as an ordered pair:

$$DZ^{CST} = \langle \alpha^{CST}, Z^{CST} \rangle, \tag{4}$$

   where:

4.2.4 Cyber Resilience Strategy

Today, companies face much more difficult questions than ever before from third parties about their cyber resilience strategies. Increasingly, regulators and major customers are demanding evidence that companies' cyber resilience strategies are working. Questions went beyond those on the Business Continuity Plan (BCP) and Disaster Recovery (DR) approach. Today's questions are:

- How do companies reduce the likelihood of service interruption?
- What will companies do if their systems are down for five days?
- How will businesses continue to operate and process transactions – manually if necessary – when systems are down for extended periods of time?
- How will companies recover effectively in a timely and well-controlled manner?

An effective cyber resilience strategy relies on several operational activities: business continuity (BC), disaster recovery (DR), incident response, and cybersecurity plans. The goal is to ensure that the organization can resume operations as soon as possible after a successful cyberattack. A cyber resilience strategy is a step towards enterprise cybersecurity maturity. The basic components of a cyber resilience strategy are shown in Fig. 2.

There are several essential steps to follow to achieve effective cyber resilience:

1 Get input – ask senior leaders across the organization to define what cyber resilience means for them. Their input helps identify the most important business activities to management.

2 Identify essential operational activities – using the input gathered, identify what is needed for the organization to produce its product and services; perform a business impact analysis (BIA) to identify mission-critical business processes and the people, processes, technologies and facility resources needed to enable those processes, as well as the potential impact to the organization if a cyber attack interrupts those processes.

3 Perform a risk analysis – a risk analysis based on a BIA can identify the most likely internal and external cyber threats to the organization's ability to conduct business. Use threat and vulnerability analyzes to identify weaknesses – such as an insecure network perimeter – that could increase the risk of a cyber attack. Conduct penetration tests to identify potential cyber vulnerabilities.

Fig. 2. Basic components of a cyber resilience strategy (Source: own elaboration).

4 Prevent cyber attacks – determine strategies, based on the above activities, to minimize the likelihood of a cyber attack occurring. In addition, evaluate additional strategies, such as training IT staff about the most effective ways to deal with cyber attacks and educating employees about the dangers of an attack. Also, ensure backup copies of all mission-critical assets are available.

5 Build and enact a cyber-resilience plan – structure an effective cyber-resilience strategy to ensure critical operational activities can be recovered and returned to normal operation. ("Build a strong cyber resilience strategy with existing tools") Use cybersecurity and DR plans to maintain IT resilience and prevent unauthorized access by cyber threat actors. Bring back mission-critical systems to full operation as quickly as possible.

6 Document and ensure – verify procedures to recover operational availability components are documented, stored in secure locations and available electronically for maximum speed of access. Also, confirm general operational procedures for mission-critical systems are available in case primary operators are not available and other employees must step in to operate these systems.

7 Test and update – ensure cyber-resilience procedures governing mission-critical company assets and business operations are regularly tested and updated. Update plans and procedures based on changes in business operations and the results of exercises. Also, patch applications and systems when recent updates or patches are available. In addition, be sure to regularly brief senior management on the state of the organization's cyber resilience.

Achieving a viable cyber-resilience strategy depends on the smooth collaboration of several preventive, detective, and responsive plans, and their associated program activities. ("Why cyber resilience is so important – ATC") Maintenance of up to date, documented and regularly exercised cyber-resilience plans can be achieved using a balanced program of activities, including cybersecurity plans, BCDR plans, incident response plans, periodic BIA/risk analysis initiatives and senior management support.

4.2.5 Technology and Work Tools in the Field of Cyber Resilience

Now that we've considered the possible types of actors, the preferences relevant to cyber resilience, and what we mean by cyber resilience, let's move on to consider cyber resilience technologies, techniques, and tools. An entity such as a company may have some cybersecurity resources and can decide how to allocate resources to Cyber Resilience. Another entity, such as a branch of government, may receive funding from a budget dedicated to resilience in general and cyber resilience. The use of resources requires competence in what technology and tools to acquire and what strategies to choose. The choice of technologies and tools can extend or limit the available strategies. Some tools can be expensive and have some advantages to increasing cyber resilience, but also disadvantages or limitations that potentially threaten cyber resilience. The choice of technologies and tools, and how they are updated over time, express adaptation to changes, which in turn affect cyber resilience over time. Practice shows that the excessive use of disconnected techniques and tools creates complex environments that can have a negative impact on an organization's ability to detect, prevent, contain, and respond to cybersecurity incidents.

Below are some sample technologies, frameworks, best practices, and most popular cyber resilience tools:

1. EDR technology (Endpoint Detection and Response is an efficient threat detection and removal technology. However, the cost alone may deter business owners or decision-makers from implementing it. "The key to building cyber resilience is to focus less on technology and more on people." ("Building a Cyber Resilience Framework – Trixter") After all, the power of data and the latest technologies can only be harnessed when we have a qualified team to oversee security operations. The set of tools supporting EDR technology include: #1.OSSEC – open-source software that offers HIDS, HIPS, log analysis, real-time Windows registry monitoring, and other EDR functions. #2. The Hive Project is a "response platform for security incidents, #3.osQuery – open-source software for handling devices licensed by Apache, which increases the visibility of connected devices, #4. Nessus vulnerability scanner – a tool for scanning communication ports useful for detecting vulnerabilities in system security – entry points that can be exploited by malicious actors #5 SNORT is a reliable open-source intrusion prevention software that allows the user to identify e-threats by analyzing real-time packet logging and network traffic #6. Ettercap Project is a cross-platform EDR tool open source that simulates ARP Poisoning and Man-in-the-attacks Middle to LAN; #7. Infection Monkey is a free and open-source cybersecurity health assessment tool that simulates the system and APT (Advanced Persistent breaches Attacks); #8. Cuckoo Sandbox is an open-source sandbox environment that allows

the user to quarantine, parse, and analyze files that exhibit malicious behavior, #9. GRR – Rapid Response is an Apache-licensed open-source incident response platform used in remote live forensics, #10. Mozilla's MIG is a free forensics platform for remote endpoints. It enables – log analysis, memory inspection, file and network inspection, full system inspection, vulnerability management, and much more.

2. Technologies needed to control, manage, and recover efficiency after cyberattacks – Automation, artificial intelligence, machine learning, and cloud services are some of the leading ways to increase cyber resiliency. Automation can improve not only operational efficiencies, but also the morale of security teams, which will be able to focus their talents on more meaningful activity and be freed of "scut work."
3. Cybersecurity framework -- A cybersecurity framework provides a common language and set of standards for security leaders across countries and industries to understand their security attitudes and those of their suppliers. ("Cybersecurity Frameworks") With an established framework, it is much easier to define the processes and procedures an organization must undertake to assess, monitor, and mitigate cybersecurity risk. ("7 Cybersecurity Frameworks To Reduce Cyber Risk – Bitsight") Popular cyber security framework is; NIST, ISO 27001 and ISO 27002, SOC2, NERC-HIPAA, GDPR, and FISMA. A cybersecurity framework provides a useful (and often mandated) basis for integrating cybersecurity risk management with third party security performance management and risk management strategies. With the platform as a beacon, you'll gain vital insight into where the greatest security risk lies and feel confident communicating to the rest of the organization that you're committed to security excellence.
4. Best Practices – There is no single method or strategy that guarantees that an organization will not become a victim, but there are best practices that organizations can take to achieve greater cyber resilience. The 5 cyber resilience best practices for 2022 are:

   - Educating employees about the dangers of clicking unfamiliar links in corporate email, as well as in their personal email and social media accounts.
   - Security tools update
   - Collaboration between IT manager, CISO and backup administrator. A solid understanding of responsibilities, strategies and processes is needed.
   - Analytics – Implementing analytics that look deep inside your data can detect more sophisticated attacks and minimize data loss. Constantly checking critical data and infrastructure to understand how data changes over time can help determine if your data center has been infiltrated and is under ransomware attack.
   - Investing in a cyber recovery strategy and in new technologies to keep backups hidden, immutable and smart enough to deal with future cyberattacks.

5. Education – By growing talent within the company with ongoing training, you keep your staff engaged. In return for offering a platform that facilitates personal and professional growth, you cultivate a loyal workforce of highly skilled security professionals. Cyber resilience should not be left to the security team alone. Instead, C-suite members must work hard to establish a strong culture that promotes peer learning, open discussion, and ongoing training on the latest incident response tools and cyber resilience strategies.

6. Most Popular Risk Management Tools – #1) LogicGate: GRC in the cloud, #2)Fusion Risk Management: Resilience meets efficiency, #3)MindBridge: Automate risk discovery, #4) SpiraPlan by Inflectra, #5) A1 Tracker, # 6) Risk Management Studio, #7) Isometrix, #8) Active Risk Manager, #9) CheckIt, #10) Isolocity, #11) Enablon, #12) GRC Cloud, #13) iTrak, #14) Analytica.

4.2.6 Cyber Resilience Indicators

There is no fixed rule for selecting key cyber resilience metrics. These metrics will depend on your industry, your organization's needs, regulations, guidelines, best practices, and ultimately your cyber risk appetite. ("14 Cybersecurity Metrics + KPIs You Must Track in 2023") Below are examples of clear cyber resilience metrics that can be tracked and presented to cyber resilience stakeholders:

1. Readiness level – how many system components are fully patched and up to date? Vulnerability scanning and vulnerability management is one of the core CIS controls that can reduce the risk of exploiting vulnerabilities.
2. Intrusion Attempts – How many times have bad actors tried to gain unauthorized access? ("14 Cybersecurity Metrics + KPIs You Must Track in 2023") You may need to refer to the cyber resilience logs to collect this information.
3. Own security assessments such as: #1 Cyber resilience reconfiguration lead time; #2 Recovery Time Objectives (RTOs); #3 Incident response time; #4 Completeness of documentation.
4. Security incidents – how many times has the attacker breached information assets or network components?
5. Mean Time to Detect (MTTD) – how long do security threats go undetected? MTTD measures how long it takes the cybersecurity team to get information about breach rates and other security threats.
6. Mean Time To Resolution (MTTR) – what is the average response time of the cybersecurity team to a cyberattack once they become aware of it? an excellent measure of the quality of the incident response plan implementation.
7. Average Retention Time (MTTC) – How long does it take to shut down identified attack vectors across all endpoints?
8. Patching cadence – How long does it take for the cybersecurity team to deploy application security patches or mitigate high-risk vulnerabilities.
9. Vendor Patch Rate – This metric includes how many threats your third-party vendor has and how many critical vulnerabilities are yet to be fixed.
10. Average supplier response time to incidents – A security incident is not only a successful cyberattack, attempts to break into suppliers can mark your organization as a potential target. The longer it takes providers to respond to incidents, the greater the chance that you will suffer from a third-party data breach. In fact, some of the biggest data breaches are the result of supplier mismanagement.
11. Cyber Resilience Awareness Level – Define and implement a Cyber Resilience Awareness Plan. The point is to know if the organization is promoting a cyber-resilience culture to reach all employees. This plan includes any initiatives to raise awareness of cyber resilience.

Policies, standards, and procedures should be complete in terms of covering all scenarios and should be regularly reviewed, updated, and approved by management.

5 Conclusion and Implication

5.1 Implication

Acting entities/Companies that want to achieve an acceptable level of cyber resilience should have an effective cyber resilience system and a cybersecurity incident response plan (CSIRP). This plan should be based on continuous monitoring and event correlation using data collected by sensors on network devices and/or endpoints as detection, investigation, and automatic response mechanisms and/or managed by security experts or threat hunters. In addition, enterprises with a high level of cyber resilience should have, as part of the external SOC team, highly qualified cybersecurity personnel as part of the incident response plan.

In this article, the focus is on defining and analyzing two concepts: cyber resilience and cyber resilience system. To compare cyber resilience with cyber resilience and cyber security, we used several aspects: actor, object of action, goal, strategy, approach, technologies and tools, and a set of indicators for measuring cyber resilience. In each of these aspects, there is a difference in the approach to cyber resilience and the cyber resilience and security system. Further work includes the characterization and analysis of the proposed components of the resilience system to address cyber resilience aspects. Finally, we outlined a set of ground rules that can be used as a guide for initial work on a cyber resilience system, as well as laying the groundwork for defining the term and detailing its core components.

5.2 Conclusions

To ensure cyber resilience, a new approach to cybersecurity must address at least the following:

1. Managing cybersecurity as an enterprise risk management issue, not as an IT issue, and adopt a "cycle of resilience" attitude.
2. Reinforcement and maintenance key pillars: 1) People, processes, technologies, politics; 2) A secure approach to digitalization; 3) Corporate cybersecurity culture.
3. Constantly adapting to new techniques and tactics of hackers and other attackers; being resilient means that this adaptation must be done in the minimum amount of time, at maximum speed, even in real time.
4. Prioritizing and mitigating risks at all levels of the entity.
5. Cyber risk management through comprehensive and collaborative management.

This paper reviews and assesses the emerging role of cyber resilience:

- First, common definitions of cyber resilience are refined and expressed as an actor's ability to resist, respond to, and recover from cyber incidents to ensure the entity's operational continuity.

- Secondly, cyber resilience encompasses many different actors divided into non-threat, cybercriminal, and hybrid actors.
- Third, the actors that affect and are affected by cyber resilience have been identified. These entities have the resources, competencies, technology, and tools of various kinds, which in turn influence the strategies and cyber resilience of all entities.
- Fourth, the expected role of cyber resilience in the future is outlined. Cyber resilience is linked to the IoT, which can be expected to simplify many aspects of life, for example by offering artificial intelligence and machine learning. Cyber resilience is also related to, for example, the future use of robots by firefighters attacking a data center fire.

6 Summary

1. Why is cyber resilience important? It is essential because it is an organization-wide system or strategy that actively monitors and responds to risks, threats, and vulnerabilities. It is a collaborative approach involving everyone in the organization and includes customers, suppliers and partners who have a good understanding of critical information assets. To achieve a sufficient level of cyber resilience for an organization, the first step is to understand the information it holds and prioritize what needs to be protected, and the final step is to implement an effective cyber resilience system. ("What Is Cyber Resilience and Why Is It Important? | RSI Security")
2. Cyber resilience is gradable, starting with classic business continuity and disaster recovery, starting with good data backup; progresses to recoverability of processes, technologies, and their dependencies (including persons and third parties); and forever-on transition, self-healing services, critical role resiliency, and failover for critical third parties. The most resilient organizations promote integration between IT, business managers and security professionals.
3. Superior resiliency includes design for resiliency from the start, secure change management, and detailed fault isolation.
4. Cyber resilience is just one scenario of a good all-risk planning program. With the rise of cyber threats and the growing importance of combining cybersecurity with resilience, the relationship of the Chief Information Security Officer (CISO) with the enterprise resilience program is becoming stronger.

References

1. Almomani, A., Gupta, B.B., Wan, T.-.C., Altaher, A., Manickam, S.: Phishing dynamic evolving neural fuzzy framework for online detection "Zero-Day" phishing email. Indian J. Sci. Technol. **6**(1), 3960–3964 (2013)
2. Bier, V., Gutfraind, A.: Risk analysis beyond vulnerability and resilience – characterizing the defensibility of critical systems. Eur. J. Oper. Res. **276**(2), 626–636 (2019)
3. Bostick, T.P., Connelly, E.B., Lambert, J.H., Linkov, I.: Resilience science, policy and investment for civil infrastructure. Reliab. Eng. Syst. Saf. **175**, 19–23 (2018)
4. Ed-daoui, I., ElHami, A., Itmi, M., Hmina, N., Mazri, T.: Resilience assessment as a foundation for systems-of-systems safety evaluation: application to an economic infrastructure. Saf. Sci. **115**, 446–456 (2019)

5. Fang, Y.P., Zio, E.: An adaptive robust framework for the optimization of the resilience of interdependent infrastructures under natural hazards. Eur. J. Oper. Res. **276**(3), 1119–1136 (2019)
6. Flammini, F.: Resilience of Cyber-Physical Systems: From Risk Modelling to Threat Counteraction, 1st edn. Springer International Publishing, Springer, Cham (2019)
7. Gisladottir, V., Ganin, A.A., Keisler, J.M., Kepner, J., Linkov, I.: Resilience of cyber systems with over- and underregulation. Risk Anal. **37**(9), 1644–1651 (2017)
8. Google Scholar Search for "Cyber Resilience": http://scholar.google.se/scholar?q=%22cyber+resilience%22. Last accessed 7 Dec 2022
9. Gupta, B.B., Agrawal, D.P., Wang, H.: Computer and Cyber Security: Principles. Applications, and Perspectives. CRC Press, Algorithm (2018)
10. Gupta, B., Agrawal, D.P., Yamaguchi, S. (eds.): Handbook of Research on Modern Cryptographic Solutions for Computer and Cyber Security. IGI Global (2016)
11. Matteson, A., Harrop, W.: Cyber resilience: a review of critical national infrastructure and cyber security protection measures applied in the UK and USA. J. Bus. Contin. Emer. Plan. **7**(2), 149–162 (2013)
12. Herrington, L., Aldrich, R.: The future of cyber-resilience in an age of global complexity. Politics **33**(4), 299–310 (2013)
13. Hult, F., Sivanesan, G.: What good cyber resilience looks like. J. Bus. Contin. Emer. Plan. **7**(2), 112–125 (2013)
14. Jiang, F., et al.: Deep learning based multi-channel intelligent attack detection for data security. IEEE Trans. Sustain. Comput. **5**(2), 204–212 (2020)
15. Kaufmann, M.: Cyber-resilience in the EU. Int. Polit. **71**(2), 274–283 (2013)
16. Kishor, S., Kim, D.S., Gosh, R.: Resilience in computer systems and networks. In: Proceedings of the International Conference on Computer-Aided Design, pp. 74–77 (2009)
17. Kott, A., Linkov, I.: Cyber Resilience of Systems and Networks, 1st edn. Springer International Publishing, Springer, Cham (2019)
18. Kaufmann, M.: Cyber-resiliens i EU. Internasjonal Politikk **71**(02), 274–283 (2013)
19. Linkov, I., Eisenberg, D., Plourde, K., Seager, T., Allen, J., Kott, A.: Resilience metrics for cyber systems. Environ. Syst. Decis. **33**(4), 471–476 (2013)
20. Linkov, I., et al.: Tiered approach to resilience assessment: tiered approach to resilience assessment. Risk Analysis **38**(9), 1772–1780 (2018)
21. Linkov, I., Palma-Oliveira, J.M. (eds.): Resilience and Risk: Methods and Application in Environment, Cyber and Social Domains. Springer Netherlands, Dordrecht (2017)
22. Linkov, I., Roslicky, L., Trump, B.D.: Resilience and Hybrid Threats: Security and Integrity for the Digital World. IOS Press, Incorporated, Amsterdam (2020)
23. Memos, V.A., Psannis, K.E., Ishibashi, Y., Kim, B.-G., Gupta, B.B.: An efficient algorithm for media-based surveillance system (EAMSuS) in IoT smart city framework. Future Gener. Comput. Syst. **83**, 619–628 (2018)
24. National Institute of Standards and Technology. Framework for Improving Critical Infrastructure Cybersecurity, V1.1. https://nvlpubs.nist.gov/nistpubs/CSWP/NIST.CSWP.04162018.pdf (2018)
25. Plageras, A.P., Psannis, K.E., Stergiou, C., Wang, H., Gupta, B.B.: Efficient IoT-based sensor big data collection–processing and analysis in smart buildings. Future Gener. Comput. Syst. **82**, 349–357 (2018)

26. Roege, P.E., et al.: Bridging the gap from cyber security to resilience. In: Linkov, I., Palma-Oliveira, J.M. (eds.) Resilience and risk. NSPSSCES, pp. 383–414. Springer, Dordrecht (2017). https://doi.org/10.1007/978-94-024-1123-2_14
27. Stergiou, C., Psannis, K.E., Kim, B.-G., Gupta, B.B.: Secure integration of IoT and cloud computing. Future Gener. Comput. Syst. **78**, 964–975 (2018)
28. United States Department of Homeland Security. NIPP 2013 – Partnering for Critical Infrastructure Security and Resilience (2013)

Simulation, Sensorization and Testing of a Hybrid Inspection Robot

Erika Ottaviano[1](✉), Pierluigi Rea[2], and Lorenzo Miele[1]

[1] University of Cassino and Southern Lazio, Via Di Biasio 43, 03043 Cassino, Italy
ottaviano@unicas.it

[2] University of Cagliari, Via Marengo 2, 09123 Cagliari, Italy

Abstract. In this paper, simulations and tests for a hybrid inspection robot are addressed. The proposed system can be used for indoor and outdoor surveys, and thanks to its compactness and ability in overpassing obstacles is extremely efficient in confined spaces, for which it is difficult or almost impossible the use of other solutions. Robots can replace operators in dangerous or unsafe operations, and in the context of structures and infrastructure inspection, autonomous or tele-operated systems can be used for monitoring purposes for securing the access to industrial sites or confined spaces. Indeed, the hybrid robot can be used for the inspection of indoor sites, which are difficult or almost impossible to access, and for which tele-operated survey by mobile robots can be an effective solution. In addition to the sensors used for the inspection, such as camera or thermal imaging camera, it can be of great interest to adopt additional sensors, named as internal sensors' suite, in order to get additional information for aiding the pilot during the survey. In this paper, we report the simulation and experimental tests for a hybrid rover developed and equipped with suitable internal sensors to accomplish an inspection task.

Keywords: Mobile Robots · Simulation · Robotic Inspection · Structural Health Monitoring · Experimental tests

1 Introduction

Robotics and Automation are widespread and commonly applied to any sector of production and quality assessment in industrial sector; but since last decades, they are not limited to conventional applications. Robots are employed for service [1], search and rescue [2], manufacturing, remote exploration [3], entertainment, agriculture [4], and last but not the least, inspection and monitoring [5] of structures and historical sites [6]. It is of great interest of taking advantage of the increased level of sophistication of intelligent systems for easing daily life activities and increasing security, resilience, and reliability of the adopted solutions. In this regard, in the next future, it is expected an enormous growth of the use of Robotics for many sectors. In particular, the development of on-site application of automation and robotics for inspection is related to the concept of smart buildings, smart, sensors, cyber-physical systems. In the future, the use of robotic maintenance can reduce up to a 95% in energy consumption and carbon dioxide emissions by replacing helicopters with suspended robots for inspecting transmission

A. Burduk et al. (Eds.): ISPEM 2023, LNNS 790, pp. 93–105, 2023.
https://doi.org/10.1007/978-3-031-45021-1_7

lines [7]. Several solutions have been developed for inspection robots, as reported in the schematic representation of Fig. 1. Table 1 reports an overview of possible solutions of ground robotic systems considering advantages and limitations related to each kind of locomotion solution.

In the last decades, the use of UAV (Unmanned Aerial Vehicles) has enormously increased because of their characteristics, namely rapid and cost-effective deployment, reliable wireless information transmission, but being aware of their disadvantages mainly related to the limited energy supply that reduces the flight distance and operation range. UAVs are equipped with camera and suitable sensors making them very effective for survey and inspection. Although they are the preferred solution for outdoor inspections, in specific circumstances it is necessary to consider Unmanned Ground Vehicles (UGV), namely in case of indoor inspection, confined areas, harsh environmental conditions, which make not possible the applications of UAVs.

Taking into account the great availability of UGVs, the most used solutions among legged [8], wheeled [9] and hybrid solutions [10, 11] are the last ones, since their main characteristics of great adaptability to different types of terrain and obstacles surpassing. The combination of legs and wheels/tracks is the most efficient kind of locomotion, also from the energetic point of view, since it affects the operation autonomy of the inspection robot.

In this paper, the proposed hybrid rover is simulated and tested for inspection purposes. More specifically, the focus is on the so-called internal sensors suite [12], which are mounted on board for improving the robot's performance during the motion. In particular, in addition to the camera used for moving the robot in the environment, other sensors can be used for aiding the pilot in understanding the environment and take decisions. These sensors are compact and are very effective for indoor survey, in confined spaces, such as in historical sites or architectural sites of interest.

Table 1. Classification of mobile robots based on locomotion type.

	Type	Characteristics
Wheeled	3 or more wheels (active and/or passive)	Efficient motion on almost flat surface Low energy consumption, Fast motion
Legged	No. of legs: 1, 2, 4, 6, 8 DOFS for the legs: 1, 2, 4	Efficient motion on uneven terrain High energy consumption High number DOFs – Complex to control
Hybrid	Combination of legs and wheels/tracks either active and/or passive	Combination of locomotion types Efficient use in complex unstructured environment (exploration, nuclear, search and rescue)
Serpentine	Hyper redundant modular robots with wheels or legs	Efficient motion on uneven terrain Small size solution well suited for inspection
Climbing	Magnetic or pneumatic	Efficient for vertical climbing inspection

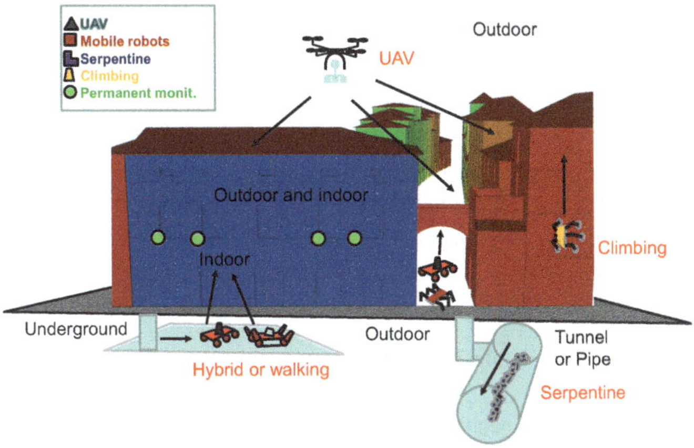

Fig. 1. Overview of robotics applied to the inspection of structure and infrastructure.

2 Hybrid Rover

The hybrid rover and its 3D model are shown in Fig. 2. The rover is composed by commercial parts. Tracks and legs are used for negotiating with obstacles and a large variety of terrains. Table 2 reports the rover specifications. In particular, it has been designed and built having a compact with size of 355 × 300 × 407mm (L × H × W) for being used in confined spaces. The roves has a 3 DOFs (Degrees Of Freedom) two of them for the tracks and a single motor for the legs, it has a payload of up to 7 kg with a mass of 7.5 kg, batteries included. Max speed is about 0.5 m/s.

Inspection robots are usually designed to operate in complex and harsh environmental situations; therefore, they may have high requirements for the operating systems. Because the great variability of the scenarios and unstructured environment, in many cases the operation is not autonomous, but tele-operated. Indeed, the greater number of information will be available, the best or easiest operation for the pilot will be. Very often vision-based systems like cameras and/or infrared cameras, are used together with other sensors, named as internal [12] because they can be used to aid the control system (if autonomous navigation is pursued), or the personnel that operates the robot (if tele operated navigation is used). Table 3 reports the internal sensors' suite that is considered for the operation of the rover.

In this context, together with a camera mounted on board for tele-operated navigation, two types of solution for internal sensors are tested, as depicted in Fig. 3.

The first solution is the Wit-Motion Bluetooth 2.0 Multi-Connect BWT901CL 9 Axis having IMU Sensor Angle Inclinometer + Acceleration + Gyro + Mag on PC/Android/MCU [13], the sensor is able to get high precision Pitch Roll Yaw rotations about X, Y, and Z axes, acceleration, angular velocity, angle, magnetic field output.

The module integrates high precision gyro accelerometer and geomagnetic field sensor, it has Wireless Bluetooth serial port communication, Bluetooth connection range $\leq$ 10m, is embedded in-built battery, also can connect with cable.

Table 2. Hybrid rover specifications

Item description	Specification	
Robot system	Size (L × H × W)	355 × 300 × 407 mm
	Mass	7.5 kg
	Max speed	Up to 0,6 m/s
	Actuation (tracks)	24VDC 122 rpm torque = 1.47 Nm
hybrid robot	Actuation (legs)	24VDC 47 rpm torque = 1.96 Nm
	DOFs	3 (4)
	Payload	up to 7 kg

The sensor is commercially available and used in industrial environment in order to develop solutions that are affordable and with robust features.

The second device tested is an IOS smartphone sensor, for which it is possible to use the following, accelerometer, gyroscope sensor, gravity sensor, magnetometer, linear acceleration taking advantage of the smartphone-integrated sensor.

The two devices have in common compactness, lightweight, and have been installed on board of the hybrid rover for experimental tests. In Fig. 3, the experimental set-up is shown with the reference systems, in order to get the correct information from the sensors, as it will be detailed in the next Sections.

Table 3. Sensor specifications.

Internal Sensor				
Item description			Specification	
Type of sensor	Model Type	Resolution	max Peak	Power Cons
Accelerometer	QTI Ver.2	0.002 m/s2	156 m/s2	1.2 mA
Gyroscope sensor	BMI160 – Bosh	0.00106 rad/s	17.45 rad/s	0.9
Gravity sensor	QTI Ver.2	0.00239 m/s2	156,9 m/s2	1.07 mA
Magnetometer	AKM-09918-Ver1	0.149 μT	4911.99μT	1.1 mA
Lin. Accel. Sensor	QTI Ver.2	0.00239 m/s2	156,9 m/s2	1.07 mA
External Sensor				
Item description	Specification			
Thermal camera (FLIR)	48MP + 5MP Thermal Imagery FLIR			
Front – Rear camera (Sony)	48MP + 19MP type			
Communication				
Item description	Specification			
WiFi router (Ethernet/Lan)	TP-LINK Model TL-WN821N			
AR6210 DSMX	Spektrum Receiver Mk610			
Control station				
Item description	Specification			
HMI Interface /Remote-Controller	Tablet 10″ – Samsung			

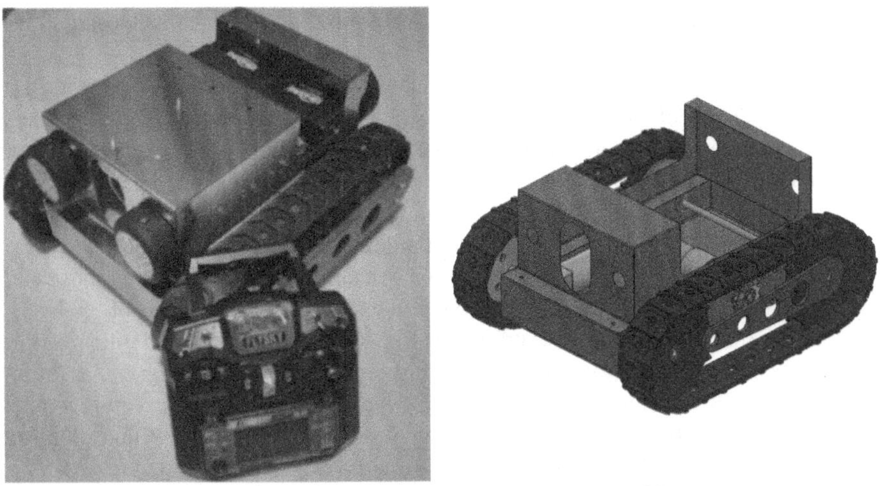

a) b)

Fig. 2. The hybrid rover: a) the prototype; b) 3D model for simulations.

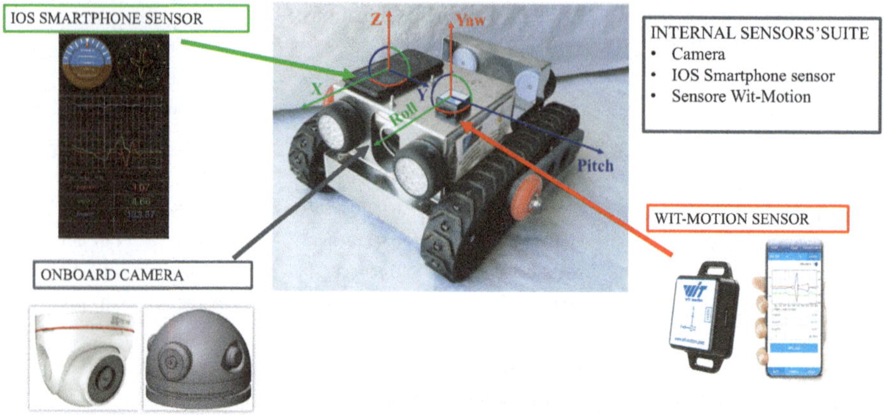

Fig. 3. Internal sensors' suite and reference frames on the hybrid robot.

3 Simulations

Simulations are carried out by using stepfield pallet reported in Fig. 4. In the context of testing robot performances, stepfield pallet have been introduced [14] for preparing robust and repeatable tests being carried out both numerically and experimentally. Within this method is possible of revealing unexpected inconsistencies between the simulation and real behavior of a mechatronic system, because it is repeatable and deterministic, [15]. The used CAD software with suitable tool is Motion allows the realistic simulations, the used PC is a workstation ASUS.

The hybrid rover behavior has been simulated, as reported in Figs. 5, 6 and 7, in negotiating with a stepfield pallet designed according to Fig. 4. In order to have repeatable tests, the rover has been operated by using the tracks only.

Simulation parameters can be compared with physical parameters, i.e., to 1) verify the correctness of the model; 2) use the simulation to get results in different scenarios; 3) use the real data together with simulation models to create alert systems to correct the operation of the system, as it will reported in the following sections. In particular, in this paper, we have focused the attention on the pitch and roll motions of the hybrid rover. Simulation tests have been run for 20 s, Fig. 8 reports the pitch and roll angles, in degrees, during the simulation, measured according to the scheme in Fig. 3. According the stepfield palled designed in Figs. 4 and 5, the first two obstacles require a pitch motion of the hybrid rover, while the last two obstacles require the roll motion of the rover, at it is visible in the simulations and then in the experimental tests.

Fig. 9 reports the Yaw angle and the Center Of Gravity (COG) velocity in the sagittal plane, the above-mentioned results are related to one of the simulations that have been run at different conditions.

Since the rover is teleoperated, it is not possible to start the legs at a given instant, therefore, the obstacle overpassing is obtained by using the tracks only, this will allow the repetition of the test experimentally for a validation of the model and evaluation of the internal sensors' behavior.

0	0	0	0	0	0	0	0	0	0	0	0	0	0	0	0	0	0	0	0	0	0	0	0	0	0	0	0	0	0	0	0	0	0	0	0	0	0	0	0	0	0	0	0	0
0	0	0	0	0	1	3	3	3	1	0	0	0	0	0	0	1	3	3	3	1	0	0	0	0	0	0	0	0	0	0	0	0	0	0	0	0	0	0	0	0	0	0	0	0
0	0	0	0	0	1	3	3	3	1	0	0	0	0	0	0	1	3	3	3	1	0	0	0	0	0	0	0	0	0	0	0	0	0	0	0	0	3	3	3	3	0	0	0	0
0	0	0	0	0	1	3	3	3	1	0	0	0	0	0	0	1	3	3	3	1	0	0	0	0	0	0	0	0	0	0	0	0	0	0	0	0	0	0	0	0	0	0	0	0
0	0	0	0	0	1	3	3	3	1	0	0	0	0	0	0	1	3	3	3	1	0	0	0	0	0	0	0	0	0	0	0	0	0	0	0	0	0	0	0	0	0	0	0	0
0	0	0	0	0	1	3	3	3	1	0	0	0	0	0	0	1	3	3	3	1	0	0	0	0	0	0	3	3	3	3	0	0	0	0	0	0	0	0	0	0	0	0	0	0
0	0	0	0	0	1	3	3	3	1	0	0	0	0	0	0	1	3	3	3	1	0	0	0	0	0	0	0	0	0	0	0	0	0	0	0	0	0	0	0	0	0	0	0	0
0	0	0	0	0	0	0	0	0	0	0	0	0	0	0	0	0	0	0	0	0	0	0	0	0	0	0	0	0	0	0	0	0	0	0	0	0	0	0	0	0	0	0	0	0

Fig. 4. Stepfield palled used for simulation and experimental tests.

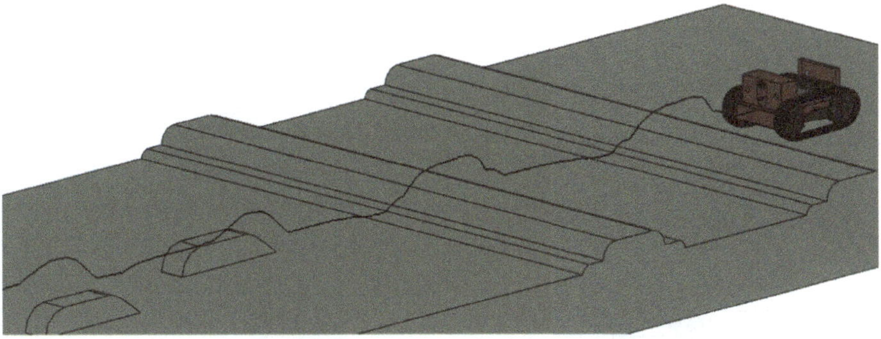

Fig. 5. A simulation of the hybrid rover in overpassing the designed stepfield pallet in Fig. 4.

4 Experimental Tests

Experimental tests have been carried out to verify the operation of the hybrid rover in overpassing obstacles. The snapshots of the tests are reported in Figs. 10 and 11.

In order to obtain repeatable measurements, the stepfield pallet in Fig. 4 has been reproduced in laboratory environment, particularly the roll and pitch motions of the rover have been analyzed experimentally and results are shown, by using the IOS smartphone sensor, and results are shown in Fig. 12, and the Wit-Motion sensor, and results are reported in Fig. 13.

It is worth to point out that only qualitative comparison can be made, for the following reasons, the hybrid rover is tele-operated, and indeed, the velocity of the prototype cannot be set by prescribing a motion law, as it is done for the simulations.

The WIT Motion sensor offers the possibility of setting the scan rate, filtering and many other parameters that can be pre-defined for the experimental tests. The IOS smartphone sensor offers less possibility in setting the acquisition parameters, but does not require calibration procedures before tests. The Wit-Motion sensor seems to be much more sensible to the environment and needs precise calibration procedures that the IOS smartphone do not require. By qualitative comparison between numerical and experimental tests, and according to the real behavior of the robot, the IOS smartphone gives better results in terms of accuracy and reliability of the results.

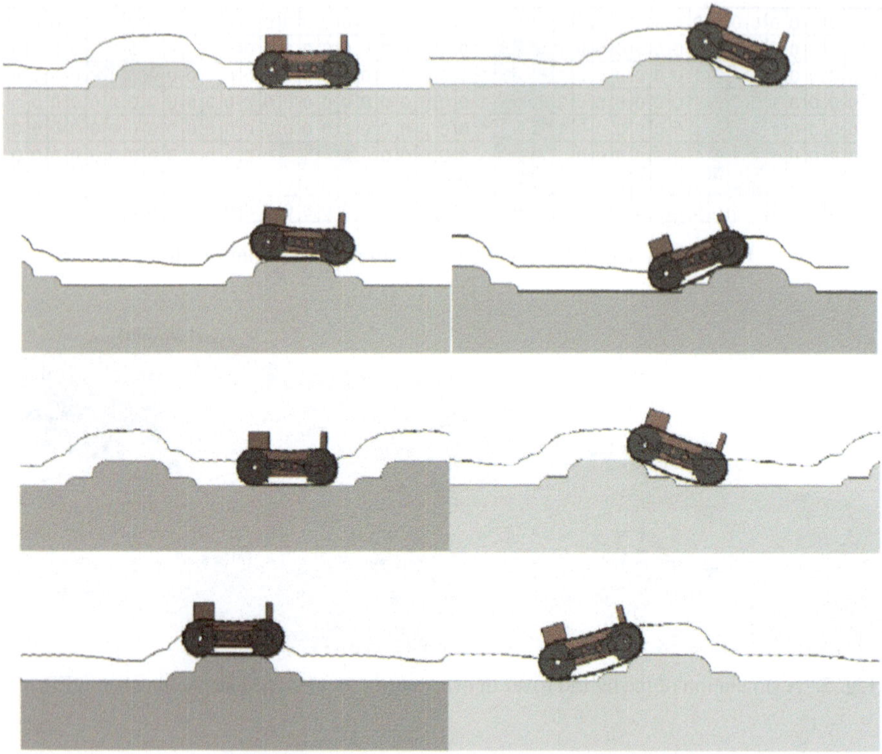

Fig. 6. Snapshots for the pitch motion.

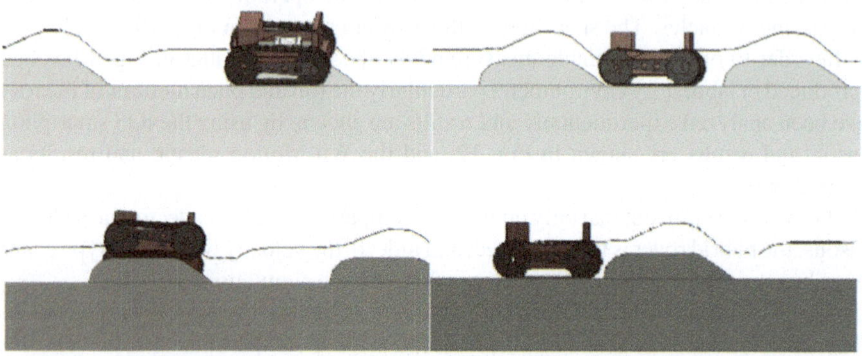

Fig. 7. Snapshots for the roll motion tracing the COG (Center Of Gravity) trajectory.

The experiments have been repeated several times, also varying the velocity of the test. These results are promising and let space for testing the robot in a realistic environment, such as infrastructures or historical sites. Both proposed solutions reveal good behavior of the sensors and the possibility to extract data for being used as additional

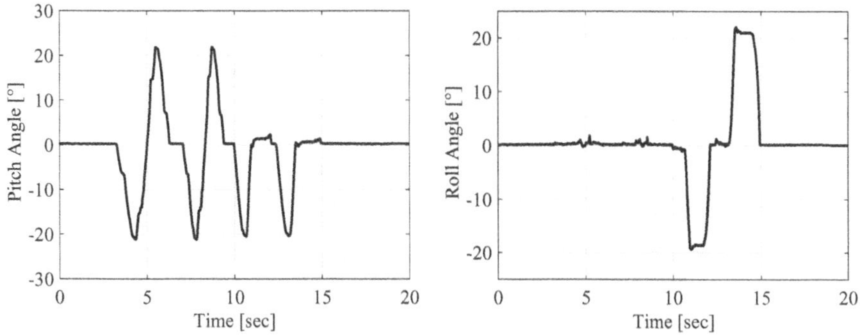

Fig. 8. Numerical results of the simulation in Fig. 4: a) pitch; b) roll.

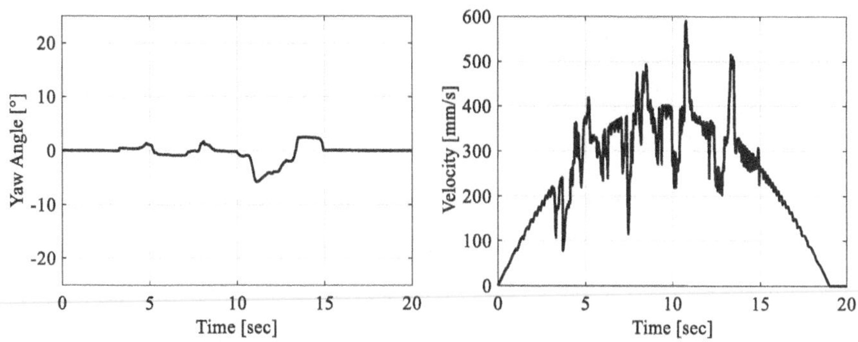

Fig. 9. Numerical results of the simulation in Fig. 4: a) Yaw; b) COG velocity.

information. Both sensors are lightweight and relatively low-cost, making them effective solutions. The information acquired by the so-called internal sensors' suite, as those reported in this Section, can be used to develop or aid the control system for decision making, by using machine learning techniques, as described in [16].

The proposed laboratory test-bed has been used in order to simulate the behavior of the system, but it can be efficiently and effectively used for building and validating the Digital Twin of the mechatronic system, which can be used remotely, as reported in [17] and [18]. The development of the so-called Digital Twin of the real system, creates a powerful tool that can be used to simulate large-scale systems, for which is very difficult to realize prototypes, such as applications in Civil Engineering [19, 20], or rehabilitation or assistive technologies [21]. For these cases, the introduction of suitable sensors can be used to effectively feed the model to adjust/reconfigure/enhance the models making them closer to the reality or following the evolution of the reality The results presented in this paper represent a step forward the use of sensors for obstacle avoidance [22], in which a review of enhancement techniques for control of mobile robots was proposed, and preliminary concept related were reported in [23].

Fig. 10. Snapshots for the pitch motion.

Fig. 11. Snapshots for the roll motion.

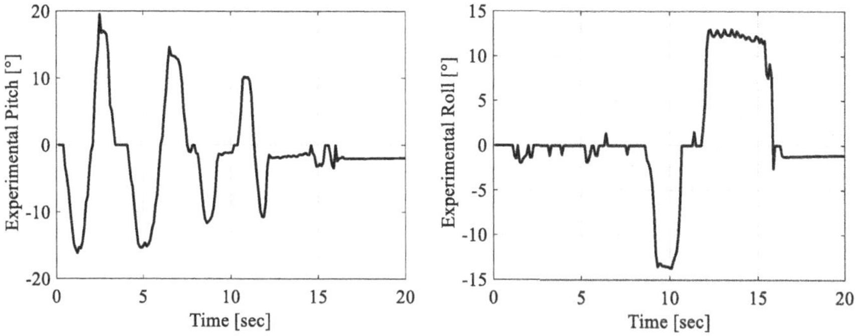

Fig. 12. IOS smartphone results: pitch and roll.

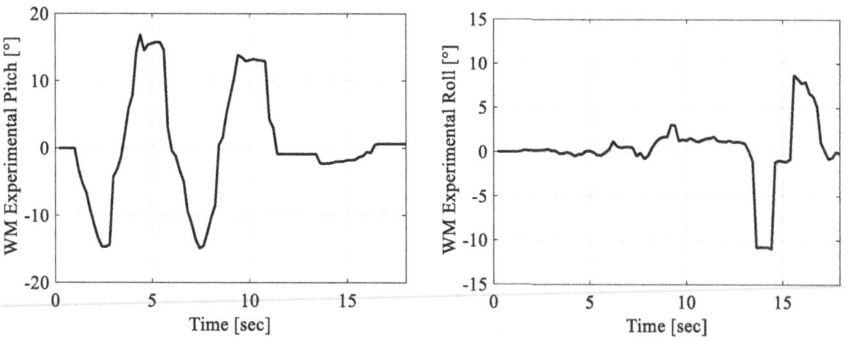

Fig. 13. Wit-Motion sensor results: roll and pitch.

5 Conclusions

In this paper, simulation and experimental tests are proposed for a hybrid robot being used for inspection of site of interest, such as structure, infrastructure and historical sites. Suitable 3D model permits the development of the simulation of a realistic scenario and operation planning. The integration of internal sensors' suite permits of gaining much inside of the survey, allowing the pilot or the control system to have more information on the environment. Furthermore, these signals may be used also for anticipating possible failures acting as alerts. Future developments consist by the integration of the proposed sensors to mobile robots or cable driven robots. The use of data from internal sensors suite can be further used to be gathered in a control system for decision making to realize an autonomous system for inspection.

Acknowledgments. This work is a part of the research Project ERIS "Estensimetri nanocaricati collocati da Robot per Il monitoraggio delle Strutture monumentali" supported by LAZIO INNOVA (n. G09493 – PO FESR LAZIO 2014/2020).

References

1. Sprenger, M., Mettler, T.: Service robots, business & information. Syst. Eng. (2015). https://doi.org/10.1007/s12599-015-0
2. Habibian, S., Dadvar, M., Peykari, B., et al.: Design and implementation of a maxi-sized mobile robot (Karo) for rescue missions. Robomech. J. **8**, 1 (2021)
3. Bellingham, J.G., et al.: Robotics in remote and hostile environments. Science **318**, 1098 (2007). https://doi.org/10.1126/science.1146230
4. Acaccia, G.M., Michelini, R.C., Molfino, R.M., Razzoli, R.P.: Mobile robots in greenhouse cultivation: inspection and treatment of plants. In: Proceedings of ASER 2003, 1st International Workshop on Advances in Service Robotics, Bardolino (2003)
5. Pelliccio, A., Ottaviano, E., Rea, P.: Digital and mechatronic technologies applied to the survey of brownfields. In: Brusaporci, S. (ed.) Handbook of Research on Emerging Digital Tools for Architectural Surveying, Modeling, and Representation, pp. 813–829. IGI Global (2015). https://doi.org/10.4018/978-1-4666-8379-2.ch027
6. Rea, P., Pelliccio, A., Ottaviano, E., Saccucci, M.: The heritage management and preservation using the mechatronic survey. Int. J. Architect. Heritage **11**(8), 1121–1132 (2017). https://doi.org/10.1080/15583058.2017.1338790
7. Nagarajan, B., Li, Y., Sun, Z.Y., Qin, R.: W: A routing algorithm for inspecting grid transmission system using suspended robot: enhancing cost-effective and energy efficient infrastructure maintenance. J. Clean. Prod. **219**, 622–638 (2019)
8. Rodriguez, A.G., Rodriguez, A.G., Rea, P.: A new articulated leg for mobile robots. Indust. Robot: An Int. J. **38**(5), 521–532 (2011). https://doi.org/10.1108/01439911111154090
9. SmpRobotics: https://smprobotics.com/ (2023)
10. Ottaviano, E., Rea, P., Castelli, G.: THROO: a tracked hybrid rover to overpass obstacles. Adv. Robot. **28**(10), 683–694 (2014). https://doi.org/10.1080/01691864.2014.891949
11. Bruzzone, L., Baggetta, M., Nodehi, S.E., Bilancia, P., Fanghella, P.: Functional design of a hybrid leg-wheel-track ground mobile robot. Machines **9**, 10 (2021). https://doi.org/10.3390/machines9010010
12. Rea, P., Ottaviano, E.: Design and development of an inspection robotic system for indoor applications. Robot. Comput. Integrated Manuf. **49**, 143–215 (2018)
13. WitM: https://www.wit-motion.com/9-axis/witmotion-bluetooth-2-0-mult.html (2023)
14. Jacoff, A., Messina, E., Weiss, B.A., Tadokoro, S., Nakagawa, Y.: Test arenas and performance metrics for urban search and rescue robots. IEE Int. Conf. Int. Robot. Syst. **4**, 3396–3403 (2003)
15. Jacoff A., Downs A., Virts A., Messina E.: Stepfield pallets: repeatable terrain for evaluating robot mobility. In: PerMIS Workshop Performance Metrics for Intelligent Systems, pp. 29–34 (2008). https://doi.org/10.1145/1774674.1774680
16. Antosz, K., Jasiulewicz-Kaczmarek, M., Paśko, Ł, Zhang, C., Wang, S.: Application of machine learning and rough set theory in lean maintenance decision support system development. Eksploatacja i Niezawodność – Maintenance and Reliability **23**(4), 695–708 (2021). https://doi.org/10.17531/ein.2021.4.12
17. Leão, C.P., et al.: Web-assisted laboratory for control education: remote and virtual environments. In: Uckelmann, D., Scholz-Reiter, B., Rügge, I., Hong, B., Rizzi, A. (eds.) ImViReLL 2012. CCIS, vol. 282, pp. 62–72. Springer, Heidelberg (2012). https://doi.org/10.1007/978-3-642-28816-6_7
18. Barros, C., Leão, C.P., Soares, F., Minas, G., Machado, J.: RePhyS: A multidisciplinary experience in remote physiological systems laboratory. Int. J. Online Biomed. Eng. (iJOE) **9**(S5), 21 (2013). https://doi.org/10.3991/ijoe.v9iS5.2756

19. Ceccarelli, M. Ottaviano, E., Galvagno, M.: A 3-DOF parallel manipulator as earthquake motion simulator. In: Proceedings of the 7th International Conference on Control, Automation, Robotics and Vision, ICARCV 2002, pp. 944–949 (2002)
20. Rinaldi, C., Di Sabatino, U., Potenza, F., Gattulli, V.: Robotized inspection and health monitoring in the gran Sasso national laboratory. Struct. Monitor Maintenance **8**(1), 51–67 (2021)
21. Rea, P., Ottaviano, E.: Functional design for customizing sit-to-stand assisting devices. J. Bionic Eng. **15**(1), 83–93 (2018). https://doi.org/10.1007/s42235-017-0006-4
22. Moniruzzaman, D., Rassau, A., Chai, D., Shamsul Islam, S.M.: Teleoperation methods and enhancement techniques for mobile robots: a comprehensive survey. Robot. Auton. Syst. **150**, 103973 (2022)
23. Rea, P., Ottaviano, E., Machado, J., Antosz, K.: A general overview of e-maintenance and possible applications. In: Rea, P., Ottaviano, E., Machado, J., Antosz, K. (eds.) Design, Applications, and Maintenance of Cyber-Physical Systems, pp. 196–218. IGI Global (2021). https://doi.org/10.4018/978-1-7998-6721-0.ch009

Two Kinds of New Flexible Eddy Current Sensors for Curved Surface Defect Detection

Xinhua Zhou(✉), Bin Wu, Xiucheng Liu, Yujue Wang, and Cunfu He

Faculty of Information Technology, Beijing University of Technology, Beijing 100124, China
wb@bjut.edu.cn

Abstract. The surface of special equipment is prone to crack after long-term service bringing great potential danger to the safe operation of the equipment. The detection of cracks and the distinction of the orientation of cracks are useful for analyzing the causes of cracks and then optimizing the structural design of special equipment components. This paper proposes two kinds of new flexible eddy current sensors. The sensors are designed based on double-layer flexible PCB circuits, all of which have 8 circular detection coils and specific array element arrangements. They can characterize the orientation of the defects while detecting the surface defects of the specimen through corresponding algorithm processing. Experiments are carried out using two eddy current sensors on curved specimens containing defects with different orientations to verify the sensors' ability. They can detect the defects and identify their orientation.

Keywords: eddy current · multi-channel · flexible sensor · nondestructive testing

1 Introduction

Curved structure is used in transport and mechanical drives which plays a vital role, especially in special equipment. During long-term service, the surface of the equipment is prone to cracks and other defects which can pose a safety hazard to the operation of the equipment. It is therefore essential to develop a non-destructive testing technique for the detection of curved structure surfaces. In addition, information on the orientation of defects can be obtained to help analyzing the causes of cracks and prevent the structures from damage in advance. However, there is currently a lack of a sensing technology that can quickly obtain information about defects on the surface of curved structures.

Traditional methods of detecting defects in curved structures include ultrasonic, magnetic flux leakage, and eddy current methods [1–4]. Ultrasonic have the advantages of a wide scanning range, fast detection speed, and high detection efficiency. To detect the defects of curved surface, Liu *et al.* [5] designed a new squirrel-cage magnetostrictive sensor using ultrasonic guided wave technology for small diameter bend pipe. It could effectively detect defects at the bend by exciting L(0,1) modal guided waves.

Magnetic flux leakage (MFL) detection is a magnetic non-destructive testing method used on the surface and inside the specimen. The principle is to use electromagnets, permanent magnets, *etc.* to saturate the surface of the specimen and obtain the leakage

A. Burduk et al. (Eds.): ISPEM 2023, LNNS 790, pp. 106–119, 2023.
https://doi.org/10.1007/978-3-031-45021-1_8

signal at the location of the defect using a magnetic sensing element. Professor Lee *et al.* [6] had designed a MFL sensor for the detection of surface defects in welded areas, in which the electromagnet was used to excite the test piece and a magnetic sensing element (GMR) was employed to detect the defects. Kim *et al.* [7] built an inspection system for the detection of curved structures in the F-5 supersonic aircraft, using a C-shaped electromagnet to excite the test piece and a hall element to detect the defects. Although both ultrasound and MLF could detect curved specimens, both methods were less universal for specimens with different curvatures or even variable curvatures.

The eddy current method, a common electromagnetic non-destructive testing method, is widely used in the inspection of conductive materials [8, 9]. The basic working principle is that the excitation coil generates a magnetic field (the primary magnetic field), which produces eddy currents on the surface of the test piece. The eddy currents create a secondary magnetic field that is reversed from the primary field. This field changes the magnitude and phase of the current in the induction coil and the electrical signal through the coil reflects information about the surface of the specimen. In particular, Zhang *et al.* [10] designed a multi-channel flexible eddy current sensor for surface defects on steel balls inside the bearing. Zhang *et al.* [11] designed a differential flexible eddy current sensor, which could effectively detect defects of 0.2 mm*1 mm*0.4 mm on the surface of CFRP materials.

This paper presents two multi-channel flexible eddy current sensors [12]. With different array element arrangements and correction algorithms, both sensors can detect the presence of defects on the surface of curved conductive materials and further characterize the orientation of the defects.

2 The Design of Eddy Current Sensors

To ensure that the sensor fits well on the curved surface during the detection process and to reduce the impact due to the lift-off, the size of the flexible eddy current sensor is designed to be relatively small, as shown in Fig. 1. Two types of sensors are designed according to the array element arrangement, called staggered eddy current sensor and floral eddy current sensor respectively [13–15].

The staggered eddy current sensor is composed of two identical flexible circuit boards, each wired on its top and bottom layers respectively. In order to excite the magnetic field generated by the coil as large as possible, it through the holes connected to the upper and lower layers of the circuit. The coil winding of the upper and lower layers of the circuit in opposite ways, where the coil wire diameter and width are 0.1 mm. When the sweeping orientation is parallel to the long side of the sensor, the sweeping resolution is improved and the probability of missing defects is reduced. Each board has 8 coils with the excitation and induction function. Based on the eddy current perpendicular mutual inductance detection principle (Fig. 2a), the two circuit boards are placed vertically on top of each other. Each coil works independently and does not affect the other.

The floral eddy current sensor consists of a flexible circuit designed to be wired on the top and bottom layers of the circuit board. With the coils on the top and bottom layers of the circuit wound in opposite ways and with a wire diameter and width of 0.1 mm. Based on the detection principle of planar mutual inductance (Fig. 2b), the circuit board

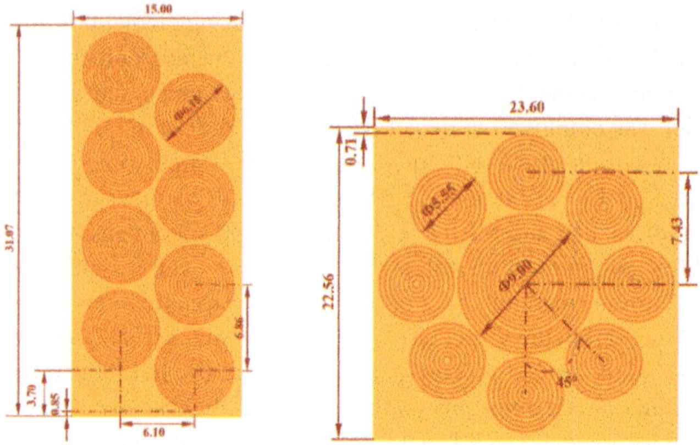

(a) (b)

Fig. 1. Dimensional drawing of two eddy current sensors

contains 9 coils, including one excitation coil and eight receiver coils. The excitation coil is placed in the center of the circuit to generate. The eight receiver coils are evenly spaced around the excitation coil to form a "flower" structure that captures the secondary magnetic field generated by the eddy current. Each coil works independently of the other without interference [16, 17].

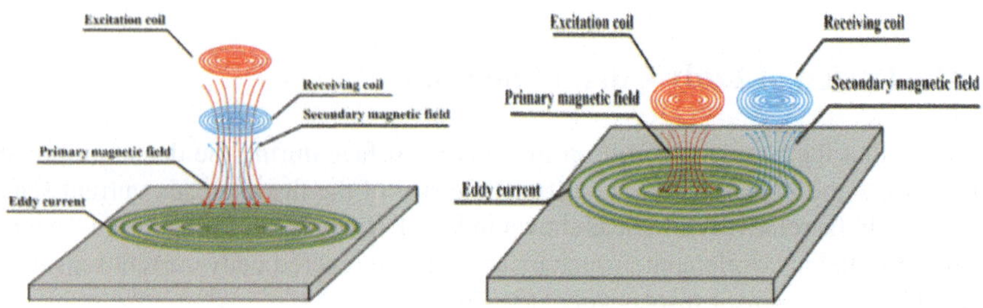

(a) (b)

Fig. 2. Schematic diagram of (a) vertical and (b) planar eddy current detection principle.

3 Sensor Detection Principle

Due to the different array arrangements of the two sensors, different sensors have distinct ways of discriminating the orientation of defects. When the staggered eddy current sensor performs a uniform sweep, the orientational information of the defect can be obtained by calculating the time (number of sampling points/sampling frequency) of the detection results of the dual-row (4 elements per row) arrays and correcting the detection results of each array element.

There are two ways of discriminating the angle of a defect by the floral eddy current sensor. In addition to reflecting the angle of the defect through time calibration according to the moment of each array element, it can also reflect by the amplitude information of results. When the sweeping object is a ferromagnetic material, the detection result of the sensor is affected by both the eddy current field and magnetic leakage field, as shown in Fig. 3. When the sweep orientation is 90° to the defect orientation, and the crack is parallel to the induced eddy current orientation, the eddy current effect is weakest and the leakage effect plays a major role in the detection, and the signal amplitude showing a negative peak of large amplitude. Since there is a difference in the angle of mutual inductance from the excitation coil to the different receiving coils, the orientation of the defect can be deduced from the magnitude of the individual array elements.

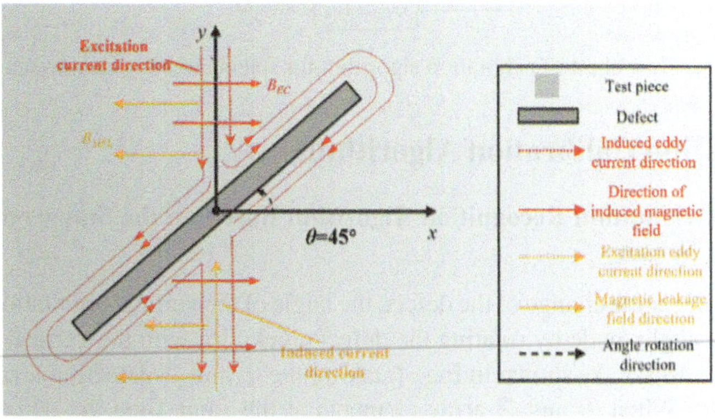

Fig. 3. Superposition of eddy current fields and leakage fields on the surface of ferromagnetic materials

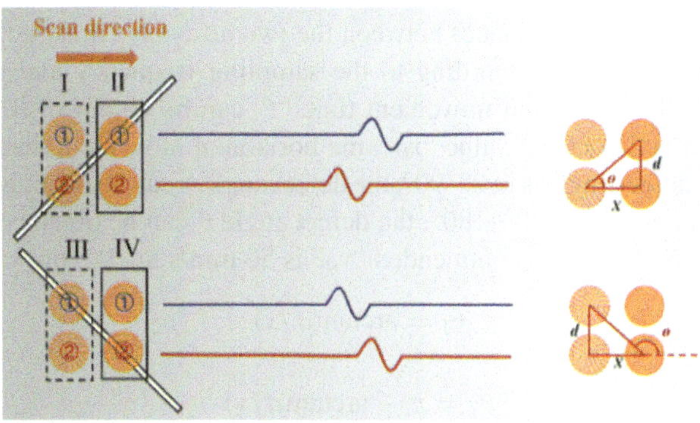

Fig. 4. Staggered eddy current sensor defect orientation identification method

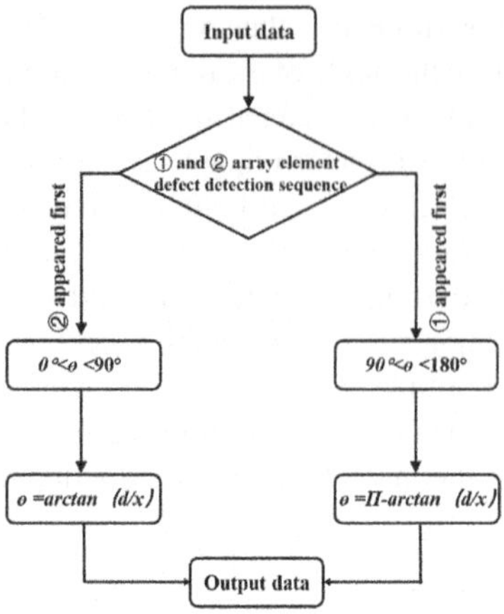

Fig. 5. Flow chart of calibration algorithm for staggered eddy current sensor

4 Orientation Calibration Algorithms

4.1 Defect Orientation Recognition Algorithm Based on the Staggered Eddy Current Sensor

According to the angle change of the defect, the angle of the vertical orientation is defined as 0°, and the angle made by rotating the defect clockwise with the vertical orientation is θ ($0° \leq \theta < 180°$). As shown in Fig. 4, taking the ① and ② detection array elements as an example. When ① and ② array elements at the same time the defect detection results, indicating that at this time the defect angle is 0°. When the sweep orientation perpendicular to the sensor long side, if the lower array element detects the moment of the defect earlier than above (process I to II), it is proved that the defect angle range is $0° < \theta < 90°$and *vice versa* defect angle is $90° < \theta < 180°$ (processes III to IV). The distance "*d*" (vertical distance) between the two arrays is 7mm and the horizontal movement distance is "*x*". According to the sampling frequency and the number of sampling points, the horizontal movement time "*t*" can be obtained. If the horizontal movement speed is a constant value "v_0", the horizontal movement distance is x ($x = v_0 * t$). When the defect angle is in 0°–90°, the defect angle θ can be obtained from Eq. (1). When the defect angle is in 90°–180°, the defect angle θ can be obtained from Eq. (2), as shown in Fig. 5. And the recommended "v_0" is 50 mm/s to 200 mm/s.

$$\theta_1 = \arctan(d/x) \tag{1}$$

$$\theta_2 = \pi - \arctan(d/x) \tag{2}$$

4.2 Defect Orientation Recognition Algorithm Based on the Floral Eddy Current Sensor

The floral sensor can determine the orientation of the defect based on the peak value of the defect detection. And the peak value of the detection signal is the largest when

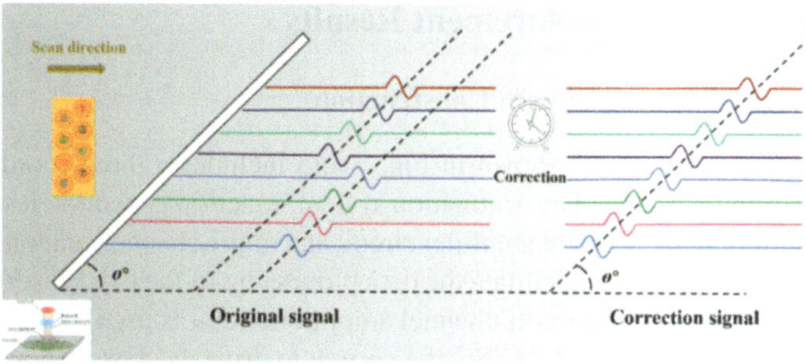

(a)

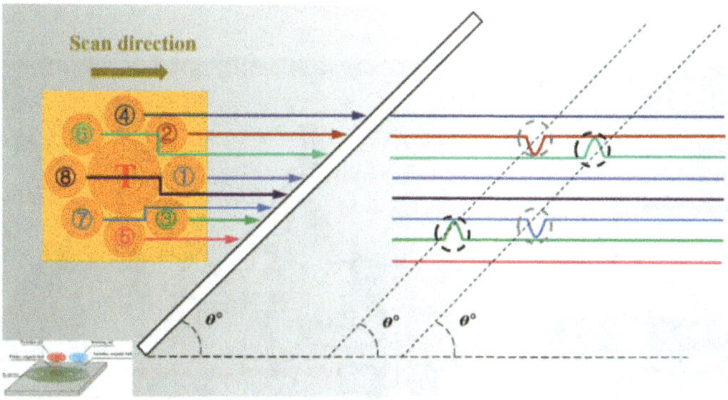

(b)

Fig. 6. Schematic diagram of (a) staggered and (b) floral eddy current signal correction

the mutual induction orientation is parallel to the angle of the defect. The orientation of the defect needs to be determined based on the peak value of the sensor. Figure 6 array element arrangement order as an example. Same definition of the angle of the vertical orientation is 0°, clockwise rotation of the defect and the angle made by the vertical orientation is θ ($0° \leq \theta < 180°$). If the ④ array element detection result is a positive peak, the defect angle is 0°. If the ② array element detection result is a positive peak, the defect angle is 45°. If the ③ array element detection result is a positive peak, the defect angle is 135°. If the only ① array element detection result appears, the defect angle is 90°. In addition, the detection results of ⑤, ⑥, ⑦ and ⑧ can be combined with their symmetrical position detection results to prevent false detection. This sensor method has the advantage of fast recognition and high detection efficiency, but the resolution of the defect angle is low.

5 Experiment and Measurement Results

5.1 Experimental Testing System Construction

The experimental test setup is shown in Fig. 7 and includes a three-coordinate scanning platform, multi-channel data acquisition system, function generator, test piece, and sensor. The multi-channel current amplifier circuit and multi-channel quadrature demodulation circuit are designed to facilitate the data processing of the two sensor excitations as well as the test results. The multi-channel amplifier circuit is used to divide the excitation signal into 8 channels and amplify the current to drive the 8 excitation coils of the staggered eddy current sensor.

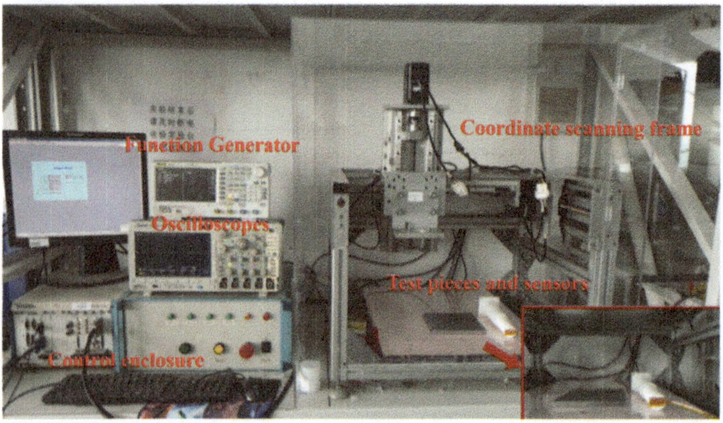

Fig. 7. Experimental system construction based on eddy current sweeping

Firstly, the two sensors are swept separately for the experiment. The scanning process involves sweeping the two sensors at different frequencies against the cracked specimen shown in Fig. 8 The specimen has a size of 150 mm*150 mm*10 mm and the crack of 80 mm*0.5 mm*10 mm is opened at the center of the specimen. The sensor excitation frequency is increased in 100 kHz increments from 100 kHz to 2 MHz, and the optimum sweep-frequency of the sensor is determined according to the magnitude of the peak value when passing through the defect. The sensor is placed on the surface of the material and passes over the specimen at a uniform speed from left to right, and the detection results are displayed and recorded in the LABVIEW software. As shown in Fig. 8 for the typical frequency of the sensor sweep results, it can be seen that the best detection frequency of the staggered eddy current sensor is about 600 kHz, the best detection frequency of the floral eddy current sensor is about 1.1 MHz. And the frequency as the excitation frequency of the sensor detection experiment (see Fig. 9).

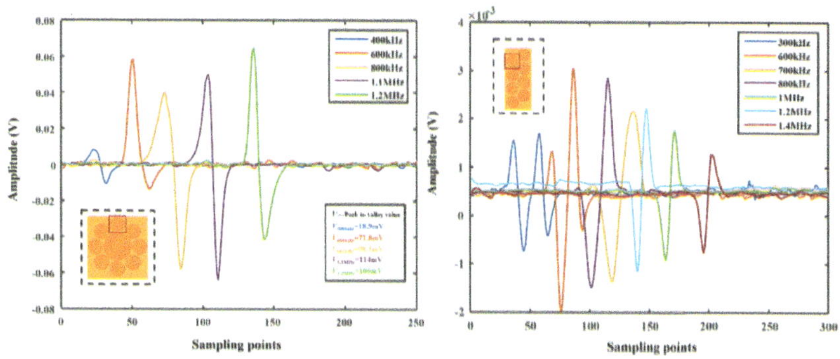

(a) (b)

Fig. 8. Sweep-frequency result of (a) staggered and (b) floral eddy current

5.2 Experimental Procedure and Analysis

5.2.1 Curved Surface Defect Detection Tests

(1) Staggered eddy current sensor

To further verify the sensor's ability to detect curved parts, a special curved part, the reducer cam was designed. The radial face, end face, and surface of the specimen were machined to simulate defects of 5 mm in length, 0.1 mm in width, and 0.1 mm in depth, respectively. To verify the detection capability of the sensors on the curved components. A sine excitation amplitude of 10 V, a frequency of 600 kHz, and a sampling frequency of 10 kHz were set in the upper computer, and the sensor was fitted to the surface of the specimen during the sweep. The test results are shown in Fig. 10 (see Table 1).

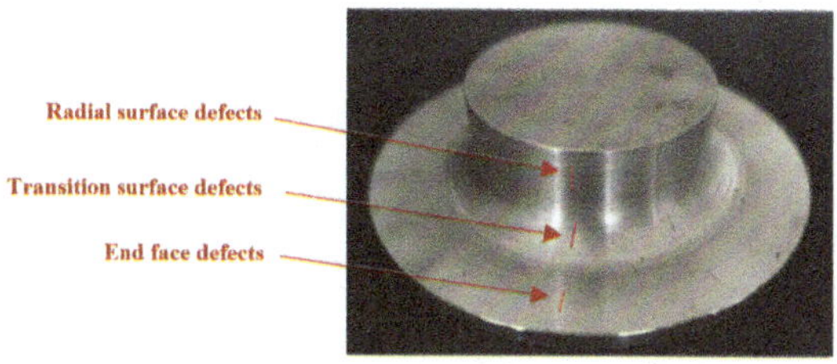

Fig. 9. Reducing boss test piece

The length of the sensor is parallel to the length of the defect and is swept from shallow to deep depending on the depth of the defect. As only part of the array passes over the defect during the sweep, it is deduced that the defect length of the specimen is less than the length of the long side of the sensor, which is in line with the actual situation. The results for the radial and end surfaces show that the sensor has four separate channels

Table 1. Reducing boss defect size

Defective markings	Defect size(length × width × depth)mm^3
1	5 × 0.1 × 0.2
2	5 × 0.1 × 0.4
3	5 × 0.1 × 0.6
4	5 × 0.1 × 0.8
5	5 × 0.1 × 1.0
6	5 × 0.1 × 1.2
7	5 × 0.1 × 1.4
8	5 × 0.1 × 1.6

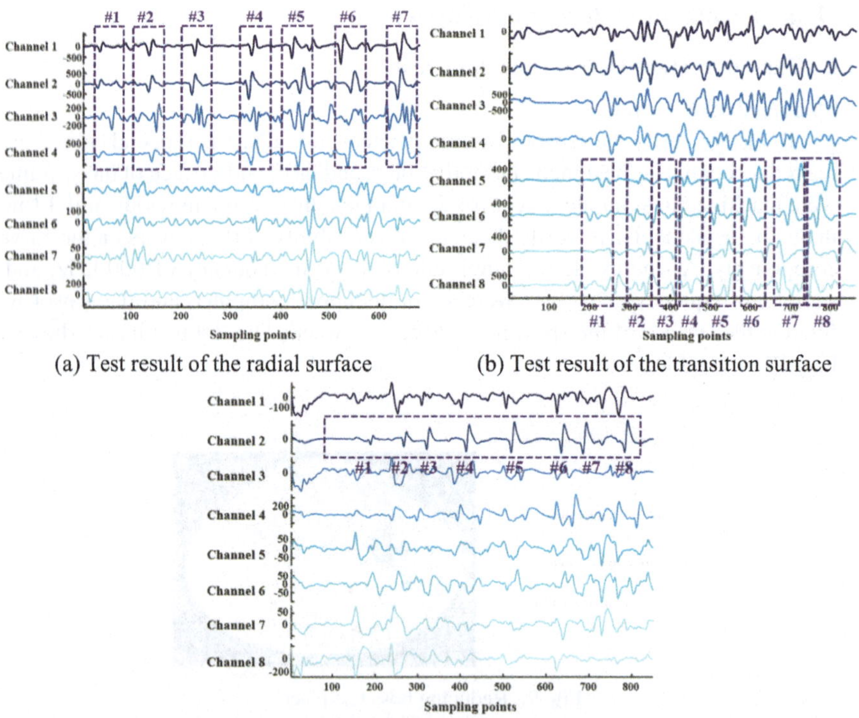

(a) Test result of the radial surface

(b) Test result of the transition surface

(c) Test result of the end face

Fig. 10. Test results of the reducing boss

that can detect defects. But only one can detect defects in curved surfaces due to the difficulty of accessing the sensor array elements in curved surfaces. It is also shown that the sensor can detect surface defects with a minimum depth of 0.2 mm in curved surfaces (see Fig. 11).

(2) Floral eddy current sensor

Using the floral sensor to detect 3 defects on the axial surface of the round bar, the diameter of the round bar specimen is 40 mm, 150 mm high, and its material is Q235. The 3 defects of different lengths and the same width and depth are designed along its axial orientation. The experiment will be the upper computer set sinusoidal excitation amplitude of 10 V, frequency of 1.1 MHz, the sampling frequency of 10 kHz. The orientation of excitation coil pointing to ① receiving coil orientation perpendicular to the orientation of defect # 1 long side. The detection results are shown in Fig. 12. The floral sensor can detect 3 defects on the surface of the round bar, which verify its ability to detect defects on the surface of the curved surface (see Table 2).

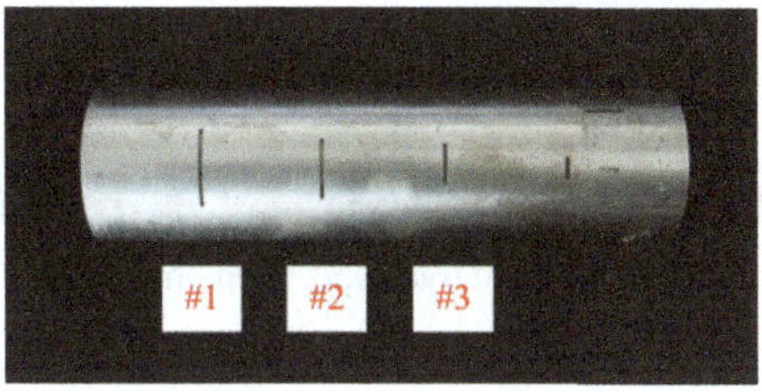

Fig. 11. Round bar specimen

Table 2. Round bar defect information

Defective markings	Defect size(length × width × depth)mm^3
1	20 × 1 × 1
2	15 × 1 × 1
3	10 × 1 × 1

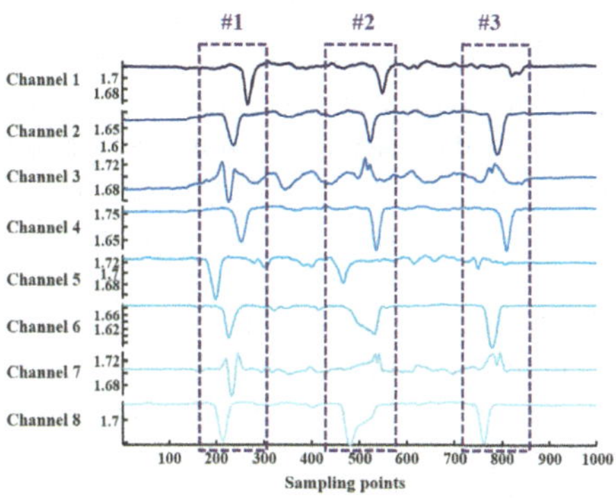

Fig. 12. Test results of round bar specimen

5.2.2 Curved Surface Defect Orientation Recognition Detection Testings

The concave track specimen is designed according to the testing requirements, the specimen size is 200 mm*50 mm*15 mm and the material is aluminum. In its width orientation, the design radius is 5.85 mm quarter arc groove, and in the arc groove along the surface of the design of four 40 mm interval defects. The width and depth of the defects are 1 mm. The defect orientation with the width of the test specimen angle are 0°, 15°, 30°, and 45°. A physical drawing of the specimen is shown in Fig. 13 and the specific dimensional information is given in Table 3.

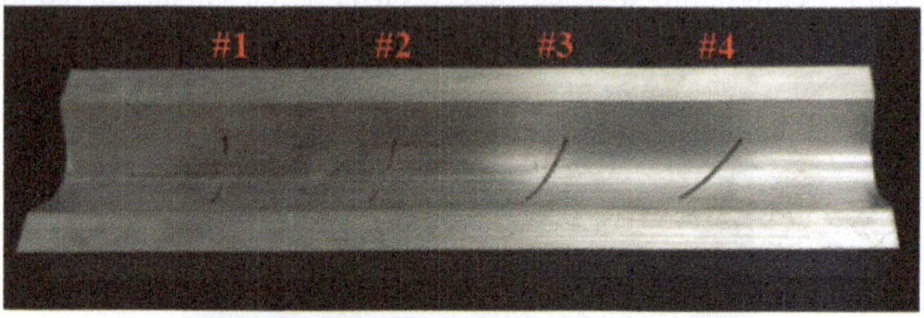

Fig. 13. The concave track specimen physical picture

Table 3. Concave track defect information

The number of defects	Defect size(width*depth)mm^2	Angle with width orientation(°)
1	1 × 1	0
2	1 × 1	15
3	1 × 1	30
4	1 × 1	45

The defects in the inner concave track specimen are swept using two separate sensors, with the orientation of the sweep perpendicular to the orientation of the long side of the sensor. First of all, the three-coordinate scanning platform carries the staggered sensors to scan the specimen. And the results of each array element will be "time-shifted correction", as shown in Fig. 14a is the corrected two-dimensional graph line results. From the graph line results can be seen in the sensor elements can detect the four defects in the specimen. To better characterize the angular information of the defects, the two rows of four channels of the detection array element signal by the algorithm processing and two-dimensional imaging, the calibration results are shown in Fig. 14b. From the detection results can be seen, the calibration results can characterize the orientation of different defects. And indicating that the sensor is more sensitive to the angular information of the defects.

Using the floral eddy current sensor to test the above specimen, the test results are shown in Fig. 15. As the material of the test piece is aluminum, there is no influence of magnetic leakage and the detection results only show a one-way negative peak, and the peak changes with the angle of the defect, proving that the eddy current sensor can

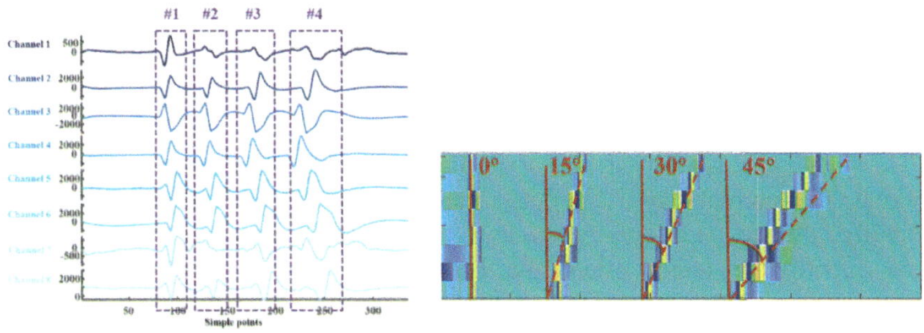
(a) (b)

Fig. 14. Sweeping (a) 2D graph and (b) image result of staggered eddy current

detect the presence of 4 defects in the test piece and infer the orientation of the defect through the peak size of the defect. In addition to depend on the moment at which each array element passes over the defect, the results are imaged by the algorithm. It can be seen that the sensor is also able to characterize defects at different angles.

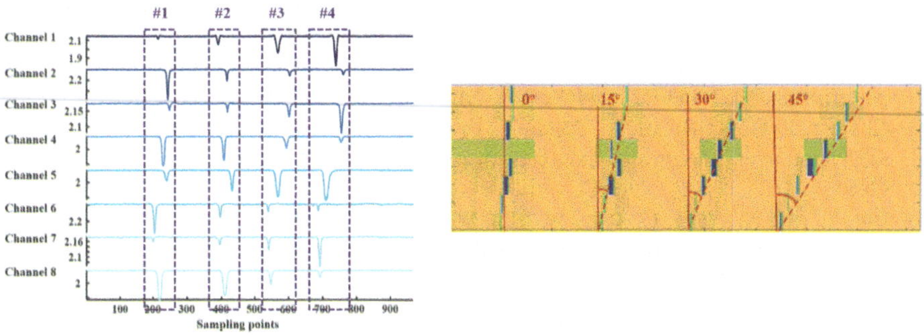
(a) (b)

Fig. 15. Sweeping (a) 2D graph and (b) image result of floral eddy current

6 Conclusion

In this paper, two flexible eddy current sensors for the detection of defects on curved surfaces are designed. Through algorithmic adjustment and amplitude analysis, both sensors are able to detect surface defects on curved parts and to identify their orientation of the defect at the same time. The main conclusions of the study are as follows:

(1) Two sensors are based on a flexible PCB circuit, and each coil contains 8 receiving coils. The staggered eddy current sensor have 8 exciting coils and the floral eddy current sensor have only one exciting coil.

(2) The staggered eddy current sensor can identify the orientation of defects by means of a time-calibrated algorithm; The floral eddy current sensor are not able to identify the orientation of the defect by means of a time calibration algorithm, but can also identify by the magnitude of the defect.
(3) The staggered eddy current sensor can detect the defects on reducing boss and the floral eddy current sensor can detect the defects on round bar, which verify their ability of detection of curve surface. Both of sensors can detect the concave track and identify the defects orientation, but the sensitivity of staggered eddy current sensor is stronger than the floral eddy current sensor.

Acknowledgements. This study was supported by the National Key R and D Program of China (2018YFC0809003).

References

1. Borsboom, C.: Eddy currenta inspection. In: Fibre Metal Laminates, pp. 499-512. Springer, Dordrecht (2001)
2. García-Martín, J., Gómez-Gil, J., Vázquez-Sánchez, E.: Non-destructive techniques based on eddy current testing. Sensors **11**(3), 2525–2565 (2011)
3. Sun, Y., Kang, Y., Qiu, C.: A new NDT method based on permanent magnetic field perturbation. NDT and E Int. **44**(1), 1–7 (2011)
4. Chotard, T., Gimet-Breart, N., Smith, A., et al.: Application of ultrasonic testing to describe the hydration of calcium aluminate cement at the early age. Cem. Concr. Res. **31**(3), 405–412 (2001)
5. Liu, Y., Liu, X., Yang, C., Guo, W., Bin, W., He, C.: Longitudinal guided wave inspection of small-diameter elbowed steel tubes with a novel squirrel-cage magnetostrictive sensor. J. Sens. **2019**, 1–9 (2019)
6. Lee, M., Kim, J., Kim, J., et al.: Nondestructive testing of moisture separator reheater tubing system using hall sensor array. Nondestr. Test. Eval. **33**(1), 35–44 (2018)
7. Kim, J., Le, M., Lee, J., et al.: Eddy current testing and evaluation of far-side corrosion around rivet in jet-engine intake of aging supersonic aircraft. J. Nondestr. Eval. **33**(4), 471–480 (2014)
8. Tan, Y., Wang, X., Moreau, R.: An innovative contactless method for detecting defects in electrical conductors by measuring a change in electromagnetic torque. Meas. Sci. Technol. **26**(3), 035602 (2015)
9. Chady, T., Spychalski, I.: Eddy current transducer with rotating permanent magnets. In: XXII International Workshop on Electromagnetic Nondestructive Evaluation (2017)
10. Zhang, W., Wang, C., Xie, F., et al.: Defect imaging curved surface based on flexible eddy current array sensor. Measurement **151**, 107280 (2019)
11. Zhang, H., Ma, L., Xie, F.: A method of steel ball surface quality inspection based on flexible arrayed eddy current sensor. Measurement **144**, 192–202 (2019)
12. Gilles-Pascaud, C., Decitre, J.M., Vacher, F., et al.: Eddy current flexible probes for complex geometries. AIP Conf. Proc. AIP **820**(1), 399–406 (2006)
13. She, S., He, Y., Chen, Y., et al.: Flexible floral eddy current probe for detecting flaws in metal plate. IEEE Sens. J. **20**(18), 10521–10529 (2020)
14. Zhang, N., Peng, L., Tao, X., et al.: Flexible ECT probe with front-end differential setting for inspection of curved CFRP structures. Compos. B Eng. **227**, 109404 (2021)

15. Rocha, T.J., Ramos, H.G., Ribeiro, A.L., et al.: Magnetic sensors assessment in velocity induced eddy current testing. Sens. Actuators, A **228**, 55–61 (2015)
16. Gloria, N., Areiza, M.C.L., Miranda, I., et al.: Development of a magnetic sensor for detection and sizing of internal pipeline corrosion defects. NDT and E Int. **42**(8), 669–677 (2009)
17. https://www.ndesensors.swri.org
18. Gardner, W.E.: Glossary – Improving the Effectiveness and Reliability of Non-Destructive Testing. 328(4), pp. 239–241. Northern Research Laboratories, Warrington (1992)

Determining the Range of Image Base of ARM Firmware

Jinmiao Wang[1,2], Ruijin Zhu[3], Xiaofei Cao[1,2], and Yuejin Zhang[4(✉)]

[1] Run Technologies Co., Ltd. Beijing, Beijing 100192, China
[2] Beijing Cyberspace Data Analysis and Applied Engineering Technology Research Center, Beijing 100192, China
[3] China Information Technology Security Evaluation Center, Beijing 100085, China
[4] Beijing Union University, Beijing 100101, China
herryyj707@vip.sina.com

Abstract. Disassembling is a necessary step for vulnerability mining or deep understanding of the operating mechanism of firmware. When disassembling firmware, the image base of the firmware need to be determined first, but the image base usually cannot be obtained directly. Given the widespread use of ARM processors in embedded systems, a method is proposed to determine the range of the firmware image base by targeting ARM firmware. First, the addresses loaded by the LDR instruction in the firmware are obtained, and then these addresses are deduplicated, sorted, and segmented to calculate the cumulative frequency. Then, the rapidly rising segment in the cumulative frequency curve is considered as the range of image base. Experimental results demonstrate that the proposed method can effectively determine the range of image base for firmware that uses LDR instruction to load address.

Keywords: Image base · ARM firmware · LDR instruction · Disassemble

1 Introduction

Firmware is an important component of embedded systems and has a significant impact on their functionality and security. There are two types of firmware based on the presence of an operating system [1]: user-space firmware and blob firmware. User-space firmware runs on some general-purpose operating systems, such as routers and network cameras, usually running a stripped-down version of Linux. Blob firmware is usually a binary image that runs directly on the processor, such as MP3 players and solid-state drives.

Currently, there are frequent attacks and security incidents targeting embedded system firmware. For example, hackers discovered four attack methods by analyzing the firmware of a brand of shared bicycles' smart locks and gained control of them. Researchers have also found vulnerabilities in the IoT ecosystem by reverse-engineering firmware [2]. Based on reverse engineering of DJI firmware, researchers designed and implemented a decoder for DroneID which is DJI's proprietary tracking protocol [3].

A. Burduk et al. (Eds.): ISPEM 2023, LNNS 790, pp. 120–132, 2023.
https://doi.org/10.1007/978-3-031-45021-1_9

Through reverse engineering and fuzz testing of Google's security chip firmware, multiple vulnerabilities were found in the chip firmware [4]. Disassembly is a common method used for security assessment and vulnerability mining in firmware analysis.

For user-space firmware, as most of the functionality is implemented by user-level programs, the definitions of system calls and program entry points are relatively standardized. Therefore, analyzing this type of firmware is similar to analyzing ordinary user-level programs. However, for blob firmware, since there is no underlying operating system, the image base and execution entry point of the firmware are usually unknown. When reverse-engineering this type of firmware, disassembly tools such as IDA Pro and Ghidra need to obtain the processor type and image base of the firmware's operating environment [5]. The processor type can be obtained by viewing product manuals [6] or physical examination [7], but the image base cannot be directly obtained [8]. The image base is the starting address of the firmware mapped into memory. A correct image base allows disassembly tools to establish accurate cross-references, including string references and function references, which are important for firmware analysts to understand the firmware. At the same time, the correct image base is helpful for firmware analysts to understand the firmware's memory layout as a whole, while an incorrect image base leads to addressing errors for instructions that reference immediate addresses. Therefore, getting the correct image base is a key step in reverse-engineering firmware.

At present, the methods for locating the image base can be divided into two types: manual determination and automatic determination.

Manual determination methods mainly rely on the experience and intuition of reverse engineers. [9] found some regular bytes at a certain offset in the firmware, and inferred that these bytes were the addresses of some symbol names, and then manually determined the image base. [10] provided an example of using the decompressed code in MIPS firmware to manually reason the image base. [11] pointed out that using the correct image base can enable disassembly tools to correctly establish cross-references of absolute address, thereby disassembling firmware more accurately. They also provided an example of using the initialization code in the BSS section to manually deduce the image base of firmware.

Automatic determination methods mainly calculate the image base based on the characteristics of firmware compilation. [12] studied the generation rules and characteristics of function entry tables, and determined the image base using the function entry table in the firmware. [13] studied the storage method and encoding rules of strings in firmware, proposed an algorithm for identifying the text pool containing string addresses in firmware, and then determined the image base using the string and text pool information in the firmware. [14] proposed a method for calculating the image base using string addresses and string offsets in firmware. [15] studied the length of string storage and loading methods in ARM firmware, and based on the fact that compilers usually store strings referenced in adjacent code in a string block according to the reference order, proposed an algorithm for locating the image base.

In the existing methods, manual determination methods highly rely on the experience and skills of reverse engineers, which requires high demands on reverse engineers; while automatic determination methods can only target firmware with certain characteristics. In this paper, we propose a method for determining the range of firmware image base.

Firstly, based on the characteristics and encoding rules of the LDR instruction, the LDR instructions in the firmware are identified and the addresses loaded by the LDR instruction are obtained. Then, these addresses are deduplicated, sorted, and their cumulative frequencies are calculated to obtain a cumulative frequency curve. The rapidly rising segments in the curve are the range of the image base. Experimental results demonstrate that the proposed method can effectively determine the range of image base for firmware that uses LDR instruction to load address.

2 Method for Determining the Range of Image Base

2.1 Identify LDR Instructions and the Loading Addresses

The ARM processor is a low-power RISC processor that supports two instruction sets: the ARM instruction set and the Thumb instruction set. The ARM instruction set is 32 bits, while the Thumb instruction set is 16 bits. The Thumb instruction set is a functional subset of the ARM instruction set. Compared to equivalent ARM instructions, the Thumb instruction can save storage space while retain all the advantages of 32-bit instructions. Because Thumb instructions have the advantage of high code density, they are used in many systems, so an ARM firmware may contain both ARM and Thumb instructions.

In ARM firmware, the LDR instruction is typically used to load addresses into registers when operating on absolute addresses. Take the firmware file iAUDIO10_FW.BIN of iAudio 10 MP3 as an example, Fig. 1 shows its disassembled code (assuming the image base is 0). Using the LDR instruction at offset 0x001B8DB8 as an example, the machine code for the LDR instruction is 40 10 9F E5. Since the firmware is in little-endian mode, the actual machine code is E5 9F 10 40.

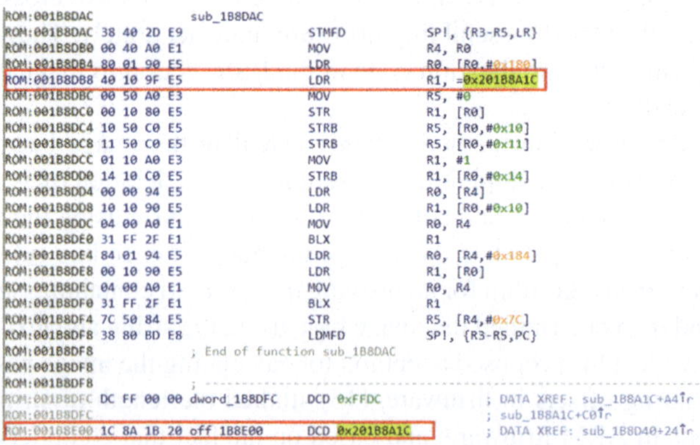

Fig. 1. An instance loaded by LDR instruction (assuming the image base is 0)

The LDR instruction has multiple syntax format, among which the encoding format for loading immediate values into register in ARM state is shown in Fig. 2(a).

By analyzing the encoding format of the LDR instruction and the machine code of the LDR instruction in Fig. 1, we can get that Rd = (0001)2 = R1, immed_12 = (0000

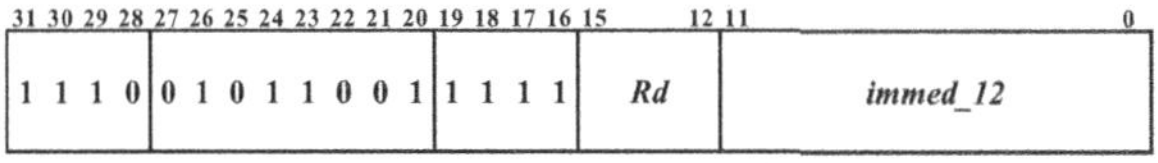

(a) ARM-based encoding format

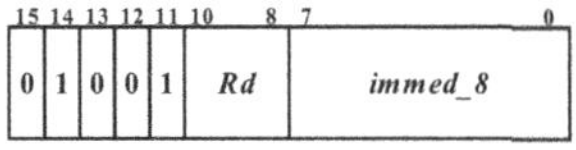

(b) Thumb-based encoding format

Fig. 2. Encoding format of LDR instruction

0100 0000)2 = 0x40. The addressing address of the LDR instruction under the ARM state is (PC&0xFFFFFFFC) + immed_12. Since the ARM processor adopts a 3-stage pipeline technology, the value of PC is PC = Current + 8 in the ARM state. Therefore, the addressing address of LDR is:

$$
\begin{aligned}
address &= (PC \,\&\, 0xFFFFFFFC) + immed\_12 \\
&= ((current + 8) \,\&\, 0xFFFFFFFC) + immed\_12 \\
&= ((0x001B8DB8 + 8) \,\&\, 0xFFFFFFFC) + 0x40 \\
&= (0x001B8DC0 \,\&\, 0xFFFFFFFC) + 0x40 \\
&= 0x001B8DC0 + 0x40 \\
&= 0x001B8E00
\end{aligned} \tag{1}
$$

Thus, the addressing address of LDR instruction is 0x001B8E00. As shown in Fig. 1, the 4 bytes starting at memory address 0x001B8E00 are 1C 8A 1B 20. Since the firmware is in little-endian mode, the actual value is 0x201B8A1C, which is the address that the LDR instruction is to load. The manually analysis shows that memory address 0x201B8A1C is the entry address of the binary function sub_201B8A1C.

The syntax and instruction loading method of LDR instruction in Thumb state are similar to those in ARM state, and are not further illustrated here. The instruction encoding of LDR in Thumb state is shown in Fig. 2 (b), which indicates that the encoding of LDR instruction in Thumb state starts with $(01001)_2$, and this encoding feature can be used to quickly identify LDR instructions in firmware.

Based on the above analysis and the encoding characteristics of LDR instruction, we introduce the FIND-LDR algorithm to identify LDR instructions in firmware files in ARM/Thumb states and calculate their loading addresses [14]. Depending on the different syntax and loading method of LDR in ARM and Thumb state, the second input of the algorithm is either ARM state or Thumb state. The FIND-LDR algorithm is shown in Algorithm 1. The time complexity of the FIND-LDR algorithm is O(*fileSize*), where *fileSize* is the size of the firmware file.

Algorithm 1: FIND-LDR algorithm

```
Imput : binary file binaryFile、ARM/Thumb state state
Output : the loading addresses of LDR instruciton in ARM or
Thumb states in the binary file
function FindLDR (binaryFile, state)
   bin[fileSize] ← binaryFile
   offset ← 0
   if (state == STATE_THUMB)
      while (0 ≤ offset < fileSize) do
         opcode ← bin[offset+1]
         opcode ← opcode & (11111000)2
         if (opcode == (01001000)2)
            PC ← offset + 4
            immed_8 ← bit[7,...,0]
            address ← (PC & 0xFFFFFFFC) + (immed_8 * 4)
            Rd ← Memory[address, 4]
            Output: Rd
         end if
         offset ← offset + 2
      end while
   else
      if (state == STATE_ARM)
         while (0 ≤ offset < fileSize) do
            if (bin[offset + 2] == 0x9F && bin[offset+3] ==
            0xE5)
               PC ← offset + 8
               immed_12 ← bit[11,...,0]
               address ← PC&0xFFFFFFFC + (immed_12)
               Rd ← Memory[address, 4]
               Output: Rd
            end if
            offset ← offset +4
         end while
      end if
   end if
end function
```

2.2 Algorithm for Determining the Range of Image Base

The image base of firmware refers to the starting position of the firmware mapped in memory. The schematic diagram of the firmware mapping is shown in Fig. 3. In ARM firmware, when operating on absolute addresses, the LDR instruction is usually used to load the absolute address into a register, and these absolute addresses usually point to the memory mapping range of the firmware. Using this principle, we can obtain a multi-set of absolute addresses loaded by LDR instructions in ARM firmware. After removing duplicates and sorting the absolute addresses in ascending order, we can obtain a set of absolute addresses. Then, the elements in the absolute addresses set are grouped, and the frequency of each group is calculated. When a group's frequency is significantly larger than the previous group's frequency for the first time, the group is the range where LDR instruction loading addresses appear in large quantities. Since the addresses loaded by the LDR instruction are within the memory mapping range of the firmware, the image base is within this group.

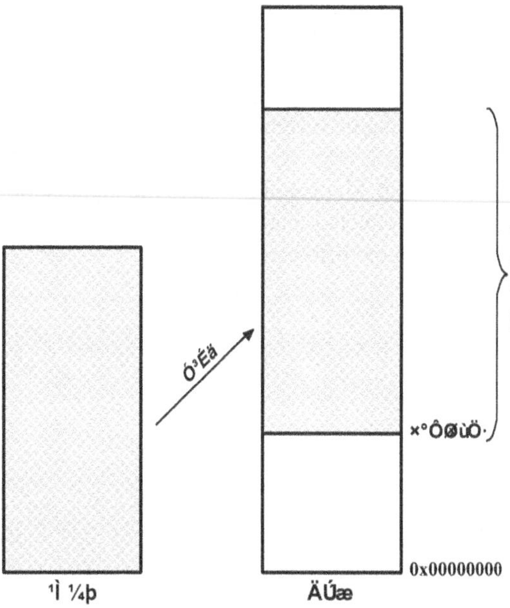

Fig. 3. Diagram of firmware loaded in memory

The algorithm for determining the image base range of ARM firmware consists of four steps:

(1) Search for LDR instructions in the firmware and output the loaded addresses.

Absolute addresses in ARM firmware are generally loaded by LDR instructions. The FIND-LDR algorithm is used to identify LDR instructions in the firmware and output the loaded addresses, which are saved as a multi-set M.

(2) Remove duplicates and sort the addresses in set *M*.

Perform a loop operation on *M* to remove duplicate elements and obtain all addresses loaded by LDR instructions. Then, sort all addresses in ascending order to obtain a set *N*.

(3) Group the elements in set N and calculate the cumulative frequency.

Let the size of the firmware file be *fileSize*, the minimum and maximum values of the elements in set *N* be min and max, respectively. The range of set *N* is [*min*, *max*]. Group all elements in set *N* with a group length of *fileSize*. The number of groups is (*max-min*) / *fileSize*. Count the number of elements in each group, and then calculate the frequency and cumulative frequency of each group.

(4) Draw the cumulative frequency graph and determine the range of image base.

Based on the cumulative frequency of the LDR loading addresses obtained by step (3), draw a cumulative frequency graph with memory addresses as the *x-axis* and cumulative frequency as the *y-axis*. When the frequency of a group is significantly larger than the frequency of the previous group for the first time, it is considered that the image base is within this group. This is shown in the curve as a rapid rise. That is, if there is a rapid rise in the cumulative frequency curve, then the image base is usually in this rapid rise range. If the rapid rise curve contains several groups, the image base of the firmware is usually in the first group.

The above steps are shown in Algorithm 2, where the input is the firmware file and the output is the cumulative frequency and cumulative frequency graph of the LDR instruction loading addresses in the firmware.

Algorithm 2: cumulative frequency of addresses loaded by LDR instruction

```
Input : firmware Firmware
Output : the cumulative frequency and cumulative frequency
graph of addresses loaded by LDR instrucion
function CalcCumulativeFrequency (Firmware)
    M ← FindLDR (Firmware, ARM)
    M ← FindLDR (Firmware, Thumb)
    N ← Ø
    for all mi ∈ M do
        if mi ∉ N do
                N ← mi
        end if
    end for
    sort(N)
    max ← getMax(N)
    min ← getMin(N)
    count ← getCount(N)
    numOfGroup ←⌈ (max - min) / fileSize ⌉
    i ← 0
    j ← 0
    for all i ∈ numOfGroup do
        while (i*fileSize <= N[j] < (i+1) * fileSize)
            group[i]++
            j++
        end while
        frequency[i] ← group[i] / count
        cumulativeFrequency[i]←cumulativeFrequency[i-1]
        +frequency[i]
        Output: cumulativeFrequency[i]
    end for
    drawChart(numOfGroup,cumulativeFrequency[numOfGroup])
end function
```

3 Experimental Results and Analysis

3.1 Experimental Results

There is currently no publicly available test set for determining the image base range of ARM firmware. Therefore, we collected multiple embedded device's firmware from Internet as the test set. The hardware environment for the experiments was an Intel Core 2.30 GHz processor with 8 GB of memory, and the software environment was Microsoft Windows7 SP1.

To verify the effectiveness of the proposed algorithm in this paper, we performed Algorithm 2 on the collected ARM firmware, and the results are shown in Table 1. The experimental results show that the actual image base of the firmware in Table 1 is within the range of image base determined in this paper. The image base range obtained by this method can be cross-verified with the results in references [12–15].

Table 1. Experimental Results

Embedded Device	Firmware	LDR	Range of image base	Actual image base	Validity
iAudio 10 MP3	iAUDIO10_FW.BIN	12656	0x1FD20C89–0x2028B585	0x20000000	✓
ABB NETA-21	uImage	16723	0xBFE5CC60–0xC02B4880	0xC0008000	✓
Pebble	tintin_fw.bin	3152	0x7FE9811–0x8058818	0x08010000	✓
Samsung Gear Fit	wingtip_in.bin	4334	0x7F00000–0x80FC000	0x08004000	✓
Sony AS30 DV	vmlinux.bin	9676	0xBFEAB880–0xC014A420	0xC0018000	✓
Sony NEX-7	av-cam.bin	26656	0xC324C5E0–0xC37A50BC	0xC3421000	✓

3.2 Case Study

Take the firmware iAUDIO10_FW.BIN of the iAudio 10 MP3 player as an example for detailed analysis and calculation of the image base range of the firmware.

First, the FIND-LDR algorithm is performed to identify 28,191 LDR instructions in the iAUDIO10_FW.BIN firmware, which means that 28,191 absolute addresses (which contains duplicate addresses) were loaded by LDR instructions. After removing duplicates, we obtained 12,656 addresses, which were then sorted in ascending order. The minimum value was 1, and the maximum value was 4294967295.

Since the file size of the iAUDIO10_FW.BIN firmware is 5,679,356 bytes (5.41 MB), which means that the group size is 5,679,356, the number of groups is (4294967295–1)/5679356 = 756.2, i.e. there is 757 groups.

Next, the 12,656 addresses are grouped and the cumulative frequency for each group is calculated. And then we plotted the cumulative frequency curve, which is shown in Fig. 4. It can be seen that there is a segment rising rapidly in the curve, and the range of the first segment is [0x1FD20C89, 0x2028B585], which means that the image base range of the firmware is [0x1FD20C89, 0x2028B585].

It was manually verified that the image base of the firmware was 0x20000000, which is within the range of [0x1FD20C89, 0x2028B585]. Therefore, the effectiveness of the method proposed in this paper is verified.

The cumulative frequency curves of firmware from Pebble, Samsung Gear Fit and Sony AS30 DV in Table 1 are shown in Figs. 5, 6 and 7. The image base range determined by the algorithm proposed in this paper is [0x7FE9811, 0x8058818], [0x7F00000, 0x80FC000], and [0xBFEAB880, 0xC014A420], respectively. By manually determination, all of the actual image base is within the respective image base range, which also verifies the effectiveness of the method proposed in this paper.

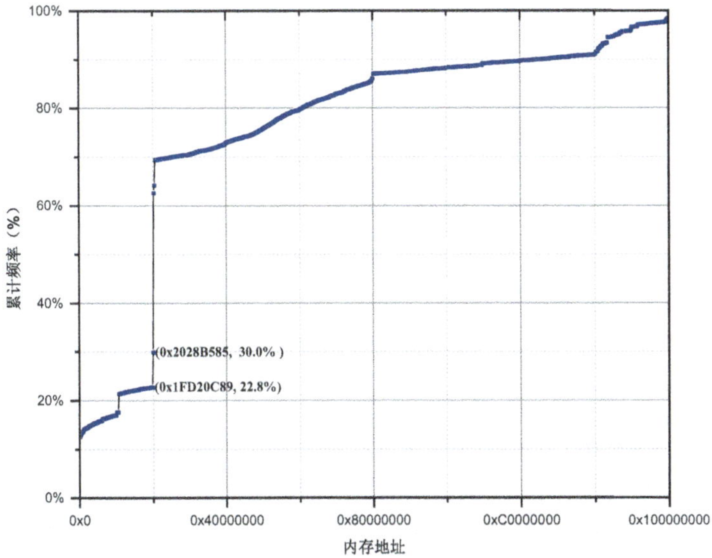

Fig. 4. Cumulative frequency curve of iAudio 10 MP3

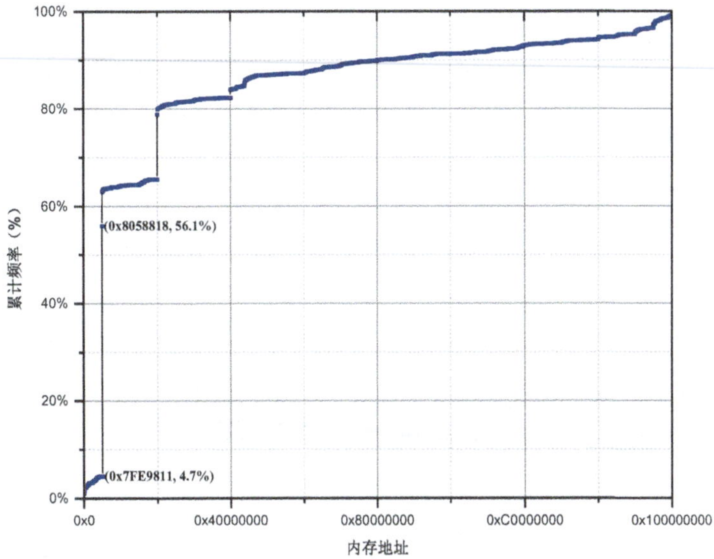

Fig. 5. Cumulative frequency curve of Pebble

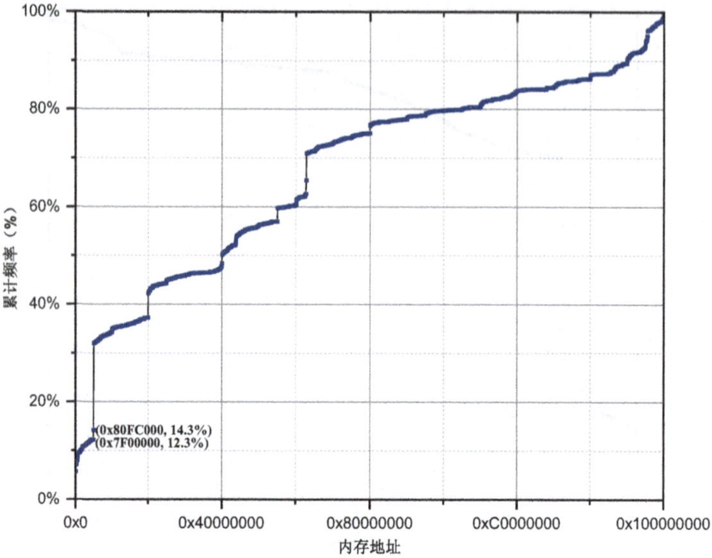

Fig. 6. Cumulative frequency curve of Samsung Gear Fit

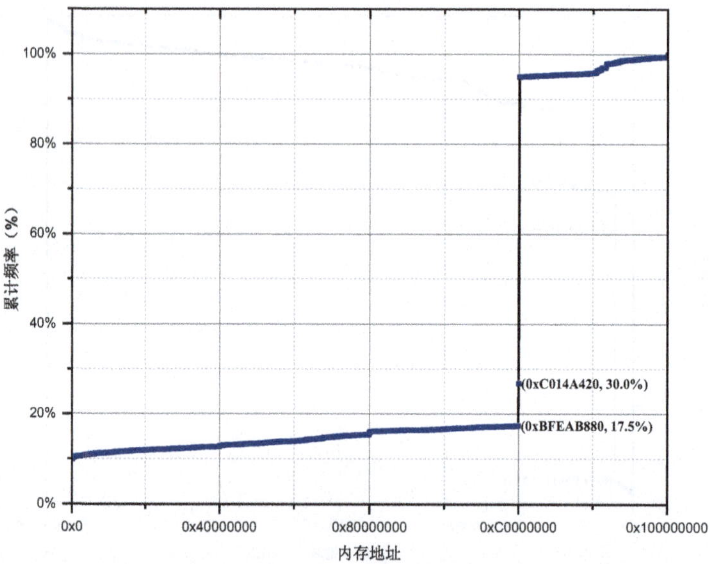

Fig. 7. Cumulative frequency curve of Sony AS30 DV

4 Conclusions

Reverse engineering is a necessary step in analyzing the security of ARM firmware. At the beginning of reverse engineering, it is necessary to determine the firmware's image base. However, most ARM firmware cannot obtain the image base directly, which poses a significant obstacle to disassembly work. In this paper, we studied the format and storage rules of the LDR instruction in firmware and proposed a method for determining the firmware's image base range using the address loaded by LDR instruction. Experimental results show that the method proposed in this paper can correctly determine the range of the firmware image base.

For future work, we will explore the encoding rules of ARM instructions and propose methods for determining the image base range of other types of firmware. We believe that these automated approaches can effectively reduce the difficulty of reverse analysis and bring convenience to the reverse engineer.

References

1. Yan, S., Wang, R., Hauser, C., Kruegel, C., Vigna, G.: Firmalice – automatic detection of authentication bypass vulnerabilities in binary firmware. In: Network and Distributed System Security Symposium, Network and Distributed System Security Symposium (2015)
2. Giese, D., Wegemer, D.: Reversing iot: Xiaomi ecosystem (2018)
3. Schiller, N., et al.: Drone security and the mysterious case of DJI's DroneID. In: Network and Distributed System Security Symposium (NDSS) (2023)
4. Melotti, D., Bellom, M.R.: Attack on titan m, reloaded: vulnerability research on a modern security chip. In: blackhat USA 2022, blackhat USA 2022, Las Vegas, USA (2022)
5. Eagle, C.: The IDA pro book: the unofficial guide to the world's most popular disassembler, No Starch Press (2008)
6. Basnight, Z., Butts, J., Lopez, J., Dube, T.: Analysis of programmable logic controller firmware for threat assessment and forensic investigation. In: Proceedings of the 8th International Conference on Information Warfare and Security: ICIW 2013, Proceedings of the 8th International Conference on Information Warfare and Security: ICIW 2013, p. 9 16. Academic Conferences Limited, Denver, Colorado, USA (2013). http://toc.proceedings.com/17592webtoc.pdf
7. Basnight, Z., Butts, J., Lopez, J., Jr., Dube, T.: Firmware modification attacks on programmable logic controllers. Int. J. Crit. Infrastruct. Prot. **6**(2), 76–84 (2013). https://doi.org/10.1016/j.ijcip.2013.04.004
8. Basnight, Z.H.: Firmware counterfeiting and modification attacks on programmable logic controllers, Ph.D. thesis, Air Force Institute of Technology (2013)
9. Peck, D., Peterson, D., Leveraging ethernet card vulnerabilities in field devices. In: SCADA Security Scientific Symposium, SCADA Security Scientific Symposium, pp. 1–19. Digital Bond Press (2009)
10. Heffner, C.: Reverse engineering vxworks firmware: Wrt54gv8 (2011)
11. Heffner, C.: Reversing belkin's wps pin algorithm (2015)
12. Zhu, R., Tan, Y.-A., Zhang, Q., Wu, F., Zheng, J., Xue, Y.: Determining image base of firmware files for arm devices. Trans. Inf. Syst. **99-D**(2), 351–359 (2016). https://doi.org/10.1587/transinf.2015EDP7217
13. Zhu, R., Tan, Y.-A., Zhang, Q., Li, Y., Zheng, J.: Determining image base of firmware for arm devices by matching literal pools. Digit. Investig. **16**, 19–28 (2016). https://doi.org/10.1016/j.diin.2016.01.002

14. Zhu, R., Zhang, B., Mao, J., Luo, Y., Tan, Y.-a., Zhang, Q.: Determining image base of arm firmware based on matching string addresses. Acta Electron. Sin. **45**(06), 1475–1482 (2017) 11–2087/TN
15. Zhu, R., Zhang, B., Mao, J., Zhang, Q., Tan, Y.-A.: A methodology for determining the image base of arm-based industrial control system firmware. Int. J. Crit. Infrastruct. Prot. **16**, 26–35 (2017). https://doi.org/10.1016/j.ijcip.2016.12.002

Supporting the Selection of Quality Tools Using Neural Networks

Beata Starzyńska[1] and Izabela Rojek[2(✉)]

[1] Faculty of Mechanical Engineering, Poznan University of Technology, M. Skłodowska-Curie Square 5, 60965 Poznan, Poland
beata.starzynska@put.poznan.pl
[2] Faculty of Computer Science, Kazimierz Wielki University, Chodkiewicza 30, 85064 Bydgoszcz, Poland
izabela.rojek@ukw.edu.pl

Abstract. Quality management tools are well-grounded in the management of enterprises regardless of an adopted quality management concept. A crucial problem to be solved is the provision of support for the selection of these tools in such a way as to choose the most useful one. The article presents an overview of traditional solutions for the selection of quality tools and their computer-aided choice. The analysis of the source literature showed a research gap regarding solutions for automatic support for the selection of quality tools. In order to resolve this problem, neural networks were used, specifically a feedforward multilayer network with backward propagation of errors. Data were prepared in the form of learning examples and many classification models based on the selected neural network were developed. As a result, the best model with the highest classification effectiveness was selected. Such a classification model can be placed in an expert system, which can then support a less experienced employee in the selection of quality tools (e.g. in the quality assurance department in an enterprise).

Keywords: Quality Tools · Manufacturing Processes · Classification · Neural Networks

1 Introduction

Quality tools (or more precisely quality management tools) are well-grounded in the management of enterprises regardless of an adopted quality management concept.

In managing a company according to J. Oakland's TQM model, they create a "tangible" process management infrastructure, oriented towards meeting customer requirements and beneficial collaboration with cooperators [1]. Total Quality Management emphasizes widespread employee involvement; within this concept, quality tools used are primarily an instrument for building this commitment (tools used in teamwork). The equally commonly used Six Sigma concept is based mainly on the measurement of effectiveness and efficiency of activities. In relation to measurable goals set in enterprises, it is quantitative tools, in particular statistical ones, that are more often used. In turn, the so-called normative approach in quality management [2] directs decision-makers to use

A. Burduk et al. (Eds.): ISPEM 2023, LNNS 790, pp. 133–145, 2023.
https://doi.org/10.1007/978-3-031-45021-1_10

systems and tools recommended in relevant standards. A manifestation of this approach is the establishment of quality management systems in enterprises in accordance with the requirements of the ISO 9001 standard [3]. It is a widely recognized instrument for building the effectiveness of management activities in enterprises (also as a component of integrated management systems). Depending on the nature of the products offered by the company, the requirements of industry standards (e.g. ISO 13485 [4]) are also implemented as well as indications of standards that explicitly refer to the use of quality tools (e.g. ISO 13053 [5]).

Regardless of their aim and scope of application, quality tools have some features in common: they are all used to collect and process quantitative and qualitative data into useful decision-making information. They are also characterized by simplicity – they can be described by a simple algorithm, program or instructions for use. As such, they are used first and foremost in enterprises' operational activity.

In the source literature, the concept of quality tools is often combined with techniques (QT&T). According to [6–8], tools and techniques are practical ways, skills, means, or mechanisms that can be used in specific tasks; including those that facilitate change or improvement.

Quality tools can be used in areas such as *industrial processes* and *non-industrial processes* [9]. In the former case, a clear division of the use of quality tools can be seen with reference to the stages of: design and development of new products (in this context they complement tools and systems of engineering design) and the production process, with manufacturing activity encompassing a wide range of product classes. In the latter group, quality tools are used in processes of other business lines than manufacturing (e.g. construction, trade) and in services [10–12].

Higher prestige, quality of products, teamwork skills, employee commitment and responsibility and increasing a company's revenue, reducing costs and the number of complaints are just some of the benefits that a company can draw thanks to the effective implementation of quality management tools and techniques [13].

The effective use of quality tools in manufacturing enterprises' activity is hindered by their diversity. This state of affairs results not only from the development of various quality management concepts, but also from the formation of new, hybrid management concepts (Human Six Sigma, Lean Six Sigma, etc.), which is accompanied, among others, by the phenomenon of "mixing" tools. This phenomenon is difficult to accept from the perspective of carrying out ongoing operational activities, because it hampers the process of selecting quality tools adequate to a problem.

The importance of the accurate selection of quality tools was indicated, among others, by [8, 14, 15]. The authors emphasize that failures in the use of quality tools (including quality techniques QT&T) do not result from their ineffectiveness *"but rather because of a lack of clear understanding about when, where and how to apply them"* [8, pp. 69]. In [16, p. 577], it was also noted that *"while solving problems and improving enterprise processes and products, there is often a significant problem connected with fast, easy and flexible choice of appropriate tools for solution activities*".

This cognitive gap can be filled by solutions that can be described as supporting the selection of quality tools. The issues raised in the article are justified in the context of considerations on Industry 4.0. Namely, Quality 4.0 is a new term representing a new featured approach to quality management [17]. "It has also been argued, that Q-4.0 should be considered as a development of traditional quality tools, with more emphasison connectedness, intelligence and automation for improving performance and making timely, data-driven decisions in an end-to-end scenario, involving all stakeholders and providing visibility and transparency." [18]. In the article [19], in turn, an overview of Industry 4.0 technologies that can support the Lean Six Sigma improvement methodology was made. The research results allow to pair quality tools with selected technologies (including neural networks), supporting their use in practice.

2 Literature Review

2.1 Traditional Solutions for the Selection of Quality Tools

The continuous development of the area of improvement results in the development of quality management tools, principles, techniques and methods. This results in an increased number of available tools. A user is not able to keep up to date with all emerging novelties, and thus may "miss" a tool that could potentially help solve a given situation or problem.

The starting point for organizing knowledge of quality tools are their classifications proposed in the source literature. These are often dichotomous divisions: one can single out tools that are traditional and new, quantitative and qualitative, for individual work and teamwork, etc.

The proper classification of a tool can be the basis for its proper selection and use. From the point of view of practitioners, it is interesting to assign quality tools to individual phases in the product life cycle, stages of the production process, depending on the method of visualization of the result or even the level of user skills [20].

From the point of view of practical applications, it is also important to use divisions that place individual quality tools in problem-solving methodologies (e.g. [21]) or in improvement methodologies (e.g. [9]). Examples of methodologies whose stages are equipped with sets of tools recommended for use in their framework are DMAIC and PDCA methodologies [22, 23].

In previous studies (source literature, guidebooks, training materials and others), the knowledge of quality tools (and techniques) is presented in a tabular (matrix) form or in the form of general selection schematics [24]. Therefore, it may be difficult to use them, which is not beneficial from the perspective of enterprises, as well as the challenges of industry 4.0 [25].

Traditional solutions offered so far require a lot of time from a user to choose and acquire knowledge of a tool. The condition for using such a form of assistance is often the need for a decision maker to have an appropriate level of preliminary knowledge and experience (e.g. [20]).

2.2 Computer-Aided Selection of Quality Tools

A large number of tools generated a need for creating support systems for their selection, which will allow to adapt a tool to a specific situation, and thus extend user perspectives with the possibility of new, subsequent applications.

Obviously, when designing solutions in the area of computer-aided selection of quality tools, one needs to be aware of who will be the recipient of a proposed solution. The range of users of quality tools includes people who are just becoming familiar with these tools as well as those who already have experience resulting from their use; operational employees and quality staff in an enterprise; "self-taught people" and people using training in this field; people making decisions about how to solve a problem (in connection with the indication of the necessary set of tools) and people occasionally using individual quality tools in operational activities.

One of only a few computerized systems serving this purpose [26] is the Intelligent Decision Support System called eCIFOD [27]. The authors of the program, with the help of experts, mapped various quality management tools and techniques, as a result of which the features of tools were identified and combined with the categories found in the CIFOD model (Fig. 1).

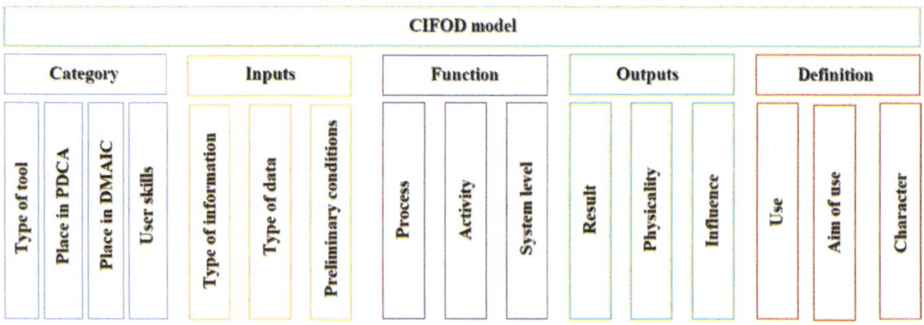

Fig. 1. Organizing tool attributes within the CIFOD model category.

In order to search for a tool, its identification features must first be specified. Therefore, when looking for the right quality tool, one needs to choose attributes from each group (see Fig. 1).

The data were then entered into an information processing system, which allowed to develop eCIFOD, a program supporting the selection of quality tools and operating in a web browser, e.g. Google Chrome and Mozilla Firefox.

The implementation of the CIFOD model helps a practitioner with limited experience and knowledge to choose a helpful tool. The model was used as an input to an intelligent and online decision support system for choosing the right quality management tool. On its basis, features were identified and assigned to individual tools contained in the database. An algorithm was created and SQL database settings were designed.

Based on the selected features, a user receives a list of tools that meet the requirements. Such a solution helps to eliminate a decision maker's bias, because it eliminates the influence of the popularity of a particular tool and can be used by a person who does not have extensive experience in this field [20, 27].

The attribute-based approach used to describe quality tools has already been used in another application. One of the first systems supporting the selection of quality tools is a system called Excellence Toolbox [16].

The target group of users of this system are enterprise employees, e.g. operational employees, junior project engineers, junior improvement specialists. These are people who have basic knowledge of quality management tools, but do not have enough experience to decide which tool to use in a particular situation – or in what sequence.

As mentioned above, from the point of view of practitioners, in such solutions it is valuable to "assign" quality tools to individual phases in the product life cycle; the most valuable are tips related to which of these tools should be used in which stages of the production process, etc. The presented system also meets these requirements.

The development of the system therefore required the formalization of knowledge of a wide range of quality tools described in the source literature and used in practice [28].

The main assumption for the development of a formalized record of knowledge of quality tools in the system database was the "transition" from an unstructured description of tools contained in the rich source literature to their definition through a finite set of features and their states (see Fig. 2). Each of the distinguished features describing quality tools adopts one or more values – e.g. in relation to the type of input data tools can have the attribute of a quantitative or qualitative tool. In the case of a feature related to the "place" of tool application in the PDCA cycle, four states of this feature were distinguished, etc.

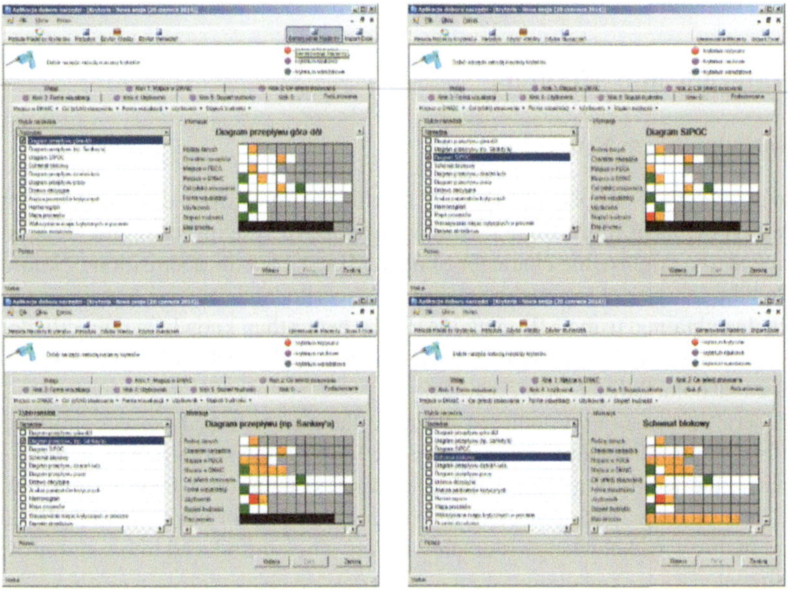

Fig. 2. The working interface of the program with a visible record of the features of example tools in the matrices.

The so-called criteria pattern method implemented in the system allows a user to find a suitable tool that meets the selection criteria, represented in the program by the defined features of tools and their states. A user goes through the consecutive steps of the tool selection procedure by selecting appropriate question tabs defined in the program. The main aim of the application is to help an inexperienced user in choosing the right tool by answering a series of questions that specify his requirements regarding the required tool.

By indicating several properties, a user receives a list of tools that match the specified criteria. Search results are given as a percentage, depending on the extent to which they match the indicated features in the form of a ranking list of tools. One can obtain even several tools that offer 100% matching with the requirements. At this point, a built-in, 'handbook' description of each tool comes to help.

The application was designed in such a way as to provide the greatest possible flexibility of knowledge stored in the system database, understood as the possibility of: constantly expanding the system knowledge with new instruments, expanding the selection criteria and others.

The application was tested in manufacturing enterprises, both by quality experts and by people who do not have extensive knowledge of quality management instruments. It was recognized as a useful tool to support the selection of quality tools, especially in SMEs.

The Excellence Toolbox program made it possible to develop a set of learning examples for the new implementation of the program. The preparation of a learning set consisted in encoding possible combinations of features of particular tools and placing them in appropriate tables.

3 Materials and Methods

3.1 Introduction to Neural Networks

Neural networks are one of the methods of machine learning (ML). Learning systems that are built on the basis of machine learning are adaptive systems, changing their internal parameters to recognize the character of data. ML enables the acquisition of knowledge based on the analysis of experts' behaviour or experimental data, i.e. learning examples. Knowledge obtained using ML methods can be better than knowledge directly deduced by people. Sometimes it is also the only way to build models if knowledge is not possessed or cannot be obtained [29, 30].

Modern information systems do not have the ability to self-correct simple errors. Therefore, the development of systems in the form of advisory systems in many fields continues – the acquisition of knowledge for expert systems or systems with a knowledge base. The overriding goal of machine learning is to search for information in the "flood" of data. In this way, knowledge is created which, when introduced into a system, makes it intelligent.

In general, it can be said that an intelligent system can change in two ways:

- a system acquires new knowledge from external sources (e.g. databases, experts),
- a system modifies itself by using knowledge it already possesses more effectively.

Therefore, the basic methods of machine learning are divided into: inductive learning and the so-called skill acquisition or speedup learning. Inductive inference uses a set of facts (observations) and a possible set of a priori hypotheses about these facts, and as a result gives the most likely generalization explaining these facts. On the other hand, deduction consists in drawing conclusions that are a logical consequence of information we have.

The main research paradigms distinguished in machine learning include: modeling of neural networks and theoretical-decision techniques, symbolic acquisition of concepts and learning in a narrow field with the intensive use of knowledge. Approaches differ in terms of: a priori knowledge possessed by a system and the way of representing and modifying knowledge in a system.

One of ML methods are artificial neural network systems. Such a system consists of a network of interrelated simple elements performing a certain processing function. It is a system that learns incrementally by modifying connections between elements. Neural network learning begins with little initial knowledge (related to the selection of elements and the structure of connections in the network). It is also very important for correct network learning to use network parameters that change continuously during learning. A neural network processes on numerical examples [31, 32].

In the article, a neural network was used in classification, which is the selection of quality tools. Classification involves the process of dividing a set of objects (examples) into classes (categories, concepts), where a class means a set of objects characterized by certain common properties. It also assumes that the structure of classes is initially known, and the goal is to search for a function/rule that allows to classify new objects, i.e. assign them to existing categories [29].

A neural network (feedforward multilayer with backward propagation of errors – MLP can be placed in an expert system, which can then support a less experienced employee in the selection of quality tools (e.g. in the quality assurance department in an enterprise).

3.2 Data Preparation

The database of tool selection examples was prepared based on interviews with experts and analysis of the work of the company's employees. All necessary data for the selection of quality tools have been specified. Examples of quality tool selections include the following input parameters: type of data (e.g. 1_1 – numerical), type of a tool (e.g. 2_1 – creative), place in PDCA (e.g. 3_1 – plan), place in DMAIC (e.g. 4-1 – define), aim of use (e.g. 5_2 – indicating dependencies), form of visualization (e.g. 6_1 – diagram), user (e.g. 7_3 – individual/team), degree of difficulty (e.g. 8_1 – easy) and an output parameter: name of a quality tool (e.g. diagram_systematics_100).

Figure 3 shows a fragment of a learning file with examples of the selection of quality tools. The example file contained 410 examples. It was divided into a learning file (80%), a test file (10%) and a validation file (10%).

3.3 Classification Models in the Form of Neural Networks

Many models of neural networks were built in experiments. The 10 best neural networks with different structures and learning with different parameters were selected (see Table 1). The constant element in the models was the input layer whose size resulted

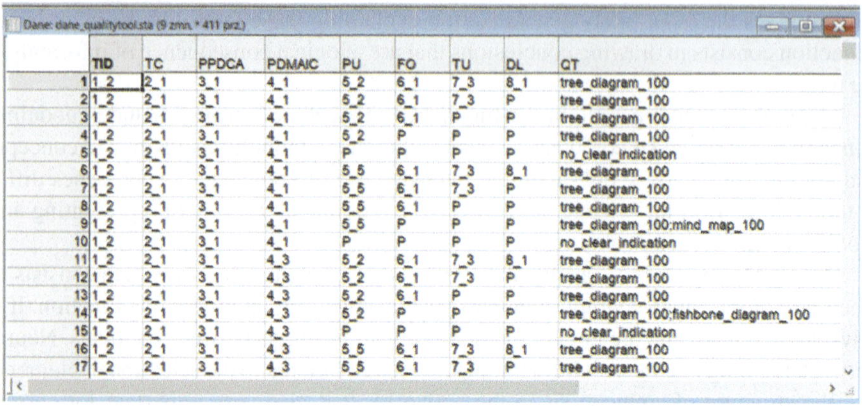
Dane: dane_qualitytool.sta (9 zmn. * 411 prz.)

	TID	TC	PPDCA	PDMAIC	PU	FO	TU	DL	QT
1	1_2	2_1	3_1	4_1	5_2	6_1	7_3	8_1	tree_diagram_100
2	1_2	2_1	3_1	4_1	5_2	6_1	7_3	P	tree_diagram_100
3	1_2	2_1	3_1	4_1	5_2	6_1	P	P	tree_diagram_100
4	1_2	2_1	3_1	4_1	5_2	P	P	P	tree_diagram_100
5	1_2	2_1	3_1	4_1	P	P	P	P	no_clear_indication
6	1_2	2_1	3_1	4_1	5_5	6_1	7_3	8_1	tree_diagram_100
7	1_2	2_1	3_1	4_1	5_5	6_1	7_3	P	tree_diagram_100
8	1_2	2_1	3_1	4_1	5_5	6_1	P	P	tree_diagram_100
9	1_2	2_1	3_1	4_1	5_5	P	P	P	tree_diagram_100;mind_map_100
10	1_2	2_1	3_1	4_1	P	P	P	P	no_clear_indication
11	1_2	2_1	3_1	4_3	5_2	6_1	7_3	8_1	tree_diagram_100
12	1_2	2_1	3_1	4_3	5_2	6_1	7_3	P	tree_diagram_100
13	1_2	2_1	3_1	4_3	5_2	6_1	P	P	tree_diagram_100
14	1_2	2_1	3_1	4_3	5_2	P	P	P	tree_diagram_100;fishbone_diagram_100
15	1_2	2_1	3_1	4_3	P	P	P	P	no_clear_indication
16	1_2	2_1	3_1	4_3	5_5	6_1	7_3	8_1	tree_diagram_100
17	1_2	2_1	3_1	4_3	5_5	6_1	7_3	P	tree_diagram_100

Fig. 3. Fragment of a file with examples from the Statistica Data Mining program.

from the amount of input data entering the neural network (8 inputs) and the output layer (1 output). Neural network models were parameterized by different sizes. Neural networks were built with 1 hidden layer, which contained a different number of neurons (in the range of 5 to 25). The number of learning epochs in the BFGS algorithm (from 10 to 120) and the error function (SOS function and Entropy), the activation function in the hidden and output layer (functions: Linear, Logistic, Exponential, Tanh, Softmax) were also changed.

Table 1. Neural networks for the selection of quality tools.

Id	MLP	Quality (learning)	Quality (test-no)	Quality (validation)	Learning algorithm BFGS	Error function	Activation function (hidden)	Activation function (output)
1	8-20-1	96.53	90.16	86.89	26	Entropy	Linear	Softmax
2	8-17-1	91.32	95.08	85.25	13	Entropy	Tanh	Softmax
3	8-29-1	94.79	91.80	91.80	14	Entropy	Tanh	Softmax
7	8-6-1	86.46	77.05	85.25	85	SOS	Linear	Exponential
8	8-27-1	98.26	86.89	80.33	63	Entropy	Logistic	Softmax
9	8-9-1	82.29	80.33	73.77	42	SOS	Logistic	Tanh
16	**8-29-1**	**96.18**	**93.44**	**88.52**	**25**	**Entropy**	**Linear**	**Softmax**
18	8-5-1	94.79	90.16	90.16	28	Entropy	Linear	Softmax
19	8-28-1	95.49	90.16	83.61	23	Entropy	Logistic	Softmax
20	8-17-1	95.49	90.16	86.89	17	Entropy	Tanh	Softmax

The best neural network is network No. 16 MLP 8-29-1 with a layer with 29 hidden neurons, 25 learning epochs, the Entropy error function, the activation function in the hidden Linear layer and the activation function in Softmax output layer with 92.72% effectiveness.

The worst network is network No. 9 MLP 8-9-1 with a layer with 9 hidden neurons, 42 learning epochs, the SOS error function, the activation function in the hidden Logistic layer and the activation function in the Tanh output layer with 78.80% effectiveness (see Fig. 4).

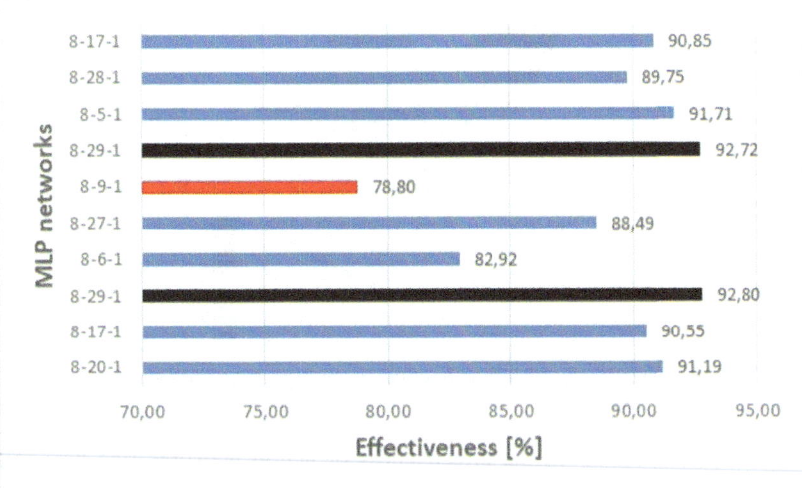

Fig. 4. Effectiveness of neural network models [%].

In order to evaluate the neural network models, a graph for the error matrix was prepared (see Table 2).

Table 2. Neural network error matrix

Neural Network	Instances	QT – All tools
9.MLP 8-9-1	Total	410
	Correct	331
	Incorrect	79
16.MLP 8-29-1	Total	410
	Correct	388
	Incorrect	22

Table 2 shows the error matrix for the two selected networks. The best network 16. MLP 8-29-1 made 22 mistakes, and the worst network MLP 8-9-1 79 mistakes. Figure 4 shows the accuracy of classifying neural networks. Network No. 16 has a much better accuracy than network No. 9, which is presented in detail in Table 3. The next figure (see Fig. 5) presents a surface graph of the best and worst neural network accuracy.

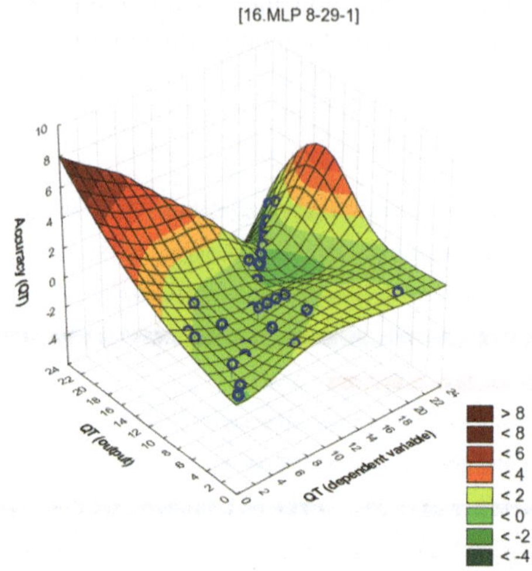

Fig. 5. Surface graph of the best and worst neural network accuracy.

In order to further check the effectiveness of neural networks (the best and the worst), the validation stage of these networks was carried out, i.e. their behaviour was examined for new input values that did not participate in the learning and testing stage of these networks. Table 3 shows the prediction of neural networks for 9 new values. This is the ultimate test of the effectiveness of neural networks. Out of 9 validation examples, network No. 9 made two mistakes, while network No. 16 correctly predicted all validation cases.

Table 3. Validation of neural networks.

Id	Prediction of the worst NN (No. 9) MLP 8-9-1	Prediction of the best NN (No. 16) MLP 8-29-1	Benchmark responses
1	tree_diagram_100	tree_diagram_100	tree_diagram_100
2	tree_diagram_100	tree_diagram_100	tree_diagram_100
3	tree_diagram_100	tree_diagram_100	tree_diagram_100
4	tree_diagram_100	tree_diagram_100; mind_map_100	tree_diagram_100; mind_map_100
5	no_clear_indication	no_clear_indication	no_clear_indication
6	tree_diagram_100	tree_diagram_100	tree_diagram_100

(*continued*)

Table 3. (*continued*)

Id	Prediction of the worst NN (No. 9) MLP 8-9-1	Prediction of the best NN (No. 16) MLP 8-29-1	Benchmark responses
7	tree_diagram_100	tree_diagram_100	tree_diagram_100
8	tree_diagram_100	tree_diagram_100	tree_diagram_100
9	tree_diagram_100	tree_diagram_100; fishbone_diagram_100	tree_diagram_100; fishbone_diagram_100

A neural network (feedforward multilayer with backward propagation of errors – MLP can be placed in an expert system, which can then support a less experienced employee in the selection of quality tools.

4 Conclusions

Quality management tools and techniques help employees analyze and solve problems, providing methodologies required to detect a problem, determine its originating cause, plan and develop solutions, verify their effectiveness, and monitor their implementation. The application of quality management tools and techniques improves communication between managers, engineers and operators as well as helps in understanding processes and detecting possible causes of their variability.

The correct use of quality management tools and techniques can bring many benefits to an organization. In addition to financial benefits (increasing a company's revenue, reducing costs and the number of complaints, etc.), there are many non-financial advantages that largely determine a company's position on the market and its further development. These are: increasing an organization's prestige, product quality, teamwork skills, employee commitment and responsibility [33].

The use of an ANN-based model to support the selection of quality tools is a very promising instrument to automate this task. A neural network (feedforward multilayer with backward propagation of errors – MLP can be placed in an expert system, which can then support a less experienced employee in the selection of quality tools (e.g. in the quality assurance department in an enterprise).

References

1. Oakland, J.: Leadership and policy deployment: the backbone of TQM. Total Qual. Manag. Bus. Excell. **22**(5), 517–534 (2011)
2. Hamrol, A.: Quality Management and Engineering. With a Look into Reality 4.0 (in Polish), PWN, Warszawa (2023)
3. ISO 9001:2015 — Quality management systems — Requirements
4. ISO 13485:2016 — Medical devices — Quality management systems — Requirements for regulatory purposes

5. ISO 13053–2:2011 — Quantitative methods in process improvement — Six Sigma — Part 2: Tools and techniques
6. McQuater, R.E., et al.: Using quality tools and techniques successfully. TQM Mag. **7**(6), 37–42 (1995)
7. Chen, S.H.: Integrated analysis of the performance of TQM tools and techniques: a case study in the Taiwanese motor industry. Int. J. Prod. Res. **51**(4), 1072–1083 (2013)
8. Castello, J., De Castro, R., Marimon, F.: Use of quality tools and techniques and their integration into ISO 9001: a wind power supply chain case. Int. J. Qual. Reliab. Manage. **37**(1), 68–89 (2020)
9. Bamford, D.R., Greatbanks, R.W.: The use of quality management tools and techniques: a study of application in everyday situations. Int. J. Qual. Reliab. Manage. **22**(4), 376–392 (2005)
10. Sharma, V., Grover, S., Sharma, S.K.: An integrated AHP-GTA approach for measuring effectiveness of quality tools and techniques. Int. J. Syst. Assur. Eng. Manage. **11**(1), 54–63 (2019)
11. Fonseca, L., Lima, V., Silva, M.: Utilization of quality tools: does sector and size matter? Int J. Qual. Res. (IJQR) **9**, 605–620 (2015)
12. Uddin, M.M.: Improving product quality and production yield in wood flooring manufacturing using basic quality tools. Int. J. Qual. Res. (IJQR) **15**(1), 155–170 (2020)
13. Abd-Elwahed, M.S., El-Baz, M.A.: Impact of implementation of total quality management: an assessment of the Saudi industry. S. Afr. J. Ind. Eng. **29**(1), 97–107 (2018)
14. Tarí, J.J., Sabater, V.: Quality tools and techniques: are they necessary for quality management? Int. J. Prod. Econ. **92**(3), 267–280 (2004)
15. Fotopoulos, C.B., Psomas, E.L.: The impact of "soft" and "hard" TQM elements on quality management results. Int. J. Qual. Reliab. Manage. **26**(2), 150–163 (2009)
16. Starzyńska, B., Hamrol, A.: Excellence toolbox: decision support system for quality tools and techniques selection and application. Total Qual. Manag. Bus. Excell. **24**(5–6), 577–595 (2013)
17. Sader, S., Husti, I., Daroczi, M.: A review of quality 4.0: Definitions, features, technologies, applications, and challenges. Total Qual. Manage. Bus. Excellence **33**(9–10), 1164–1182 (2022)
18. Turner, M., Oakland, J.: Defining Quality 4.0. Qual. World | Summer, 25–31 (2021)
19. Pongboonchai-Empl, T., Antony, J., Garza-Reyes, J. A., Komkowski, T., Tortorella, G.L.: Integration of industry 4.0 technologies into lean six sigma DMAIC: a systematic review. Prod. Plann. Control 1–26 (2023)
20. Santos, G., et al.: New needed quality management skills for quality managers 4.0. Sustainability **13**(11), 6149 (2021)
21. Tague, N.R.: The Quality Toolbox. ASQ Quality Press, Milwaukee (2004)
22. Hagemeyer, C., Gershenson, J.K., Johnson, D.M.: Classification and application of problem solving quality tools: a manufacturing case study. TQM Mag. **18**(5), 455–483 (2006)
23. Nedra, A., et al.: A new lean six sigma hybrid method based on the combination of PDCA and the DMAIC to improve process performance: application to clothing SME. Industria Textila **70**(5), 447–456 (2019)
24. Shahin, A., Arabzad, M.S., Ghorbani, M.: Proposing an integrated framework of seven basic and new quality management tools and techniques: a roadmap. Res. J. Int. Stud. **17**(17), 183–195 (2010)
25. Saifuddin, I., Rizal, H.S.: A study of quality tools and techniques in the context of industrial revolution 4.0 in Malaysia. What's new? Calitatea **21**(174), 88–96 (2020)
26. Hasan, M.S., et al.: Decision support system classification and its application in manufacturing sector: a review. Jurnal Teknologi **79**(1), 153–163 (2017)

27. Amran, M.M., et al.: Development of intelligent decision support system for selection of quality tools and techniques. Int. J. Mach. Learn. Comput. **9**(6), 893–898 (2019)
28. Starzyńska, B.: Practical applications of quality tools in Polish manufacturing companies. Organizacija **47**(3), 153–164 (2014)
29. Russell, S., Norvig, P.: Artificial Intelligence: A Modern Approach, 4th edn. Financial Times Prentice Hall, London (2019)
30. Rojek, I.: Classifier Models in Intelligent CAPP Systems. In: Cyran, K.A., Kozielski, S., Peters, J.F., Stańczyk, U., Wakulicz-Deja, A. (eds.) Man-Machine Interactions. Advances in Intelligent and Soft Computing, vol. 59, pp. 311–319. Springer, Berlin, Heidelberg (2009)
31. Tadeusiewicz, R., Chaki, R., Chaki, N.: Exploring Neural Networks with C#. CRC Press, Bo-ca Raton (2017)
32. Rojek, I., Dostatni, E., Hamrol, A.: Ecodesign of technological processes with the use of decision trees method. In: Pérez García, H., Alfonso-Cendón, J., Sánchez González, L., Quintián, H., Corchado, E. (eds.) SOCO/CISIS/ICEUTE -2017. AISC, vol. 649, pp. 318–327. Springer, Cham (2018). https://doi.org/10.1007/978-3-319-67180-2_31
33. Górski, F., et al.: Virtual reality training of hard and soft skills in production. In: Proceedings of the 23rd International ACM Conference on 3D Web Technology, pp. 1–2. ACM, Poznan (2018)

Information Design Management of Machining Parts on Metal Cutting Machines

Vira Shendryk[1](✉), Petro Pavlenko[2], and Justyna Trojanowska[3]

[1] Sumy State University, Sumy 40007, Ukraine
v.shendryk@cs.sumdu.edu.ua
[2] Zhejiang Ocean University, Zhoushan 316022, People's Republic of China
[3] Poznan University of Technology, 60965 Poznan, Poland

Abstract. This article deals with the problem of technological production preparation, as the task of choosing the optimal machining method on machine tools with numerical control. The task of finding the optimal solution is considered in the context of Industry 4.0 and 5.0 paradigms. The analysis of the possibilities for increasing the technological readiness of enterprises made it possible to identify qualitative and quantitative criteria for assessing the flexibility of production. Also determined the main compensation resource that allows you to provide adaptability. The optimality of solutions is determined based on their economic efficiency and cognitive complexity. The proposed approach allows organizing the process of forming all possible decision alternatives and performing their preliminary assessment according to several optimality criteria in the automated system of technological preparation of production. The possibility of interaction between information systems and decision-makers is shown in the information management of the design processes of the technology for machining parts on metal-cutting machines. The article presents the results of testing the proposed method, proving its effectiveness.

Keywords: Optimal Solution · Information Design Management · Decision Making

1 Introduction

From the beginning of the development of various types of production and their transformation into established industries, there is a constant transformation of the role of the human. The industry is the driver of economic and social relations development, starting with the first industrial revolution. The use of machines in production displaced humans from the role of a single productive force. At the same time, the problem of analyzing human-machine interaction arose for the first time. Subsequent industrial revolutions did not come close to solving this problem. The achievements of Industry 4.0, namely the development of automation of production processes, ensuring their implementation by productive cyber-technical systems, have increased production productivity [1].

In modern conditions, automated management systems perform information support of organizational and production processes of industrial enterprises of various types.

A. Burduk et al. (Eds.): ISPEM 2023, LNNS 790, pp. 146–158, 2023.
https://doi.org/10.1007/978-3-031-45021-1_11

Functions of information support for technological preparation of production are currently performed by various integrated information systems for production purposes, such as CAM systems, MES systems, and others. Their positive contribution is especially useful for enterprises with established nomenclature and production technology. At discrete-type enterprises, such automated systems provide automation of the processes of control, diagnosis, and regulation of the processes of machining of various parts on Machines with Computer Numerical Control (CNC).

At the same time, the dynamics of economic development, the competitiveness of enterprises, and limited resources require the ability to quickly and flexibly reconfigure technological processes, to respond accordingly to all changes in sales markets. Automation of management processes has removed humans from the management cycle, transferring them to the components of automated systems. At the same time, the possibility of the adaptive influence of people on production processes, and, accordingly, the possibility of flexible reconfiguration of technological processes, was lost. The complexity of automated systems has significantly increased the information pressure on people who manage processes and make decisions [2, 3].

At the same time, there are potential opportunities for improvement in matters of ensuring the necessary production productivity with existing resources and limitations. According to the concept promoted by Industry 5.0, there is an approach that provides benefits to industry, workers, and society. This approach puts the worker at the center of the production process by adopting a human-centric approach to digital technology. It empowers workers, promotes circular production models, and supports technologies that make resource use more efficient.

That is why this paper is aimed at solving the problem of increasing the productivity of mechanical processing of parts, within the limits of existing production resources and limitations. The key element in this approach should be the improvement of the professional work of the engineer-technologist and operator of the machine with CNC.

2 Literature review

Modern realities of production development form important strategic challenges, which consist of the ability to quickly introduce, adapt and manufacture a new product. The adaptability of production processes is one of the main pillars of Industry 4.0. The concept of the Industry 5.0 paradigm requires adaptability and flexibility not only from production and information systems but also from personnel involved in production [2].

In [2], several types of flexibility related to the context of production and production processes are defined, and their classification and priority are also given. From the point of view of the development of the concepts of Industry 4.0 and 5.0, it is interesting to consider some of the identified types of flexibility:

- Flexibility of machines - that is, the ability of a machine (industrial equipment, machine tools) in a specific production system to perform various types of operations without excessive effort to switch from one operation to another [4–11]. This flexibility is the most common in production systems and provides the basic structure for other types of flexibility [7]. This type of flexibility, while focusing primarily on the physical ability of the machine to handle a new operation, is also related to the capabilities of software and software applications (i.e., the cyber part of manufacturing systems). This type of flexibility can be measured by the number of different operations that can be performed without requiring additional resources for a major overhaul of the machine [8].

- Routing flexibility – this type of flexibility consists of the ability to produce a part by alternative routes in the system [4, 6, 12, 13]. Routing flexibility should be considered alternative routes that can be used in different machines or operations. Routing flexibility can increase the efficiency of part processing planning and allow for a better balance of machine loads. The effectiveness of this flexibility is affected by the supply of parts to machines or lines. Paper [14, 15] presents a construction methodology for a modern approach to tool selection and solving the problem of assigning optimal cutting parameters for specific production conditions. The authors obtained mathematical formulation determining the extreme values of the technological process optimality criteria. Quantitatively, this type of flexibility can be evaluated by such a criterion as the average number of possible ways in which a certain type of part can be processed in each system [14, 16, 17].
- Process flexibility is the flexibility that characterizes the ability to vary the range of types of parts (groups of products) that the system can produce without major adjustments [4, 6, 18]. This type of flexibility is also known as optionality and short-term flexibility. Such flexibility allows you to reduce the size of parts batches and inventory costs [19]. Realization of such flexibility is possible only due to operational planning. In [20, 21] the problem of sequencing jobs on a single machine to minimize total cost was considered. The authors applied constructive heuristics to find suitable solutions for the single-machine scheduling problem. The quantitative criterion that characterizes such flexibility is the size of the batch of parts that the system can produce without long-term reconfiguration [22, 23].
- Human polyvalence is a person's ability, as part of the production process, to perform unfamiliar actions or operations and to learn other actions in a minimum time. It should be noted that a person is able to adjust faster than other components of the production system, material processing technologies, or components of the information system. This type of flexibility can be considered compensatory for other types of flexibility. It is thanks to the quick adaptation of a person that it is possible to reduce production time and compensate for the inflexibility of other processes. This type of flexibility can be developed in the context of Industry 5.0, as the focus is more on a human-centered system, and therefore this type of flexibility becomes very relevant. However, there is some limitation which is in the cognitive aspects. Therefore, the underlying complexities of this type of flexibility should be explored more thoroughly. Efforts aimed at reducing the cognitive load (quantitative routines in processing and perceiving information, the duration of cognitive processes, and the decision-making process) can be considered one of the criteria for quantifying this flexibility.
- Interoperability of IT systems is the ability to exchange data and information with other information systems and/or decision support programs. This type of flexibility consists in organizing quick access to data and information for persons responsible for organizing production processes. The difficulty with this type of flexibility is organizing the ability to share IT program and/or system data with others, especially

presenting the information in an understandable, prepared form for decision-makers. This is a serious problem since the lack of common approaches to processing and sharing information can lead to data loss. With the loss of data, the collapse of information processes can occur. IT compatibility can be measured in terms of the time spent and the cost of efforts to prevent such threats.

Thus, summarizing the analysis of flexibility, according to [3], it can be stated that "flexibility is the system's ability to adapt and find solutions to certain and/or uncertain changes without large investments and in a relatively short time." Ensuring the flexibility of production processes becomes an opportunity to improve production. Because the flexibility of the production system can significantly speed up the search for effective solutions. The level of flexibility also strongly influences the development of the production environment and strategic goals. Moreover, considering the modern short life cycle of products, the increase in the variety of products, and the acceleration of technological processes, the flexibility of production becomes an important condition for competitiveness.

In order to adapt to the demands of the highly competitive global market, manufacturing companies need to rely on a new strategy that would maximize their profits, forcing manufacturing companies to be more flexible and proactive. Therefore, the concept of flexibility is directly related to the concept of economic efficiency, and the effectiveness of measures to increase flexibility can be measured quantitatively, accordingly, the task of increasing flexibility can be considered as an optimization task. At the same time, the introduction of new products, with greater functional diversity, better quality, and lower prices, form some basic customer expectations that generate several constraints and are uncertain. The possibilities of information support for production processes and information support for personnel in decision-making are also uncertain. The use of data and information, in addition to the possibility of reconfiguration (self-adjustment) of the production system, provides an opportunity to use human potential and increase the level of several types of flexibility, which allows you to meet specific customer requirements, respond to variations in demand, provide product personalization and efficiently use available resources more actively. In this context, making decisions related to the design of technological processes has become very difficult, because there is no method of justifying the decision in a specific context that is rapidly changing. Thus, reaction and decision-making time becomes shorter. In the context of personalized mass production, the need for flexibility and optimization of production processes with an emphasis on people-centricity is becoming more and more relevant.

3 Problem Statement

The conducted preliminary studies showed that despite the development and involvement of modern systems of automated control and regulation of the processes of machining of parts on metal-cutting machines, the tasks of optimization and management of the design processes of the technology of such processing remain poorly formalized and unresolved.

Therefore, the purpose of this study is the development and practical testing of a method of optimal management of the process of technological design of the technology of machining of parts, which ensures the necessary productivity of production with existing resources and limitations and at the same time reduces the information load on people who make decisions in the management of technological processes.

The object of the study is the process of managing the design of the optimal technology for the machining of parts on metal-cutting machines.

The subject of the research is models, methods, and tools for designing the technology of machining parts on metal cutting machines.

We will consider the management process's formalization and the management method's development in the example of the technological production preparation (TPP) of a machine-building enterprise. At the same time, it is necessary to consider a large number of factors and connections between them, which change both in space and in time. There is a task of choosing the most rational variant of the implementation of machining from a possible set according to the given quality criterion, i.e., solving the optimization problem. By optimization, we will understand the process of finding the extremum of some quantitative value (parameter) of the designed object, presented in the form of a function.

We will consider the optimal solution to be the best of a certain set of solutions that provides the given product quality characteristics in the most productive way at minimum costs, that is, the fulfillment of two main criteria – maximum production productivity and minimum cost (optimization of costs).

4 Materials and Methods

Modern metal cutting machines and mechanized technological complexes have a large number of various systems of adaptive regulation and control over the process of machining parts. During machining, a large number of various parameters of the processing process, temperature modes of cutting, vibration, etc. are taken into account. At the same time, the productivity of machining, its cost, and other characteristics of the efficiency of the production process depend, first of all, on those design solutions that are generated and adopted by the technological engineer at the stage of designing technological operations of machining of parts of any complexity. The quality of these decisions depends on the experience and intuition of the relevant specialists. Thus, the process of choosing solutions is highly subjective, depends on the human factor, and is not guaranteed to be error-free. In addition, the person making the decision has to process a large volume of data that characterizes the technological process.

The development of information technologies provides an opportunity to automate the process of choosing the optimal solution, reducing the burden on the decision-maker. It is possible to achieve this by transferring to an automated system of technological preparation of production the process of forming all possible decision alternatives and performing their preliminary assessment according to several criteria of optimality.

The concept of "optimality of the design solution" at the design stage is always relative. Considering the economic feasibility, as a rule, it is necessary to ensure the minimum cost with maximum productivity, given the conditions of the quality model and the limitations imposed by the existing resource capabilities. In this case, at the waste disposal stage, it is necessary to consider various input data, and the choice of the quality function to solve the optimization problems of technological design depends on the specific production problems. The general form of such an optimization problem, from the point of view of optimization of production costs, can be reflected by the integral function of solid waste in the form of minimization of product production costs [20]:

$$\sum_{n>>1}^{n} M_t^{ДОЛ} + \sum_{n>>1}^{n} З_t^{ДОЛ} + \sum_{n>>1}^{n} A_t^{ДОЛ} \to \min ,$$

where $M_t^{ДОЛ}$ is the share technological material capacity of manufacturing the *t*-th product for the *n*-th operation; $З_t^{ДОЛ}$– equity technological capacity for wages; $A_t^{ДОЛ}$– share technological amortization capacity of production of the *t*-th product for the *n*-th operation.

There is a large number of publications in which various options for solving such problems are offered [20]. However, in the practice of real technological design, they are usually not used. For use at the stage of technological design, it is necessary to have relatively simple mechanisms for the implementation of optimization tasks. Such a solution can be especially relevant when designing technological operations for multi-purpose processing centers. In this article, on the example of the design of the technology of machining body parts on multi-purpose processing centers, we will consider the solution to the task of optimal management of the process of technological design.

In general, the machine processing time of a part is described by the expression:

$$T_m = t_p + t_x, \tag{1}$$

where t_p is the working time of the tool; t_x is the idle time of the tool.

In turn,

$$T_p = L_p / S_{mp}, \tag{2}$$

$$t_x = \frac{L_x}{S_{Mx}} + t_M = \frac{l_p}{V_p} + \frac{l_x}{V_x} + t_з + t_{поз} \tag{3}$$

where L_p, L_x – the length of the trajectory of the tool movement during the working and idle strokes, respectively;

S_{Mp}, S_{MX} – minute feeds of working and idle movement of the tool (tool feed rate); $t_з$ – tool replacement time; $t_{поз}$ – positioning time of the table-satellite of the machine. When sequentially processing one part with several tools, we will get the total machine processing time:

$$t_M = \sum_{g=1}^{G}\sum_{r=1}^{R}\sum_{j=1}^{J}\sum_{i=1}^{I}\left(\frac{l_p}{V_p} + \frac{l_x}{V_x} + t_з + t_{поз}\right)_{grji}, \tag{4}$$

where l_p, l_x are the length of the trajectory of the tool during working and idling, respectively; V_p, V_x – speed of movement of the tool during working and idle strokes, respectively; $t_з$ – tool replacement time; $t_{поз}$ – time for positioning the table-satellite of the machine; i, I – index and number of working moves, respectively; j, J – index and number of idle runs, respectively; r, R – index and number of sides of the product, respectively; g, G – index and number of positions during product processing, respectively; $t_{\sum}$ – total processing time of the side of the part; S is the number of parts during one positioning.

It is known from the theory of machine productivity [21] that the real productivity of machining of a part for a continuous process is determined by the expression:

$$\Pi_{\Phi} = \frac{1}{T_M + T_X + T_{\Pi}},$$

where T_X is the average idle time; T_{Π} – is idle time.

That is, it is possible to control the performance of machining of the part by changing the components of the machining time of T_M processing, since the T_{Π} time cannot be controlled at the technology design stage, and the T_X time is taken into account in the designs of multipurpose machines.

At the same time, it is obvious that the machine processing time of the product t_M forms a linear homogeneous space $t \in t_M$ in the set of indices of working, idle runs, and technological transitions (positionings), which will be presented in the form of a system of equations by the corresponding indices i, b, l:

$$t_{P_{is}} = \sum_{i=1}^{I} \frac{l_{P_{is}}}{V_{P_{is}}}, \quad i = \overline{1, I}, \quad s = \overline{1, S};$$

$$t_{X_{js}} = \sum_{j=1}^{J} \frac{l_{X_{js}}}{V_{X_{js}}} + t_{з_{js}}, \quad j = \overline{1, J}; \tag{5}$$

$$t_{\Sigma_{rq}} = \sum_{r=1}^{R} \frac{l_{X_{rq}}}{V_{X_{rq}}} + t_{з_{rq}} + t_{поз_{rq}}, \quad r = \overline{1, R}, \quad g = \overline{1, G}, \tag{5}$$

where $t_{\sum}$ is the total processing time of the side of the part;

j, J – index, and a number of idle runs, respectively.

After combining expressions (4) and (5) and intermediate calculations, we obtain the sequence of partitioning the space t_M into subsets that create a σ-algebra in t_M such that

$$\frac{1}{\Pi_{\phi}} = \sum_{l=1}^{L} \sum_{b=1}^{B} \sum_{i=1}^{J} (t_M)_{lbi} \le t_M \le t_M^{з}, \tag{6}$$

where $t^{з}_M$ is the specified machine processing time of the part.

From the classical theory of optimal dynamic control, there are known options for ensuring the extreme value of Π_{ϕ}[22]. Let us denote the vector of changes in the time characteristic of the future planned technological operation $y = f1(t)$, and the vector of controlled parameters necessary for control as $U = f2(tp(x), Vp(x))$.

Taking into account the postulates of the theory of optimal dynamic control, we get a classic control problem. In this case, the management of the design process of the technology of processing parts on metal cutting machines can be reduced to finding the minimum functionality:

$$\Phi = f_3\{U(y), t_M[U(y)]\} \rightarrow \min, \tag{7}$$

where $y = f_1(t)$ is the vector of changes in the time characteristics of the technological process; $U = f_2(t_{p(x)}, V_{p(x)})$ is a vector of controlled parameters such as $t_{p(x)}$ – time, and $V_{p(x)}$ – speed of movement of the tool during working and idling, respectively.

It is clear from the expressions (4–7) that in order to obtain an optimal design solution, the technological engineer has the opportunity to adjust such components of the technological operation (positions, transition, etc.) $L_{p_{xj}}$, $S_{M_{pxj}}$ та $L_{x.x_i}$, $S_{M_{x.xi}}$. Within the planned T_K processing time.

Thus, the structural elements of the technological process, such as the operation, the processing position, the transition will be optimally designed only if the probability of this event is in a given range according to the classical theory [22].

To move from the general (7) to the specific statement of the problem, we will construct a generalized functional that would characterize the σ-algebra given by the system of Eqs. (4–6) and calculate the probability P_1^t of an event in which the technological process will be designed for the given σ-algebra in space t_M as, that processing time $t_M \leq t_M^з$. Thus, technological modes (parameters) t_p and t_x will be used, which will ensure the maximum value of probability P_1^t. The above description of actions is a formal description of the problem of optimal dynamic control [22].

For its practical solution, we first assume that P_1^t is probability that the parameters of the space t in some structure of the designed technological process are in the interval $\mu(t_1, t_M^з) = |t_1 - t_M^з|$. Then,

$$P_1^t = \frac{\mu(t_M^з, t_1)}{\mu(t_M^з, t_k)}, \tag{8}$$

where μ is the symbol of measure in space $t = \{t_p;\ t_x\}$, and $\mu(t_M^з, t_k) = t \leq t_M^з$; t_k, , – the time of machining of a part at a given technological operation (position or transition), which are calculated and specified during the design of the technological operation as a whole. During the next stage of technological design of the structure of the technological operation itself, the technological engineer uses all possible variants of the components $U = f_2(t_{p(x)}, V_{p(x)})$ to obtain a specific variant that is optimal in the given production conditions.

Based on the postulates of the theory of controlled random processes [23], it is possible to state that the technological operations of machining of parts of any complexity designed by the method described above will have the greatest efficiency only when:

$$h_i = 1 - P_i^t = 1 - \frac{\mu(t_M^з, t_i)}{\mu(t_M^з, t_k)}. \tag{9}$$

And for the general view, we have the following expression:

$$h_k = 1 - P_M^t = 1 - \frac{\mu(t_{k-1}, t_k)}{\mu(t_M^з, t_k)}. \tag{10}$$

The resulting sequence $\{h_{\mathrm{K}},\ k = \overline{1, \mathrm{K}}\}$ in this case is a measure that reflects the ultimate goal of the work of a technologist, namely the quality of the process of managing the design process of the machining technology of any part. In this way, it is possible to assert the achievement of the final goal of technological design due to the occurrence of the event of completion of the optimal design management process and the probability of this event P^3. Then, using the well-known theorem on the sum of probabilities [24], the generalized characteristic P^3, which describes the quality of the design process of the technology of machining of the part that is optimal in terms of performance, can be calculated by the formula:

$$P_n^з = h_1 + h_2 + \ldots + h_k + \sum_{f=1}^{k} (-1)^{f-1} \sum_{n_1 < n_2 < \ldots n_f}^{k} h_{n_1} \ldots h_{n_f}, \tag{11}$$

where $n = \overline{1, K};\ f = \overline{1, K}$.

Therefore, the functional (11) is a function of its arguments and can be considered as a criterion that characterizes the σ-algebra given on t by relations (4) and (5). Thus, a technological engineer, develops a structure of technological operations or a control program for a Machine with Computer Numerical Control, by changing the parameters l_p, l_x and V_p, V_x, has the opportunity to ensure the maximum value of the function $P^з$. The set of ratios (5), (6) and (11) determine the optimal model of technological design management for each specific case. There is a single solution of the obtained ratios [25, 26]. The search for this solution is carried out by sorting through variants of the values.

$L_{PX_j}, S_{M_{PX_j}}$ and $L_{X_i}, S_{M_{X_i}}$. By generating variants of the control parameters, we obtain the maximum value of the functional (11) and, thereby, the technological operation or other structural elements of the technological process that is optimal in terms of its performance. At the same time, various methods of solving the optimization problem (4), (5), and (7) can be used. The obtained part processing performance is optimal under the given limitations and real production conditions. Thus, the method presented in the article provides management of the process of optimal design of the technology of machining of parts on metal cutting machines and provides informational support for project decisions.

5 Results

For experimental testing and practical operation in real production conditions, the OPTIMUM computer program has been developed and tested, which performs all the necessary calculations and automates the management of the design process of the technology of machining of parts. The input of initial data $u = f_2(t_{P(X)}, V_{P(X)})$ and t_x, $t_{noз}$the number of tools, and the required number of variations are assumed. The program automatically calculates the machining processing time of the part t_M and the actual processing performance $\Pi_\phi$using the entered equipment parameters and the number of variations of the variable parameters. As an example, a process is considered in which three tools and two positionings are used to process one part, and the equipment parameters (lengths of tool movement on working and idle strokes, feed rate of working and idle tool movements) are selected with reference to a specific tool and multi-purpose machining center IR500MP4.

The calculated values must be compared with the value of the specified time of machining of the part from the point of view of fulfillment of the condition $t_M \leq t_M^{з}$, and in case of non-fulfillment of the condition, it is necessary to vary the parameters of the equipment and technological process. The results of the calculations are shown in Table 1 and on the diagram (Fig. 1).

When changing the number of tools and positioning, it is necessary to expand the fields of the existing table. The OPTIMUM software allows you to calculate the processing intensity and manage the design process of machining parts of any complexity.

The developed method of managing the optimal process of designing the technology of machining of parts and the OPTIMUM software was included as one of the main component subsystems in the automated system of technological preparation of production [27].

Table 1. Data for optimal design management of the technological operation of machining of the body part

Equipment parameters	Tool 1			Tool 2			Tool 3			Positioning time	
	1	2	3	1	2	3	1	2	3		
The length of the trajectory of movement during the working stroke, mm	40,00	40,00	40,00	25,00	25,00	25,00	10,00	10,00	10,00	20	1
	30,00	30,00	30,00	25,00	25,00	25,00	15,00	15,00	15,00	15	2
Speed of movement during the working stroke, mm/s	2,00	2,50	3,00	1,00	1,50	2,00	3,00	3,50	4,00	20	1
	2,00	2,50	3,00	1,00	1,50	2,00	3,00	3,50	4,00	15	2
The length of the trajectory of movement at idle, mm	40,00	40,00	40,00	25,00	25,00	25,00	10,00	10,00	10,00	20	1
	30,00	30,00	30,00	25,00	25,00	25,00	15,00	15,00	15,00	15	2
Speed of movement at idle, mm/s	5,00	6,00	7,00	10,00	11,00	12,00	5,00	6,00	7,00	20	1
	5,00	6,00	7,00	10,00	11,00	12,00	5,00	6,00	7,00	15	2
Tool replacement time, s	15,00	15,00	15,00	20,00	20,00	20,00	10,00	10,00	10,00	20	1
	15,00	15,00	15,00	20,00	20,00	20,00	10,00	10,00	10,00	15	2
Time of machining of a part with a tool, s	79,00	69,67	63,33	95,00	77,88	69,17	33,33	31,31	29,82		
Variations	1			2			3				
The total time of machining of the part, s	242,33			213,85			197,32				
Part processing productivity, 1/s	0,0041			0,0047			0,0051				
Increase in processing productivity in percent, %				13,32%			22,81%				

The system has successfully passed experimental and industrial operations and is used by the technological services of MotorSich PJSC (Zaporizhia, Ukraine).

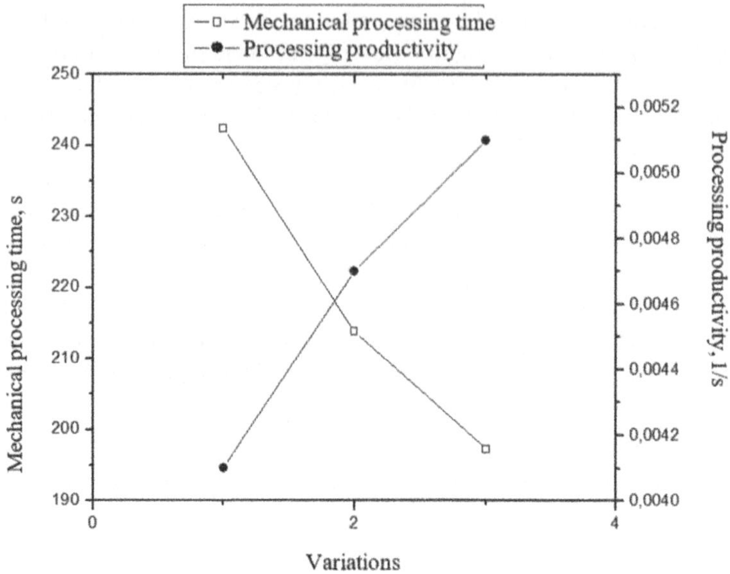

Fig. 1. Processing time and performance

6 Conclusion

The study considers the problem of optimizing the process of technological preparation of products in the context of the achievements of Industry 4.0 and 5.0. It is shown that the efficiency of dynamic development and competitiveness of the enterprise directly depends on the adaptability of production and the flexibility of technological processes. Different types of flexibility are considered. It was determined that in production processes, human polyvalence combined with IT interoperability creates a synergistic effect and can be a compensatory type of flexibility for other processes. It is also determined that transferring the generation and evaluation of possible variants of technological processes and processing routes to information systems significantly reduces the information pressure on the decision-maker and allows the development of algorithms for optimal dynamic management of the functions of technological preparation of production. Flexibility indicators were evaluated in quantitative characteristics such as time, resource consumption, and productivity. This made it possible to link these indicators and formalize the task of optimizing the process of technological preparation of production. Methods of determining the profitability of using the appropriate technological regime are also proposed, which allowed performing a preliminary assessment of the possibility of implementing specific technological operations. Generalized cost-effectiveness is a criterion that describes the quality of the design process of the optimal performance technology of machining of the part.

The method is quite simple from the point of view of numerical implementation and is not critical to the choice of the optimization procedure. The method is implemented in OPTIMUM software [27]. The corresponding software product was integrated into the automated system of technological preparation of production as one of the main constituent subsystems. Approbation of the method of optimal dynamic management of the functions of technological preparation of production, implemented in the lost application, was performed for the multi-purpose processing center IR500MP4. Experimental verification confirmed the success of the proposed method.

Prospects for further research in this direction lie in the development of theoretical and applied foundations of information management of human polyvalence, both at the level of a specific specialist and at the level of a team (for example, a project department [28]).

Acknowledgment. The research was partially supported by the Polish National Agency for Academic Exchange within the project "Strengthening the scientific cooperation of the Poznan University of Technology and Sumy State University in the field of mechanical engineering" (agreement no. BPI/UE/2022/8–00).

References

1. Ghobakhloo, M., Fathi, M., Iranmanesh, M., Maroufkhani, P., Morales, M. E.: Industry 4.0 ten years on: a bibliometric and systematic review of concepts, sustainability value drivers, and success determinants. J. Clean. Prod. **302**, 127052. Elsevier BV (2021)
2. Xu, X., Lu, Y., Vogel-Heuser, B., Wang, L.: Industry 4.0 and industry 5.0—inception, conception and perception. J. Manuf. Syst. **61**, 530–535. Elsevier BV (2021)
3. Chehami, A., Baboli, A., Einabadi, B., Ebrahimi, M., Rother E.: Determination and prioritization of flexibility types in the context of industry 4.0: a use case in automotive industry. In: 2022 IEEE International Conference on Industrial Engineering and Engineering Management (IEEM) Proceedings, pp. 1529–1535. IEEE, Kuala Lumpur, Malaysia (2022). https://doi.org/10.1109/IEEM55944.2022.9989757
4. Boyer, K.K., Leong, G.K.: Manufacturing flexibility at the plant level. Omega: Int. J. Manage. Sci. **24**(5), 495–510 (1996)
5. YounesSinaki, R., Sadeghi, A., Mosadegh, H., Almasarwah, N., Suer, G.: Cellular manufacturing design 1996–2021: a review and introduction to applications of industry 4.0. Int. J. Prod. Res. **61**, 5585–5636 (2022). https://doi.org/10.1080/00207543.2022.2105763
6. Sethi, A.K.: Flexibility in Manufacturing: A Survey. Kluwer Academic Publishers (1990)
7. Sokolov, O., Hosovsky, A., Ivanov, V., Pavlenko, I.: Movement monitoring system for a pneumatic muscle actuator. J. Eng. Sci. **10**(1), A1–A5 (2023). https://doi.org/10.21272/jes.2023.10(1)
8. Priyadarshini, J., Gupta, A.K.: Mapping and visualizing flexible manufacturing system in business and management: a systematic review and future agenda. J. Modell. Manage. (2023) https://doi.org/10.1108/JM2-02-2022-0035
9. Ranta, J.: The impact of electronics and information technology on the future trends and applications of CIM technologies **260**, 231–260 (1989)
10. Gan, Z.L., Musa, S.N., Yap, H.J.: A review of the high-mix, low-volume manufacturing industry. Appl. Sci. **13**(3), 1687 https://doi.org/10.3390/app13031687
11. Brill, P., Mandelbaum, M.: On Measures of Flexibility in Manufacturing Systems? (1987)
12. Renna, P., Materi, S., Ambrico, M.: Review of responsiveness and sustainable concepts in cellular manufacturing systems. Appl. Sci. **13**(2), 1125 (2023). https://doi.org/10.3390/app13021125

13. Stecke, K.E., Suri, R.: Proceedings of the second ORSA/TIMS (1986) Conference on Flexible Manufacturing Systems – Operations Research Models and Applications, Held at the University of Michigan, Ann Arbor, MI, USA, 12–15 August 1986, vol. 2. Elsevier Science Limited
14. Kegg, R.L.: Flexible manufacturing trends and data base requirements (1984)
15. Ivchenko, O., et al.: Method for an effective selection of tools and cutting conditions during precise turning of non-alloy quality steel C45. Materials **15**, 505 (2022). https://doi.org/10.3390/ma15020505
16. Chung, C.H., Chen, I.J.: A systematic assessment of the value of flexibility for an FMS. In: Proceedings of the Third ORSA/TIMS Conference on FMS, Cambridge, Massachusetts, pp. 27–34 (1989)
17. Buzacott, J.A., Mandelbaum, M.: Flexibility and productivity in manufacturing systems. In: Proceedings of the Annual IIE Conference, pp. 404–413 (1985)
18. Ranta J., Alabyan, A.K.: Interactive analysis of FMS productivity and flexibility (1988)
19. Ancelin, B.: Quels critères de performance pour les nouveaux ateliers? Rev. française Gest. Ind. **8**(1), 65–84 (1989)
20. Varela, M.L.R., et al.: Collaborative paradigm for single-machine scheduling under just-in-time principles: total holding-tardiness cost problem. Manage. Prod. Eng. Rev. **9**(1), 90–103 (2018). https://doi.org/10.24425/119404
21. Gerwin, D., Tarondeau, J.-C.: International comparisons of manufacturing flexibility. Manag. Int. Manuf. 169–185 (1989)
22. Mikhalevich, V.S., Volkovich, V.L., Voloshin A.F.: Algorithms for sequential analysis and variant screening in discrete optimization problems. Cybernetics **3**, 76–84 (1980)
23. Kuhnle, A., Kaiser, J.P., Theiß, F., et al.: Designing an adaptive production control system using reinforcement learning. J. Intell. Manuf. **32**, 855–876 (2021). https://doi.org/10.1007/s10845-020-01612-y
24. de Lacalle, L.N.L., Lamikiz, A.: Machine tools for high performance machining. Springer, London (2010). https://doi.org/10.1007/978-1-84800-380-4
25. Propoi, A.I.: Elements of the Theory of Optimal Discrete Processes. Nauka, Moscow (1973)
26. Gihman, I., Skorohod, A.: Controlled Stochastic Processes. Springer, New York, NY (2012) https://doi.org/10.1007/978-1-4612-6202-2
27. Pavlenko, P.M., Khlevny, A.O., Zaritskyi, O.V., Khlevna, Y.L., Tretyak, V.V.: Certificate No. 60620 on registration of copyright, 07/14/2015. Computer program Technology of automated management of project works of technical preparation of aviation and machine-building production ("U TPV"). Statement 05/14/2015, published 14.07.2015
28. Pavlenko, P., Zaritskyy, O., Khlevnyy, A.: Information technology for performance management of industrial production. Eastern-Eur. J. Enterp. Technol. **1**(2), 24–30 (2015)

Virtual Training at ELLA Platform – The Case Study

Ľuboslav Dulina[1(✉)], Beáta Furmannová[1], Mariana Máchová[1,2,3], Michaela Bodingerová[1], and Tomáš Michulek[2]

[1] Faculty of Mechanical Engineering, Department of Industrial Engineering, University of Žilina, Žilina, Slovak Republic
{luboslav.dulina,beata.furmannova,mariana.machova, michaela.bodingerova}@fstroj.uniza.sk
[2] Edgecom a. S, Žilina, Slovak Republic
tomas.michulek@edgecom.eu
[3] Institute of Competitiveness and Innovation, University of Žilina, Žilina, Slovak Republic

Abstract. The article describes the ELLA Platform as the result of research and development. The platform is used as a tool for creating virtual training. In the introductory part of the article, the ELLA Platform is presented as the result of our development. Ella is an integrated software ecosystem on which multiple software applications are built. These applications focus on the design and analysis e. g. of internal logistics, monitoring, and simulation of automotive logistics assets, and safety systems, etc. This article is focused on virtual training at the ELLA Platform. Furthermore, the article provides an example of virtual training that has already been carried out, and the ELLA Platform has been used in the preparation.

Keywords: Industrial engineering · Virtual training · Augmented reality · ELLA Platform

1 Introduction

The initial perception of training and development activities did not consider them as contributors to a company's value addition or competitiveness. However, there has been a gradual shift in this perspective today. Employing innovative training and development practices allows companies to achieve superior financial outcomes compared to their non-adopting counterparts. Moreover, these practices aid in the societal development of essential human resources required to meet competitive challenges.

Presently, numerous companies endorse the belief that learning through training, development, and knowledge management empowers employees to enhance and expand their skill sets. This, in turn, enables the improvement and creation of novel products, innovative processes, and top-notch customer service. Concurrently, fostering development activities and career management ensures that employees are prepared for potential managerial and leadership positions, thereby motivating them to stay with the company across all hierarchical levels. The focus on education through training, development, and

A. Burduk et al. (Eds.): ISPEM 2023, LNNS 790, pp. 159–173, 2023.
https://doi.org/10.1007/978-3-031-45021-1_12

knowledge management has transcended the realm of mere bonuses; it is now considered indispensable for gaining a competitive advantage and fulfilling employee expectations [8].

As businesses strive to retain competitiveness in the global market amid rising demand for skilled labour, they face the challenge of incorporating effective training methods. Technologies such as social media, tablets, and computers have alleviated certain costs associated with staff management, training, and onboarding. However, the key challenges lie in ensuring [1]:

- The inclusion of crucial elements such as practice, feedback, variable pace, etc., in these training methods.
- Implementing a blended learning approach that balances self-learning technologies with methods fostering interpersonal interaction.
- Comprehensive coverage of all aspects relevant to future staff assignments within the training process.

2 Possibilities of a Virtual Reality in Education

Teaching methods aimed at imparting complex knowledge and skills often draw upon managed social constructivist education theories. These theories emphasize personalized and realistic learning experiences that revolve around mastering authentic tasks. Immersive technologies present an opportunity to enhance such educational approaches by fully engrossing the learner in a virtual environment. By leveraging technologies like headsets, these immersive experiences can generate richer sensory stimuli, thus deepening the psychological immersion and evoking a sense of virtual presence, i.e., the illusion of being present in that virtual place [2].

Virtual reality (VR) facilitates sensory immersion by combining audio-visual cues with haptic elements. Users can interact with the virtual environment, while the VR system seamlessly responds to their movements, preserving the illusion of presence. While VR learning offers significant advantages, it also comes with certain drawbacks. Therefore, understanding the fundamental principles and appropriately evaluating the suitability of this technology's application is essential to align it with the requirements of the educational process. When implemented effectively, virtual reality can successfully translate training into practical skills and knowledge [9].

3 ELLA Platform for Creating Virtual Training

The ELLA Platform was developed through an in-house research initiative as an integrated software ecosystem, serving as the foundation for multiple software applications. These applications are primarily focused on the design, analysis, monitoring, and simulation of internal logistics in the automotive industry, including virtual training capabilities. The platform enables individual solutions to exchange data on virtual worlds and objects, as well as to process real-time information about industrial operations. This integration facilitates the delivery of unique solutions to users (customers) with enhanced added value.

The development of the platform involved the utilization of various technologies, including C++, Python, OpenGL, PhysX, and others. A significant portion of the system logic is scripted in Python using a visual programming module (Figs. 1 and 2).

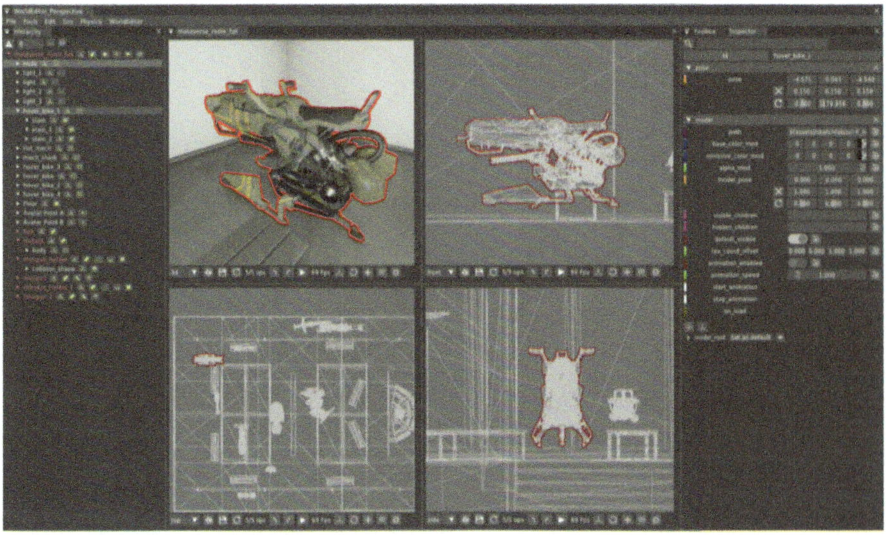

Fig. 1. Example of work in the ELLA Platform.

```
namespace ella
{
    namespace userInterface
    {
        class CameraViewPolicy;
        class CameraViewPolicy_Ortho;
        class CameraViewPolicy_Perspective;

        class EllaToolsExport Camera : public PtrBase
        {
            //==============Default methods start==================================================
        public:

            static Sptr<Camera> New();
            static Sptr<Camera> Alloc(const String & typeName = "");
            static Sptr<Camera> ConstructObject();
            ~Camera() override;

        protected:

            Camera();
            void Construct();

        public:

            static void FillClassDefinition(const Sptr<ClassDefinition> & def);
            void Serialize(const Sptr<serialization::OutputArchive> & ar) override;
            void Serialize(const Sptr<serialization::InputArchive> & ar) override;
            //===============Default methods end===================================================
        private:

            Double44 _pose;
            Sptr<CameraViewPolicy> _viewPolicy;

            Sptr<utilities::DelegateSignal> _onSendEvent;
```

Fig. 2. Scripting tree in the ELLA Platform.

Ella encompasses a comprehensive suite of tools intended for the development of industrial applications and software targeting the domains a simulation and emulation of logistics processes, design of enterprises adhering to the principles of Industry 4.0, data collection, analysis, and real-time evaluation, control and navigation of logistical assets, creation of specialized training levels within virtual and augmented reality environments, server solutions optimized for non-graphical support, 3D web interfaces.

A visual depiction of the developed ELLA core, along with its associated tools and products, is presented in Fig. 3.

Fig. 3. The scheme of the ELLA Platform.

Hence, the ELLA Platform serves as the foundational framework for constructing virtual training programs. It has been meticulously developed and designed to enable efficient and straightforward programming using a visual programming approach. This means that all applications and software solutions can be programmed within the ELLA Platform utilizing its integrated visual programming tool. Such visual programming empowers users to build diverse systems, ranging from real-time vehicle control to intricate virtual world simulations, leading to enhanced productivity when working with the platform (Fig. 4).

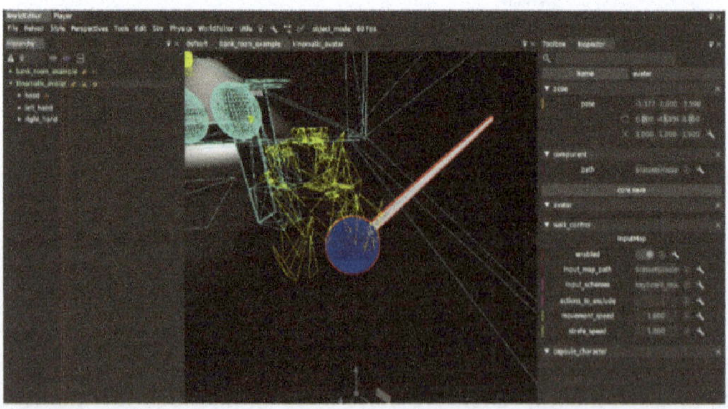

Fig. 4. Example of visual programming in the ELLA Platform environment.

This software platform facilitates the seamless integration of a wide array of modules within a virtual reality environment. This compatibility is mainly supported by the following features: full compatibility with Windows operating systems, support for modules created in the C++ language, graphical core that supports OpenGL 3.3 and above, built-in support for module serialization a 3D user interface and support for multi-monitor systems and touch control.

The ELLA Platform offers the flexibility to create training programs in both a simpler 2D variant and a fully immersive virtual reality variant. These variants are accompanied by different minimum software configurations, ensuring the smooth creation and execution of virtual training sessions. These configurations are outlined in Table 1.

Table 1. Proposed configurations.

	2D variant of virtual training	VR variant of virtual training
Processor	i5 7300	i7 7700
Memory	8GB	8GB
Graphic Card	NVIDIA GTX	NVIDIA GTX 1080
Virtual Glasses	–	HTC Vive Pro

The versatility of use and user specific requirements are also supported by the wide variability of the operating system and programming languages shown in Fig. 5.

4 System Toolkit Based on ELLA Platform

The Virtual Training System Toolkit constitutes a pivotal component within the ELLA Platform, specifically designed for the development of virtual training systems. This toolkit comprises a comprehensive collection of interfaces and components that facilitate the efficient creation and interaction with the virtual world.

Fig. 5. ELLA Platform – support of systems and languages.

Hence, the Virtual Training System Toolkit empowers users to engage in training sessions within a virtual reality environment. Its key constituents include the following main elements, which together enable standardized rendering of the virtual scene, seamless navigation within the virtual environment, and interaction with the elements present within the environment.

4.1 Ensuring the Best Performance

The described module serves as a visualization tool capable of rendering complex design data, rectifying frame rate fluctuations (jumps and decreases in FPS). Additionally, it automatically adjusts the level of detail in case the standard 90 FPS is compromised, ensuring smooth virtual reality performance.

The module responsible for movement within the virtual environment enables users to navigate through a virtual space that exceeds the standard dimensions for virtual glasses. The movement is facilitated through a teleport system, and the module restricts movement to predefined zones, thereby managing permitted transport destinations.

The controllers' interaction module facilitates user interactions with virtual objects, allowing for seamless engagement and manipulation.

Moreover, a performance optimization module is integrated into the Virtual Training System Toolkit, enhancing virtual reality performance on newer series of graphics chips.

This toolkit also comprises a variety of components, such as Avatar 2D/3D, inventory management, interactive zones for object capturing and interaction within the virtual world, teleportation zones, an assembly toolkit featuring screws, diagrams (Virtual Training Diagrams), and materials. It includes support for various language mutations.

5 ELLA Platform – The Use in Education

The creation of virtual training consists of five distinct steps and an illustration of its implementation for training with a robotic system is depicted in Fig. 6.

Part no. 1 – Input data collection

- 3D laser scanning to capture the physical environment;
- Gathering production process information, factory layout and equipment data;
- Incorporating safty-related rules and guidelines;
- Utilizing existing 3D models of the factory or equipment, if available.

Part no. 2 – 3D Modelling

- Importing 3D models from various formats (stp, jt, fbx…);
- Creating 3D models based on photo documentation and drawings;
- Constructing 3D models of robots, accessories, control panels and environment;
- Preparing and optimizing CAD data using tools like PIXYZ;
- Applying custom materials to the models.

Part no. 3 – Inverse kinematics

- Establishing connection for serial chain of rigid links in the robotic system;
- Solving the inverse kinematics problem using closed-form solutions for robots;

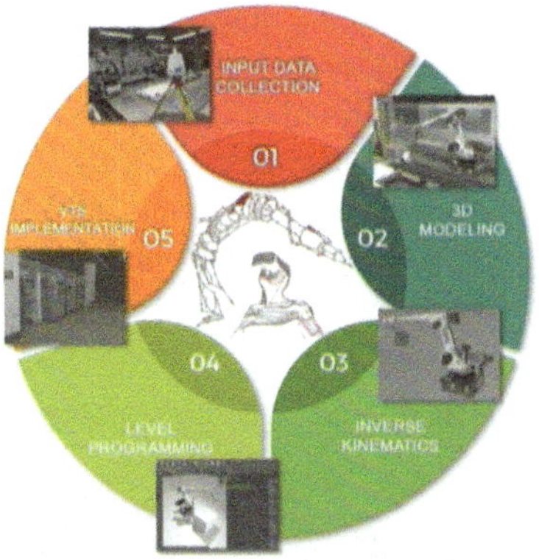

Fig. 6. ELLA Platform.- education circle.

- Accommodating inverse kinematics calculations for 3 to 7 axis robotic systems.

Part no. 4 – Level programming

- Creating a virtual factory model that represents the training environment;
- Developing a custom editor module to facilitate efficient content creation;
- Designing customer-specific training modules tailored to their needs.

Part no. 5 – Virtual training system implementation

- Implementing user management and statistics tracking functionalities;
- Incorporating virtual charts to visualize training progress and performance;
- Integrating a learning management system to facilitate effective training organization and administration.

6 Current Applications of Virtual Training in the Industry

The current rapid development and gradual decrease in virtual reality's investment costs help implementation in various industries. Businesses use the ability of virtual reality to faithfully simulate the real process and situations that are dangerous and difficult to imitate. For this reason, it creates a suitable environment for safe and effective training. Many companies from different industries have adopted virtual training as part of their corporate education system.

In the current market, some companies offer virtual training solutions as a product for other businesses. One example is training solutions based on ELLA Platform. The training solutions offered include [4]:

- Virtual maintenance-oriented training;
- Virtual training focused on assembly training and sequencing;
- Virtual training oriented to quality control;
- Logistics-oriented virtual training.

Virtual maintenance-oriented training allows the training of employees to master maintenance operations of industrial robots, CNC and other machines. Training guides the user through all necessary actions, monitors his progress and warns of errors. Figure 7 shows a sample of the robotic cell maintenance process. The training uses an entirely virtual environment with a high degree of interaction and a high level of immersion and VR headset with the necessary sensors and controllers for smooth movement. It uses a teleport as a type of different movement.

Fig. 7. Edgecom Virtual Maintenance Training Solution.

One of the other solutions offered is virtual training of the assembly process. Complex assembly tasks are suitable for virtual reality training. Repeated virtual reality training streamlines employee performance and prevents costly errors that could affect production. The employee is introduced to the assembly process step by step. This virtual process corresponds to its real copy using the same tools and movements. Training provides the necessary instructions for each operation [3].

There is also a great interest in virtual training in the field of internal logistics. This type of training makes it possible to use the virtual environment and the faithful functional virtual model of the selected means of transport with identical control elements and driving characteristics of a real means of transport. (see Fig. 8).

Fig. 8. Virtual assembly training solution [3].

Logistics training allows you to set different levels. Specific events in the level can be activated or deactivated for specific training. For example, if a worker will pass a component interfering with the path of a forklift, this component can be activated, deactivated, or moved to another level. As a result, the trained employee will not know if and where this attention testing event will occur. These random events are specified during the process of collecting data on the real situation in the workplace. These can be falling crates, obstructions in the road, another intersection for forklift trucks, etc.

7 The Case Study – Virtual Trainer for Painting Car Body Part

The Ella software platform was also used to create a virtual training program for painting a car body part for a passenger car manufacturer in the Czech Republic. For the needs of the specific training, it was necessary to take into account all aspects of the painting work activities so that the training replaces the lengthy and inefficient training of painting on real body parts directly in the paint shop. For this it was necessary:

- To use the same spray gun for the virtual training as is used in the real operation;
- To create a virtual paint shop environment (software);
- Create a virtual painting process with all the parameters of the painted surface to match the real process (software).

Hardware Platform

Based on the company's request, a real spray gun was modified into a "virtual spray gun" for the needs of virtual painting. A SATAjet 4000 B HVLP spray gun was used, which was interfaced with sensors (see Fig. 9). Based on the defined spraying parameters, the following was selected:

- Spray and flow control sensing (circular or flat);
- Material flow sensing;
- Trigger sensing (the main control for the spray gun);
- Sensing of paint flow rotation and the air pressure in the spray gun.

Fig. 9. Spray gun for a virtual training [3].

The spray gun also included a sensor for sensing external devices for HTC Vive virtual glasses powered by a battery with a minimum of 6 h, which is fully sufficient for the needs of training. The hardware platform thus defined was called the Virtual Training Systems Hardware Platform Spray Gun (VTSHWSG).

Software Platform

Using the ELLA Platform, a software module called the Virtual Training System Painting Module (VTSPAINT) was developed for painting. This module contains a component that can be wrapped to any mash and thus simulate painting equipment with the following functions:

- Interfacing with real hardware devices (a spray gun);
- Physical simulation of painting processes;
- Module usable for both 2D and 3D imaging technologies;
- Connection to virtual reality;
- Output statistics in the form of 2D and 3D graphs.

The virtual training system includes virtual training modules created on the ELLA Platform, whose structure and characteristics are presented in Table 2.

Implementation of the Virtual Training System Level

As part of the implementation, the company needed to deliver a comprehensive level of virtual painting solution that included 3 modules: learning, training, testing.

The prerequisite for completing the training was that the workers already had basic knowledge of painting, i.e., they were qualified. The aim of the painting training was to improve skills and experience. Thus, the learning focused on familiarization with the virtual system, especially on the operation of a particular spray gun, and consisted of the following areas:

- Theoretical knowledge of the manual painting process;
- A virtual tour of the spray gun with a detailed description;
- Virtual breakdown of the components of the spray gun;
- Virtual description of the main settings of the spray gun.

In addition, this module focused on learning how to spray properly with the SATAJet spray gun in the next step. The learning module further included:

- Proper spray gun holding;
- The correct setting of spraying parameters:
  - Circular/flat spray control;
  - Flow control and material flow control;
  - Trigger control – main control for the spray gun;
  - Paint flow rotation control;
  - Spray gun air pressure control;
- Maintaining distance of the spray gun from the object to be sprayed with vibration;
- Correct application and layering of paint.

After familiarization with the theoretical knowledge, the training module followed. This module focused on training the knowledge acquired in the learning module. In this module, the trainee was asked to spray three types of objects:

- Simple flat sheet;
- A more complicated sheet metal shape;
- A complicated sheet with holes and folds.

This module was finally evaluated using statistics that will take into account:

- Spraying time;
- Correct parameter settings;

Table 2. Modules on the ELLA Platform created for the painting training program.

Software module name			Software module description
ESC	ELLASC	ELLA Platform System Core	A software platform module that allows a wide range of modules to coexist in a virtual reality-based environment: • Compatibility with Windows OS 8, 10 and later; • Compatibility with modules created in C++ languages; • Graphics core supporting Open GL 3.3 and higher; • Integrated support for module serialization; • 3D user interface; • Support for multi-monitor systems and touch control.
ELC	VTSLCM	Virtual Training System Level Control Module	User interface for managing the levels of the virtual simulator allowing: • Control of individual levels (start, level, level stop, level restart); • Editing the theoretical slides of each level; • Editing of video sequences of each level.
ESM	VTSSTM	Virtual Training System Statistic Module	User interface for managing virtual trainer level statistics: • Evaluation of time levels (tracking of final training time, tracking of penalty time, tracking of real training time, comparison of expected vs. real training time); • Database of penalties of achieved levels; • Graphical evaluation of results; • Level evaluation spreadsheet; • Export of statistics to xlsx files; • Tracking and evaluation of level subtasks.

(continued)

Table 2. (*continued*)

 	VTSUMM	Virtual Training System User Management Module	User interface for managing users and accounts of the virtual simulator: • Database using structured search and query language SQL; • Database with administrator accounts (trainers); • Database of trainees; • Creation of new training users; • Editing of existing training users; • Tracking basic information
 	VTSVRM	Virtual Trainer System Virtual Reality Module	A module that allows you to view training sessions in a VR environment. It contains the main parts that will allow standardized rendering of scenes in VR, movement around the virtual environment and interaction with the environment. • Dynamic Quality Management; • Module for movement in the virtual environment; • A module for interaction with controllers; • A module for performance optimization in VR.
 	VTSPAINT	Virtual Training System Painting Module	Module designed for physical simulation of painted devices. This module contains a component that can be wrapped to any mash, thus simulating a paint device with the following functions: • Interfacing with real hardware devices; • Physical simulation of painting processes; • Module usable for both 2D and 3D imaging technologies; • Connection to VR; • Output statistics in the form of 2D and 3D graphs.

- Adherence to spraying distances;
- Layering.

These statistics were supported by a 3D graphical map using visualization using intensities (see Fig. 10). After the implementation of the training, it was possible to proceed to testing in the virtual space: (see Fig. 11). This module was aimed at training the painting process using a SATAjet spray gun in real plant conditions.

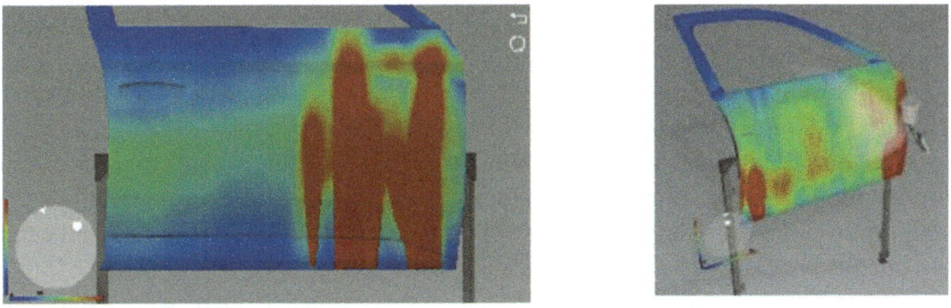

Fig. 10. 3D graphical map using intensity visualization [4].

In the testing, the trainer or the system, using automatic generation, selected one real component from the database, which had to be painted correctly by the trainee based on the levels of learning and training passed by the trainee. The component consisted of a real car model produced in the enterprise. This module also included a digitized real workplace of the enterprise and its accessories. Complete training statistics were also evaluated within this level. The activities selected for training were:

- Which took place in a real paint cell area;
- Whose task was to paint three different shaped components;
- With a level without auxiliary indicators;
- With statistical evaluation of the manual painting process.

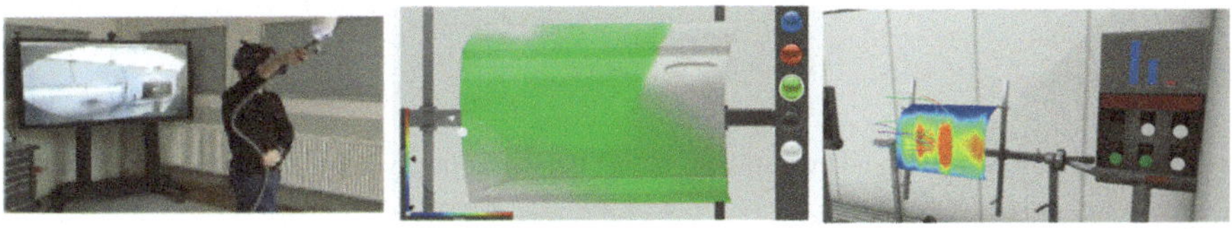

Fig. 11. Virtual space testing and 3D graphical map.

8 Conclusion

The ELLA Platform serves as a comprehensive framework for integrating software and hardware components essential for virtual training. In education, virtual training operates at two primary levels. The first level involves the hardware and software aspects, while the second level pertains to the design and implementation of the education scenarios. These two levels are interlinked, and their successful integration relies on principles from didactics, psychology, and sociology.

The hardware and software sector in virtual training is continuously evolving and challenging to predict for the near future. In contrast, the field of education scenarios is dependent on the development of industry-driven work activities, such as Industry 4.0 and Industry 5.0, and the current demands of enterprises for their employees' knowledge, skills, and competencies. The convergence of these two levels is a subject of ongoing research aimed at developing procedures that facilitate an efficient and effective combination of hardware-software security and education scenarios.

Virtual training implemented with the ELLA Platform has been primarily focused on industrial companies, particularly within automotive production, logistics, and EHS (Environment, Health, and Safety) domains. While tailoring this training to meet specific customer requirements, the creators also took into account additional considerations such as employee turnover, safety and ergonomics applications (both in training and actual work conditions), and industrial engineering principles.

Developing such solutions would be more streamlined if there were established procedures that apply to various forms of training, along with guidelines on creating training in different domains. These procedures would constitute a methodology that harmonizes the two education levels mentioned earlier. The authors of the research are currently working towards the creation of such a methodology as their primary objective. VR in industry offers opportunities to decrease design and production costs, maintain product quality and reduce the time needed to go from product concept to production.

Due to its interactive nature, VR is also used for simulations of training sessions to communicate knowledge on assembly and maintenance processes. So far, there are hardly any scientific findings on the effectiveness of virtual learning and training for industrial applications. It is also unclear how training can be improved in virtual environments to maximize the learning outcome [10].

Acknowledgment. This work was supported by the KEGA Agency under the contract no. 032ŽU-4/2021.

References

1. Allen, W.C.: Overview and evolution of the ADDIE training system. Adv. Dev. Hum. Resour. **8**(4), 430–441 (2006)
2. Bailenson, J., et al.: The use of immersive virtual reality in the learning sciences: digital transformations of teachers, students, and social context. J. Learn. Sci. **17**(1), 102–141 (2008)
3. EDGECOM Homepage. Maintenance focused virtual training. https://www.edgecom.eu/virtual-training-systems. Accessed 05 Mar 2020
4. Liu, D., Dede, C., Huang, R., Richards, J.: Virtual, Augmented, and Mixed Realities in Education, 1st edn. Springer, Singapore (2017)
5. Gola, A., Plinta, D., Grznar, P.: Modelling and simulation of reconfigurable manufacturing system for machining of casing-class parts. In: Malinovska, L., Osadcuks, V. (eds.) 20th International Scientific Conference Engineering for Rural Development, pp. 1563–1568. Latvia University of Life Sciences and Technologies, Jelgava (2021)

6. Grznar, P., Gregor, M., Gola, A., Nielsen, I., Mozol, S., Seliga, V.: Quick workplace analysis using simulation. Int. J. Simul. Model. **21**(3), 465–476 (2022)
7. Noe, R.A., Kodwani, A.D.: Employee Training and Development, 5th edn. McGraw-Hill Education, New York (2010)
8. Noe, R.A., Kodwani, A.D.: Employee Training and Development, 7th edn. McGraw-Hill Education, New York (2017)
9. Rossou, M.: A VR playground for learning abstract mathematics concepts. IEEE Comput. Graph. Appl. **2**(5), 82–85 (2009)
10. Wolfartsberger, J., Zimmermann, R., Obermeier, G., Niedermayr, D.: Analyzing the potential of virtual reality-supported training for industrial assembly tasks. In: Grabot. B. (ed.) Computers in Industry, vol. 147, p. 10. Elsevier, The Netherlands (2023)

Web Application for the Learning of Emotions in Children with Autism Spectrum Disorder

Christophe Magalhães[1], João Sena Esteves[1,2], Ana Paula Pereira[3](✉), Vinícius Silva[2], and Filomena Soares[1,2]

[1] Department of Industrial Electronics, School of Engineering, University of Minho, Guimarães, Portugal
a73770@alunos.uminho.pt, {sena,fsoares}@dei.uminho.pt
[2] R&D Centre Algoritmi, Guimarães, Portugal
a65312@alunos.uminho.pt
[3] Department of Psychology of Education and Special Education, CIEC, University of Minho, Braga, Portugal
appereira@ie.uminho.pt

Abstract. Individuals with Autism Spectrum Disorder (ASD) tend to have a difficulty in interpreting and even expressing basic emotions such as joy or sadness, difficulties that can be mitigated with the support of tools that help those children understand and demonstrate emotions. The main goal of the presented work is to promote the learning of emotions in children with ASD, in a playful, didactic and simple way, through the use of a developed web application. Several games with different objectives were developed in order to optimize the acquisition of specific skills. The application gathers these games and makes them accessible in different contexts. A pilot study with seven children aged between six and eight years, three with ASD and four with typical development, was performed to validate the developed games, by analysing their suitability for the target audience. The results obtained allowed to verify that all players, with and without ASD, showed positive evolutions throughout all sessions, as they showed higher levels of interpretation regarding the game requests and completed the game with fewer errors and in less time.

Keywords: Autism Spectrum Disorder (ASD) · Children · Emotions · Web Programming

1 Introduction

Emotion recognition promotes a child's social communication development and learning. However, not all people find it easy to communicate through gestures and emotions. A case in point are individuals with Autism Spectrum Disorder (ASD), who tend to have a difficulty in interpreting and even expressing basic emotions such as joy or sadness. These difficulties can be mitigated with the support of tools that help those children understand and demonstrate emotions. This paper concerns the development and application of interactive technologies to support and promote new adaptive teaching-learning

A. Burduk et al. (Eds.): ISPEM 2023, LNNS 790, pp. 174–187, 2023.
https://doi.org/10.1007/978-3-031-45021-1_13

approaches for children with ASD. Serious games are more intended for educational purposes than entertainment. In this case, the focus is on helping users with ASD in the learning of emotions. They may be a useful tool to help developing social-emotional skills in children with ASD.

1.1 Affective Computing

The ability to recognize and express affection is one of the most important characteristics of the human being. Affective computing addresses the ability of computers to process behavioural cues and to interact with human users in a way that is similar to human interaction. Without the ability to process this information, one cannot expect a computer to be able to interact with humans in a harmonious way. Affective computing concerns multidisciplinary knowledge of areas such as psychology, physiology and computer science [1].

Systems and devices that can recognize, interpret, process and simulate human behavioural cues have been in development for a considerable time. The recognition of emotional information requires extracting meaningful patterns from the obtained data. To perform an automatic analysis of facial expressions, it is pertinent to divide this process into three stages, namely the detection and face tracking, feature extraction (such as AUs), and classification of expressions (usually the six basic expressions described by Paul Ekman) [2].

1.2 Emotions and ASD

ASD is characterized as a neurological disorder associated with impairment of communication and social interaction and restrictive and repetitive behaviour. Individuals with ASD tend to have greater difficulty in interpreting and even expressing basic emotions, such as happiness or sadness. It is difficult to carry out an analysis based on the look, which complicates the interpretation of what the someone else is trying to express. However, it is possible to mitigate these difficulties these obstacles teaching these people to express and interpret emotions [3].

There is a big difference in the social communication development of children with ASD compared to typically developing children. Different approaches, methodologies and tools are needed to help in the intervention process with children with ASD [4–6]. More recently, the use of affective computing has been explored in order to facilitate the processes of learning, and this work aims to help in this process, in a didactic way.

1.3 Serious Games and ASD

With the constant technological evolution, new tools and new approaches have been proposed in the teaching-learning process. Serious games are included in the special category of games, with specific objectives and purposes, in which the player uses his knowledge to solve problems or to know new problems and practice tasks.

Serious games are applied to simulate critical situations that involve some kind of risk, taking decisions or even to develop specific skills, which is the fundamental point of this dissertation, which focuses more specifically on the development of the ability to recognize emotions. For teaching-learning purposes, situations can be simulated in which the use of a knowledge is necessary for the evolution in the game. In some cases, teaching and training may be combined to simulate situations where you learn something to use in everyday life.

The purpose of these games, when aimed at teaching-learning, can be divided into three categories: awareness, knowledge building, and training [7].

With the constant evolution of technology, the gaming area has also gained from this and, in this way, it is possible to explore areas that were previously of little relevance. More and more serious games are used to help with everyday problems, and new methods are explored to optimize learning as can be seen in Fig. 1 [8].

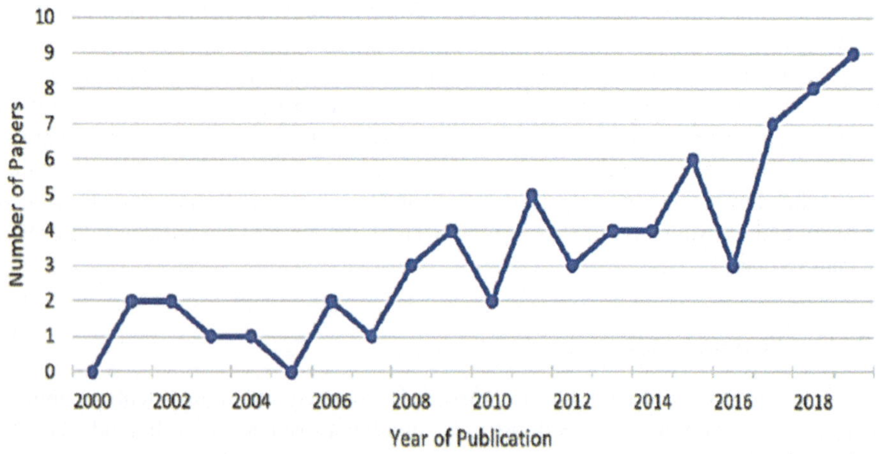

Fig. 1. Total Number of Papers that Presented a Serious Game for PEA, Intervention or Evaluation in Each Year [8].

In the period between 2019 and 2021, 12 articles were published about interventions with serious games and entertainment games in ASD [9].

Madsen [10] developed an interface called Emotion Bubbles, in which each displayed emotion has a particular colour. Positive emotions were assigned colours with a good feeling, such as green, blue and purple, indicating a productive state, and negative emotions were assigned warm colours, such as red, orange and yellow, indicating that the user must be aware of a possible impediment in the course of the conversation. Figure 2 shows an example of the face analysis system and the bubble interface of the emotion detected [10].

The main goal of the work presented in the following sections is to promote the learning of emotions in children with ASD, in a playful, didactic and simple way, through the use of a web application gathering several games developed with different objectives, aimed at optimizing the acquisition of specific skills. The application makes the games accessible in different contexts.

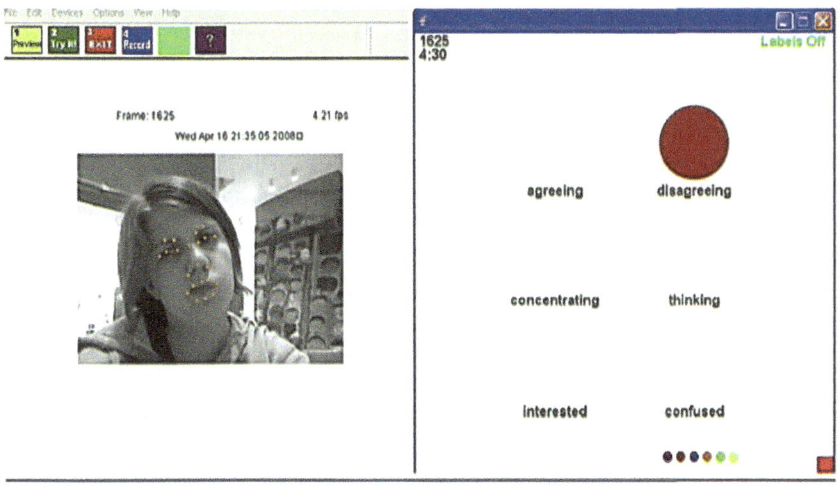

Fig. 2. Example of the face analysis system (on the left) and the developed Emotion Bubbles interface (on the right) [10].

2 Games and Application Development

The JavaScript programming language was used to develop a web application. It was used in front-end programing, which is responsible for the "visual" part of the application. Two early languages, HTML and CSS, were also used. Furthermore, it was possible to create a complete web application using Node.Js along with a library that helps in the task of creating a web server, and through the use of some front-end frameworks and a database.

As the games were developed for children with ASD, special care had to be taken not to make it too difficult or too confusing for the users, simplifying the graphical interface presented. In all games, several images are presented in the Help menu, with the aim of helping a player to assimilate even better the different emotions.

After successfully completing the two developed games, the player will find it easier to interpret and associate different emotions with certain characteristic features of a human face.

It is possible to create an account, consisting of name, age, email and password. It is account will be responsible for containing the information of each player and, mainly, for storing the failed attempts in each game, so that whoever plays or monitors can have a better perception about what you are doing wrong and reflecting on what to do to rectify your reasoning.

In this work, two serious games were developed. The first one, Emotions Game, aims to build a facial expression through the use of various facial parts such as the eyes, eyebrows, nose and mouth. The second game, Recognition of Facial Expressions, prompts the user to replicate a requested emotion, but the success of the task is verified by means of a camera and a system that validates whether the emotion is correct.

2.1 Emotions Game

The purpose of the Emotions Game is to create a face that represents the requested emotion. This game has four emotions: Happiness, Sadness, Anger and Surprise. It uses various parts that make up a human face, such as the mouth, nose, eyebrows and eyes. As can be observed on the left side of Fig. 3, a set of parts to drag is presented towards the incomplete face to create the emotion of Anger. On the right side, three images make up the Help menu, which presents the most pronounced features of each emotion in order to help the player find the correct face part.

The menu containing the parts of the face is updated with each passing level, and the positions are randomly changed to make the player's task more difficult. The face shown in Fig. 4 was correctly completed. When all emotions are correctly completed, a warning shows the message "Game completed successfully!". A child has to drag a part of an incomplete face, according to a requested emotion, from the options menu that is on the right side and must interpret which parts correctly correspond to the requested emotion. Figure 4 shows the emotion of Sadness completed successfully.

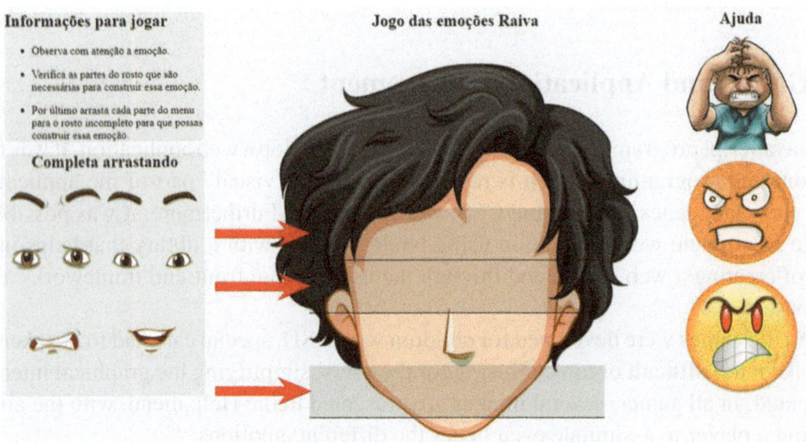

Fig. 3. General aspect of the Emotions Game

Furthermore, whenever the player hits or misses a constituent part of the face to be completed, a message is displayed on the screen, as can be seen in Figs. 5 and 6.

In addition to the visual notification, a sound notification is also given through an audio that says: "Good! You got it!" as a positive reinforcement to the player.

The same way that a sound notification is given when a player gets it right, when he makes a mistake, he hears an audio saying: "Oh! Try again". Thus, the player can realize that he/she was right or wrong.

The colours around and inside the text boxes, green and red, are in accordance with the ideas that people associate with right and wrong, respectively.

Fig. 4. Correctly filled face for the emotion of Sadness

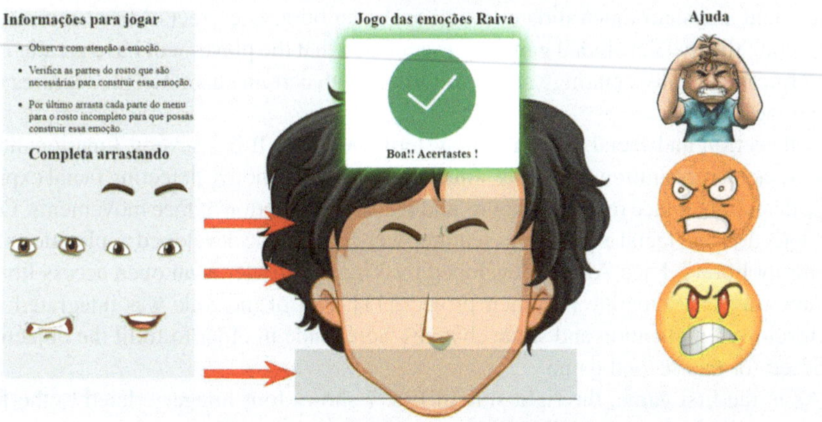

Fig. 5. Visual notification to alert if the face part was chosen correctly

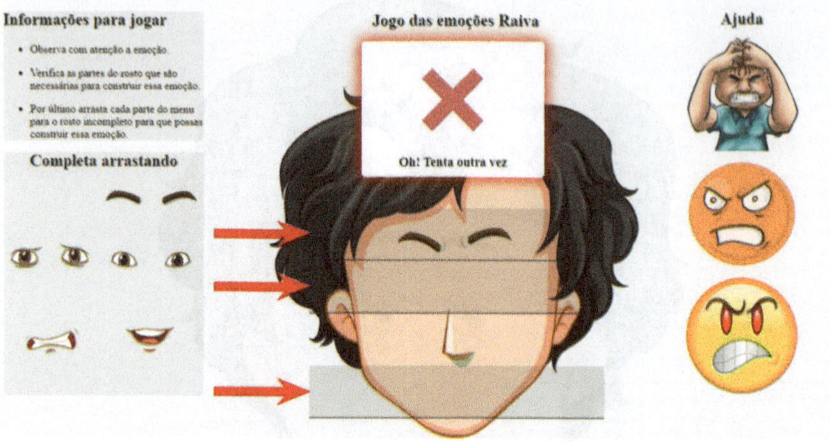

Fig. 6. Visual notification to alert if the face part was chosen incorrectly

2.2 Recognition of Facial Expressions

The second game is called Recognition of Facial Expressions. It prompts the user to replicate a requested emotion, but the success of the task is verified by means of a camera and a system that validates whether the emotion is correct. After successfully completing the two developed games, it is expected that the player will have greater ease in interpreting and associating various emotions with certain characteristic features of a human face.

A detection and facial recognition system was used. It has several functionalities, such as: facial recognition, checking similarities between photos, detecting facial expressions, detection of face reference points and real-time detection of face movements. Only face detection and facial expressions features were used in the developed application. The JavaScript library, Face-API.js, developed by Vincent Mühler is an open access library that any user can integrate into their project [11]. Part of the code was integrated into the developed application and some changes were made in order to fulfil the objectives proposed for this second game.

As in the first game, the right side of Fig. 7 shows four images related to the four emotions used in this game: Happiness, Anger, Surprise and Sadness. Key features of each emotion, such as the mouth and the eyebrows, are highlighted. On the left side, with a yellow underline, the emotion to be performed is requested, and the player should imitate this emotion in the best possible way.

Figure 8 shows the detection of multiple emotions, such as: Sadness, Joy, Surprise and Anger. When the system performs the validation with a probability above 70%, another emotion is asked until the game is completely finished.

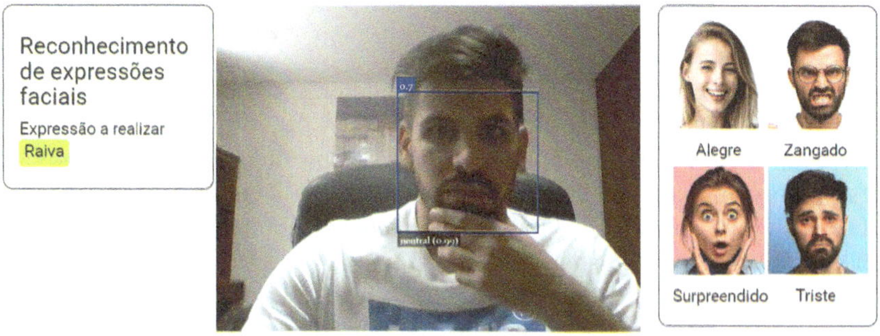

Fig. 7. Neutral state

Fig. 8. Multiple emotion detection: Sadness, Happiness, Surprise and Anger

3 Pilot Study

Five test sessions were carried out with seven children aged between 6 and 8 years old attending the elementary school with the goal to verify the evolution of the game for all of them and, also, whether the application allows responding to the functioning specificities and special needs of children with ASD. Three children have a diagnosis of ASD (Mateus, Josué and Joaquim) and four are children with normative/typical development (Madalena, Ricardo, Miguel and Fernando). To protect the identity and confidentiality of children, fictitious names are used, and the children's parents signed an informed consent in which they were notified about the objectives and methods applied in the research. Figure 9 illustrates the setup of a test session with a child.

Fig. 9. Setup of a test session with a child

4 Results

In order to analyse the evolution of each player, a comparative analysis of the results obtained in the first and the fifth test sessions was carried out. Table 1 presents the data of the seven children.

Table 1. Percentage of correct answers from the players for each emotion in the 1st session and the 5th session (%)

Percentage of correct answers by emotion of all players between the 1st and 5th session								
Fictitious Name	Happiness		Anger		Surprise		Sadness	
	1st	5th	1st	5th	1st	5th	1st	5th
Mateus*	75	100	75	100	75	100	75	100
Josué*	100	100	75	100	100	100	75	100
Joaquim*	60	100	75	100	100	100	60	75
Madalena	100	100	75	100	100	100	100	100
Ricardo	100	100	10	100	100	100	100	100
Miguel	100	100	75	100	100	100	60	100
Fernando	100	100	75	100	100	100	100	100

*Children with ASD

Using the correct answers from all players in the five sessions, an average of all sessions and standard deviation were computed for each emotion, in order to find which emotions were more difficult for children to interpret and had lower percentages of correct answers. This data is shown in Table 2. It is clear that the emotion of Sadness and, soon after, the emotion of Anger were the most difficult to interpret.

Table 2. Overall average of correct answers and standard deviation for each emotion in the five sessions performed

Emotion	Average of all sessions (%)	Standard deviation
Sadness	92.29	13.79
Surprise	97.14	8.07
Anger	92.43	12.39
Happiness	96.71	9.54

In the first sessions, most of the children got the eyebrows wrong for the emotion of Sadness, and this lowered the average throughout the sessions, in comparison with the other emotions.

Regarding the emotion of Anger, the most common mistake was getting the eyebrows and eyes wrong. The children mainly confused the eyes of the emotion of Happy with the eyes of the emotion of Anger and, sometimes, the eyebrows, too. So, the average in this emotion is lower in comparison to the other emotions.

In order to carry out a complete analysis of the evolution between sessions, a comparison of the completion time for each emotion was carried out. Table 3 shows which were the emotions in which the players had more difficulties, according to the time they took to complete it.

Table 3. Comparison of completion times for each emotion

	Completion time for each emotion in the 1st and 5th session (min)							
Fictitious Name	**Happiness**		**Anger**		**Surprise**		**Sadness**	
	1st	**5th**	**1st**	**5th**	**1st**	**5th**	**1st**	**5th**
Mateus*	1:03	0:19	1:52	0:21	1:08	0:17	1:44	0:23
Time Reduction	0:44		1:31		0:51		1:21	
Josué*	0:34	0:14	0.32	0.32	2:43	0:13	0:34	0:50
Time Reduction	0:20		0:00		1:30		- 0:16	
Joaquim*	1:55	0:13	1:59	0:24	1:11	0:38	1:58	0:27
Time Reduction	1:42		1:35		0:33		1:31	
Madalena	0:40	0:10	0:48	0:18	0:28	0:13	0:27	0.18
Time Reduction	0:30		0:30		0:15		0:09	
Ricardo	0.29	0:10	1:05	0:09	0:41	0:14	1:07	0:19
Time Reduction	0:19		0:56		0.27		0:48	
Miguel	0:34	0:17	0:55	0:13	0:40	0:15	1:06	0:13
Time Reduction	0:17		0:42		0:25		0:53	
Fernando	0:50	0:30	0:20	0:15	0:17	0:09	0:18	0:12
Time Reduction	0:20		0:05		0:08		0:06	

*Children with ASD

Comparing times from the first and the last sessions, it is possible to observe a positive evolution in the players with ASD, Mateus, Josué and Joaquim. The emotions of Sadness and Anger have higher completion times than the rest, because they were the emotions in which all players had greater difficulties. Josué's results are different, as he has a longer completion time in the fifth session, compared to the time used in the first session for the emotion of Sadness. He holds a constant time in the emotion of Anger. However, he presents a positive evolution in the emotion of Surprise, where he managed to reduce the execution time in 1:30 min. It is possible to conclude that the players easily adapted to the game and demonstrated greater levels of confidence and motivation throughout the sessions.

The players without ASD (Madalena, Ricardo, Miguel and Fernando) also showed a positive evolution throughout the sessions, and became faster and faster in the choice of the parts of the human face (eyebrows, eyes and mouth) that characterize each of the emotions, presenting more doubts, namely in the emotions of Sadness and Anger.

All the completion times were recorded and are presented in Table 4. A comparison was made between the completion time of the game in the first and the fifth sessions. To complete the game, it is necessary to successfully complete the four indicated emotions.

Table 4. Time for game conclusion in the 1st and 5th session (min)

	Fictitious name	Time for game conclusion in the 1st and the 5th session (min)	Time reduction
Children with ASD	Mateus	5.52–1.22	4:30
	Josué	4:43–1:49	2:54
	Joaquim	7:05–1:55	5:10
Children with typical development	Madalena	2:24–1:01	1:23
	Ricardo	3:23–0:52	2:31
	Miguel	3:21–0:58	2:23
	Fernando	1:47–0:51	0:56

By analysing the data in Table 4, it is possible to conclude that the players with PEA were the ones who had a better evolution in terms of completion times. These players were not completely comfortable in the first sessions, and took longer to decide which options were correct. Throughout the sessions, they improved in all aspects, and this was later reflected in the total time needed to complete the game.

It is important to emphasize that all players, with or without PEA, showed positive evolutions throughout all sessions, as they showed higher levels of interpretation regarding the requests of the game and completed it with fewer errors and in less time.

5 Conclusions and Future Work

This paper presented a web application gathering several games developed with different objectives, aimed at optimizing the acquisition of specific skills. The application makes the games accessible in different contexts and was developed in order to promote the learning of emotions in children with ASD in a playful, didactic and simple way.

Through all the data recorded in the five sessions carried out in an exploratory study, it was possible to make several comparisons and verify the evolution of each player. The emotions in which the children had more difficulties were Sadness and Anger, with 92.29% and 92.43% of correct answers, respectively. This result can be explained by the fact that these are less frequent emotions in a child's daily life, compared to the emotions of Joy and Surprise. This lesser use of the emotions of Sadness and Anger may have resulted in a greater number of errors in their identification, and in greater confusion with parts of the face of other emotions, namely with regard to the eyebrows and the mouth.

Analysing the results, it is possible to conclude that all children evolved throughout the sessions, and that they all demonstrated the need for a greater amount of time to finish the game in the first and second sessions, potentially due to the novelty factor. From the second session onwards, all children demonstrated greater confidence and speed in executing emotions. The obtained results may constitute important indicators for the promotion and development of this type of games in the context of emotions recognition and imitation skills, in children with or without ASD who attend the first cycle of basic education. The social relevance of the type of application developed for schools in the first cycle of basic education and for families of children with and without ASD is reinforced.

The application offers the possibilities of creating an account with the data of each player and saving the results of each game session in its database, for future studies. As future work, some improvements can be made in terms of the confidence interval and validation of the system, and also at the aesthetic level, in order to reinforce the child's motivation for the game. Emotions such as Fear and Disgust can also be introduced. These two emotions, although more difficult to interpret when compared to the emotions already used in the application, could provide greater complexity in the game and the need for collaborative work between peers. This collaborative work promotes interactions between children, with or without ASD, and the development of problem-solving skills. Finally, an exploratory study could be carried out, with a larger sample, in order to validate the behaviours of children with or without ASD regarding the recognition and imitation of emotions.

Acknowledgments. This work has been supported by national funds through FCT – Fundação para a Ciência e Tecnologia within the R&D Units Project Scope: UIDB/00319/2020. The authors thank the teachers and students of the Elementary School of Gualtar (EB1/JI Gualtar) in Braga for the participation.

References

1. Tao, J., Tan, T.: Affective computing: a review. In: Tao, J., Tan, T., Picard, R.W. (eds.) Affective Computing and Intelligent Interaction, pp. 981–995. Springer, Berlin Heidelberg, Berlin, Heidelberg (2005)
2. Vaigai College of Engineering, Institute of Electrical and Electronics Engineers. Madras Section, and Institute of Electrical and Electronics Engineers. In: Proceedings of the 2017 International Conference on Intelligent Computing and Control Systems (ICICCS) (2017)
3. Uljarevic, M., Hamilton, A.: Recognition of emotions in autism: a formal meta-analysis. J. Autism Dev. Disord. **43**(7), 1517–1526 (2013). https://doi.org/10.1007/s10803-012-1695-5
4. Chen, J., Wang, G., Zhang, K., Wang, G., Liu, L.: A pilot study on evaluating children with autism spectrum disorder using computer games. Comput. Hum. Behav. **90**, 204–214 (2019). https://doi.org/10.1016/j.chb.2018.08.057
5. Dantas, A.C., do Nascimento, M.Z.: Recognition of emotions for people with autism: an approach to improve skills. Int. J. Comput. Games Technol. **2022**, 6738068 (2022). https://doi.org/10.1155/2022/6738068
6. Mihova, P., Stankova, M., Andonov, F., Stoyanov, S.: The use of serious games for developing social and communication skills in children with autism spectrum disorders – review. In: Lim, C.P., Vaidya, A., Chen, Y.-W., Jain, V., Jain, L.C. (eds.) Artificial Intelligence and Machine Learning for Healthcare: Vol. 2: Emerging Methodologies and Trends, pp. 181–196. Springer International Publishing, Cham (2023). https://doi.org/10.1007/978-3-031-11170-9_7

7. Dos, L., Machado, S., Ronei, I., de Moraes, M.: Serious Games Baseados em Realidade Virtual para Educação Médica Serious Games Based on Virtual Reality in Medical Education (2011). www.abragames.org
8. Hassan, A., Pinkwart, N., Shafi, M.: Serious games to improve social and emotional intelligence in children with autism. Entertain. Comput. **38**, 100417 (2021). https://doi.org/10.1016/j.entcom.2021.100417
9. Silva, G.M., et al.: Interventions with serious games and entertainment games in autism spectrum disorder: a systematic review. Dev. Neuropsychol. **46**(7), 463–485 (2021). https://doi.org/10.1080/87565641.2021.1981905
10. Madsen, M., el Kaliouby, R., Goodwin, M., Picard, R.: Technology for Just-in-Time in-Situ Learning of Facial Affect for Persons Diagnosed with an Autism Spectrum Disorder. In: Proceedings of the 10th International ACM SIGACCESS Conference on Computers and Accessibility, in Assets 2008, pp. 19–26. Association for Computing Machinery, New York, NY, USA (2008). https://doi.org/10.1145/1414471.1414477
11. Vincent, M.: face-api.js – JavaScript API for Face Recognition in the Browser with tensorflow.js (2018). https://itnext.io/face-api-js-javascript-api-for-face-recognition-in-the-browser-with-tensorflow-js-bcc2a6c4cf07

Lean Manufacturing Knowledge Correlation Model

Iwo Podloch[1,2] and Krzysztof Nowacki[2,3](✉)

[1] Silesian University of Technology, 44-100 Gliwice, Poland
iwo.podloch@polsl.pl
[2] Magna International, 47-225 Kędzierzyn-Koźle, Poland
krzysztof.nowacki@polsl.pl
[3] Silesian University of Technology, 40-019 Katowice, Polska

Abstract. This article describes the results of a statistical study based on data collected in a survey. The questionnaire survey conducted at one of the Polish plants of an international automotive corporation was designed to investigate the level of knowledge, attitudes and awareness towards issues related to production management systems. Statistical analysis derives the correlations occurring in the "knowledge" subgroup of the data, and a substantive expert study was conducted based on the results. In the study, in addition to describing the correlations captured, a description is undertaken of how to improve the performance of areas with the weakest knowledge level scores using the correlations detected.

Keywords: management systems · Lean Manufacturing · statistical analysis

1 Introduction

Implementation of standards-based systems is driven by the requirements of the market, customers or suppliers, while so-called Lean systems are driven by the desire or necessity to optimize processes and reduce the cost of managing businesses. To be able to improve performance and meet the requirements of standards and customers, teams of modern manufacturing plants must demonstrate their knowledge of Lean systems tools. In plants with Lean systems, the autonomy at their foundation is important. Every employee in the company must show commitment to developing and optimizing not only those issues for which he or she is directly responsible, but also by participating in workshops, improvement circles or even by being a member of a root cause analysis team or completing a Near miss report. In order to be part of a system based on leveraging the intellectual potential of each employee, each employee, starting with the operator, must demonstrate knowledge first. In Lean-based production systems, the training system plays a very important role. Employees must understand what the Lean system is all about, they must know the requirements of issues related to quality management systems, customer requirements, and know how to use optimization tools or those related to daily operations such as TPM or 5S.

A. Burduk et al. (Eds.): ISPEM 2023, LNNS 790, pp. 188–199, 2023.
https://doi.org/10.1007/978-3-031-45021-1_14

Implementation of systems based on Lean Manufacturing brings very positive results for companies, and as a result, a large number of manufacturing plants are choosing to do so. In addition to the term Lean Manufacturing, the Lean concept is also increasingly popular under the name Lean Management, which proves its increasing use in non-manufacturing industries. The effect of introducing Lean systems into plants is a significant improvement in production indicators, as confirmed by numerous studies: [1–10].

1. cost reduction in the range of about 10–15%
2. reduction in workload time of about 28%
3. inventory reduction of about 10%–95%
4. reduction in warehouse space of about 90%
5. increase in productivity about 60–180%
6. reduction in employment approx. 50–66%
7. reduction of production area about 50–70%
8. reduction in lead time by about 97%
9. reduction in production cycle length approx. 70–86%
10. increase in work safety by about 55%
11. reduction of changeover time by up to 90%

Effective introduction of systems based on Lean, in addition to the aforementioned issues as shown, among others, by the analysis conducted by the Lean Enterprise Institute Poland, allows to shorten the product transition time (lead time), improve quality, communication, increase flexibility against order fluctuations, shorten the implementation time of new products, improve financial liquidity, improve safety and reduce the likelihood of environmental hazards. [2, 3, 11–25].

2 Research Methodology

The purpose of the research was to determine the feasibility of building a model for correlating Lean Manufacturing knowledge, among employees working in a manufacturing plant. Such a model can form the basis of activities to sustainably improve the level of knowledge among plant personnel.

The research was conducted using a proprietary questionnaire containing 16 questions on knowledge in the area of Lean Manufacturing, which is presented in Table 1. The set of questions was developed in cooperation with the Lean department. Questions were answered on a 5-point Likert scale, where 1 meant definitely no, and 5 meant definitely yes.

The results were subjected to scale reliability analysis using the α-Cronbach test. The Nunali criterion was used to verify the reliability of the scale. Using the stepwise method, questions were removed from the scale, after removing which the.

α-Cronbach's value increased. After obtaining a satisfactory value of α-Cronbach's a (min. 0.7), using the Pearson's r test, a correlation analysis was performed between pairs of questions. In each case, the level of significance was assumed to be $\alpha = 0.05$. Based on the used correlation coefficients for pairs for which $p < \alpha$, the LM knowledge model was built. [21–25, 27–47].

Table 1. Questions included in the research questionnaire

Question marking	Content of the question
q1 (p7)	How would you rate your knowledge of Lean principles?
q2 (p14)	Do you think the operator is responsible for the condition of the machine he/she works on?
q3 (p15)	Do you think keeping records of machine parameters, activities, etc. on MAFACT boards is important?
q4 (p60)	In your opinion, should technical employees be trained in Lean Manufacturing?
q5 (p61)	Do you think that performing and updating Standardization of Changeovers can bring tangible benefits?
q6 (p62)	Should part of the responsibility for machine maintenance lie with the operator?
q7 (p64)	Should the process engineer and/or production engineer analyze the layout of machines and stations and the number of operators in their area?
q8 (p65)	Is 100% Pull production possible or does it depend on the type of production?
q9 (p66)	Should Maintenance be involved in detailed failure analysis and the use of advanced systems to schedule parts replacement?
q10 (p67)	Is Continous Flow always feasible?
q11 (p69)	According to you, should Finance departments be trained in Lean and MAFACT principles?
q12 (p70)	Should machines be grouped according to their application? E.g., all CNC machines next to each other?
q13 (p71)	Do you think machines should be arranged in equal rows turned in the same direction?
q14 (p72)	Is it possible to achieve full replacement of operators? That is, situations where all employees can be freely dedicated to other positions?

3 Research Results

The research was conducted at a manufacturing plant located in Poland. The plant is part of a global automotive concern. LM principles are implemented at the plant, and the following LM tools are used on a daily basis:

- 5S
- TPM
- Autonomous maintenance
- Kaizen
- PDCA
- 5 times Why
- Standardization of work

- Visual management
- One piece flow
- Continuous flow
- VSM - as a document that is the basis for making changes in routing

and related tools like:

- BBS
- Near Miss
- Raport 8D / One Pager

et al.

The survey included 60 middle management employees. The choice of the research group was dictated by the direct impact of the respondents on the organization of production processes, for which comprehensive knowledge is necessary. Structure of the research group: women 31.4%, men 68.6%, age 24–55, higher education 72.2%, secondary education 27.8%, seniority in manufacturing companies applying ISO standards and Lean principles 2–21 years. (22–40).

Stepwise analysis of the scale's reliability, in the third step, selected a homogeneous scale consisting of 14 questions. The α-Cronbach's value for the scale was 0.727. Questions p7, p14, p15, p60, p61, p62, p64, p65, p66, p67, p69, p70, p71, p72 were selected from the scale for further analysis (questions p8–13, p16–59 p63, p68, p73 were rejected from further analysis). A pairwise correlation analysis was performed for the scale. The results of the analysis are shown in Table 2. Correlations for which $p < \alpha$ in Table 2 are in bold.

Based on the analysis, 12 statistically significant pairs were selected and used to build the LM knowledge model.

Table 2. Correlations between pairs of scales

	Q2	Q3	Q4	Q5	Q6	Q7	Q8	Q9	Q10	Q11	Q12	Q13	Q14
Q1	−0.042	0.134	0.492	0.087	0.261	0.027	0.140	0.155	−0.21	0.284	−0.185	−0.246	−0.141
Q2		0.119	0.163	−0.061	0.340	0.216	0.282	0.085	0.12	0.295	0.001	0.143	0.019
Q3			0.266	0.173	0.381	−0.076	0.191	0.298	0.204	0.174	0.279	−0.028	−0.007
Q4				0.467	0.196	0.242	0.310	0.28	0.017	0.220	−0.120	−0.110	0.123
Q5					0.282	0.241	0.329	0.153	0.224	−0.016	0.114	0.026	0.291
Q6						0.315	0.299	0.235	0.350	0.471	0.15	0.122	0.181
Q7							0.239	0.104	0.010	0.122	0.318	0.374	−0.267
Q8								0.149	0.535	0.067	0.042	0.069	0.365
Q9									0.198	0.204	0.134	0.303	0.018
Q10										−0.018	0.306	0.255	0.605
Q11											−0.145	−0.010	−0.086
Q12												0.778	0.079
Q13													0.083

4 LM Knowledge Model

The graphical construction of the knowledge model began with the pair of questions with the strongest co-correlation, questions 70 and 71 where r = 0.778. The model was then supplemented with additional questions that correlate with those already included. The result was the graphical version of the model shown in Fig. 1. Of the pairs of questions that were selected to build the LM model, each has a positive r correlation, meaning that, that is, by raising awareness in one area, we can simultaneously positively influence another related area. (47).

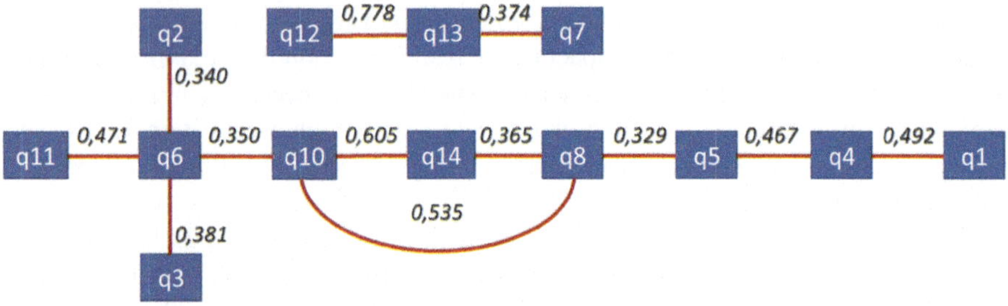

Fig. 1. LM knowledge model

The developed LM knowledge model requires expert verification and determination of the possibility of indirect influence on individual questions through the formation of knowledge in the range corresponding to the individual pairs, which are the components of the model. A content analysis of the individual correlations between questions was developed with the values of the individual correlations starting with the largest. The strongest correlation (r = 0.778) was determined for question pair q12 and q13:

– q12 - q13 (r = 0.778)

- questions relate to the analysis and planning of LayOuts;

– q1 - q4 (r = 0.492)

- questions verify the respondent's level of knowledge;

– q6 - q11 (r = 0.471)

- o questions are interrelated by addressing the issue of engagement;

– q4–q5 (r = 0.467)

- questions verify the respondent's level of knowledge;

– q3–q6 (r = 0.381)

- questions are related to issues of TPM nature;

– q7–q13 (r = 0.374)

- questions are related to the form of layout indicating the issue of responsibility for its analysis and optimization;

– q2–q6 (r = 0.340)

- ask directly about autonomous maintenance;

– q5–q8 (r = 0.329)

- questions verify the respondent's level of knowledge;

– q6–q10 (r = 0.350)

- questions verify the respondent's level of knowledge;

– q10–q14 (r = 0.605)

- Continous Flow - Flexible Manpower;

– q8–q10 (r = 0.535)

- Pull System - Continuous Flow;

– q8–q14 (r = 0.365)

- Pull System - Flexible Manpower;

In this case, we observe a correlation between the question of Continous Flow type and Pull production scheduling system and the question of substitutability and flexibility of production stations, a correlation that may be due to the combined level of difficulty of these questions, surpassing the understanding of such basic Lean tools as 5S or TPM. In contrast, a direct factual relationship between these questions does not appear to exist.

The most significant correlations relate to questions related not so much to Lean tools, which can be implemented in a straightforward relationship, but to questions about planning management systems, flow design and production lines. These issues are difficult to explain with a presentation without providing examples. It is most advisable to carry out a simulation demonstrating the principles of their operation and proving their advantages over traditional models or the lack of a system.

Another issue noticeable on the model is the issue of autonomous maintenance and as such the issue of responsibility and commitment. Understanding of these issues also needs to be done by presenting real-life examples, but in this case, it would be necessary to work on two tracks developing "soft skills," reinforcing the sense of agency and the importance of each employee's role in the efficient operation of the production plant.

5 Possible Applications of the LM Knowledge Model

The results presented here confirm the intersection of the different areas of knowledge, providing a basis for finding additional opportunities to shape the knowledge of employees during training. These opportunities should manifest themselves with planning to address topics related to the individual areas, the correlation of which is known.

For the company analyzed in the article, the lowest average rating was obtained for q8 (2.47 points). Such a rating applied to a group of engineers employed in operational states. Based on the developed LM knowledge model, it can be concluded that an increase in knowledge in the area of q8 can result from an increase in knowledge in the areas of q10 ($r = 0.535$), p72 (r - 0.365) and p61 ($r = 0.329$). Accordingly, training courses that cover Continous Flow (q10), interchangeability of machine operator competencies (q14) and standardization of changeovers should cover issues related to the Pull system (q8). The low level of understanding of the Pull System in the group of operational specialists and engineers is not surprising, and in practice the lack of understanding of the methodology's system tools issues is a common phenomenon in this group.

The maximum sum of answers to all 16 questions (90 points) that are the pre-subject of the analysis indicates the level of knowledge of respondents in the area of knowledge. For the purpose of the analysis, the criteria used to divide the respondents (gender, education) determined the average values of the final ratings, which are shown in Fig. 2. The summary shows that the lowest rating is obtained by employees employed in operational positions (49.6 points) and those with higher education (50.2 points.) Thus, the employer should put more emphasis on training these groups in lean knowledge.

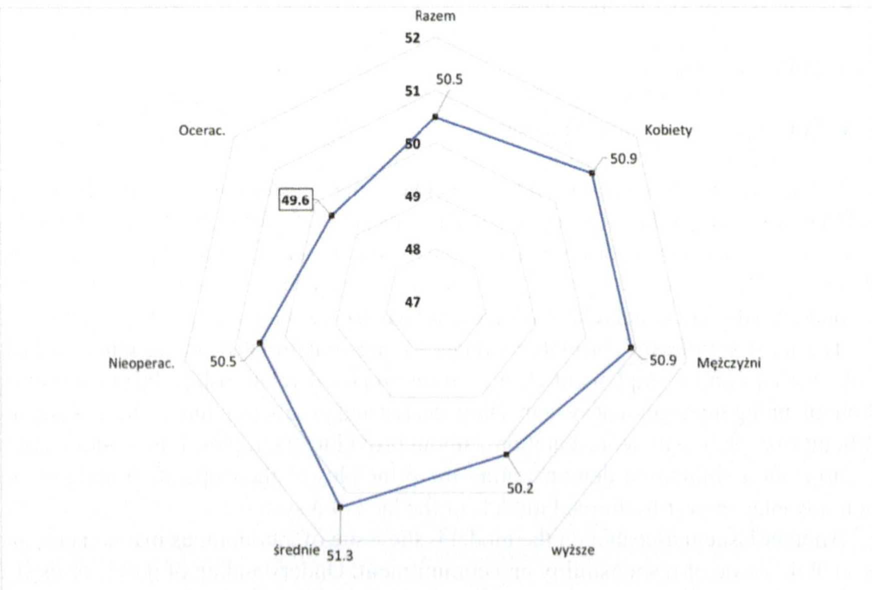

Fig. 2. Evaluation of LM knowledge by the analyzed employee groups (max 90 points)

The standard deviation and coefficient of variation were compared for each group according to the adopted criteria (Table 3). The coefficient of variation (standard deviation mean value) is very comparable for most groups except for executives where the score obtained was the highest (Fig. 3). Comparing the results of departments, the lowest were captured in the group of people not directly employed in Lean-related cells and, more specifically, in the group "specialist/engineer in an operational position."

Table 3. Results of evaluation, standard deviation and coefficient of variation of individual groups.

variable	evaluation	standard deviation	coefficient of variation
total	44,6	7,23	0,162
women	44,9	6,77	0,151
men	45,0	7,24	0,161
managers	46,5	4,97	0,107
specialists/engineers	44,2	7,73	0,175
higher education	44,0	6,99	0,159
secondary education	46,3	7,97	0,172
non-operational staff	43,9	6,67	0,152
operational staff	44,0	6,96	0,158

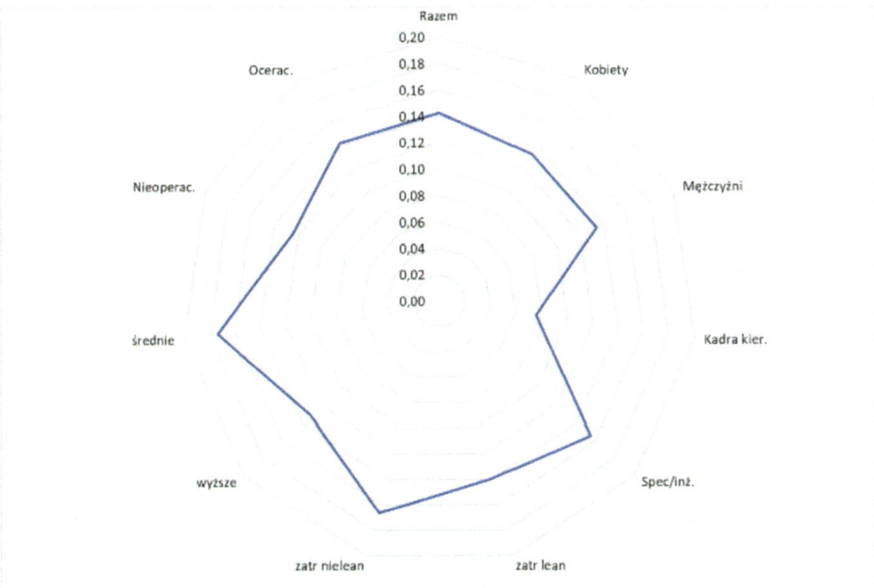

Fig. 3. Graph showing the coefficient of variation result for each group.

We can observe that in the "specialist/engineer in an operational position" group, the worst rating is given to question q8: Is pull production 100% implementable or does it depend on the type of production?

The low level of understanding of the Pull System in the group of operational specialists and engineers is not surprising, and in practice, the lack of understanding of the methodology's system tool issues is a common phenomenon in this group (Fig. 4).

Referring to the model built earlier, the correlation of question p64 on the Pull system with question q10 on Continuous Flow was presented, both questions are related as mentioned to more advanced Lean tools, which are very difficult to understand without the use of simulation. Correlations are also present with questions on operator substitutability at the workstations (q14) and labor standardization (q5), converging questions whose common denominator is flow planning and production planning with "flexible manpower" at the line level.

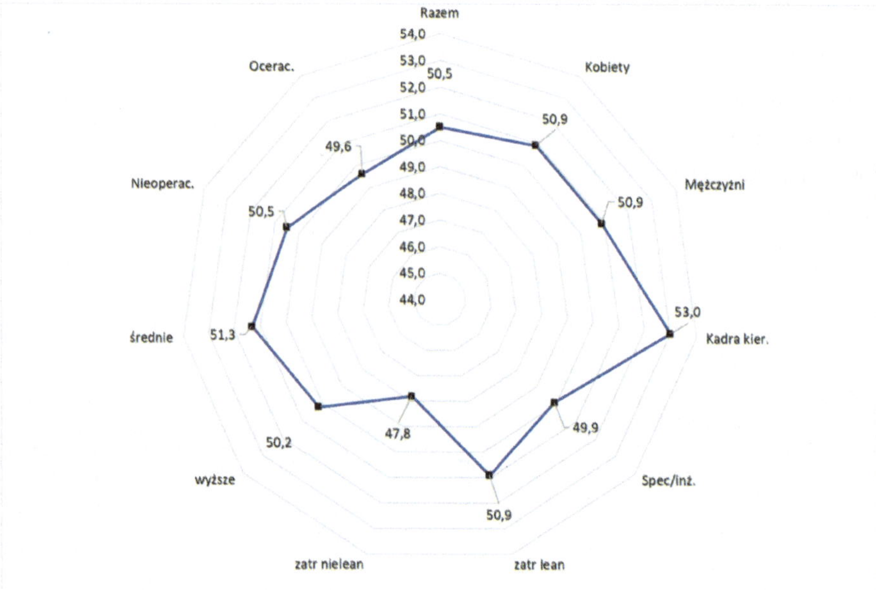

Fig. 4. A chart showing the result of the survey evaluation for each group.

6 Summary

6.1 Model Analysis

Based on the research, it was proven that the developed model can be used to study correlations, and conduct further statistical analysis. By determining the level of knowledge in terms of individual questions as well as defined groups, as presented on the basis of the model, it can possibly be used to identify employee groups requiring more attention. The results of the questionnaire evaluation in the individual groups of the adopted research population are comparable-no less, however, differences in the overall and area evaluation may be observed in subsequent studies.

6.2 Case Stage Analysis

From the analysis, it is clear that improving the level of Lean knowledge and thus production efficiency at the analyzed plant can be achieved by intensifying training illustrating (simulations) flow design, production planning, differences in planning systems and ways to implement parts flow at the plant-wide level, as well as planning and analysis of optimization potential from the line level (LayOut and line balance) for operational specialists and engineers. This group is also crucial in connection with its function as a bridge between office and production staff, sometimes it is the attitude of this very group that has a significant impact on the perception of Lean system tools by leaders and operators. The issue of the transfer of attitude to the shop floor by this very group is also important in this case, because the level of knowledge of managers in the analysis looks very good, and if this attitude had the opportunity to be transferred to production it would be possible to raise the level of Lean knowledge throughout the plant.

References

1. Prońko, J., Soboń, A., Zamiar, Z.: Zarządzania produkcją, Wydawnictwo Uniwersytetu Humanistyczno-Przyrodniczego w Kielcach, Kielce s. 252 (2008)
2. Sahoo, A.K., Singh, N.K., Shankar, R., Tiwari, M.K.: Lean philosophy: implementation in a forging company. Int. J. Adv. Manuf. Technol. **36**, 451–462 (2008)
3. Hill, A.V.: The Encyclopedia of Operations Management, vol. 195. FT Press, Upper Saddle River (2011)
4. Bo, Z., Zhanwen, N., Feng, L., Wang, X.: Evaluation system for lean knowledge management ability based on improved Gray correlation analysis. Int. J. Industr. Eng.: Theory Appl. Pract. 2008, Manag. Res. News (2020)
5. Sim, K.L., Rogers, J.W.: Implementing lean production systems: barriers to change
6. Bhasin, S., Burcher, P.: Lean viewed as a philosophy. J. Manuf. Technol. Manag. **17**, 61 (2006)
7. Nicholas, J.: Lean Production for Competitve Advantage, vol. 2. CRC Press, New York (2011)
8. Lewis, M.A.: Lean production and sustainable competitive advantage. Int. J. Oper. Prod. Manag. **20**, 960 (2000)
9. Holweg, M.: The genealogy of lean production. J. Oper. Manag. **25**, 428 (2007)
10. Murman, E., et al.: Lean Enterprise Value: Insights from MIT's Lean Aerospace Initiative, vol. 6, p. 137. Palgrave, New York (2002)

11. Murphy, M: Organisational change and firm performance. OECD Science, Technology, and Industry Working Papers 2002/14, p. 9. OECD, Paris (2002)
12. Ward, P., Zhou, H.: Impact of information technology integration and lean/just-in-time practices on lead-time performance. Decis. Sci. **37**, 177–189 (2006)
13. Belekoukias, I., Garza-Reyes, J.A., Kumar, V.: The impact of lean methods and tools on the operational performance of manufacturing organisations. Int. J. Prod. Res. **52**, 5346–5366 (2014)
14. Alpenberg, J., Scarbrough, D.P.: Exploring communication practices in lean production. J. Bus. Res. **69**, 4959–4963 (2016)
15. Belekoukias, I., Garza-Reyes, J.A., Kumar, V.: The impact of lean methods and tools on the operational performance of manufacturing organisations. Int. J. Prod. Res. (2014). Int. J. Oper. Prod. Manag.
16. Bhamu, J., Sangwan, K.S.: Lean manufacturing: literature review and research issues
17. Möldner, A.K., Garza-Reyes, J.A., Kumar, V.: Exploring lean manufacturing practices' influence on process innovation performance. J. Bus. Res. **106**, 233–249 (2020)
18. Koch, T.: Minione 10 lat ruchu Lean w Polsce: wnioski i perspektywy, X Międzynarodowa Konferencja Lean Manufacturing: materiały konferencyjne, Wrocław 2010, Lean Enterprise Institute Polska. za: Jak stosować metody Lean Manufacturing (Oszczędnego Wytwarzania) do wprowadzania innowacji, materiały, Warszawa (2011)
19. Prońko, J., Soboń, A., Zamiar, Z.: Zarządzania produkcją, Wydawnictwo Uniwersytetu Humanistyczno-Przyrodniczego w Kielcach, Kielce, p. 253 (2008)
20. Emiliani, B.: Origins of lean management in America. J. Manag. Hist. **12** (2006)
21. Radeka, K.: The Mastery of Innovation: A Field Guide to Lean Product Development, vol. 3. CRC Press, Boca Raton (2013)
22. Snee, R.D.: Lean six sigma – getting better all the time. Int. J. Lean Six Sigma **1**, 26 (2010). 11
23. Hafey, R.B.: Lean Safety: Transforming Your Safety Culture with Lean Management, pp. 5–6. CRC Press, New York (2010)
24. Liker, J.K. (ed.): Becoming Lean: Inside Stories of US Manufacturers. Productivity Press, New York (1998), pp. 186–187
25. Forza, C.: Surveys, research methods for operations management (2016)
26. Tice, J.: Lean production and EMSs: aligning environmental management with business priorities. Environ. Qual. Manag. Winter (2005). 3
27. Jasińska, S., Żurek, M., Wyrwicka, M.K.: Analiza efektywności wdrożenia Lean Manufacturing. Studium przypadku, Economics and Management – 1/2015, Politechnika Poznańska (2015)
28. Aczel, A.D.: Statystyka w zarządzaniu, Warszawa (2020)
29. Nowacki, K.: Modelowanie Bezpieczeństwa w Przemyśle, Wydawnictwo Politechniki Śląskiej, Gliwice (2019)
30. Womack, J., Jones, D., Roos, D.: The Machine That Changed the World, Warszawa (2007)
31. Krafcik, J.: Sloan Management Review (1988)
32. Shook, J., Rother, M.: Learning to see, lean enterprise institute (2017)
33. Graupp, P., Wrona, R.: Podręcznik TWI, lean enterprise institute (2010)
34. Johansson, P.E.C., Lezama, T, Malmsköld, L, Sjögren, B, Moestam Ahlström, L.: Current state of standardized work in automotive industry in Sweden. In: Forty Sixth CIRP Conference on Manufacturing Systems (2013)
35. Holweg, M.: The genealogy of lean production. J. Oper. Manag. **25**(2), 420–437 (2007)
36. Garvin, D.A.: Quality problems, policies, and attitudes in the United States and Japan: an exploratory study. Acad. Manag. J. **29**(4), 653–673 (1986)
37. Dudek, M.: Szczupłe systemy wytwarzania, Difin, Warszawa (2016)

38. Nogalski, B., Czerska, J., Klimek, A.: Wykorzystanie metod teorii masowej obsługi i lean management w usprawnianiu procesów w zarządzaniu organizacją publiczną, "Współczesne zarządzanie", nr 1/2010, pp. 62–63 (2010)
39. Bednarek, M.: Doskonalenie Systemów Zarządzania, nowa droga do przedsiębiorstwa lean, Difin sp. z o.o., Warszawa (2007)
40. Koch, T.: Jak stosować metody Lean Manufacturing (Oszczędnego Wytwarzania) do wprowadzania innowacji, Politechnika Wrocławska,"E-narzędzia i technologie generatywne jako szybka ścieżka do innowacji", nr 1, Warszawa (2011)
41. Antosz, K.: Lean Manufacturing doskonalenie produkcji, Oficyna Wydawnicza Politechniki Rzeszowskiej, Rzeszów (2015)
42. Womack, J., Jones, D.T.: Lean Thinking: Banish Waste and Create Wealth in Your Corporation, FreePress (2003)
43. Pomietlorz, M.: Istota koncepcji Lean Manufacturing
44. Walentynowicz, P.: Zakres zastosowania Lean Management w przedsiębiorstwach produkcyjnych Konferencje KZZ 2015, zebrane artykuły (2015)
45. Wolniak, R.: Metody i narzędzia Lean production i ich rola w kształtowaniu innowacji w przemyśle Konferencje KZZ 2013, zebrane artykuły (2013)
46. Żebrucki, Z., Kruczek, M.: Doskonalenie struktury łańcucha dostaw z wykorzystaniem koncepcji lean
47. Forghani, M.A., Tavasoli, A.: Investigating the relationship between knowledge management dimensions and organizational performance in lean manufacturing. Int. J. Manag. Account. Econ. (2017)

Assessment of the Effectiveness of Six Sigma Methodology Implementation - A Literature Review

Paula Kolbusz[1(✉)] and Katarzyna Antosz[2]

[1] PhD School of Engineering and Technical Sciences, Faculty of Mechanical Engineering and Aeronautics, Rzeszow University of Technology, Al. Powstancow Warszawy 12, 35-959, Rzeszów, Poland
d585@stud.prz.edu.pl

[2] Faculty of Mechanical Engineering and Aeronautics, Rzeszow University of Technology, Al. Powstancow Warszawy 12, 35-959, Rzeszów, Poland
katarzyna.antosz@prz.edu.pl

Abstract. Within the application of Six Sigma philosophy, the DMAIC methodology (D-Define, M-Measure, A-Analyse, I-Improve, C-Control) to analyse existing processes is used. The aim of this paper is to analyse the current state of the art of assessing the effectiveness of the implementation of manufacturing process improvement projects using this methodology. The methodology of the work was carried out according to the following steps: 1) identification of relevant data sources and keywords; 2) definition of the timeframe; 3) search of databases; 4) analysis of the review and summary of the results obtained. The following databases were analysed as part of the study: Science Direct, Web of Science and Scopus from 2017 to 2022. The result of the re-search is the identification of performance indicators used in Six Sigma projects, which were related to the different phases of the DMAIC methodology. The result of the analyses is the identification of groups and areas of indicators that would allow the effectiveness of improvement activities to be assessed at each phase of a Six Sigma project using the DMAIC methodology.

Keywords: Six Sigma · Efficiency · Indicator · DMAIC Methodology

1 Introduction

The requirements of ever-changing markets condition the continuous adaptation of companies' offers. The development and continuous improvement of management systems is aimed at anticipating these changes and therefore fully satisfying the needs and expectations of each partner (customers, employees and other stakeholders), as well as maintaining a competitive advantage. One way to achieve operational excellence is to implement various quality improvement initiatives such as Total Quality Management, Agile & Lean Manufacturing, etc. [1]. Nowadays, there is tremendous pressure on organizations to improve customer satisfaction and quality in the organization, while

A. Burduk et al. (Eds.): ISPEM 2023, LNNS 790, pp. 200–214, 2023.
https://doi.org/10.1007/978-3-031-45021-1_15

reducing inefficiencies and errors. Businesses are adopting newer and newer solutions to attract and retain existing customers, as this is a key driver of the economy [2]. One way to achieve excellent business performance involves understanding the process and subsequent improvement. Philosophies such as Kaizen, Total Quality Management, Six Sigma or the use of dynamic models are among the innovative approaches to improving process quality. Their idea of optimizing processes, as well as continuously improving the knowledge of their employees, makes it easier to deal with global changes, but also process disruptions and extraordinary circumstances. Effective measurement of service quality cannot be limited to a specific moment, so it is necessary to ensure continuous quality monitoring throughout the entire process [3]. In addition, nowadays the efficient use of energy has taken on great importance in the industrial sector. Furthermore, stringent environmental impact regulations are prompting companies to increase energy efficiency in the short term to avoid penalties [4]. Another reason for the development of the Six Sigma methodology is the variety of products, which is rapidly increasing in large manufacturing companies, production lines are beginning to face the problem of time wastage. In order to reduce various types of waste, more and more companies are using Lean Management, or Six Sigma, as a business management tool, as well as using information systems such as SAP as an ERP system for enterprise resource planning [5].

In addition, Industry 4.0 and Lean Six Sigma solutions complement each other, resulting in a global supply chain management model. There is an algorithm for predicting the robustness of a manufacturing process to synthesise tolerances, which is an introductory step to expand the use of Six Sigma in process improvement in the Industry 4.0 vision. In 2017, an approach to organise the data collected using Industry 4.0 technology was proposed to help the organisation make strategic decisions. In 2018, data handling methods related to Lean Six Sigma phases were described in detail in relation to Industry 4.0 data collection technologies [6]. Based on the literature reviewed, significant interest in the subject of Six Sigma has been perceived in recent years, as well as its increasing combination with other methodologies and areas of application. Further research in this topic is important, as the improvement of production processes is an important issue from the point of view of the efficiency of the production process and the costs allocated to production. In addition, it is important nowadays to continuously improve production processes in order to develop a competitive advantage in the global market.

2 Six Sigma

Six Sigma is a systematic and structured, customer-driven approach that aims to improve the performance and quality of processes, products and services using statistical techniques [7]. Six Sigma involves a structured process improvement strategy that places processes on a continuous improvement path [5]. The main objective of the Six-Sigma approach is to achieve customer satisfaction by producing products with less variability. This is very important in the manufacturing process, as products that do not meet customer requirements are eliminated and accepted for re-manufacturing, and only those that are free of defects can be sold [8]. The Six Sigma philosophy, using the application of the most innovative digital technologies, seeks to reduce inefficient use of resources and non-optimized service processes [6].

Six Sigma provides structured methods, such as DMAIC (Define, Measure, Analyse, Improve and Control) and DFSS (Design for Six Sigma), by which individuals can organise their creative ideas to improve project performance [9]. DMAIC stands for Define-Measure-Analyse-Improve-Control. This method is based on process improvement according to the Deming cycle. It is the improvement of processes in many different areas of the company. The DMAIC cycle consists of five stages that are interlinked [2]. The costly correction of products late in the product life cycle causes the development of the methodology to focus on the product design stages to eventually link it to the Six Sigma quality level. Hence, this methodology is called Design for Six Sigma (DFSS). It is recognised that the Six Sigma programme aims to improve the existing process. Nevertheless, DFSS focuses on product design, especially in the early stages of the product life cycle. The use of DFSS incorporates design thinking and preventive problem solving and intends to achieve breakthrough innovation. DFSS is a universal design method, combining quality management and Six Sigma. DFSS not only optimises design goals, but also reduces the sensitivity of constraints and goals to fluctuations in uncertainty, i.e. improves system resilience [10]. In contrast, DMAIC focuses on statistical thinking and analytical problem solving and results in incremental innovation [11]. Although, Six Sigma has wide application in various industries [2–11], the evaluation of its effectiveness and the use of various metrics to assess it is still very limited.

The purpose of this article is to identify the state of the art in the applied performance indicators of the Six Sigma method in manufacturing process improvement using DMAIC. The article is structured as follows. Section 3 presents the methodology of the analyses conducted, while Sect. 4 discusses the results of the analysis of literature sources. The paper concludes with a summary and description of directions for further research.

3 Research methodology

The work was carried out in the following stages:

1. Identification of relevant data sources and keywords.

   The main method of data collection was the collection of secondary publications, where data can be obtained from published sources and can be further analyzed for additional interpretations or conclusions. Articles in languages other than English, conference proceedings, textbooks, journal reports, masters and doctoral theses and unpublished working papers were not included. Three scientific databases were analyzed: Science Direct, Scopus and Web of Science. The search began by selecting a set of keywords and possible combinations that might be relevant to the topic. The Six Sigma concept covers a wide range of aspects, so the search field had to be limited accordingly. The databases were searched considering the following keyword combinations:

   - "six AND sigma";
   - "six AND sigma AND indicator";
   - "six AND sigma AND efficiency AND assessment".

2. Defining the time frame.
   The systematic approach of the 2017 - 2022 literature review used in this article allows for the identification of Six Sigma developments, identifies the areas in which it is applied, highlights the performance indicators used.
3. Searching databases.
   The search methodology was based on the following principles: only selected timeframes were included, suggested keyword combinations were used in the 'keywords' field to narrow down the group of publications to those most relevant and adding value to this work, only articles in the area of production processes were included.
4. Analysis of the review and summary of the results obtained.
   Identified after the keywords, the articles in each year allowed the identification of the indicators used to assess the effectiveness of improvement projects in production processes.

4 Results of the Research

4.1 Number of Scientific Publications Analysed

The scientific publications analyzed are divided in Table 1 according to the following criteria: database, search years and keyword combination. Figure 1 shows the breakdown of sources considering keyword combination: "six AND sigma AND indicator", "six AND sigma AND efficiency AND assessment" and database, while Fig. 2 shows keyword combination: "six AND sigma AND indicator", "six AND sigma AND efficiency AND assessment" and search years. The literature search carried out mainly analyzed works on the evaluation of the effectiveness of the implementation of improvement projects and the related indicators used in the DMAIC methodology. The established

Table 1. Results of database analysis.

Keywords	Database	Review articles						
		Search years						
		Overall	2017	2018	2019	2020	2021	2022
six AND sigma	Science Direct	558	74	82	86	90	114	112
	Web of Science	967	127	138	177	161	197	167
	Scopus	1388	191	206	220	232	256	283
six AND sigma AND indicator	Science Direct	29	5	2	4	4	7	7
	Web of Science	7	0	2	0	0	4	1
	Scopus	35	1	3	6	4	9	12
six AND sigma AND efficiency AND assessment	Science Direct	7	0	1	1	2	2	1
	Web of Science	0	0	0	0	0	0	0
	Scopus	8	1	1	3	0	1	2

search strategy initially identified a total of 86 publications for the three databases analyzed. Each paper was analyzed in detail to exclude repetitive ones and those outside the scope of the work topic. Taking these constrains into account, the number of articles was reduced to 39.

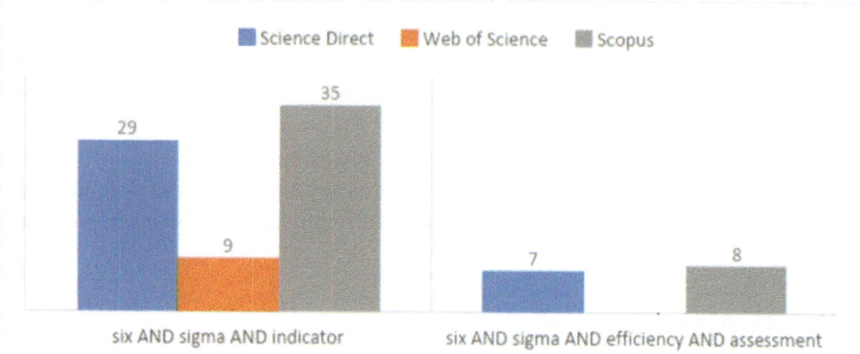

Fig. 1. Number of publications by keyword combination and databases searched.

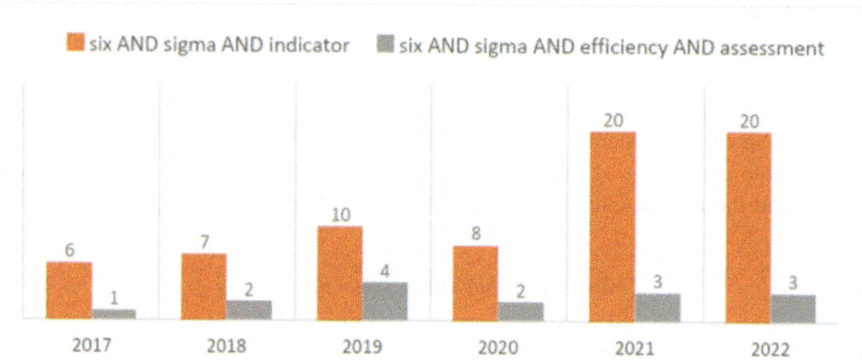

Fig. 2. Number of publications by keyword combination and search years.

4.2 Six Sigma Performance Evaluation Indicators

In the case of performance indicators, we are dealing with a measure of the relationship of the resources involved to the result achieved. Six Sigma statistics provide added value to the research because they make it possible to detect processes that need to be improved. Table 2 shows the detailed classification of the indicators, with a citation of the literature in which they were proposed and the stage of the Six Sigma project in which they were determined.

From a search of the databases, it can be concluded that an upward trend in the number of publications per year has been noted in each version of the keywords used.

Table 2. Detailed classification of indicators, for each phase of the DMAIC methodology.

Indicator	Formula / definition	Stage of the DMAIC SS methodology	Paper ID
VOP	Total System's performance: $\frac{2VOC+VOB}{3} = VOP$, where: VOC-voice of customer, VOB-voice of business	D	[12]
VOC	Voice of customer - summary of customer satisfaction	D	[13]
Importance Index II	II = (F1 × criticality level) + (F2 × cost level) + (F3 × performance level); The success of a Six Sigma project can be measured by the amount to which we reduce the value of II in the project	D	[40]
% service level performance	% of satisfied customers	D	[15]
% productivity effort	Indicator calculated on the basis of the stocking of products in the storage area, taking into account the distribution between floors	M	[15]
Compliance with 5S performance	% compliance of the layout with the introduced 5S principles	M	[15]
RPN (Risk Priority Number)	Risk priority number: RPN = SEV*OCC*DET, where: SEV - severity OCC - occurrence DET-detection	A	[15, 41]

(*continued*)

Table 2. *(continued)*

Indicator	Formula / definition	Stage of the DMAIC SS methodology	Paper ID
Picking time	Time from placing order by customer to completion of order	I	[15]
Order cycle time	Time from customer order to delivery	I	[15]
Indicator of work-off	Daily operating time of the machine in relation to the quantity of products produced	C	[14]
Cpk	Actual process capability index	C	[1, 15–17]
DPMO	Defects per Million Opportunities	C	[1, 18–22]
PPM	Number of defects per million possibilities of their occurrence	C	[7, 16, 17, 23, 26]
Sigma level	Six Sigma Level (SSL) - standard deviation of the variable	C	[7, 15, 19–22, 24–28]
CS	Customer Satisfaction - the sum of 30% quality control (QC) and 70% time to market (TTM)	C	[23]
Apt	Actual production time - efficiency is defined as meeting the planned production time target	C	[6]
Rq	Rework quality - effectiveness is defined as meeting the planned target for the quality of the reworked product	C	[6]
Sq	Scrap quality - efficiency is defined as meeting the planned product quality target	C	[6]
Non-Value-Added Activity	Reducing non-value added activities	C	[29–31]
Performance improvement indicator	% of available space (free space), % reduction in waste, increase in operator productivity	C	[32]

(continued)

Table 2. (*continued*)

Indicator	Formula / definition	Stage of the DMAIC SS methodology	Paper ID
A waste of time	Excess of time needed to perform a given operation	C	[33]
Productivity	The ability of the workforce to produce a specified number of units of goods/services in a specified period of time	C	[31, 34]
Lead time	Time including value-added activities, non-value-added activities and inventories are evaluated in terms of their time contribution to process execution time	C	[31, 35]
Machine cycle time	Indicator of achievement of machine cycle time target	C	[16, 17, 31, 36, 37]
KPI (Key Performance Indicator) - Maintenance	Elimination of: breakdowns, changeovers, downtime, reduced speed, errors and corrections	C	[2]
KPIs - call centre	Update rate, improvement of the process of catching errors before they turn into defects, reduction of customer waiting times, fast resolution of requests	C	[38]
IpB	Incidents per Billion	C	[23]
DPU- Defects per unit	Total number of defects / total number of products	C	[19]
DPO - Defects per opportunity	Total number of defects / total number of opportunities	C	[19]
Number of complaints	Number of reports of product defects from customers	C	[31, 39]
% of product returns	Number of returned products per year	C	[15, 31]

(*continued*)

Table 2. *(continued)*

Indicator	Formula / definition	Stage of the DMAIC SS methodology	Paper ID
Yp	Combine the concepts pf process yield and process loss to propose a process yield index Yp $Y_p = \int_{LSL}^{USL}(1 - \lvert\delta\rvert)dF(x)$ $\delta =$ $\left\{\frac{(X-T)}{D_u}, T \le X \le USL\frac{(X-T)}{D_l}, LSL \le X \le TD_u = USL - T, D_l = T - LSL\right\}$ where: X - quality characteristic of product, T - process loss	C	[42]
Sigma metric	$Sigmametric = \frac{TAE-bias}{CV}$ Where: $\%bias = \frac{labvalue-referencevalue}{referencevalue} * 100\%$ TAE - Total Allowable Error CV - imprecision	C	[43]
Q_{pkh}	$Q_{pkh} = \frac{1-\lvert\delta_h\rvert}{\gamma_h} + 1$ $\delta_h = (\mu_h - T)/d$ $\gamma_h = \sigma_h/d$ μ_h- process mean σ_h- standard deviation T = (USL + LSL)/2, d = (USL-LSL)/2, USL - upper specification limit LSL - lower specification limit	C	[44]
OEE	Overall Equipment Effectiveness	C	[45]

(continued)

Table 2. (*continued*)

Indicator	Formula / definition	Stage of the DMAIC SS methodology	Paper ID
Absenteeism	Number of absentees to total number of employees	C	[31]
Delivery performance	Percentage of orders that were delivered on time / Total number of orders	C	[28, 31]
Customer feedback	% customer satisfaction	C	[31]
Employee satisfaction	% of satisfied employees	C	[31]
Inventory levels	Metric includes the quantity of stock located in warehouses, logistics centres, and physical stores and comparison with customer orders	C	[31]
Market Share	% global market share	C	[31]
Material handling	Material transport involves moving short distances within a building or between a building and a transport vehicle	C	[31]
Profitability	The difference between a company's total revenue from the sale of products and the total costs incurred to produce those products	C	[28, 31]
Sales Turnover	The company's total amount of products or services sold over a given period of time	C	[31]
Skills	% improvement of staff skills (internal, external training, workshops)	C	[31]
Throughput	Network capacity refers to the rate at which messages are delivered over a communication channel	C	[31]
TAT	Turnaround time - the amount of time it takes to complete a process	C	[28]
% scrap	% non-conforming products in relation to all products manufactured	C	[13, 16, 17, 31, 36, 37]

(*continued*)

Table 2. (*continued*)

Indicator	Formula / definition	Stage of the DMAIC SS methodology	Paper ID
O	Number of Opportunities for defect - Potential failure of process or project activities	C	[46]
D	Number of Defects - Defects of process or project activities	C	[46]
DI	Number of Defective Items - Defective process or project activities	C	[46]
CTD	Correction Time per Defect - Time needed for repeating related activity	C	[46]
CCD	Correction Cost per Defect - cost of repeating related activity	C	[46]
IaC	Items at Complete - Number of process or project activities	C	[46]
C	Completeness of activities - What part of activities is completed or what is the completion rate of activities	C	[46]
SD	Start Date of activity in time - The date when assessed activities have started	C	[46]
PCD	Planned Completeness Date - Planned date of current completeness of assessed activities	C	[46]
ACD	Actual Completeness Date - Actual date of current completeness of assessed activities	C	[46]

At the same time, the number of publications in the field of performance evaluation and related indicators is considerably lower than that on the general application of the Six Sigma methodology.

Fifty-nine indicators were identified to verify the effectiveness of improvement activities in Six Sigma projects. These indicators were used in different phases of the DMAIC cycle, two of the indicators discussed relate to risk assessment.

The analysis showed that indicators can be divided into groups taking into account different criteria. The division of indicators can be made, among others, on the basis of the area of application: production processes, maintenance, healthcare, laboratories - each of these areas defines specific indicators. A further subdivision can be made taking into account whether an indicator sets a reference point otherwise known as zero value, or whether it can be calculated on the basis of generally accepted formulas. The benchmark in this case is the value of the variable in question before the improvement measures were implemented. The next breakdown criterion is the calculation of the indicator as time or number of deficiencies. One of the last criteria seen is the breakdown of indicators taking into account the stage of the improvement project in which it is used. A significant number of indicators are only calculated in the control phase. In addition, the indicators presented were only used in one of the project phases. The effectiveness of improvement activities is therefore influenced by indicators that can be used at each stage of the project.

5 Conclusions and Directions for Further Research

This article focuses on identifying the indicators used in the Six Sigma methodology to assess the effectiveness of improvement activities. The analysis of the literature shows that the most common phase of the DMAIC cycle during which the effectiveness of improvement activities is calculated is the Control phase. Furthermore, the results obtained indicate, the application of this philosophy is broad and can be implemented in different economic contexts, and its impact is more often studied in the industrial sector. The most commonly used indicators are Cpk, DPMO, PPM, SSL, % scrap and machine working cycle. In the bibliographic sources analyzed, there is a lack of indicators to evaluate the effectiveness of improvement activities at different stages of Six Sigma project implementation, which should be the subject of further research.

Most of the indicators listed are calculated in the control phase - after the improvement measures have been implemented. This means that the team analysing the problem can only draw conclusions from the implemented actions after they have been implemented. The Six Sigma methodology for process improvement used to date does not provide a method by which the implemented measures can be evaluated at each stage of a Six Sigma project. Therefore, it is reasonable to carry out an overall evaluation of the activities in the Six Sigma methodology implementation and then define a method to be used to evaluate the effectiveness of all activities in a process. The effectiveness of improvement activities is therefore influenced by indicators that can be used at each stage of the project. It will already be known to the problem analysis team at the project implementation stage whether the defined activities will have the intended effect.

References

1. Pugna, A., Negrea, R., Miclea, S.: Using six sigma methodology to improve the assembly process in an automotive company. Procedia Soc. Behav. Sci. **221**, 308–316 (2016)
2. Smętkowska, M., Mrugalska, B.: Using six sigma DMAIC to improve the quality of the production process: a case study. Procedia Soc. Behav. Sci. **238**, 590–596 (2018)
3. Nedeliaková, E., Štefancová, V., Kudláč, Š: Six sigma and dynamic models application as an important quality management tool in railway companies. Procedia Eng. **187**, 242–248 (2017)
4. Meo, I., Papetti, A., Gregori, F., Germani, M.: Optimization of energy efficiency of a production site: a method to support data acquisition for effective action plans. Procedia Manuf. **11**, 760–767 (2017)
5. Chen, S., Fan, S., Xiong, J., Zhang, W.: The design of JMP/SAP based six sigma management system and its application in SMED. Procedia Eng. **174**, 416–424 (2017)
6. Titmarsh, R., Assad, F., Harrison, R.: Contributions of lean six sigma to sustainable manufacturing requirements: an Industry 4.0 perspective. Procedia Cirp **90**, 589–593 (2020)
7. Costa, J.P., Lopes, I.D.S., Brito, J.P.: Six sigma application for quality improvement of the pin insertion process. Procedia Manuf. **38**, 1592–1599 (2019)
8. Ikumapayi, O.M., Akinlabi, E.T., Mwema, F.M., Ogbonna, O.S.: Six sigma versus lean manufacturing-an overview. Mater. Today: Proc. **26**, 3275–3281 (2020)
9. Hwang, G.H., Lee, K.C., Seo, Y.W.: How does Six Sigma influence creativity and corporate performance through exploration and exploitation? Total Qual. Manag. Bus. Excell. **29**(13–14), 1668–1684 (2018)
10. Yu, H., Yang, J., Ding, X., Wang, H., Wang, S.: Six sigma robust optimization method based on a pseudo single-loop strategy and RFR-DBN with insufficient samples. Comput. Struct. **257**, 106653 (2021)
11. Yang, C.C., Jou, Y.T., Lin, M.C., Silitonga, R.M., Sukwadi, R.: The development of the new process of design for six sigma (DFSS) and its application. Sustainability **14**(15), 9294 (2022)
12. Panagopoulos, I., Atkin, C., Sikora, I.: Developing a performance indicators lean-sigma framework for measuring aviation system's safety performance. Transp. Res. Procedia **22**, 35–44 (2017)
13. Essawy, M.H., Attia, E.A., El-Assal, A.M.: An integrated model for achieving business strategic objectives with a case study. Int. J. Six Sigma Compet. Adv. **11**(4), 253–276 (2019)
14. Costa, T., Silva, F.J.G., Ferreira, L.P.: Improve the extrusion process in tire production using six sigma methodology. Procedia Manuf. **13**, 1104–1111 (2017)
15. Daza-Moran, F.J., Ramirez-Alva, A.S.D., Quiroz-Flores, J.C., Collao-Diaz, M.F.: Improving service level performance by implementing lean six sigma in SMEs of the gaming peripherals industry in Peru: a case study. S. Afr. J. Ind. Eng. **33**(2), 168–184 (2022)
16. Singh, J., Ahuja, I.P.S., Singh, H., Singh, A.: Development and implementation of autonomous quality management system (AQMS) in an automotive manufacturing using quality 4.0 concept-a case study. Comput. Industr. Eng. **168**, 108121 (2022)
17. Bottani, E., Montanari, R., Volpi, A., Tebaldi, L., Di Maria, G.: Statistical process control of assembly lines in a manufacturing plant: process capability assessment. Procedia Comput. Sci. **180**, 1024–1033 (2021)
18. Vincent, A., Pocius, D., Huang, Y.: Six Sigma performance of quality indicators in total testing process of point-of-care glucose measurement: a two-year review. Pract. Lab. Med. **25**, e00215 (2021)
19. Mendes, F.S., Soares, J.M.: Financial KPI analysis in the implementation of DMAIC in a pharmaceutical organization - a case study. Int. J. Qual. Res. **15**(1) (2021)

20. Lemke, J., Kijewska, K., Iwan, S., Dudek, T.: Six sigma in urban logistics management-a case study. Sustainability **13**(8), 4302 (2021)
21. Tasdemir, C., Gazo, R.: Validation of sustainability benchmarking tool in the context of value-added wood products manufacturing activities. Sustainability **11**(8), 2361 (2019)
22. Fontalvo-Herrera, T.J., Herrera, R., Zambrano, J.: Three-phase method to assess the logistics service using six sigma metrics, Hotelling's T-square control chart and a principal component capacity indicator. Int. J. Prod. Qual. Manag. **35**(1), 17–39 (2022)
23. Pereira, M.T., Bento, M.I., Ferreira, L.P., Sá, J.C., Silva, F.J.G., Baptista, A.: Using six sigma to analyse customer satisfaction at the product design and development stage. Procedia Manuf. **38**, 1608–1614 (2019)
24. Jingjing, Z.H.U., Xiaojun, W.A.N.G., Zhang, H., Yuwen, L.I., Ruixing, W.A.N.G., Zhiping, Q.I.U.: Six sigma robust design optimization for thermal protection system of hypersonic vehicles based on successive response surface method. Chin. J. Aeronaut. **32**(9), 2095–2108 (2019)
25. Ibrahim, I., Sultan, M., Yassine, O.G., Zaki, A., Elamir, H., Guirguis, W.: Using Lean six sigma to improve timeliness of clinical laboratory test results in a university hospital in Egypt. Int. J. Lean Six Sigma **13**, 1159–1183 (2022)
26. Zorbozan, N., Zorbozan, O.: Evaluation of pre-analytical and postanalytical phases in clinical biochemistry laboratory according to IFCC laboratory errors and patient safety specifications. Biochemia Medica **32**(3), 357–365 (2022)
27. Wang, H., Zhang, G., Zhou, S., Ouyang, L.: Implementation of a novel six sigma multi-objective robustness optimization method based on the improved response surface model for bumper system design. Thin-Walled Struct. **167**, 108257 (2021)
28. Bhat, S., Gijo, E.V., Rego, A.M., Bhat, V.S.: Lean six sigma competitiveness for micro, small and medium enterprises (MSMEs): an action research in the Indian context. TQM J. **33**(2), 379–406 (2021)
29. Priya, S.K., Jayakumar, V., Kumar, S.S.: Defect analysis and lean six sigma implementation experience in an automotive assembly line. Mater. Today: Proc. **22**, 948–958 (2020)
30. Abbes, N., Sejri, N., Xu, J., Cheikhrouhou, M.: New lean six sigma readiness assessment model using fuzzy logic: case study within clothing industry. Alex. Eng. J. **61**(11), 9079–9094 (2022)
31. Yadav, N., Shankar, R., Singh, S.P.: Cognitive aspects of lean six sigma. Qual. Quant. **56**(2), 607–666 (2022)
32. Bastos, N.M., Alves, A.C., Castro, F.X., Duarte, J., Ferreira, L.P., Silva, F.J.G.: Reconfiguration of assembly lines using lean thinking in an electronics components' manufacturer for the automotive industry. Procedia Manuf. **55**, 383–392 (2021)
33. Kochov, A., Argilovski, A.: Case study: six sigma project for reducing manual handling of materials in real manufacturing company. Tehnički glasnik **14**(4), 499–506 (2020)
34. Ifrim, A.M., Biţan, G.E., Maier, D., Fogoroş, T.E.: Improving the performance of organizational innovation processes by applying the six sigma methodology. In: Proceedings of the International Conference on Business Excellence, vol. 14, no. 1, pp. 1098-1108 (2020)
35. Acero, R., Torralba, M., Pérez-Moya, R., Pozo, J.A.: Order processing improvement in military logistics by value stream analysis lean methodology. Procedia Manuf. **41**, 74–81 (2019)
36. Guleria, P., Pathania, A., Bhatti, H., Rojhe, K., Mahto, D.: Leveraging lean six sigma: reducing defects and rejections in filter manufacturing industry. Mater. Today: Proc. **46**, 8532–8539 (2021)
37. Rathi, R., Kaswan, M.S., Garza-Reyes, J.A., Antony, J., Cross, J.: Green lean six sigma for improving manufacturing sustainability: framework development and validation. J. Clean. Prod. **345**, 131130 (2022)

38. Bloj, M.D., Moica, S., Veres, C.: Lean six sigma in the energy service sector: a case study. Procedia Manuf. **46**, 352–358 (2020)
39. Castro, S., Riedel, C.: Assessment of the implementation of manufacturing excellence in a fiber-based packaging manufacturing environment. Procedia CIRP **63**, 113–118 (2017)
40. Al-Qatawneh, L., Abdallah, A.A., Zalloum, S.S.: Six sigma application in healthcare logistics: a framework and a case study. J. Healthc. Eng. (2019)
41. Aleksandrovskaya, L.N., Ardalionova, A.E., Papic, L.: Application of probability distributions mixture of safety indicator in risk assessment problems. Int. J. Syst. Assur. Eng. Manag. **10**, 3–11 (2019)
42. Wang, C.H., Chen, K.S.: New process yield index of asymmetric tolerances for bootstrap method and six sigma approach. Int. J. Prod. Econ. **219**, 216–223 (2020)
43. Sahu, P.K., Ramisetti, N.R., Cecchi, T., Swain, S., Patro, C.S., Panda, J.: An overview of experimental designs in HPLC method development and validation. J. Pharm. Biomed. Anal. **147**, 590–611 (2018)
44. Chen, K.S., Wang, C.H., Tan, K.H.: Developing a fuzzy green supplier selection model using six sigma quality indices. Int. J. Prod. Econ. **212**, 1–7 (2019)
45. Mjimer, I., Aoula, E.S., Achouyab, E.H.: Monitoring of overall equipment effectiveness by multivariate statistical process control. Int. J. Lean Six Sigma **13**(4), 847–862 (2022)
46. Csiszér, T.: The FAR model-the 'Rubik's Cube'of process and project monitoring. Acta Polytechnica Hungarica **15**(4) (2018)

Analysis of Basic Characteristics of Textile Yarn Using Image Processing Techniques

Marta Rodrigues[1], Filipe Pereira[1,2], and José Machado[1](✉)

[1] University of Minho, MEtRICs Research Centre, 4804–533 Guimarães, Portugal
jmachado@dem.uminho.pt

[2] University of Minho, Algoritmi Research Centre, 4804–533 Guimarães, Portugal

Abstract. This paper intends to explore part of the development of a simple and compact system that operates an algorithm based on image processing techniques, concerning the characterization of textile yarn. Specifically, it aims to conclude about parameters such as the diameter, the linear mass, the volume, the imperfections, the twist's pitch and direction, and the pilosity coefficient. These are considered essential regarding the yarn parametrization for quality control. The main goal of this project was to develop a practical solution based on image processing that arises as innovative in comparison to the existing equipment in the industry.

Keywords: Textile yarn analysis · Image processing · Quality yarn basic parameters

1 Introduction

The current textile industry has available many technologies that allow the analysis of textile yarn concerning its quality control. However, these methodologies don't enable the attainment of all relevant parameters and are associated with robust and high-cost equipment. To get a full characterization of textile yarn, several testing procedures are needed, which is very time-consuming.

The use of image processing techniques in the textile industry is still covered as an innovative progress. This knowledge area is promising in the sense of offering a variety of improvements applied to the measurement, analysis, and quality control tools for textile yarn production and maintenance.

This article intends to explore the development of an algorithm that acknowledges the most important parameters regarding yarn quality control, such as diameter, linear mass, specific volume, faults, pilosity coefficient, and twist direction and pitch.

This paper is divided into a total of five sections, whereas the current purpose is the introduction to the theme. Secondly, a brief context of the current market for yarn quality control is given. The third chapter lodges the methodologies behind the application of image processing techniques. Next, the results are introduced and discussed accordingly to what was expected. Finally, a few conclusions about the explored work are shown.

A. Burduk et al. (Eds.): ISPEM 2023, LNNS 790, pp. 215–227, 2023.
https://doi.org/10.1007/978-3-031-45021-1_16

2 Methods for Textile Yarn Parametrization

Many technologies allow to conclude about the yarn's homogeneity, on which its quality is based. Namely, there are a few parameters that are considered relevant to this arrangement, like linear mass, faults, pilosity coefficient, and twist. Specifically, Uster is a company that remains the main reference in the industry, as it offers a variety of tools that acknowledge different necessities. However, the prices are very high, and the equipment is heavy and complex, which encumbers its handle.

Regarding yarn's hairiness, there is a diverse range of methodologies like microscopic methods, weighting methods, and electrostatics methods. The first one consists in observing directly the yarn, so it is possible to conclude its morphology. However, this technique is time-consuming, as it takes a lot of analyzed samples to get accurate results. The weighting techniques are based on the difference in the yarn's mass before and after a process named singeing, which burns loosed fibers. For threads with a smaller length, this method is inefficient since is more challenging to access all hairs. Lastly, the electrostatic procedure involves the creation of an electric field that will lift every loosed fiber and, consequently, allow its score. Of all technologies, this one is the most produced, although it slightly alters the yarn's morphology. As for an example of an equipment, the Uster Zweigle HL400 grants the study of hairiness through the automatic recognition of the yarn's core. Subsequently, it counts every hair that protrudes from the identified core [1, 2].

The study of the yarn's uniformity is grounded by optical and capacitive measurements that study the mass variation per unit of length. The capacitive system is established by two metal plates that are positioned parallelly as a capacitor is used as a sensor. When a yarn is placed in the electrical field it occurs capacity variation. It is fundamental to ensure the yarn's morphology, its dielectric constant, and humidity content, so it maintains the stability between the capacity and the mass. Moreover, the optical system concerns a light source and photodiode detector. When the yarn mass appears in the illuminated field, partial light will be absorbed and the other will be reflected. The photodiodes measure the amount of light that is absorbed and reflected. The Uster Tester 6-C800 and the Uster Classimat 5 are two equipment references that, respectively, conclude about the mass variation and the detection of faults [3, 4].

Regarding the study of the twist, the traditional technique is based on twisting-untwisting the yarn repeatedly. The principle of this method is that the contraction of a single yarn length is equal to any number of twists. Although the procedure is simple, its disadvantage is its inaccuracy as the applied tension easily influences the results. Uster also provides an appliance that surveys this parameter, named Uster Zweigle Twist Tester 5 [5].

Overall, all the technologies cited are solely capable of analyzing one parameter. If a full characterization of the yarn for quality control is needed, it's necessary to compound several technologies and equipment, which is unfit and time-wasting. The disclosure of image processing techniques arises as an innovative solution.

3 Image Processing Methodology

This chapter uncovers the methodology developed for the application of image processing techniques to extract yarn's parameters, such as diameter, linear mass, specific volume, faults, pilosity coefficient, and twist's pitch and direction.

Image processing is defined as a scientific area that brings together a series of different techniques. These are developed in mathematical functions that allow the handling and improvement of an image to obtain a set of goals, like useful data, for example. It's possible to describe three different elements that compose the whole system: the image acquisition element, the image processing element, and the image exhibition element [6].

Thus, the first step was image acquisition for which a simple capture system was established. This prototype was composed of a low-cost USB camera – whose specifications are resumed in Table 4 –, and a light source, that was implemented into the camera itself. The camera was the chosen element for the image acquisition. Moreover, it was important to create a non-reflective environment to avoid light reflection. The achievement of this was completed by using a box that was painted internally black (Table 1).

In short, it was created as an accessible system that allowed to capture yarn samples in real-time, so later these could be submitted into the created algorithm to extract the required parameters. This system worked as a support device to test the veracity of the application of the image processing techniques.

3.1 Extraction of Yarn's Parameters

Firstly, the captured image went through a simple pre-image processing treatment. This task was supported by the development platform Visual Studio, to which the open-source library OpenCV was employed. These were the image processing elements of the elaborated system, being that it was the means to create the algorithm.

The procedure started with the acquisition of an image of colors, which was the base to create a grayscale image by conversion. This initial output was the starting point for the whole methodology. A Gauss filter was then applied to remove noise, that is characterized by sudden variations in density in an image, giving it a textured appearance.

Image segmentation established the creation of a binary image from the grayscale image, where the yarn was highlighted from the background. This division came from

Table 1. Camera's principal specifications

Camera's Specifications	
Model	USB Microscopic Camera
Auxiliary Light Source	8 LED Lights (adjustable intensity)
Resolution	640*480 Pixels
Magnification	50X – 1000X
Frame Rate	30 f/s

the definition of a threshold value that works as a reference to the pixel value conversion. If the pixel value was lower than the threshold value, the pixel assumed its value as 0 (black); if the pixel value was greater than the threshold value, the pixel assumed its value as 1 (white). Therefore, it was created two sub-groups within the same range of intensity values. In this regard, the Otsu algorithm was chosen as it automatically presumes the ideal threshold value [7]. Figure 1 illustrates, respectively, the results from the application of the Gauss filter, and the application of the Otsu algorithm.

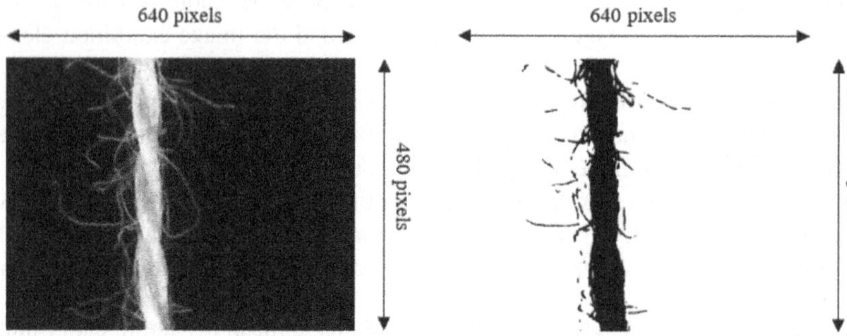

Fig. 1. Result from the application of the Gauss filter (left), and after the application of the Otsu algorithm for segmentation (right) on a sample of purple cotton yarn.

The diameter was the first parameter to be analyzed. For this, the segmented image went through a morphological operation to withdraw the representation of the loosed fibers. The closing operation was chosen as it removed the necessary pixels to individualize the yarn's core by closing any existing gaps. To ensure this result, the last filter was used to remove any residual noise. Finally, it was implemented a binary inversion to put the yarn area as the focal point of analysis. Figure 2 shows the result of this procedure, where is mainly possible to visualize the removal of the loosed fibers.

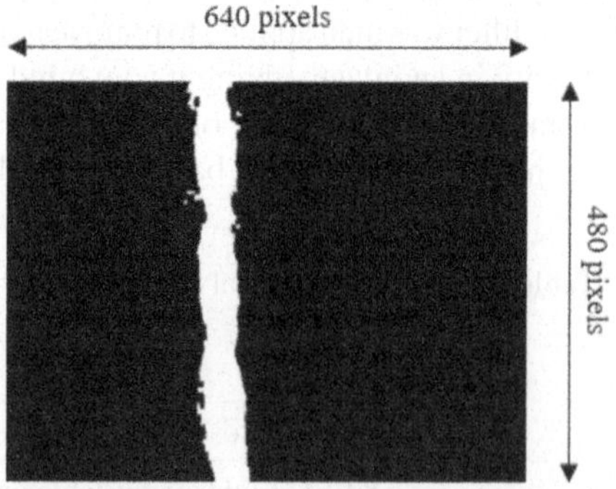

Fig. 2. Result from the application of the full procedure on a sample of purple cotton yarn.

To obtain the diameter, each pixel was wandered to conclude about its intensity value within a range between 0 (black) and 1 (white). Since the yarn is represented by white pixels, the location of each one of these was kept in a vector created by the algorithm. When the image was checked, the diameter was calculated through the mean value of every value saved into the mentioned vector. This parameter has a mathematical relationship with linear mass, which is expressed in grams per kilometer (g/km), and it's also known by the standard as Tex. The following Eq. 1 elucidates the mentioned association.

$$Tex = \phi\left(\frac{g}{km}\right) \times \rho\left(\frac{g}{cm^3}\right) \times \pi\left(\frac{d(cm)}{2}\right)^2 \times \left(\frac{10^5 cm}{km}\right) \tag{1}$$

Namely, ϕ is the porosity, ρ is the material's density, and d is the yarn's diameter.

Moreover, it was also possible to reckon on the yarn's specific volume (cm^3/gm), which is based on Eq. 2.

$$V_t = \frac{\pi \times r^2}{Tex \times 10^5} \tag{2}$$

where V_t is the specific volume and r is the yarn's radius in mm.

Next, the yarn's defects were studied to determine if there was any significant mass variation in relation to the mean value. It's possible to distinguish three different types:

- Thin Point: when the linear mass at a given point is smaller than the mean value by 50%.
- Thick Point: when the linear mass at a given point is greater than the mean value by 50%.
- Nep: when there's a severe oscillation of the linear mass with an increase equal to or greater than 100%.

Figure 3 illustrates the different types of defects that were enunciated.

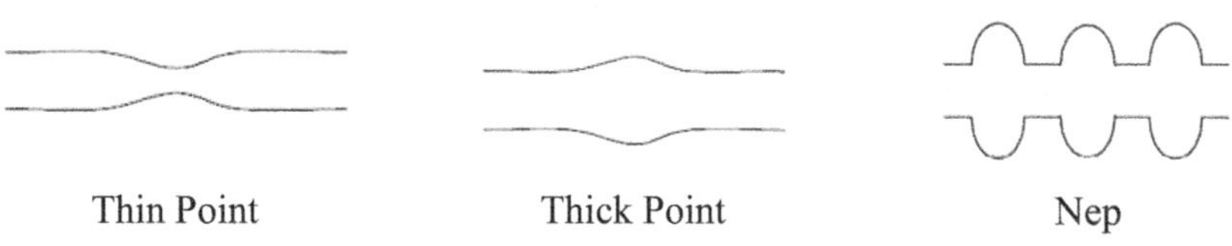

Fig. 3. Types of yarn defects [8].

The yarn sample was divided into five equal sections, as shown in Fig. 4. Each portion was analyzed, whereby the corresponding linear mass was determined. Then, the algorithm took the obtained value for each and compared it to the mean linear mass value, to calculate the variation between both values. According to the percentages of variation previously presented, it was verified whether there was any significant dissimilarity. If so, the respective defect was identified and an increment was added to a vector created by the algorithm, to account the sum of defects in every sample of textile yarn examined.

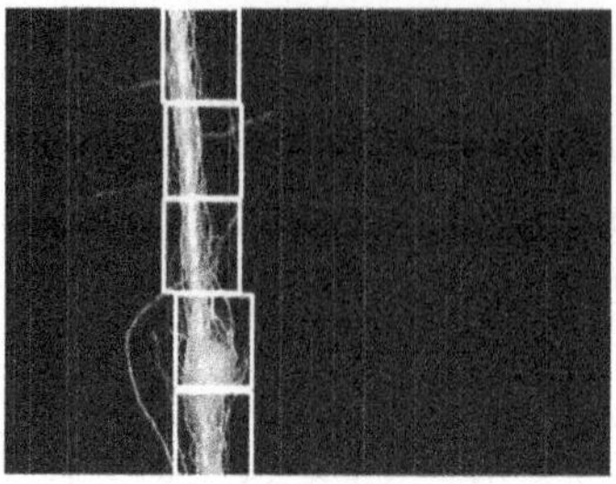

Fig. 4. Division of a sample of yellow cotton yarn to determine if there's a defect.

The study of the pilosity coefficient came after, as the study of the number of loosed fibers per unit of length, as Eq. 3 describes it.

$$H = \frac{l_H}{l_{Yarn}} \tag{3}$$

where H is the pilosity coefficient, l_H is the total length of loosed fibers (in meters) and l_{Yarn} is the total yarn's length (in meters). This coefficient characterizes the yarn's hairiness, which is composed of fibers that emerge from the yarn's core. There are two types of loosed fibers that classify this parameter (Fig. 5):

- Looped fibers: fibers whose ends are incorporated into the core of the thread, while the central part protrudes from it.
- Protruding fibers: only one end of the fibers is attached to the yarn's core, and the rest of the body is freely exposed.

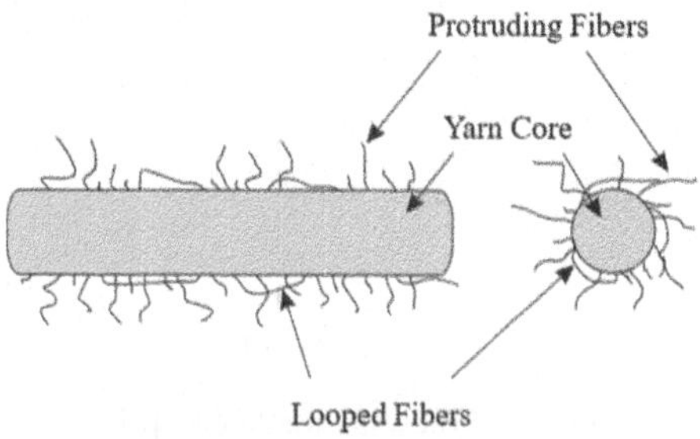

Fig. 5. Types of Loosed Fibers [2].

Regarding the image processing for the study of this parameter, it was necessary to apply an additional image treatment to illustrate again the hairiness, as it was removed previously to study other parameters. As such, the yarn's core was dilatated to enlarge its boundaries by adding white pixels. Then, the result was subtracted from the segmented image, which resulted in eliminating the yarn's core. Lastly, the result from the subtraction went through a thinning process which intends to generate the object's skeleton by decreasing its total area. Next, in Fig. 6, observe the result of this strategy.

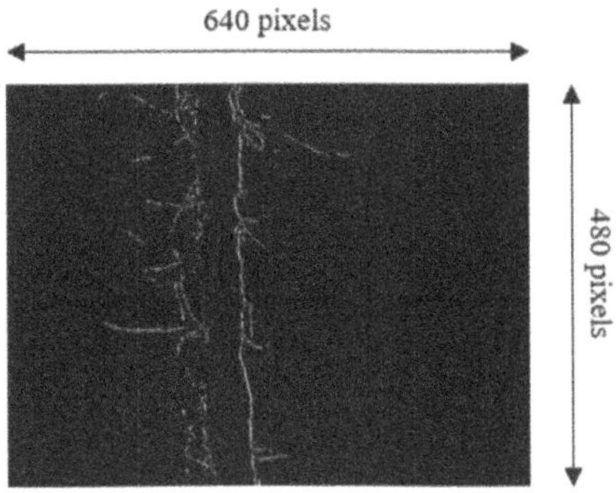

Fig. 6. Result from the skeletonization technique on a sample of purple cotton yarn.

Each white pixel from the figure above was counted and kept in a vector. Once the full examination was completed, every pixel area was calculated, and posteriorly, the main area was totalized. The result was then divided by the total yarn's length represented, to obtain the number of loosed fibers per unit of length (in millimeters).

The final parameter studied was the twist which is characterized by its pitch and direction. For this, the light source for image acquisition was adjusted to create shadows in the yarn to highlight the different twist portions. After capturing a new image, it was necessary to put it through an image processing procedure. Following the initial steps of grayscale conversion and filter Gauss application, the image segmentation took a new threshold value. While previously the threshold value was automatically configured by the Otsu algorithm, this value was configured manually after many attempts to achieve a goal. Finally, the residual noise was eliminated by bearing on an erosion operation. Next, in Fig. 7, observe the final product for this parameter's analysis.

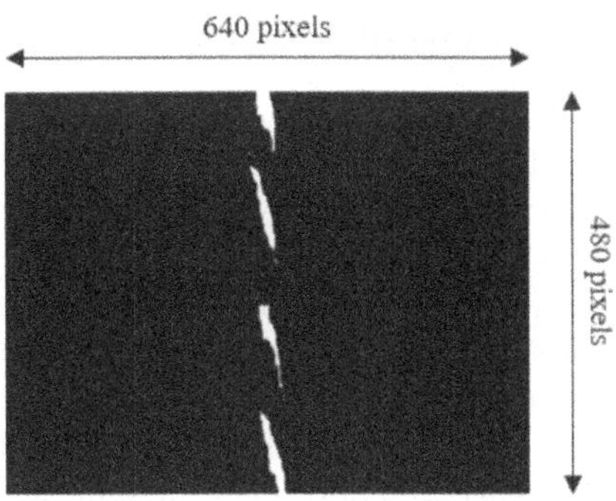

Fig. 7. Results from the application of a morphological operation of the erosion on a sample of white polyester yarn.

The individualized objects shown in Fig. 6 were identified and then labeled one by one to acknowledge the examination of the twist's direction and the twist's pitch. Firstly, the twist's direction was examined by setting all the object's orientations in space. This was determined by the Principal Components Analysis, which is a statistical method that grants a linear transformation of the input information by stipulating two eigenvectors. These indicate the orientation in which the data varies the most. The attribution of the direction type was given accordingly to the following logic:

- If the object's orientation is lower than 90º, it assumes type S.
- If the object's orientation is higher than 90º, it assumes type Z.

In turn, the twist's pitch problem was approached by the Bounding Box Rectangle method, which recognizes a set of points, and limits it with the smallest box possible. For this, each object was analyzed and limited. The height of each bounding box corresponds to the respective pitch [9, 10]. Figure 8 illustrates the conversion between the different labelled objects and their bounding boxes.

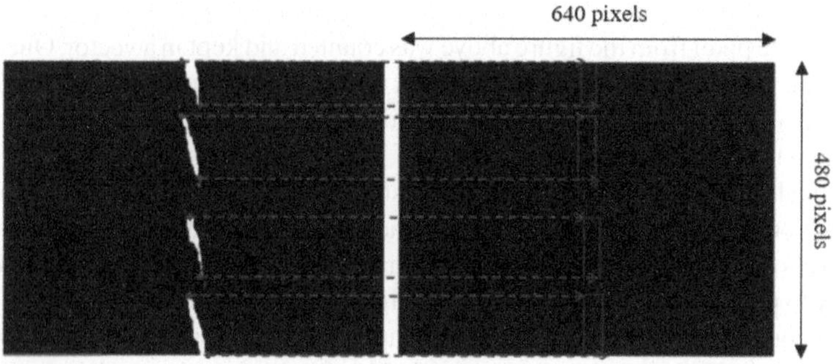

Fig. 8. Conversion of the labelled objects to its respective bounding boxes on a sample of white polyester yarn.

3.2 Testing of the Developed Algorithm and the Image Acquisition System

After the idealization of the algorithm for the yarn parameter's extraction, it was necessary to define a series of quality factors that affect the capture of an ideal digital image. As so, illumination, contrast, and yarn position consisted of crucial aspects that were considered.

Regarding the illumination, it was mentioned previously that the USB microscopic camera had a light source implemented into itself. In addition to this, it was verified that an external light source was also required. The combination of both allowed an accurate digital representation of the yarn, specifically the fibers, which offered greater resistance to being represented by the image exhibition element in the absence of a suitable light source.

The contrast was studied through the change of color of the background from where the yarn digital image was acquired. First, it was tested an ordinary background that didn't answered to any specifications. While the yarn was correctly portrayed, there was a lot of noise associated, which should be avoid for the image processing. Therefore, a black background was implemented to guarantee that the light wasn't reflected. It was also found that using bright colored yarn enhanced the contrast.

About the yarn positioning, promptly it was ascertained that the yarn should maintain a constant position during the sample taking. Various working distances, i.e., distance between the camera and the yarn, were tested to verify which was the most suitable to keep every detail of the yarn. Therefore, as the camera didn't have an excellent resolution, it was determined that the yarn should be kept close to the camera. Furthermore, while the yarn moved in front of the camera, it was also fundamental to keep it steady to avoid any blurring on the digital image.

In summary, the most advantageous scenario for image acquisition covered the following conditions:

- The textile yarn, preferably bright colored, was inserted into a closed environment. A box was used for this purpose, and it was painted internally by black to remove any possible noise on the digital image.
- The illumination was supported by two light sources: a LED ring implemented in the USB camera, and an external LED ring. Both rings allowed its adjustment manually, which granted to modify the light intensity according to the ruling goal.
- The yarn was positioned at a working distance of 10.0 mm, and it was kept constant during the whole testing procedure. The camera was operated by an 890X magnification. Furthermore, there was no need to calibrate the camera, as it was a generic one.

After the definition of the quality factors, these were applied into the system, so the image processing phase could be initiated. There were three different types of yarn that were submitted for analysis: a purple cotton yarn, a yellow cotton yarn, and a white polyester yarn. The system was designed to study only these types of yarn, since, although the material was different, the yarns were structurally similar.

4 Analysis and Discussion of Results

The study relied upon two different materials: cotton and polyester like it was stated before. A total of 550 samples were acquired for each yarn to put into practice the developed algorithm. The obtained results were compared with the Uster Tester 3, which is equipment that grants values for the linear mass and the hairiness coefficient. As such, only for these was possible to obtain a variation error. Observe Tables 2 and 3.

Table 2. Results for the linear mass.

Type of Yarn	System	Uster Tester 3	Variation Error
Cotton Purple Yarn	63.9 g/km	56.4 g/km	13.2%
Cotton Yellow Yarn	25.4 g/km	23.8 g/km	6.66%
Polyester White Yarn	21.2 g/km	20.9 g/km	1.54%

Mainly, the linear mass presented the best results, as the variation errors were lower. The most concerning aspect translated to the gotten results for the detection of loosed

Table 3. Results for the hairiness coefficient.

Type of Yarn	System	Uster Tester 3	Variation Error
Cotton Purple Yarn	5.58	5.44	2.56%
Cotton Yellow Yarn	9.91	4.77	107.7%
Polyester White Yarn	6.31	3.61	74.8%

fibers. The cotton yellow yarn and the polyester white yarn conferred very high values that went behind what was expected. During the testing, it was possible to visualize the many outcomes from the established procedure, and for that, the loosed fibers' morphology was also visible. The cotton purple yarn's hairiness was easily detected as it was less complex and smaller. These loosed fibers spread in a balanced and progressive way through the yarn's length, like seen in Fig. 9. Although the algorithm was not programmed to detect and distinguish the two types of hairiness, it was possible to discern by the naked eye lots of looped fibers.

Fig. 9. Sample of purple cotton yarn.

The white yarn and yellow yarn had a wilder aspect, as the fibers easily intertwined with each other. Nevertheless, they were also longer. Figure 10 illustrates two examples of samples for the cited yarns.

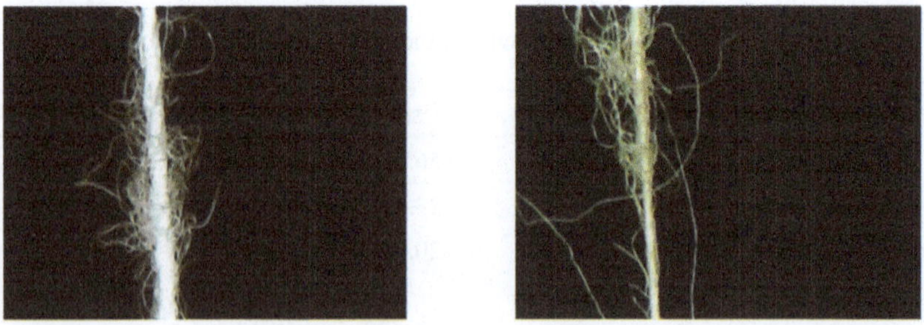

Fig. 10. Sample of white polyester yarn (left), and sample of yellow cotton yarn (right).

When comparing these to the sample of purple cotton yarn presented previously, a big difference in the respective morphology is visible. The study of the pilosity coefficient for the yellow yarn and white yarn revealed itself as more complex, as there was a bigger yarn density overall. Several samples indicated a structural irregularity in the loosed fibers.

For the analysis of the values of the average diameter and the specific volume, it was considered the equations previously displayed. Because they both derive mathematically from the linear mass, the results were found satisfactory following the conclusions taken from those.

Regarding the yarn's faults, it was viable to visualize and perceive the three different types along the samples. Table 4 presents the total of defects attained.

Table 4. Results for the counting of faults.

Type of Yarn	Thin Points	Thick Points	Neps
Cotton Purple Yarn	0	15	1
Cotton Yellow Yarn	14	31	17
Polyester White Yarn	3	13	1

The cotton yellow yarn revealed itself as the most defective, with a total of 62 faults. The other two yarns were very similar in numbers, and the only difference was that the purple yarn had no thin points. Overall, thick points were the most identified type. These values were validated by the direct visualization and confirmation of the captured digital images for each sample.

Finally, for the twist's direction, the results were also directly confirmed by observation of the yarn's pictorial representation, as it was visible to the naked eye. Only the yellow yarn was not suitable for this procedure, and for that, it was only possible to retain conclusions for the purple and white yarns. This outcome was also applied to the twist's pitch, which got the results presented in Table 5.

Table 5. Results for the twist's pitch.

Type of Yarn	System
Cotton Purple Yarn	0.9 mm
Cotton Yellow Yarn	Not identified
Polyester White Yarn	0.4 mm

The obtained results are adequate. The outcomes from the table above consist of the average values obtained for each tested sample. The white yarn's values were in a smaller range, for which these were more consistent.

5 Conclusions and Future Work

Overall, the results were optimistic, considering the created prototype was very low-cost. It allowed to achieve the stipulated goals, namely, to obtain the full characterization of textile yarn for quality control. However, it was also possible to detect a few limitations that should be paid attention to for future work.

Firstly, the most concerning aspect was based on the hairiness coefficient results, which had a higher error variation in comparison with the Uster Tester 3. The morphology of the loosed fibers for the yellow and white yarn was very different from the purple yarn, which was less complex and easily detected. Mostly, the fibers were prone to intertwine with each other, making their spotting difficult. This contrast could justify the gotten results. Furthermore, the improvement of this aspect could be treated with the introduction of a static electricity generator. This mechanism would permit the stretching of the loosed fibers and, consequently, untangle the fibers from each other, so its detection was more perceptible.

Likewise, for future work, the methodology established for the twist's analysis would also need to be reviewed. Although the results were good, the authors will focus on verifying correctly the gotten results, especially for the twist's pitch. Specifically, it would be important to have access to a piece of equipment like the Uster Zweigle Twist Tester 5, to corroborate the gotten values for the twist's pitch.

The improvement of the designed system should be prioritized as well, pointedly through its automatization, like by developing an automatic unwinding and winding system, for example. It would also be interesting to insert an interface that would allow the user to handle dependent variables, like the threshold value for the twist's study, for example. In addition, to reaffirm the settled procedure, it would be compelling to use a camera with a better resolution to acquire digital images with better quality still making valuable the obtained results.

The application of the system based on image processing in the textile industry would revolutionize the study and analysis of yarn quality control, as it offers a simpler and low-cost solution given the current situation. While there are technologies and equipment that allow the study of the cited yarn parameters, these are particularly complex to handle, time-consuming, and very high cost. Nevertheless, this area of study contains a great capacity for continuous improvement, where even artificial intelligence can play a key role in textile yarn parameterization, as verified in [11].

References

1. Wang, X.H., Wang, J.Y., Zhang, J.L., Liang, H.W., Kou, P.M.: Study on the detection of yarn hairiness morphology based on image processing technique. Int. Conf. Mach. Learn. Cybern. **5**, 2332–2336 (2010)
2. Carvalho, V., Cardoso, P., Belsley, M., Vasconcelos, R.M., Soares, F.O.: Yarn diameter characterization using two orthogonal directions. Exp. Tech. **35**(5), 7–15 (2011)
3. Sparavigna, A., Broglia, E., Lugli, S.: Beyond capacitive systems with optical measurements for yarn evenness evaluation. Mechatronics **14**(10), 1183–1196 (2004)
4. Carvalho, V., Cardoso, P., Belsley, M., Vasconcelos, R.M., Soares, F.O.: Development of a yarn evenness measurement and hairiness analysis system. In: IECON 2006-32nd Annual Conference on IEEE Industrial Electronics, pp. 3621−3626. IEEE (2006)

5. Czaplicki, Z.: A new method of measuring twist of yarn. Fibres Text. East. Eur. **14**(1), 27–29 (2006)
6. Behera, B.K.: Image-processing in textiles. Text. Prog. **35**, 2–4 (2004)
7. Bangare, S.L., Dubal, A., Bangare, P.S.: Reviewing Otsu's method for image thresholding. Int. J. Appl. Eng. Res. **10**(9), 21777–21783 (2015)
8. Pinto, R., Pereira, F., Carvalho, V., Soares, F., Vasconcelos, R.: Yarn linear mass determination using image processing: first insights. In: IECON 2019-45th Annual Conference of the IEEE Industrial Electronics Society, vol. 1, pp. 198–203. IEEE (2019)
9. Pandey, P., Singh, Y., Tripathi, S.: Image processing using principle com-ponent analysis. Int. J. Comput. Appl. **15**(4) 2011
10. Chelishchev, P., Sørby, K.: Estimation of minimum volume of bounding box for geometrical metrology. Int. J. Metrol. Qual. Eng. **11**(9) (2020)
11. Pereira, F., et al.: Intelligent computer vision system for analysis and characterization of yarn quality. Electronics **12**(1), p. 236 (2023)

Modeling the Impact of Leadership on Change Management Decisions Using System Dynamics Simulation

Manuka Lakdinu[1], M. Mavin De Silva[1,2](✉), Kavith Balachandra[1], Amila Thibbotuwawa[3], H. Niles Perera[3], Tadachika Nakayama[2], and Peter Nielsen[4]

[1] Department of Transport Management and Logistics Engineering, University of Moratuwa, Moratuwa, Sri Lanka
mavinds@uom.lk

[2] Extreme Energy-Density Research Institute, Nagaoka University of Technology, Nagaoka, Japan

[3] Center for Supply Chain, Operations and Logistics Optimization, University of Moratuwa, Moratuwa, Sri Lanka

[4] Department of Materials and Production, Aalborg University, Aalborg, Denmark

Abstract. Many firms are shifting from legacy systems to Enterprise Resource Planning (ERP) system solutions. ERP is a project requiring extensive change management to expect implementation success. It entails severe technological changes, and significant organizational behavior and business process changes. Despite this, the research has not thoroughly investigated the relationship between change management challenges and ERP deployment. This research examines and assesses how leadership factors will aid transformational decision-making in implementing ERP. Empirical research reveals that leadership responsibilities like proper communication between leaders and employees, enhancing core-team productivity, improving re-engineering process quality, and top management commitment will promote implementation success. The research problem is defined using the Relative Importance Index (RII). The dynamic behavior of ERP implementation was then analyzed and validated using the System Dynamics (SD) technique. The findings emphasize the importance of numerous leadership elements in change management decisions and how they will influence the long-term effectiveness of ERP deployment.

Keywords: ERP Implementation · Business Process Reengineering · System Dynamics · Change Management Decisions · Leadership

1 Introduction

The industry is a dynamic environment. Every day, new technologies join the industrial picture. To thrive in today's economy, technological adaptation is required. Because the third industrial revolution is focused on digitalization, automation, and electronics, where ERP has become a key component in that paradigm [1]. As a result, many organizations realized the significance of these technologies.

A. Burduk et al. (Eds.): ISPEM 2023, LNNS 790, pp. 228–242, 2023.
https://doi.org/10.1007/978-3-031-45021-1_17

When successfully implemented, an ERP system is a business management system that manages and integrates every business process of an organization [2]. ERP software handles finance and controlling, materials management, sales and distribution, production planning, human resource management, Customer Relationship Management (CRM) and Supply Chain Management (SCM) [2]. Furthermore, by lowering the operations' cycle time, ERP will improve performance, quality, and customer satisfaction [3].

Companies are struggling to realize actual value of ERP systems [4]. These failure rates, however, have little impact on the organizations that commit money to deploy ERP systems [5]. As a result, firms are willing to adapt their business processes in order to improve technology.

ERP implementations are constantly failing due to a variety of failure causes. An incorrect ERP selection, poor project management practices, inadequate business process re-engineering (BPR) activities, and lack of organizational management support can be highlighted as a few main failure factors. Most ERP system mismatches are caused by low quality in BPR activities. As a result, the company's actual business processes were not adequately re-engineered to match the ERP system [6].

Business Process Change (BPC) is the combination of evolutionary (continuous) and revolutionary (radical) management methodologies. Business Process Reengineering (BPR), Business Process Innovation (BPI), and Business Process Transformation (BPT) are subgroups of the revolutionary methodology (Fig. 1). Total Quality Management (TQM), Six Sigma, and ISO standards are all components of the evolutionary approach [7]. Therefore, the quality of the revolutionary approach activities will be important in a BPC project such as ERP deployment.

We discovered a gap between revolutionary and evolutionary approaches. Most organizations are solely concerned with the revolutionary approach, which we recognize as a fundamental disadvantage of ERP implementation. Therefore, this research will assess leadership's impact in developing change management decisions for ERP implementation success [8–10].

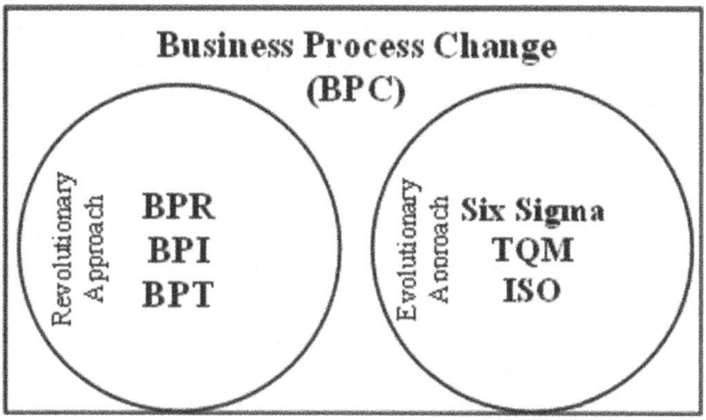

Fig. 1. Elements of Business Process Change (BPC)

The rest of the paper is organized as follows. Section 2 provides a review of studies on ERP and ERP implementation, BPR process, importance of change management, and how leadership influence change management decisions. Section 3 explains the research design and methodology. Section 4 is dedicated to examine the relationship between several components using System Dynamics (SD) technique. The paper concludes with a discussion under Sect. 5 which interprets key findings followed by a conclusion.

2 Literature Review

ERP systems have evolved into one of the most important components of every organization. ERP emerged in the 1980s as a manufacturing resource planning (MRP) development. Nowadays, ERP systems can be identified as the heart of every business [11]. ERP systems are comprehensive software packages that combine an organization's business operations and functions to deliver a comprehensive view of the entire business via a single IT platform [12].

Once, ERP was used only in capital-intensive industries such as product manufacturing, aerospace, construction and civil engineering, and marine industry. However, now they are widely used in practically every sector/industry, including hotel supply chains, retail, education, telecommunication, and finance [2]. ERP implementation projects failed due to a lack of technical understanding and issues with time and budget allocations, BPR, change management, and business cultural readiness [13].

2.1 Business Process Re-engineering (BPR)

Business Process Re-engineering (BPR) involves the radical and complete redesign of core business processes to improve cost, quality, service, and speed dramatically. BPR is based on the idea that organizations can be made simpler and more efficient by restructuring their core processes [7].

It is commonly understood that the internal and external processes would be altered during the ERP implementation. It will either be a total or partial re-engineering. These business operations, however, must be standardized and well-synchronized. As a result, these standardized, re-engineered, and optimized processes will compel organizations to implement ERP successfully [14].

There are three primary phases of the organization's re-engineering and process mapping. Those developing the "As-Is" model, the "To-Be" model, and the "Bridging the Chasm" model [15]. During the As-Is phase, consultants from the ERP software vendor visit the organization to watch, analyze, and identify how the processes work. The most crucial and critical procedures were identified during the To-Be phase. The ERP will then be expanded with new features that will benefit the organization. In the last phase, the organization must develop a change management plan to adopt the ERP properly. This program will comprise training, familiarization, customization, and testing sessions [15].

2.2 Change Management in BPC Projects

Management must constantly adapt to new trends in order to improve the organization. As a result, businesses are always in constant flux. Digitalization and technology are significant changes that every organization must deal with regularly. It evolved into an organizational driving force [16].

ERP implementation is considered a BPC project. As a result, it encompasses a transformational process from the As-Is to the To-Be phase. This transformation is performed by implementing numerous improvements in people, processes, and systems. The number of stakeholder expectations and the effort they put into achieving those expectations will determine the extent of the organizational change [9].

A careful, progressive, and bureaucratic implementation methodology combined with effective change management and network management will result in a successful ERP implementation. Seven important success factors have been discovered. They are namely: Strategic efforts, learning capacity, cultural resilience, information technology, network connections, transformation management practice, and process management practices [8].

Agile methodologies such as daily standup meetings, pair programming, and regular assessments will shrink the complexity of BPC in ERP implementation and boost quality, employee performance, and stakeholder satisfaction following ERP adoption. The best approaches to succeed in BPR are to do ongoing reviews, walkthroughs, and inspections [17].

In a study, Gill [18] outlines the importance of introducing and sustaining effective leadership while managing the transformation. To be considered an effective leader, these individuals must be cognitively, spiritually, emotionally, and behaviorally intelligent.

2.3 Impact of Leadership in Change Management Decision Making

In a dynamic corporate world, change is unavoidable for the survival of organizations. Changes in businesses will have a substantial impact on employees' morale, insecurity, and self-esteem. As a result, effective leaders must understand how changes affect employees [19].

Effective and strategic leadership must be transformational to serve the organization in changing contexts. These transformational leaders must maintain strong morale, energy, and ethics. They must also recognize employees' most prevalent issues and work collaboratively to address them [19].

According to Kavanagh and Ashkanasy [20], effective leadership requires open communication and transparent transformation processes. If not, employees will reject, oppose, or become emotionally stressed by the changes around them. Employers will be forced to accept lower productivity among their staff because of this.

Transformational leadership will have a favorable impact on innovation. Effective and charismatic leaders will impact the organization and the feeling of team identity during such transformational shifts. Employee engagement will eventually improve as a result [21].

2.4 System Dynamics

System dynamics is used to analyze and solve complicated situations by identifying the dynamic behavior of complex systems. System dynamics is also utilized to determine the link between various components. It is a modeling technique based on computer simulation. Following that, new plans and policies will be introduced based on the results [22].

Causal Loop Diagrams (CLD) depict the system's feedback structure and causal links between its variables [23]. SD theories are useful to ERP implementation because the ERP can be viewed as a system and the survey results may be viewed as variables influencing the success of ERP implementation [24].

3 Methodology

The study aims to discover leadership characteristics involved in change management decisions and how they affect the success of ERP adoption. We employed a SD methodology to accomplish our goal (Fig. 2).

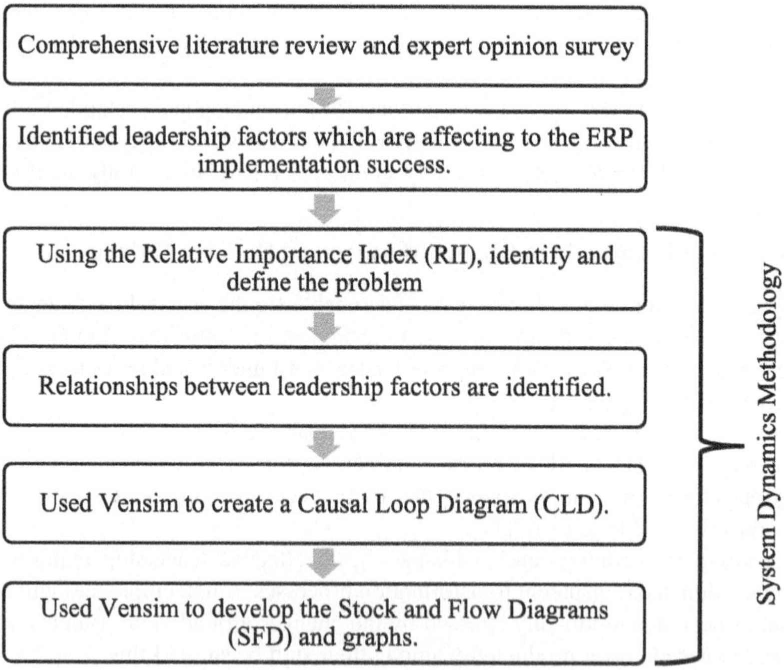

Fig. 2. Research Process Flow

3.1 Problem Definition Through RII Analysis

An RII analysis was used to identify the research challenge following the expert opinion survey. This is the SD methodology's defining the problem stage [25].

$$R = \frac{\Sigma W}{H \times N} \tag{1}$$

where;

R = RII.
W = Weight given by the interviewee to the factor.
H = Highest weight in the Likert-scale.
N = Total number of Interviewees [26] (Table 1).

Table 1. RII analysis output

Factor	RII
Correct definition of the project's scope	0.953846
Offer the necessary resources and ERP system training	0.938462
Top management support	0.923077
Train the user	0.923077
Inspire and empower end users to help them overcome the cultural change	0.907692
Consultant assistance	0.907692
Across the organization, communicate ERP implementation strategy, vision, goal, objectives, and processes	0.892308
ERP implementation processes and scenarios are identified	0.876923
Designated project manager in the client's organization (Organizational matrix. Weak matrix, Balanced matrix, Strong matrix)	0.876923
Creating a collaborative work atmosphere	0.876923
Initial success in the early stages of implementation	0.876923
Consultation sessions with the ERP vendor	0.861538
End-user participation and assistance	0.861538
Top management's commitment towards the project	0.830769
Appropriate allocation of human resources	0.815385
Continuous supervision and improvement	0.815385
Management's willingness to adapt to new technologies and change	0.8
Appropriate time, materials, and budget	0.8
Users' IT proficiency	0.8
A separate steering committee is in charge of monitoring and reviewing	0.769231

(*continued*)

Table 1. (*continued*)

Factor	RII
The workforce's ability to quickly adopt to the new technology	0.753846
The organization's collaborative learning environment	0.753846
The system implementation takes a bureaucratic approach (because of the changing environment)	0.723077
Employer and employee should have open formal and informal communication	0.692308

These factors were gathered from literature survey and expert opinion interviews conducted. Then these factors were asked to rate according to the importance by industry experts.

3.2 Conceptualize the Scenario from a CLD

Factors and their interactions were conceptualized using Vensim® PLE software in a CLD. This is the SD methodology's second stage [25].

3.3 Simulate Through an SFD

After the CLD, we developed the Stock and Flow Diagrams (SFD) to simulate the model. Even though CLDs are extremely useful to illustrate the system's interdependencies and the system's feedback structure, it cannot capture the stock and flow behavior of the system. Stocks are the aggregations created by the difference between the total inflow of a process and its outflow. Stocks can be identified as disequilibrium dynamics in the system, because of the decoupling rates of flow. The inventory of a production organization can be identified as a stock. A stock's accumulation level will depend on inflows and outflows [23].

4 Analysis and Findings

The elements linked to leadership were discovered from the prioritized components and utilized to build the causal loop diagram to identify the feedback structure and the causal relationship among those factors. The Vensim® PLE was used to create the CLD.

Figure 3 depicts the relationship and feedback structure of leadership variables that influence ERP implementation success. The polarity of feedback loops is denoted by the letters "R" and "B" in circles. The letter "R" stands for Reinforcing feedback loops (Positive loops), and the letter "B" stands for Balanced feedback loops (Negative loops). Arrows illustrate the relationship between different variables, and polarity of the relationship is denoted by "+" and "–" markings near the arrowhead.

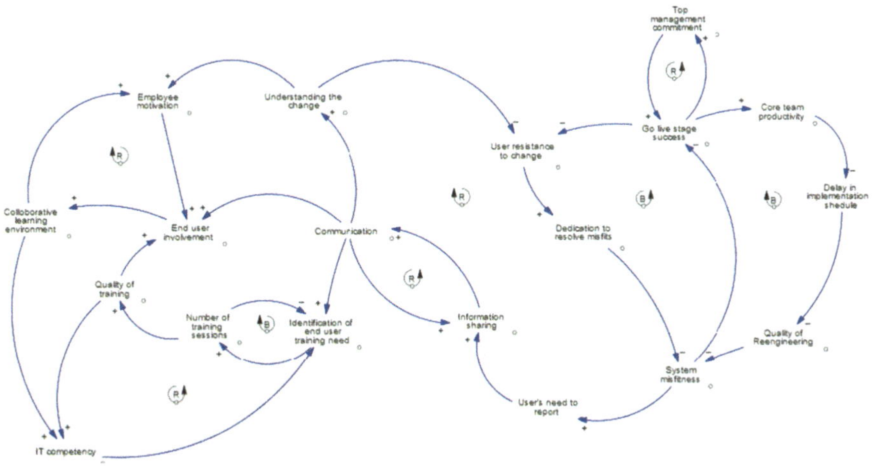

Fig. 3. Developed Causal Loop Diagram

4.1 Conceptualization Through CLD

Communication. Communication is regarded as one of the most important aspects of any BPR effort. Therefore, understanding and accepting the change is critical for success in a business process change project such as ERP implementation. So, it is the employer's responsibility to convey to employees the vision, mission, strategies, goals, and objectives of the ERP implementation. Employee comprehension of the project will improve because they are aware of forthcoming adjustments.

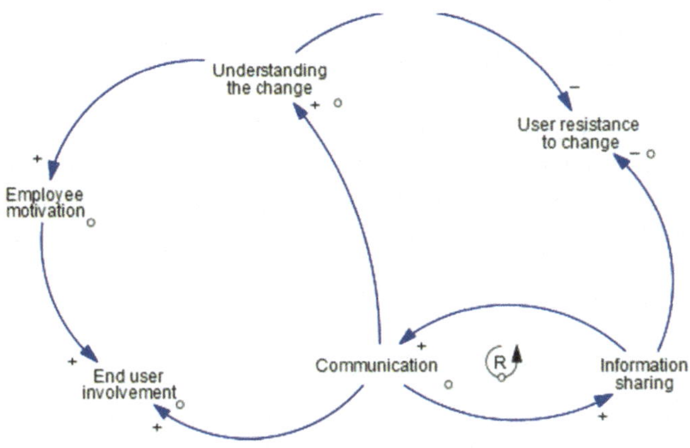

Fig. 4. Identified communication loop No. 1

Communication and information transfer, as shown in Fig. 4 and 5, boost comprehension of change and reduce resistance to change. Therefore, the quantity and quality of information will boost employee awareness. It will then increase end-user involvement in ERP software. Hence, effective communication and established goals will boost staff morale and motivation.

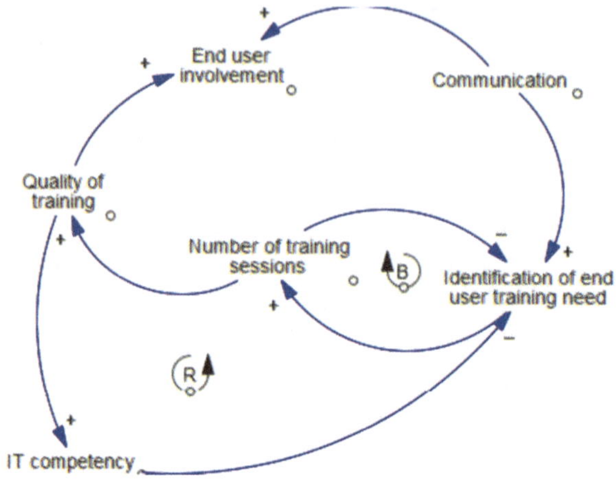

Fig. 5. Identified communication loop No. 2

In another way, effective communication will help with ERP implementation by enabling the firm to identify the workforce's training requirements. It will assist users in improving their ERP IT literacy, and as a result, they will become more involved in ERP.

Commitment of the Top Management. The organization's management must fully commit to the ERP installation process. Their actions will be important in many scenarios, as depicted in Fig. 6. Management's ideas and thoughts on the BPR process will be a significant success element in ERP adoption.

Because the organization's initial business processes were captured in the BPR stage, and business processes will overlap with ERP procedures, reducing ERP system mismatches. Therefore, if there are fewer misfits, employees' dedication to resolving them will be lessened, as would their resistance. As a result, top management and administration involvement will be critical to the success of the "Go Live and Support" stage.

4.2 Formalization Through SFD Simulation

We built SFD to simulate the problem after conceptualizing it with CLD (Fig. 7). Interviews, the internet, literature, and observations were used to collect data for the SFD. The SFD simulation model primarily focused on the organization's business requirements, business process reengineering log, and case log (case log is a log of cases sent by users on various difficulties in ERP. This is known as tickets in the SAP world [27]).

Business Requirement Stock. In this stock, we considered how the business requirement approval process hap-pens. Business modification inflow is the initial customer

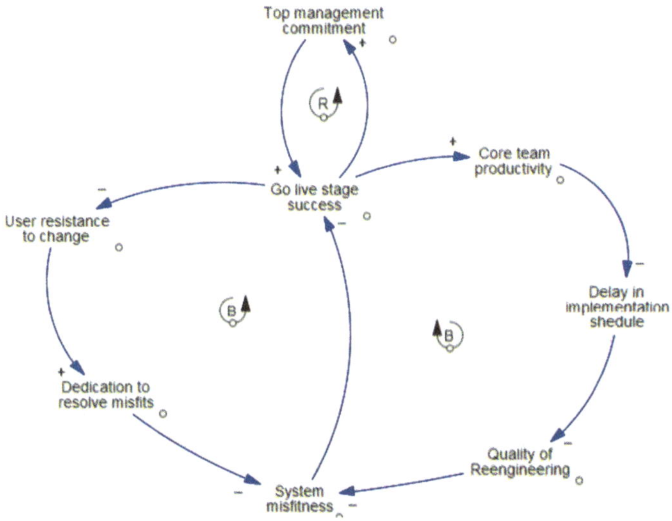

Fig. 6. Identified commitment of the top management loop No. 1

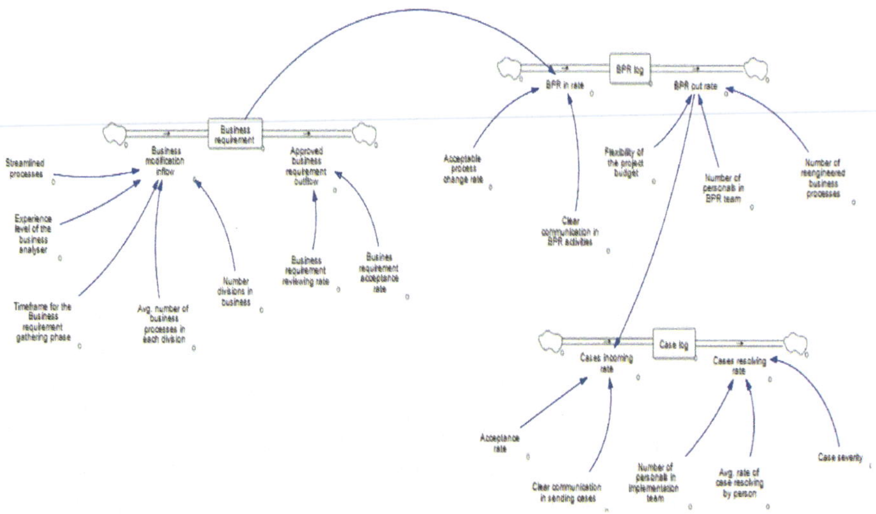

Fig. 7. Developed SFD for the ERP implementation.

requirement for ERP software. This includes the expectations of the customer. Business modification inflow is influenced by the average number of business processes in each division, the number of divisions in the business, the timeframe of the business requirement gathering phase, the experience level of the business analyzer, and streamlined (standardized) processes the organization.

Approved business requirement outflow is the number of processes the ERP-providing company will accept. This will depend on the business requirement reviewing rate and business requirement acceptance rate of the ERP vendor company. The aggregate value of the stock will be the difference between the inflow and outflow.

$$\textit{Business modification inflow} = \frac{\textit{Avg.number of business processes in each division} \times \textit{Number divisions in business} \times \textit{Experience level of the business analyser} \times \textit{Streamlined processes}}{\textit{Timeframe for the Business requirement gathering phase}} \tag{2}$$

$$\textit{Approved business requirement outflow} = \textit{Busines requirement acceptance rate} \times \textit{Business requirement reviewing rate} \tag{3}$$

$$\textit{Business requirement} = \textit{Business modification inflow} - \textit{Approved business requirement outflow} \tag{4}$$

Business Process Re-engineering Log Stock. We identified how the BPR activities are happening and what causes such activities in this stock. The BPR rate depends on the number of business processes that must be re-engineered. This rate will depend on the organization's business requirement (finalized and approved requirement), acceptable process change rate (the process change rate that the client organization is willing to change), and clear communication in BPR activities.

BPR out rate is the process re-engineering output rate of the business consultants whom the ERP vendor company deploys. This rate will depend on the project budget's flexibility, the number of re-engineered business processes, and the number of personnel in the BPR team.

$$\textit{BPR in rate} = \textit{Business requirement} \times \textit{Acceptable process change rate} \times \textit{Clear communication in BPR activities} \tag{5}$$

$$\textit{BPR out rate} = \textit{Flexibility of the project budget} \times \textit{Number of personals in BPR team} \times \textit{Number of reengineered bu}\sin\textit{ess processes} \tag{6}$$

$$\textit{BPR log} = \textit{BPR in rate} - \textit{BPR out rate} \tag{7}$$

Case Log Stock. This stock collects all user cases, issues, and bugs reported by the customer. The incoming rate is when the user reports cases about the system. Here we assumed that, when the output rate of BPR increases over time, the number of cases reported will also increase. The other variables affecting the incoming rate cases are acceptance rate, and clear communication when sending those.

Case resolving rate will be depended on the number of personnel in the implementation team, the average rate of case resolving by a person, and case severity (the depth of the case or issue).

$$\textit{Cases incoming rate} = \textit{BPR out rate} \times \textit{Acceptance rate} \times \textit{Clear communication in sending cases} \tag{8}$$

$$\textit{Cases resolving rate} = \textit{Avg. rate of case resolving by person} \times \textit{Number of personals in implementation team} \times \textit{Case severity} \tag{9}$$

$$\textit{Cases log} = \textit{Cases incoming rate} - \textit{Cases resolving rate} \tag{10}$$

4.3 Evaluation of Results

Business Requirement and BPR. When the client has additional business requirements, the re-engineering team will typically engage in more BPR activities. The findings of our simulation confirmed this connection.

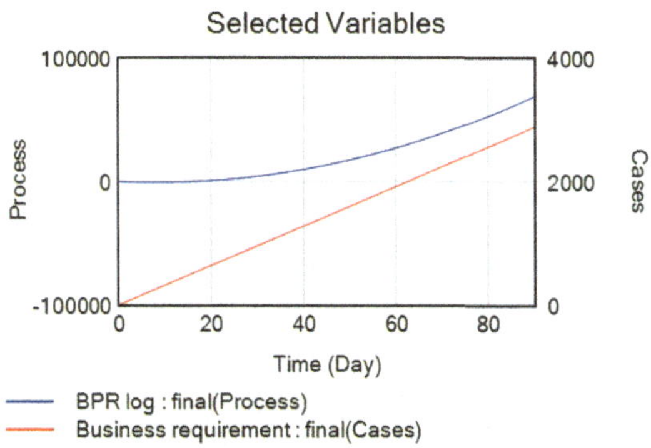

Fig. 8. Business requirements Vs. BPR log

As in Fig. 8, when the business requirements increase, BPR activities will also increase exponentially. The model runs for ninety days time period. In that period, the total accumulation in business requirement cases is around three thousand. The total accumulation in BPR activities is around fifty thousand (The number is higher because we consider activities as the measurement). Therefore, this graph illustrates how much BPR work must be done to customize one business requirement.

BPR Log and the Case Log

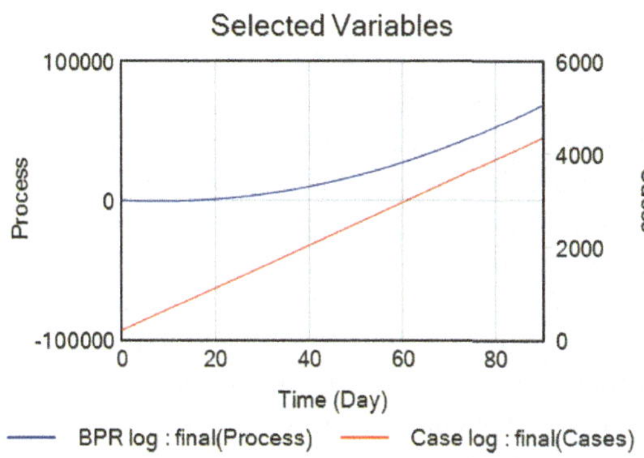

Fig. 9. BPR Log and the Case Log

As per the graph (Fig. 9), when the number of BPR activities increases, the number of cases accumulated in the case log will also increase. The reason behind these fluctuations is that BPR activities will re-engineer the existing processes, creating issues when the users are executing the new processes. So, the number of cases will be increased.

These simulation results were generated through information gathered from the internet and using hypothetical numbers. Adding more realistic and actual data can increase the model's accuracy. Nevertheless, the pattern of the results will remain the same because we will not change any of the model's equations.

5 Conclusions

This research aims to discover and assess how leadership commitments in change management decisions will enhance ERP implementation success. Industry specialists from ERP vendor companies were chosen to collect the essential data. According to the findings, the role of leadership in ERP implementation is critical [10].

In SFD, our main aim is to evaluate the contribution of leadership to developing a change management plan. Even though we did not use leadership as direct inputs for the SFD, we used certain leadership measures in SFD. For example, the flexibility of the project budget, clear communication in BPR activities, and clear communication used in reported cases can be identified as variables that the leadership of the client organization will influence [7, 24].

First, leaders must think strategically and recognize the need for ERP before developing a vision, mission, strategies, objectives, and goals. The information must then be communicated appropriately to the workforce. The entire implementation must then be continuously monitored and supervised so that leaders can identify training needs and employee perceptions of ERP. Then they can determine the appropriate type of trainings, motivation sessions, inspiration, and empowerment which will help employees to improve.

In the BPR process, it is critical to collaborate with the ERP vendor organization effectively. Leaders must facilitate all information required by consultants and work throughout the process. The initial success of the implementation will be based on proper BPR.

We identified that number of cases reported by the user will increase with the number of BPR activities. Therefore, these cases should be properly identified and resolved immediately. Otherwise, users' will be demotivated and inefficient. Proper investigation on cases reported and ensure that there will be no backlog of cases remaining is one of the major responsibilities of leaders in the implementation project.

As mentioned earlier, it is difficult to build a linear relationship between leadership and ERP implementation. This can be considered as the dynamic nature of the leadership. Therefore, in a complex adaptive system such as organization, it is inevitable that new innovations are emerging. These new innovations such as ERP will impact the organizations drastically once it is implemented and it is vital to manage these changes/dynamic behaviors as leaders in order to minimize the negative impacts to the organization [28].

The study can be expanded to determine when a company should deploy an ERP system because knowing when to execute a large investment like ERP can save much money.

Acknowledgement. The authors would like to acknowledge the financial support given by the Norwegian Program for Capacity Development in Higher Education and Research for Development (NORHED II - Project number 68085), the "Politics and Economic Governance" sub-theme, the project "Enhancing Lean Practices in Supply Chains: Digitalization", which is a collaboration between the University of Stavanger (Norway), ITB (Indonesia), and the University of Moratuwa (Sri Lanka).

References

1. Rifkin, J.: The Third Industrial Revolution, First Ed. PALGRAVE MACMILLAN®, London (2011)
2. Shehab, E.M., Sharp, M.W., Supramaniam, L., Spedding, T.A.: Enterprise resource planning: an integrative review. Bus. Process. Manag. J. **10**(4), 359–386 (2004). https://doi.org/10.1108/14637150410548056
3. Gardiner, S.C., Hanna, J.B., LaTour, M.S.: ERP and the reengineering of industrial marketing processes. A prescriptive overview for the new-age marketing manager. Ind. Mark. Manag. **31**(4), 357–365 (2002). https://doi.org/10.1016/S0019-8501(01)00167-5
4. Marnewick, C., Labuschagne, L.: A conceptual model for enterprise resource planning (ERP). Inf. Manag. Comput. Secur. **13**(2), 144–155 (2005). https://doi.org/10.1108/09685220510589325
5. Kießner, P., Perera, H.N.: Managing complexity in variant-oriented manufacturing: a system dynamics approach. In: Freitag, M., Kinra, A., Kotzab, H., Megow, N. (eds.) Dynamics in Logistics, pp. 363–375. LDIC 2022. LNL. Springer, Cham (2022). https://doi.org/10.1007/978-3-031-05359-7_29
6. Wong, A., Chau, P.Y.K., Scarbrough, H., Davison, R.: Critical failure factors in ERP implementation. In: 9th Pacific Asia Conference on Information Systems I.T. Value Created (PACIS 2005), no. January (2005)
7. Rosenberg, Z., Riasanow, T., Krcmar, H.: A system dynamics model for business process change projects, no. December (2015)
8. Motwani, J., Subramanian, R., Gopalakrishna, P.: Critical factors for successful ERP implementation: exploratory findings from four case studies. Comput. Ind. **56**(6), 529–544 (2005). https://doi.org/10.1016/j.compind.2005.02.005
9. Alghamdi, J.S., Muzaffar, Z.: Quality assurance model capturing ERP implementation facets through RVRA services. In: International Conference on Advanced Communication Technology ICACT, pp. 1452–1456 (2011)
10. Lakdinu, M., De Silva, M.M., Balachandra, K., Frank Perera, K.P.D.: Leadership in change management decisions for successful ERP implementation - a system dynamics perspective. In: 2022 Moratuwa Engineering Research Conference (MERCon), pp. 1–6 (2022). https://doi.org/10.1109/MERCon55799.2022.9906200
11. Costa, C.J., Ferreira, E., Bento, F., Aparicio, M.: Enterprise resource planning adoption and satisfaction determinants. Comput. Human Behav. (351) (2016)
12. Klaus, H., Rosemann, M., Gable, G.G.: What is ERP? Foundry Manag. Technol. **127**(8), 94 (1999)
13. Der Leu, J., Lee, L.J.H.: Enterprise resource planning (ERP) implementation using the value engineering methodology and Six Sigma tools. Enterp. Inf. Syst. **11**(8), 1243–1261 (2017). https://doi.org/10.1080/17517575.2016.1215537
14. Akyuz, G.A., Rehan, M.: Requirements for forming an e-supply chain. Int. J. Prod. Res. **47**(12), 3265–3287 (2009). https://doi.org/10.1080/00207540701802460

15. Okrent, M.D., Vokurka, R.J.: Process mapping in successful ERP implementations. Ind. Manag. Data Syst. **104**(8), 637–643 (2004). https://doi.org/10.1108/02635570410561618
16. Albach, H., Meffert, H., Pinkwart, A., Ralf, R.: Management of permanent change—new challenges and opportunities for change management (2015). https://doi.org/10.1007/978-3-658-05014-6_1
17. Baig, J.J.A., Shah, A., Sajjad, F.: Evaluation of agile methods for quality assurance and quality control in ERP implementation. In: 2017 IEEE 8th International Conference on Intelligent Computing and Information Systems ICICIS 2017, vol. 2018-Janua, no. Icicis, pp. 252–257 (2017). https://doi.org/10.1109/INTELCIS.2017.8260055
18. Gill, R.: Reading 2_Gill_change leadership. J. Chang. Manag. **3**, 307–318 (2003)
19. Belias, D., Koustelios, A.: The impact of leadership and change management strategy on organizational culture. Eur. Sci. J. **10** (2014). https://doi.org/10.1111/j.1467-8551.2006.00480.x
20. Kavanagh, M.H., Ashkanasy, N.M.: The impact of leadership and change management strategy on organizational culture and individual acceptance of change during a merger. Br. J. Manag. **17**(SUPPL. 1) (2006). https://doi.org/10.1111/j.1467-8551.2006.00480.x
21. Paulsen, N., Maldonado, D., Callan, V.J., Ayoko, O.: Charismatic leadership, change and innovation in an R&D organization. J. Organ. Chang. Manag. **22**(5), 511–523 (2009). https://doi.org/10.1108/09534810910983479
22. Poles, R.: System Dynamics modelling of a production and inventory system for remanufacturing to evaluate system improvement strategies. Int. J. Prod. Econ. **144**(1), 189–199 (2013). https://doi.org/10.1016/j.ijpe.2013.02.003
23. Sterman, J.D.: Business dynamics: systems thinking and modeling for a complex world, vol. 26, no. 1. Jeffrey J. Shelsfud (2000). https://doi.org/10.1108/sd.2010.05626aae.001
24. Cyrusm, K.M., Aloini, D., Karimzadeh, S.: How to disable mortal loops of enterprise resource planning (ERP) implementation: a system dynamics analysis, Systems (2018). https://doi.org/10.3390/systems6010003
25. Jayalath, M.M., Perera, H.N.: Mapping post-harvest waste in perishable supply chains through system dynamics : a sri lankan case study, vol. 16, no. 3, pp. 526–543 (2021)
26. Azman, N.S., Ramli, M.Z., Razman, R., Zawawi, M.H., Ismail, I.N., Isa, M.R.: Relative importance index (RII) in ranking of quality factors on industrialised building system (IBS) projects in Malaysia. In: AIP Conference Proceedings, vol. 2129, no. July (2019). https://doi.org/10.1063/1.5118037
27. SAP, "What is ERP?" (2022). https://www.sap.com/insights/what-is-erp.html
28. Lichtenstein, B.B., Uhl-bien, M., Marion, R.,Seers, A., Orton, J.D., Schreiber, C.: Complexity leadership theory : An interactive perspective on leading in complex adaptive systems. Emerg. Complex. Organ. (2006)

Using an ISO 9001 Based Framework as a Benchmark to Identify Best Practices Used by Sri Lankan Practitioners when Selecting Suppliers

Nikila Dharmadasa[1], M. Mavin De Silva[1,2](✉), Amila Thibbotuwawa[3], Tadachika Nakayama[2], and Izabela Iwa Nielsen[4]

[1] Department of Transport Management and Logistics Engineering, Faculty of Engineering, University of Moratuwa, Katubedda 10400, Sri Lanka
mavinds@uom.lk

[2] Extreme Energy-Density Research Institute, Nagaoka University of Technology, Nagaoka, Niigata 940-2188, Japan

[3] Center for Supply Chain, Operations and Logistics Optimization, University of Moratuwa, Katubedda 10400, Sri Lanka

[4] Department of Materials and Production, Aalborg University, Aalborg, Denmark

Abstract. Supplier selection is highly industry-specific, with industry-specific regulations and requirements, making gauging the quality of the process difficult. This raises the requirement for industry practitioners and quality institutes to benchmark the Supplier Selection Process (SSP) to identify where the process adapted currently stands and what the good practices within the industry are. Benchmarking is an ideal method of identifying the best practices in a process, but the usage of benchmarking to assess and identify best practices in a SSP lacks scholarly and practitioner literature. This study intends to combine benchmarking practices and ISO Standards to identify best practices in SSP by developing a structured benchmarking framework. This is achieved through a two-phase methodology. The first phase will focus on framework derivation and the second phase will focus on framework validation. The recommended Framework was applied to two organizations operating in the Consumer Goods industry, and the results show that adaptation of a structured benchmarking framework aids in identifying best practices in a structured and quantifiable manner. The authors recommend using this Framework as a base framework and a decision support tool to identify best practices within an industry or across industries and to improve the supplier selection processes further on the results achieved from the Framework.

Keywords: Supplier Selection · Supply Chain Management · Quality Standards · Process Benchmarking

A. Burduk et al. (Eds.): ISPEM 2023, LNNS 790, pp. 243–254, 2023.
https://doi.org/10.1007/978-3-031-45021-1_18

1 Introduction

Supply Chains of organizations have been evolving over time, resulting from the changing roles of supply chain stakeholders [1]. When considering the stakeholders of a supply chain, suppliers play a critical role because they are the key drivers of the upstream logistics of any organization. Additionally, the raw material requirements have become more complex over time. The conventional raw material requirement has been made more complex with the evolving packing material requirements. This has driven a drastic change of suppliers in a supply chain to be more of a strategic role [2].

With suppliers becoming increasingly strategic in the supply chain, the role of supplier selection or the procurement process should also be considered a strategic activity [3]. The supplier selection process has been considered routine clerical work earlier [4]. But in the modern-day context, with the transition of the supplier selection process towards a strategic role, aligning the process with the best practices is essential for an organization. In addition, the identification of best practices is best achieved through a standardized benchmarking approach [5].

Standardization in Benchmarking is essential when identifying best practices in supply chains because a non-standard benchmarking process can create complications when properly gauging the peer organizations under review for the benchmarking process [5]. Further, the benchmarking process should follow a structured framework, which should act as a process evaluation and improvement framework. The literature in this domain made it clear that Process Improvement and Process Evaluation must be taken through a structured framework to gain tangible results. Hence, supply chain practitioners should follow a structured, standardized framework to improve supply chain processes such as supplier selection. Furthermore, quality should be a key concern in it too.

From a structured literature review, the authors identified a knowledge gap in the SSP Evaluation and Benchmarking literature. This is further studied in Sect. 2. Scholars have explored Supplier Selection as a Multi Criteria Decision Analysis (MDCA) problem, but it has not been studied as a complete process. Further, the benefits realization of using a structured benchmarking framework in identifying best practices in Supplier Selection has not been studied in previous academic or practitioner literature. This study seeks answers to three questions to bridge this research gap.

1. What are the processes used by Sri Lankan industry practitioners for supplier selection?
2. Do these processes align with the provisions provided by ISO 9001:2015 Standard?
3. What are the best practices Sri Lankan industry practitioners use in Supplier Selection?

Answering these research questions was done by following a rigorous 2 phase research methodology which will be further explained in Sect. 3.

This study further argues that quality has become a key concern in the modern world because quality has become a key driver of two key aspects: Competitive advantage and Globalization [6]. With that as a major consideration, quality should be addressed in the early stages of the supply chain, because as per Bashan & Kordova [6], organizations can achieve their quality targets by mitigating any adverse quality issues in the early stages of the supply chain. Integrating quality into a benchmarking framework, as mentioned above, will ensure the supplier selection process improvement is achieved with quality outputs. Integrating quality for such Framework should use a standardized theoretical approach that follows the Plan-Do-Check-Act (PDCA) cycle [7]. The ISO 9001 standard is a standardized approach that follows the PDCA cycle [8]. Hence, it is visible that following a benchmarking framework based on a quality standard will aid in the needs of practitioners to improve their procurement processes and maneuver the organizational procurement function towards more of a strategic function rather than a routine activity function.

The remainder of the paper will have the following order: Sect. 2 provides a literature review that will assess the background of multiple aspects pertaining to the research domain and scope, Sect. 3 will provide a detailed overview of the research methodology, Sect. 4 will provide a detailed summary of the results and discussion, and Sect. 5 will provide Conclusions, Acknowledgements, and References.

2 Literature Review

Supplier selection has evolved throughout time from being a routine activity to a strategic activity. This is visible when assessing the past literature. Soukup [4] has identified the supplier selection process as a key routine activity of the purchasing department, further he has identified supplier selection as a key responsibility of the purchasing manager. Further, this article considers supplier selection by assessing only a few aspects, such as cost and lead time. When looking at later studies, this perspective has evolved. The study by Flynn et al. [9] identified that quality is a key dimension that should be considered in the supplier selection process, provided TQM (Total Quality Management) and JIT (Just in Time) have been key drivers in organizational strategy. Further, they suggest that supplier selection is a key operations management activity that should bring a competitive advantage to the organization through the right utilization of resources. Even though scholars have identified the importance of other factors apart from cost and lead time, a study done by Verma & Pullman [10] identified that practitioners still preferred to select suppliers based on conventional criteria such as cost and lead time. However, with time, studies have identified multiple aspects as critical criteria for short-listing suppliers. Significantly, Sarkis & Talluri [11] identified supplier selection as a key strategic activity rather than a routine activity. With time, the industries have shifted focus to strategic activities and focused on competitive advantage, which resulted in the supplier selection process converting into a sophisticated process within organizations [12]. Throughout time, the procurement function has been becoming complex. This is visible from the study by Beil [13], which thoroughly mentions that significant organizational revenues are spent on purchasing. This implies the significance of purchasing as a function. Further, the author emphasizes practitioners' importance and focus on value-based purchasing. This is visible throughout recent literature after 2010. The focus of supplier selection is a strategic activity and an activity that is solely driven based on value creation for the organization.

In previous literature, supplier selection has been considered an MCDA problem. Such literature has explored different criteria for assessing Operations Research (OR) based on mathematical and statistical models. This part of the literature review will assess the previous literature in the domain of supplier selection methods and the development of the literature throughout time. It was further evaluated by the study done by Patil [14]. In his study he explores different supplier selection criteria and different supplier selection methods that are assessed by previous literature up to 2014. Further in this study, it assesses the different criteria that previous scholars in the previous literature assess. Furthermore, it is visible that Price, Product Quality and Delivery are the most prioritized criteria in previous literature. Additionally, the same study revealed that Environmental risk, Global Factors, and being Solution Oriented are less explored or less studied in previous literature in the supplier selection methods domain. This gives an important insight as to the lack of studies into areas that are critical to supplier selection.

Table 1. SSP in OR approaches

Year	Researcher	Proposed Model
2014	Şahin & Yiğider	AHP + TOPSIS
2015	James Freeman, Tao Chen	AHP + Entropy + TOPSIS
2016	A Fallahpour, EU Olugu, SN Musa	Fuzzy DEA
2017	Segura and Maroto	PROMETHEE + MAUT
2018	Ghadikolaei and Parkouhi	Fuzzy ANP
2019	S Hosseini, AA Khaled	Hybrid ensemble + AHP
2020	Ramakrishnan & Chakraborty	TOPSIS

From Table 1, it is visible that supplier selection methods, in the perspective of OR approaches, have evolved from just OR approaches to mixes of OR approaches and Artificial Intelligence/Machine Learning approaches such as Hybrid Ensemble. The supplier selection literature has broadly assessed various factors related to supplier selection processes. How it differs based on the business scale [15], phases of the process, how it aligns with the product life cycle, different procedures, and what kind of a role suppliers play in a supply chain and how it has affected supplier selection. However, through the review, it was visible that the literature in assessing the supplier selection process is less and a possible research area. Further, it is seen through past literature that multiple methods have been assessed to assess suppliers in the selection process. Even though scholars have explored how these processes should be integrated into the supplier selection process, the use of these in assessing the supplier selection process is not visible.

3 Materials and Methods

3.1 Study Methodology

This study was designed with a phased methodology to address the research gaps identified from the literature review. Further, other related literature such as the study by Jochem & Landgraf [16] displayed the importance of having a research methodology such that it is structured in a concise and clear manner. This study employed a two-phase methodology. The initial requirements for the framework development are compiled in the first phase. A structured As-is analysis was done on industry firms to identify the current industry practices in supplier selection. Another key input was past literature to define the assessment criteria in the Framework. The remainder of the methodology is depicted in Fig. 1.

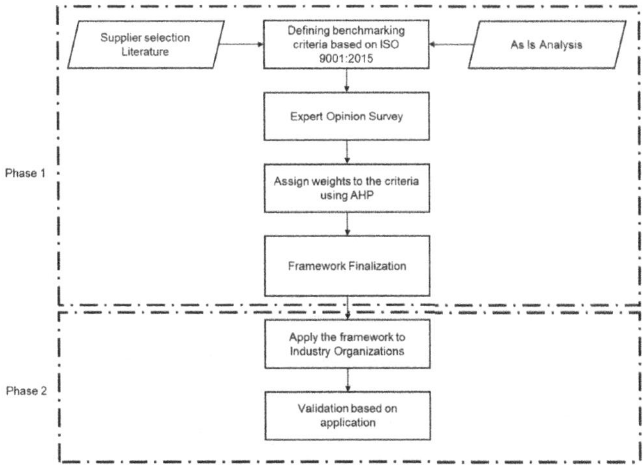

Fig. 1. Phase Breakdown of the Study

3.2 The Framework for Benchmarking

This study focuses on assessing the supplier selection process of two different organizations. If benchmarking is an important process management method, it should be done methodically [16]. Benchmarking is more successful when a theoretical framework is followed [7]. These theoretical frameworks include mainly PDCA-based frameworks. For the purpose of this study, a framework that fundamentally follows PDCA was developed based on the Framework developed by Jochem & Landgraf [16]. This Framework followed 5 steps, and for this study, it was modified to have 4 steps (See Fig. 2). The modification was made to ensure that the framework aligns with the research objectives and provides a suitable structure to evaluate the supplier selection process of the two organizations. By streamlining the framework to 4 steps, the study can focus on key aspects that are relevant to the research question and avoid any unnecessary steps that may not be applicable to the context of the study. Another requirement for this Framework to be based on PDCA is that ISO 9001 itself is based on PDCA [8]. Hence, when developing an ISO 9001 based framework, the Framework itself should fundamentally follow the PDCA cycle. In the 1st step of the Framework, a process analysis is conducted to identify the current state of the supplier selection process. As it was discovered in the literature review, any process improvement methodology/framework should follow a methodical flow [17]. The first step focuses on assessing the As-is state of the supplier selection process using the criteria defined based on ISO9001:2015 and previous literature. The Framework will allocate scores to the aspects that are assessed in the process.

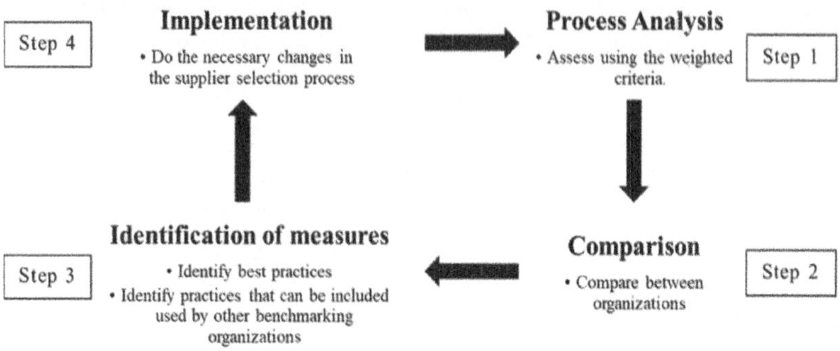

Fig. 2. Recommended Benchmarking Framework

3.3 Defining the Hierarchy of Benchmarking for the AHP

A benchmark is a comparison. Furthermore, the comparison should be made using identical criteria. Since this Framework is based on ISO 9001:2015, the main criteria were defined based on the key focus areas of ISO 9001:2015. Initially, the main criteria were identified from past studies that have studied ISO 9001 for process evaluation, ISO 9001 in Supplier selection, and organizational evaluation. Additionally, based on the findings of past literature (see Table 2) and the ISO 9001:2015 standard document, a set of main criteria were defined. They are *Quality Management System and Processes, Leadership, Planning, Infrastructure, Evaluation.* The hierarchy's sub-level criteria and relevant KPIs were defined based on the existing provisions (such as leadership inclusiveness, customer focus, etc.) and new additions to the ISO 9001:2015 standard (such as risk-based thinking).

3.4 Assigning Weights

Analytical Hierarchy Process (AHP) is a mathematical process that can mimic the human decision-making rationale [23]. It allocates a set of weights until the goal is achieved. This process of allocating weights is done through a structured mathematical process. Scholars in MDCA problems often use this method. But another use of this model is to assign weights to a set of selected criteria. This practice was visible in the study done by Banaeian et al. [24], where weights were assigned to a set of criteria using AHP as a guide. In their study, the assessment of criteria was done by 4 experts: 2 academic experts and 2 industry experts. Initially, advanced variants of the AHP were considered such as Fuzzy AHP, but the decision to use the conventional AHP method was due to the primary intention of this is to assign weights, not to find the most accurate answer to typical MDCA Problem.

Table 2. Summary of ISO 9001 based process assessment criteria

Author/Year	Study purpose	Criteria Identified
(Biazzo & Bernardi, 2003)	Process management practices Process management practices and quality systems standards Risks and opportunities of the new ISO 9001 certification	Customer Focus, Leadership, Involvement of People, Process Approach, System Approach
(Lee et al., 2003)	To develop a high-quality-supplier selection model for supply chain management and ISO 9001 system	Quality Management System Audit, Product Test, Percentage of workforce with technical qualification, Process capability index, Annual training hours per employee
(Carmignani, 2008)	To provide a structured approach to provide the best answers to the ISO 9001 requirements	Quality Policy and objectives, Strategic Planning, resource identification, Personnel
(Okwiri, 2013)	To use ISO 9001 quality management system audit as an Organizational effectiveness evaluation tool	Quality policy and Quality objectives, Management Commitment, Leadership, Process Approach, Management review
(Psomas & Pantouvakis, 2015)	To identify ISO 9001 overall performance dimensions	Product service Quality, Operational performance, Market performance, Financial Performance

Since AHP gives weights to all the criteria (main and sub), the sub-criteria weights will not make sense if they are not considered with the impact of the weight of the min criteria. Hence, a global weight is considered for the sub-criteria. The global weight is calculated by multiplying the sub-criteria weight by the main criteria weight. See Fig. 3 for the final hierarchy tree, including the allocated global weights.

4 Results and Discussion

4.1 Process Analysis and Comparison

These are the first two steps in the Framework. The benchmarking Framework was initially applied to two industry organizations in the Consumer Goods sector of Sri Lanka (Considered as Organization 1 and Organization 2). The Framework's output is a score that is based on the weights assigned for the assessment criteria and a set of KPIs, which was achieved using the equation below.

$$\text{Score allocated per main criteria} = \sum w_i * k_i \tag{1}$$

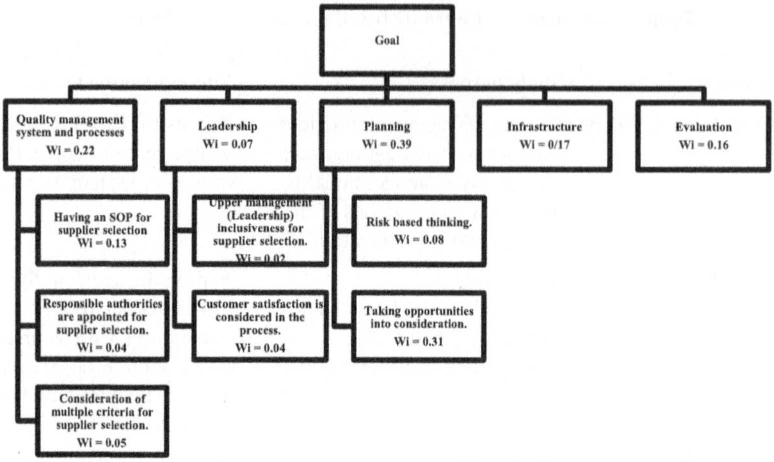

Fig. 3. Hierarchy of criteria for the AHP

w_i = *Criteria Weight*
k_i = *Respective KPI of the Assessment Criteria*

See Fig. 4 below for the respective scores of the main criteria based on the As-is state, and based on these scores, the best practices can be identified.

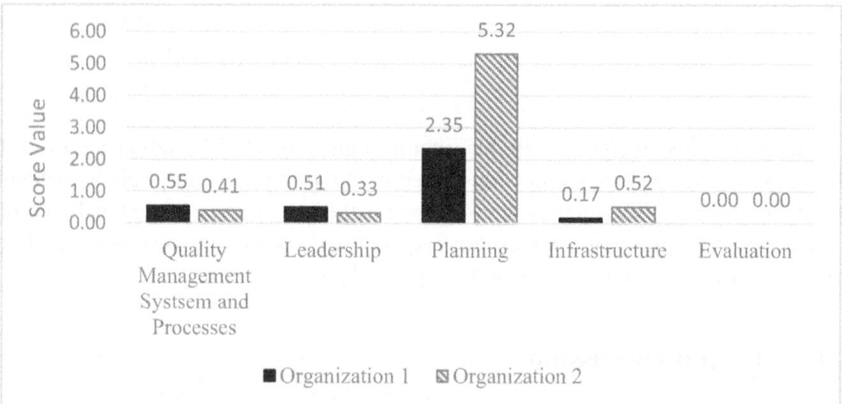

Fig. 4. Comparison of Organization 1 and Organizaiton 2

It is visible that Organization 2's supplier selection process is better in Planning aspects and Infrastructural aspects when compared with that of Organization 1's supplier selection process. Looking at the KPIs of both the organizations, it is visible that organization 2 is able to consider multiple suppliers enabling them to assess a wider range of options giving them much more of a space to assess the quality and other key parameters prior to selecting one. And they are able to do this because they possess a dedicated Decision Support System in the nature of a Supplier Database. Further, they possess multiple labs for quality testing purposes. These good practices make the Supplier Selection Process of Organization 2 comparatively better than that of Organization 1's.

4.2 Identification of Measures

This is the third step in the Framework. It is visible that Organization 2's supplier selection process is better in Planning and Infrastructural aspects when compared with that Organization 1's supplier selection process. Assessing the KPIs of both organizations, it was observed that organization 2 considers multiple suppliers, enabling them to assess a wider range of options, and giving them much more space to assess the quality and other key parameters before selecting one. Furthermore, they can do this because they possess a dedicated Decision Support System in a Supplier Database. Further, they possess multiple labs for quality testing purposes. These good practices make the Supplier Selection Process of Organization 2 comparatively better than Organization 1's.

4.3 Implementation

This is the fourth step of the Framework. This Framework acts as a problem-solving method for organizations. Because the Framework reveals the problems in the supplier selection process and from the Framework itself the problem can be solved because it reveals the best practices in the peer organizations as well. Below is a 5 why analysis (Fig. 5) done for Organization 1's supplier selection process. It is visible that the root cause was identified in just two why's, that is because the benchmarking Framework already identified the surface issue. Unless such a need could have been discovered due to a quality defect in the final finished product.

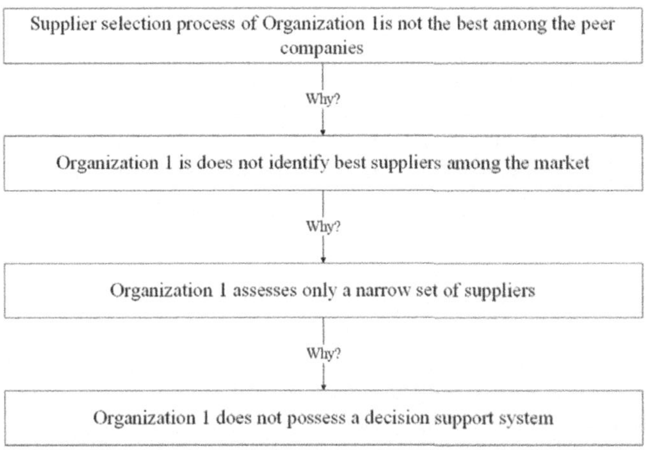

Fig. 5. Five Why analysis for Organization 1

Nonetheless, when implementing the identified best practices, the scale and agility of the organization and its processes should be considered. This is because when applying the best practices to the organization, some tweaks will have to be made for it to be in sync with the organizational aspects (i.e. scale, culture, agility, perception towards change).

4.4 Framework Validation

The framework validation was done in the form of an experiment conducted with experts' help. The experiment was to rate the states of the As-is and To-be states of derived processes from the Framework. The Paired Wilcoxon Signed Rank Test was used for this at a confidence level of 95% and for the analysis, the Null hypothesis was that there is no significant difference between the two ratings that the experts gave. And the Alternate hypothesis was that there is a significant difference in the experts' ratings. Based on the sample of 7 experts, the critical value for the Wilcoxon Signed Rank Test was 2. From the analysis, it was visible that there is a significant difference in the ratings given to the two processes. This implies that the experts validate the output from the Framework, which is the To-be state of the newly recommended supplier selection process that includes the recommendations mentioned in Sect. 4.2.

5 Conclusion

This study was initiated with the intention of developing a Framework based on ISO 9001:2015 provisions to assess and evaluate Supplier Selection Processes adopted by organizations. The sole purpose of this recommending Framework is to benchmark Supplier Selection Processes in a quantifiable way and finally identify where they stand among their peers. When developing the study, key insights were derived such as the lack of literature in the explored topic and how important having a structured framework for process evaluation efforts was. In concluding remarks, from the framework application on industry organizations, it was found that the Framework suggested in this study is applicable and provides important insights, and it was validated through a scientific experiment conducted with Experts. The uses of the recommended Framework are immense, and the applications are not only limited to the industry practitioners, such as procurement managers, but also for the regulators such as Standard makers to benchmark organizations in a specific industry.

As far as the limitations of this study are considered, it was discovered that the study being limited to two organizations would have some bias since the Framework could have been applied to multiple organizations. Further, another limitation was that the suggested Framework is a base level framework and has the risk of being not deep enough to give industry-specific insights into supplier selection processes.

The recommended future research areas are more focused on mitigating the limitations that were observed during this study. A key research area from this study is the integration of such Framework into organizational ERP systems where the organization can realize the benefits of such Framework through the ERP system itself. Further, the study could be extended to either be industry specific (since this study recommends a base-level benchmarking framework) or to be more generalized across industries by making the assessment more flexible.

Acknowledgements. The academic staff of the Department of Transport and Logistics Management of University of Moratuwa Sri Lanka guided this study. The authors also thank the Center for Supply Chain, Operations and Logistics Optimization, University of Moratuwa, for the guidance provided during the study.

References

1. MacCarthy, B.L., Blome, C., Olhager, J., Srai, J.S., Zhao, X.: Supply chain evolution–theory, concepts and science. Int. J. Oper. Prod. Manag. (2016)
2. Rajeev, A., Pati, R.K., Padhi, S.S., Govindan, K.: Evolution of sustainability in supply chain management: a literature review. J. Clean. Prod. **162**, 299–314 (2017)
3. Obura, C., Shale, N.I., Mukanzi, C.: Assessment Of Supplier Appraisal Influence On Performance Of Procurement Function (2021)
4. Soukup, W.R.: Supplier selection strategies. J. Purch. Mater. Manag. **23**(2), 7–12 (1987)
5. Cuthbertson, R., Piotrowicz, W.: Supply chain best practices–identification and categorisation of measures and benefits. Int. J. Product. Perform. Manag. (2008)
6. Bashan, A., Kordova, S.: Globalization, quality and systems thinking: integrating global quality Management and a systems view. Heliyon **7**(2), e06161 (2021)
7. Moriarty, J.P., Yasin, M.M.: A theory of benchmarking. Benchmarking Int. J. **18**, 176–209 (2011)
8. Gupta, P.: Beyond PDCA - a new process management model. Qual. Prog. **39** (2006)
9. Flynn, B.B., Schroeder, R.G., Sakakibara, S.: A framework for quality management research and an associated measurement instrument. J. Oper. Manag. **11**(4), 339–366 (1994)
10. Verma, R., Pullman, M.E.: The scholarly commons an analysis of the supplier selection process. Int. J. Mgmt Sci. **26**(6), 739–750 (1998)
11. Sarkis, J., Talluri, S.: A model for strategic supplier selection. J. Supply Chain Manag. **38**(4), 18–28 (2002)
12. Çebi, F., Bayraktar, D.: An integrated approach for supplier selection. Logist. Inf. Manag. **16**(6), 395–400 (2003)
13. Beil, D.R.: Supplier selection. Wiley Encycl. Oper. Res. Manag. Sci. (2010)
14. Patil, A.N.: Modern evolution in supplier selection criteria and methods. Int. J. Manag. Res. Rev. **4**(5), 616 (2014)
15. Waikar, A.M., Huynh, M.Q., Cope, R.F., Tate, U.S.: Evaluating key factors in supplier selection for micro–businesses: implications for buyer satisfaction. Int. J. Integr. Supply Manag. **6**(3–4), 284–309 (2011)
16. Jochem, R., Landgraf, K.: Quality management benchmarking: FDA compliance in pharmaceutical industry. Int. J. Health Care Qual. Assur. **23**(8), 690–698 (2010)
17. Siha, S.M., Saad, G.H.: Business process improvement: empirical assessment and extensions. Bus. Process. Manag. J. **14**(6), 778–802 (2008)
18. Biazzo, S., Bernardi, G.: Process management practices process management practices and quality systems standards risks and opportunities of the new ISO 9001 certification. Bus. Process. Manag. J. **9**(2), 1463–7154 (2003)
19. Lee, M.S., Lee, Y.H., Jeong, C.S.: A high-quality-supplier selection model for supply chain management and ISO 9001 system. Prod. Plan. Control **14**(3), 225–232 (2003)
20. Carmignani, G.: Process-based management: a structured approach to provide the best answers to the ISO 9001 requirements. Bus. Process. Manag. J. **14**, 454–466 (2008)
21. Okwiri, O.: ISO 9001 Quality Management System Audit As an Organizational Effectiveness Evaluation Tool, vol. 20, no. 1, pp. 15–29 (2013)

22. Psomas, E., Pantouvakis, A.: ISO 9001 overall performance dimensions: an exploratory study. TQM J. **27**(5), 519–531 (2015)
23. Saaty, T.L.: Modeling unstructured decision-making-AHP. In: International Conference on Mathematical Modeling (1977)
24. Banaeian, N., Mobli, H., Nielsen, I.E., Omid, M.: Criteria definition and approaches in green supplier selection – a case study for raw material and packaging of food industry. Prod. Manuf. Res. **3**(1), 149–168 (2015)

Requirements for Flow Control Systems for Discrete Production of Mechanical Products

Jaroslaw Chrobot(✉)

Wroclaw University of Science and Technology, 27 Wybrzeze Wyspianskiego Street, 50-370 Wroclaw, Poland

jaroslaw.chrobot@pwr.edu.pl

Abstract. Requirements regarding the quality and traceability of delivered products in some industrial sectors, such as the food, pharmaceutical and automotive industries, where the life and health of the end user depends on the quality of the product, cause production flow control systems (PFC) to become an important element of the functioning of production enterprises. PFC systems are IT systems that are generally used for recording production data (in a database) at various stages of industrial processes, mainly from PLC controllers (in this respect they perform the function of traceability systems, that enable tracking products), but also provide blocking of identifiable products, which were incorrectly processed during earlier stages of production or follow wrong technology sequence. PFC systems used to support discrete production of mechanical products are characterised by taking into consideration rework and disassembly workplaces to enable reuse of products and components. Workers have to confirm to the PFC system the positive (reuse possible) or negative result of rework/disassembly operations. Further requirements regard safe functioning by locating PFC systems' software on central IT server as well as enabling a communication with workers using remote desktops or panel PCs with dedicated IT software. The article discusses desired PFC architecture, which integrates rework, disassembly as well as remote communication. The described solution is the result of projects in which the author of this article participated, the aim of which was to develop and launch a PFC system that meets the basic industrial requirements from the area of discrete manufacturing of mechanical products.

Keywords: Production Flow Control · Data Registration · Traceability · Mechanical Products · Product Tracking

1 Introduction

Requirements regarding the quality and full traceability of delivered products in some industrial sectors, such as the food [3, 12–15], pharmaceutical [17], electronics, aerospace as well as automotive industries [11, 18, 19], where the quality and life of the end user depends on the quality of the product, make the implementation of Production Flow Control systems (PFC) a very important element of business operations. For the food industry, traceability is not only required by regulatory agencies and third party

A. Burduk et al. (Eds.): ISPEM 2023, LNNS 790, pp. 255–267, 2023.
https://doi.org/10.1007/978-3-031-45021-1_19

audit standards, but it is a key element of risk management and control of food safety and quality. A robust, documented system that provides traceability, helping companies to meet regulatory or certification standards for food safety, supports to identify products quickly in the case of contamination [3]. The already existing traceability standards define product data identification, acquisition, as well as processing [1].

Traceability is a collection of complete and chronological information about unique identified products. It allows to distinguish materials when reporting and when analyzing the entire life cycle of a given product. This means that one can thoroughly analyze a unit product from the earliest stages of its production (the ability to identify materials from suppliers that were used to manufacture the product), through the production process (location of processes, stages, operators who participated in the production, service history, as well as information about possible failures or quality problems) to the final stage, i.e. sales and customer location. "Traceability" is not only a term used in the food industry. Also it is used in the electronics industry, where traceability is of great importance in the aspect of minimizing the costs of corrective actions. With the help of an unambiguous marking, one can track the origin of the final product along the entire supply chain, up to individual components. Unambiguous identification allows, in the case of a final product defect, to quickly isolate the source of the defect and launch a precise corrective action. In this way, the costs and financial losses associated with such actions can be significantly reduced. Traceability systems have become very important management information systems within the production companies as they can efficiently control the supply chain, by minimizing the risks of the production process, and helping to enhance the customer reliability on products [4]. Also according to [7] "traceability has become a critical risk management tool for a wide range of companies, including those in the food, medical, electronics and automotive industries".

In food production, it is important to maintain the appropriate freshness of goods, temperature, packaging, storage, transport or to place reliable and true information on labels. In the case of medicines production, it is extremely important to maintain the correct proportions, expiry date, origin of raw materials and, most importantly, their impact on human life and health.

The ability to trace food brings huge benefits for quality and ecology. Currently, agriculture uses a lot of fertilizers and pesticides, which have a dramatic effect on vegetables and fruits. Manufacturers outdo each other to make their products more eye-friendly, longer fresh or better in taste, but they often forget about the main advantages, i.e. their naturalness and the content of vitamins and minerals. When buying a given product, the customer wants to be sure that its production was carried out in accordance with various safety standards and rules, and that the information on the label is true. The product tracking system makes it possible to trace, for example, the route of a given bottle of olive oil, starting from the olive orchard, ending with the retail store. What's more, the ability to trace production is also of great importance if harmful substances or bacteria are detected. The use of the traceability system makes it possible to withdraw products of a given series from stores in a sufficiently short time. Another advantage of this system is also product tracking during production processes, which prevents waste of products due to their expiration date or improves company management because it allows to check where a given product, whether a batch of products is located and what processes have already been performed, which directly affects for scheduling of production.

This is very different in the aerospace or automotive industry, where it is important to design individual parts, calculate their strength, select the right material, and then inspect the individual components and their joints. The automotive industry, where nowadays in use are globalized platforms and standardized vehicle architectures, even a single faulty part can have serious after-effects. "Faced with costly recalls, automotive manufacturers are confronted with increasingly complex requirements and stricter specifications. To be future-proof and competitive, companies therefore need powerful automation solutions paired with integrated traceability solutions such as laser markers, barcode readers, barcode verifiers and controllers that can capture and analyze data without interfering with production" [22]. In case of aerospace industry, traceability helps to locate and then report where a part has been since it was manufactured or last certified. The aerospace industry is unique, because a part must be tracked and traced even after a plane has been manufactured and delivered [23].

In case of pharmaceutical industry, companies, due to the use of their products, are burdened with a huge responsibility for human life, as well as they are subject to many control authorities. Pharmaceutical products require strict control of production processes, which involves the possibility of detailed identification of production batches. This is particularly important when, despite many required audits or certificates, there are many products on the market that are counterfeit or those whose properties differ from those promised on the packaging.

The current interest in ecology has caused that more and more people are looking for information about the companies whose products they want to buy. This also applies to the furniture industry, whose customers want to know in which country a given tree grew, what type of tree it is, what carbon footprint was created during the production of the product or what additives were used during the process. In order to provide such information to buyers, manufacturers are forced to introduce a product tracking system. In the case of wood processing, the stages to be recorded are logging, sawmill, use of the resulting sawdust, production of chipboard, production of edge-glued timber, production of veneer, production of furniture, wholesale and retail. The tagging system therefore begins during felling and there an RFID tag is applied to the log. Another issue is the marking of by-products, veneer sheets are marked with ink printing, while chips are stored in silos with RFID tags. Then, at the chipboard production stage, information from the silos is read and a tag is also placed on the boards, just like on a silo. In the case of the main product, i.e. solid wood furniture, the RFID tag is present at every stage of production, it contains not only the information that the customer expects, but also may have tips for operators.

The textile industry is currently highly stigmatized, e.g. due to illegal trade, unethical manufacturing practices and toxic agents used. Although this applies only to a part of the clothing industry, all companies are exposed to doubts from customers. Additional problems are: rapidly growing competition, the emergence of counterfeits and uncontrolled overproduction. The introduction of a "blockchain" database system enables insight into the production practices used, process location tracking, materials used, health and safety regulations, employee rights and anti-corruption regulations. The "blockchain" technology is mentioned in [16, 17, 21]. This solution allows to build trust between manufacturers and customers from around the world. Production begins with a fiber manufacturer, then continues as fabric production, and then the goods can go to the clothing manufacturer. The finished product goes to stores where it can be purchased by customers. The total supply chain often includes many subcontractors, depending on the complexity of the product: manufacturers of zippers, buttons, or producers of various types of fabrics from which one product is produced.

Production Flow Control Systems (PFC) are IT systems that are generally used to register discrete production data at various stages of industrial processes (in this respect they perform the function of traceability systems, that enable tracking products), but also provide blocking of incorrectly processed identifiable products during earlier stages of production.

PFC systems in discrete production ensure registration of information important from the customer's point of view (production date, numerical values of various parameters (e.g. screwing angle and torque, number of rivets applied, pressure, temperature, rod length, component identifier etc.), result of the operation (positive, negative)), which allow to detect defective products. PFC systems enable analyzing the reasons for these defects and defense against potential customer complaints or even a significant reduction in the number of complaints. PFC systems bring manufacturing benefits to enterprises expressed through savings of capital allocated for managing customer complaints, but above all an increase in income through increased customer confidence and the opportunity to gain new markets.

A typical discrete production system for mechanical products, controlled by the PFC system, consists of production stations managed by Programmable Logic Controllers (PLC) or PCs (in the industrial version with an external monitor) – Fig. 1.

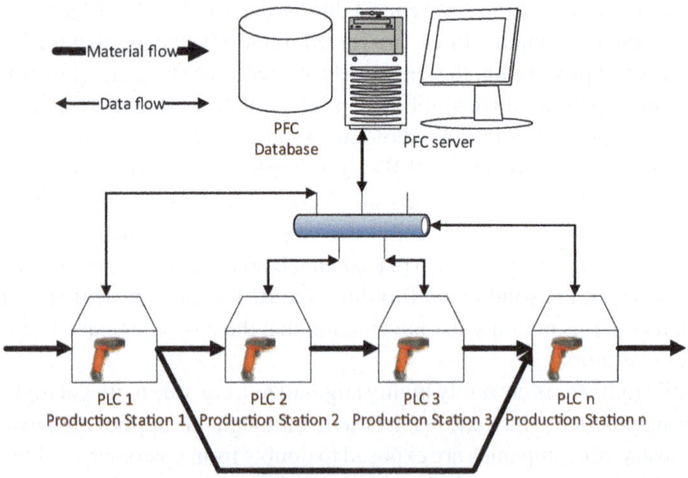

Fig. 1. Principle of PFC system operation [11]

The PFC system through a computer network and OPC server (commercially available software for communication with PLCs of various producers), or its own internal communication solution, reads information stored in the PLC/PC about the result of the operation performed and the values of various types of important parameters (e.g. angles and torques of screwing operation, length, force, number of rivets etc.). These data are saved in a database (usually it is an SQL database) and are assigned to an unique identifier read from the product label in the form of a bar code, two-dimensional code (datamatrix code) or RFID data transponder (tag). When the product enters the next production station, the PFC system additionally queries the database, whether the previous operation was successful, whether the product follows the defined technological route and whether it is compatible with the currently implemented processing program set for the designated reference (product type, e.g. manual version, electrical version, etc.) In the case of a negative response, the PFC system sends a PLC (over OPC server) or a PC the command to the production station. The PFC system can additionally support automatic setup of production stations, i.e. the reference code read at the first production station in the technological process and assigned to the product is transferred internally to subsequent production stations with a setup order (change of production station processing program). PFC systems can be integrated with ERP or MES systems by delivering data about number of pieces produced in a certain time period, to give a clear view over a status of an order fulfillment. Installation of a PFC system requires the following issues to be handled: (1) production station's PLC/PC software able to communicate with PFC server over installed computer network, (2) SQL database with a proper data structure, (3) configured PFC server for connection with production station as well with the database.

One of the main challenges in traceability systems development is that there is no one solution that will solve traceability problems for every kind of industry. Traceability needs are dependent on the context of the organization and can differ from case to case in the same organization [2].

The described PFC system solution is the result of projects in which the author of this article participated, the aim of which was to develop and launch a PFC system that meets the basic industrial requirements from the area of discrete manufacturing of mechanical products.

2 Central Server Architecture of PFC Systems

Central architecture is a safe solution for PFC systems and is characterized by both central PFC system and database server. In the architecture the concept of the Station is understood as the production single unit (included into a Production Line) controlled by PLC or PC. The PLC or PC can control a larger number of Stations as well. This architecture include groups of n Stations connected within local networks to separate them from the main network (to minimize the data flow) inside a group.

The components of the central server architecture are (see Fig. 2).

- PFC System Server – a powerful computer server with the OPC server, database and Virtual PFC System Points running on it,
- Virtual PFC System Point – a virtual computer (not a physical PC, but using hardware resources of the PFC System Server) with PFC system IT application to acquire data

from the group of n Stations. It is a virtual computer without a local database. It communicates with the database which runs on PFC System Server,
- Access Point – any physical computer with configured access to the PFC System Server and dedicated reporting application to generate production reports according to defined criteria,
- PFC System Visualization – an optional physical PC to visualize PFC system IT application for a production line and to use PFC system functionality (i.e. printing spare labels, view data flow, querying a logbook to analyse causes of problems in the data flow etc.) – to enable a communication with workers using remote desktops or panel PCs with dedicated IT software (including rework and disassembly stations).

The advantage of the architecture is possibility to include more processes into PFC system by adding new Virtual PFC System Points and configure them to be able to communicate with the database and Stations. The other advantage can be maintenance of the whole traceability software in one place (e.g. there is no need to go to PCs and upload new version of PFC system software), as well as having at disposal mechanisms of software backup using spare data discs and spare servers to avoid hardware problems. From the other side the disadvantage can be concentrating all of the PFC system software on one server and therefore necessity to have more effective (expensive) hardware. Figure 2 considers as well rework and disassembly stations included into PFC system, that support the reuse of mechanical components of a product.

PFC systems are required to be configurable to shorten time of implementation and update. Configurability is the ability to include different processes into PFC system in relatively short time. Configurable PFC systems allow the configuration of the number of production stations and their type (managed by PLC or PC), their network addresses, as well as the types of products allowed there, without changing the software itself. It is already well described in [11].

PFC systems are evolving towards more detailed information, i.e. step-by-step data acquisition and analysis within one production station, which requires significant system efficiency. Therefore, the performance of these systems in terms of data processing and communication in the Ethernet network is constantly being improved.

3 Supporting the Reuse of Products by Considering Rework and Disassembly

PFC systems used to support discrete production of mechanical products are characterized by taking into consideration rework and disassembly workplaces to enable reuse of products and components. Workers have to confirm to the PFC system the positive (reuse possible) or negative (reuse refused) result of rework/disassembly operations. PFC system checks if the result of rework operation was positive on the product that enters a production station.

3.1 Rework Station

In case of a rework station, the example could be manual welding to fix some failures of a robotized welding. The example station is equipped with RFID reader (to read RFID tag

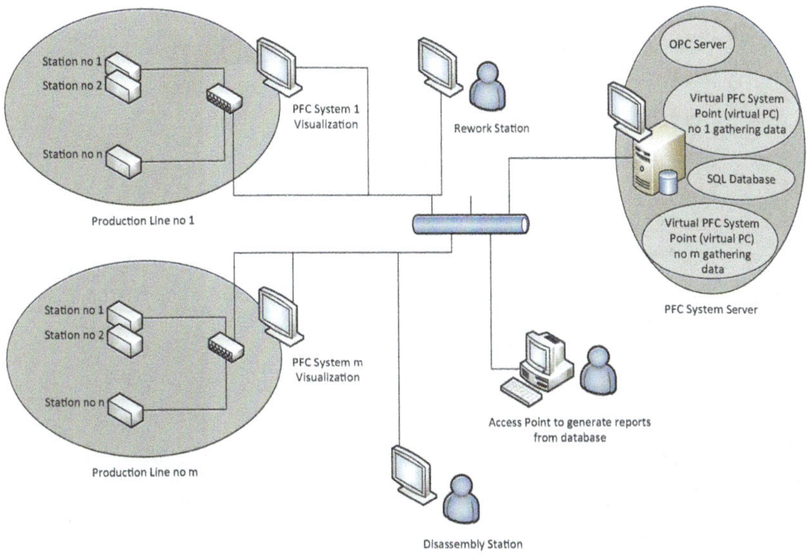

Fig. 2. Central server architecture of a PFC system [based on [11]]

(transponder), which includes product identifier) and a hardware cabinet with response buttons (see Fig. 3), which are connected with an automation device - Programmable Logic Controller (PLC). The PLC communicates directly with the PFC system via a special created communication protocol using Ethernet network. A result button puts operation status to inform the PFC system as well as shows the feedback from PFC system to the worker.

To confirm a positive or negative rework result a product in the PFC system, the worker, after the rework is finished, has to:

- ensure that the RFID tag is located in the RFID reader reading area and
- push "Start RFID Reading" white button – then the button lights constantly with white color.
- push a result button ("Welding Structure OK" – the green one or "Welding Structure not OK (NOK)" – the red one), then the button lights constantly with its color.

The PFC system gives then a feedback. The positive or negative response can be recognized as set of blinking appropriate result buttons.

Another example of a rework station is a rework of a product supported by a local IT application, which communicates with PFC system. The local IT application, installed on a local PC placed at the rework station, waits for a scan of an identification code. The scan is made by an employee (operator). An authorized operator logs into the system each time the operator wants to make a failure analysis of a product. To do this the operator has to scan the individual bar or matrix code from a label, which is fixed on the product and identifies the product in the PFC system.

The operator can manually edit the product identifier in a special text field. The operator in the "Disassembly" window manually scans the reference code (thus a diagram

Fig. 3. Cabinet with response buttons – green button pushed (Color figure online) [source: own materials]

of the correct structure appears on the screen), and then the NFP label code of the product structure (Fig. 4).

The operator clicks on the diagram of the product structure which bolted connections are defective. Marked (defective) connections are colored red, unmarked ones are considered good and remain with green. After positive finishing the rework (changing defective bolted connections into non-defective ones), the worker confirms to the PFC

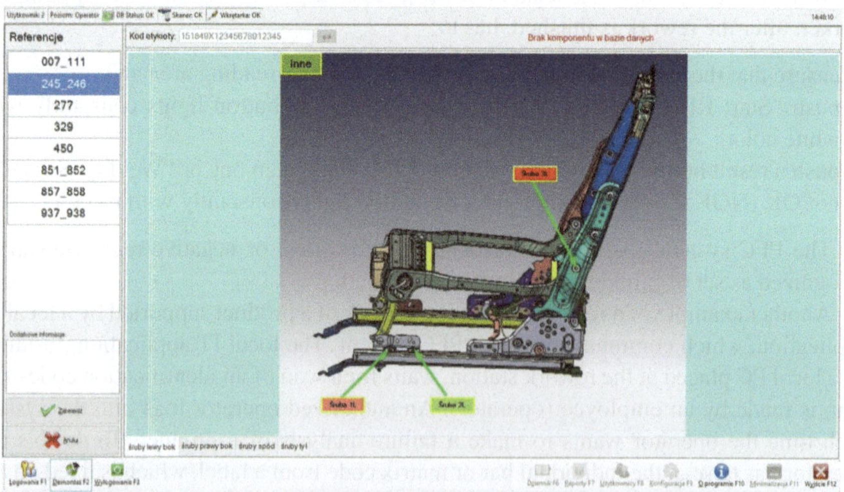

Fig. 4. The scheme of a mechanical product with erroneous bolted connections marked (red connection) (Color figure online) [source: own materials]

system with a virtual button (or scanned control code), that the product shall be allowed by PFC system to further processed on production stations.

3.2 Disassembly Station

In case of a disassembly station, the example could be manual disassembly of a car seat structure to identify and return to a production some components like electrical seat adjuster. Operator scans product unique identifier label code of a component. The label code together with the "NOK" default status is then added to the list of scanned components (Fig. 5). In case when component is permitted to be used in assembling process again, operator scans special "OK" barcode for a chosen component on the list. Operator can scan special "NOK" barcode for a chosen component in the list, in case the component is not permitted for assembling again.

In case of problems with scanner the operator can enter codes with keyboard into the "Code" text field and next accept with a virtual button. Next the operator scans „Save" special barcode to save the list of components in the database. In case of problems with scanning, operator can use the special "Save" button. If after scanning of the identifier label of a component, the PFC system identifies that identifier label duplicates or there is no identifier label code in the database, it informs with text status and with graphical status about refusing the disassembling operation.

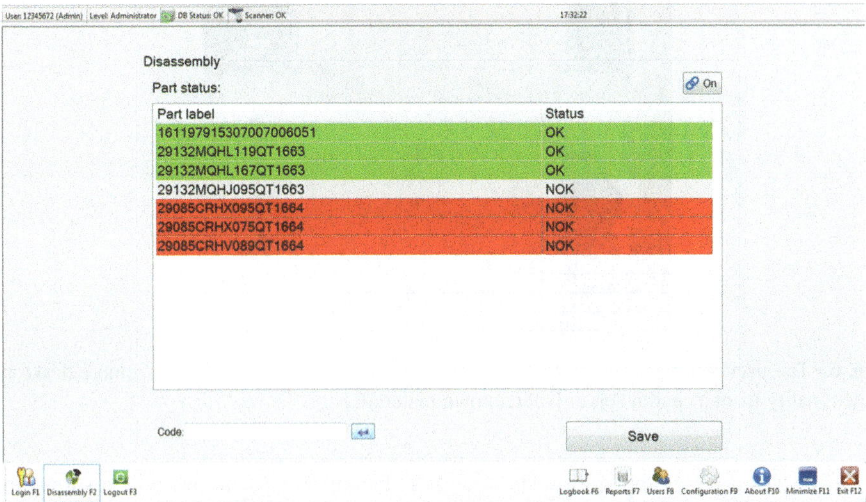

Fig. 5. Disassembling station window [source: own materials]

3.3 Visualisation and Supporting Communication

Further requirements for PFC systems with central server architecture regard enabling communication with workers using remote desktops or panel PCs with dedicated IT software. Because the main PFC system application runs on a central server, the worker have to have somehow an access to system's functionality. It can be made in two ways. The first way is to visualize its user interface (Fig. 6) on a panel using functionality of a remote desktop. In this way workers can use system's functionality after logging in using for example RFID token (eg. Euchner Key System EKS). Figure 6 shows an example user interface of a PFC system, which visualizes operations made on production stations as well as communication statuses of production stations.

Another example is visualization of a communication with the PFC system supported by a local IT application. The local IT application, installed on a local PC placed at the production station, waits for a scan of an identification code, made by an employee (operator).

Fig. 6. The user interface of a PFC system (visualization) accessible by the "remote desktop" functionality from a central server [source: own materials]

The station panel application starts with a "Produktion" window, which shows the following information (see Fig. 7):

- status section: station name, logged user, traceability status (green – connection with traceability server OK, red – connection with traceability server NOK) – "Status Traceability", traceability ON/OFF (green – traceability ON, red – traceability OFF) – "Trace EIN/Trace AUS", database status (green – communication with database OK, red – communication with database NOK) – "Datenbank", PLC status (green – communication with PLC OK, red – communication with PLC NOK) – "SPS",

- parts/hour section ("Schichtstatistik"): shows the statistics of the current shift – how many parts were done so far ("Zur Zeit"), how many parts are planned for chosen reference ("Geplant …") in this moment, how many parts are requested ("Angefordert") during the current shift; if number of parts done is less than number of parts planned, the color of the section gets red (Fig. 1a), otherwise it gets green,
- PLC section ("SPS"): this section enables to choose reference ("Referenz") and shows the status of processing step ("Status") – red status means traceability problems or PLC problems,
- parts in painting department section ("Teile in der Lackiererei"): shows the number of parts, which were welded, but not yet entered next assembly stations,
- traceability data section ("Traceabilitydaten"): this section shows the details regarding the traceability message (red color data mean traceability problems with detailed explanation of the cause) – it requires the active communication with the PFC server.

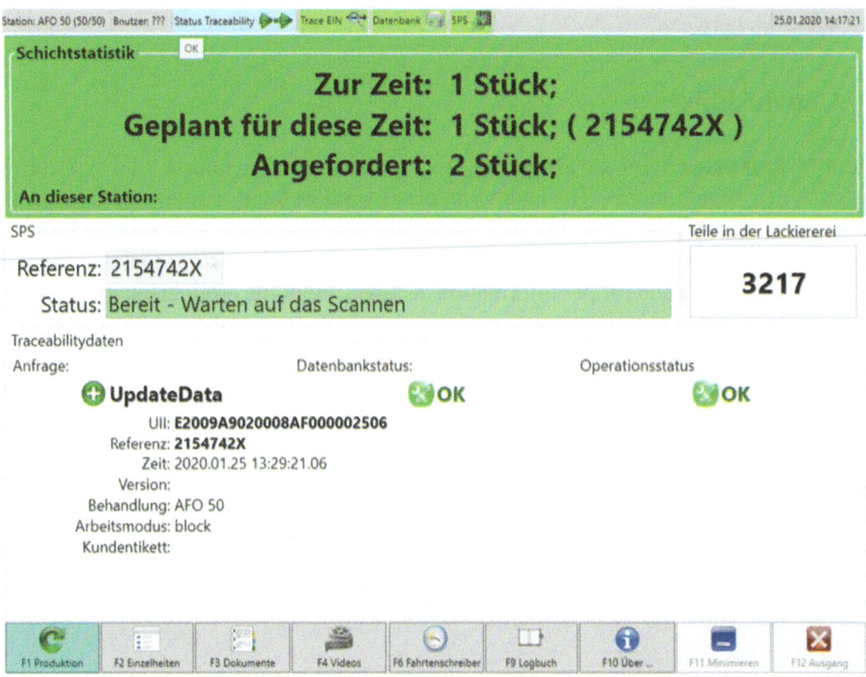

Fig. 7. IT application supporting communication between a worker and PFC system [source: own materials]

4 PFC Systems on the Market

According to the [7–9] latest reports, in case of traceability systems on the market mostly there is product marking based on barcode, QR code as well as RFID technology, which is applicable in the described PFC system. Furthermore, according to [7], "a framework is an essential part of any traceability system design as it enables a company to tailor a system to their specific products and requirements".

One of the examples of traceability systems on the market is IQMS Manufacturing Trace system. According to [10] "the IQMS web-enabled suite of Track and Trace tools provides the visibility you need to quickly find, track and report critical information related to a product's location, production status, and supplier source." Its functionality includes among others: Regulatory and Compliance Reporting (production history reporting), Detailed Search and filtering (detailed database query), as well as Real-Time Production and Quality Monitoring (gathering events' data from devices), which corresponds to the functionality of the described PFC system concept.

Another example is DelmiaWorks traceability system. According to [20] "manufacturing traceability means that you know which supplier lots and or serial numbers were used in which finished products, and who received those finished products. With a lot and serial traceability system, you have a way to track and trace goods from the supplier that sent the raw material through delivery to the customer, and all the production processes and stages in between."

5 Conclusions

Currently, manufacturers are struggling with increasing competition, and with growing customer requirements. Traceability systems support them in taking care about the quality of their products. Traceability systems are record keeping systems that show the path of a particular product from suppliers through intermediate steps to the consumers [5]. Manufacturers need proper information and data about the product location, its processing history, raw materials, etc. at each point so as to control the production process, therefore traceability has become one of the integrated parts of the supply chain logistics management [6]. The implementation of PFC systems is thus becoming a very important element of the functioning of companies, especially large organizations, or those operating in specialized market sectors that operate under high pressure of legal regulations. The ability to prepare, maintain and use the registered information minimizes the costs of withdrawing potentially harmful or unfit goods, as well as implementing preventive measures to eliminate production errors. A desirable feature of PFC systems is considering such production processes as rework and disassembly as well as visualization of PFC functionality on remote screens right in the production workshop.

References

1. GS1 Standards Document, Business Process and System Requirements for Full Supply Chain Traceability, GS1 Global Traceability Standard, Issue 1.3.0 (2012)
2. Maro, S., Steghöfer, J.P., et al.: A configurable and extendable traceability management tool. In: 2016 IEEE 24th International Requirements Engineering Conference (RE), pp. 407–408 (2016)
3. Nsf Homepage. https://www.nsf.org/newsroom_pdf/NewFood_Traceability_Oct2015.pdf. Accessed 09 Jan 2018
4. Bougdira, A., Ahaitouf, A., Akharraz, I.: An intelligent traceability system: efficient tool for a supply chain sustainability. In: AIP Conference Proceedings, vol. 1758, p. 030010 (2016).

5. Intracen Homepage. http://www.intracen.org/Traceability-in-food-and-agri-products/. Accessed 31 May 2020
6. Dharmendra, K.M.D., Henry, S., Sekhari, A., Ouzrout, Y.: Traceability as an integral part of supply chain logistics management: an analytical review. In: International Conference on Logistics and Transport (ICLT 2015), Nov 2015, Lyon, France (2015)
7. Schuitemaker, R., Xu, X.: Product traceability in manufacturing: a technical review. Procedia CIRP **93**(2020), 700–705 (2020)
8. León-Duarte, J.A., De La Re-Iñiguez B.M.: Revisión de aplicaciones de sistemas de trazabilidad. Caso de estudio industria arnesera. Estudios de Administración. 27(2), 96–112, 40–56 (2020)
9. GrandViewResearch Homepage. https://www.grandviewresearch.com/industry-analysis/track-trace-solutions-market. Accessed 25 Sep 2021
10. Iqms Homepage. https://www.iqms.com/products/quality/manufacturing-traceability. Accessed 26 Sep 2021
11. Chrobot, J.: Flexible production flow control systems. Enterp. Manage. **24**(1−2), 2–7 (2021)
12. Zheng, M., Zhang, S., Zhang, Y., Hu, B.: Construct food safety traceability system for people's health under the internet of things and big data. IEEE Access Electronic, **9**, 70571−70583 (2021). ISSN: 2169-3536
13. Conti, M.: EVO-NFC: extra virgin olive oil traceability using NFC suitable for small-medium farms. IEEE Access **10**, 20345−20356 (2022)
14. Xiao, X.: Improved traceability process for frozen tilapia waste elimination in cold chain. Cleaner Eng. Technol. **4**, 100148 (2021)
15. Agnusdei, G.P., Coluccia, B., Elia, V., Miglietta, P.P.: IoT technologies for wine supply chain traceability: potential application in the Southern Apulia Region (Italy). Procedia Comput. Sci. **200**, 1125–1134 (2022)
16. Saez, C., Karageorgos, A., Papini, R., Cerrai, D., Mier, M., et al.: Smart Pharmaceutical Manufacturing: Ensuring End-to-End Traceability and Data Integrity in Medicine Production. Elsevier, Big Data Research (2020)
17. Haq, I., Esuka O.M.: Blockchain Technology in pharmaceutical industry to prevent counterfeit drugs, Int. J. Comput. Appl. (0975–8887), **180**(25), 8−12 (2018)
18. Costa, F., do Sameiro Carvalho, M., Fernandes, J.M., Alves, A.C., Silva, P.: Improving visibility using RFID–the case of a company in the automotive sector. Procedia Manuf. **13**, 1261−1268 (2017)
19. Velandia, D.M., Kaur, N., Whittow, W.G., Conway, P.P., West, A.A.: Towards industrial internet of things: crankshaft monitoring, traceability and tracking using RFID. Robot. Comput. Integr. Manuf. **41**, 66−77 (2016)
20. DelmiaWorks Homepage. https://www.solidworks.com/product/delmiaworks/manufacturing-erp/manufacturing/manufacturing-traceability/. Accessed 07 May 2023
21. TextilSphere Homepage. https://www.textilesphere.com/2022/06/blockchain-in-textile-industy.html. Accessed 07 May 2023
22. Omron Homepage. https://industrial.omron.eu/en/solutions/blog/traceability-in-automotive. Accessed 07 May 2023
23. Frequentz Homepage. http://frequentz.com/industries/industrial/aerospace/. Accessed 07 May 2023

An Augmented System Based on Machine Learning for Boccia Assisted Gameplay

João Cruz[1], Vinícius Silva[2](✉), João Sena Esteves[1,2], and Filomena Soares[1,2]

[1] Department of Industrial Electronics, School of Engineering, University of Minho, Guimarães, Portugal
a85939@alunos.uminho.pt, {sena,fsoares}@dei.uminho.pt
[2] R&D Centre Algoritmi, Guimarães, Portugal
a65312@alunos.uminho.pt

Abstract. In order to promote the practice of sports, several approaches using technology have been employed to gamify and augment the user experience. Following this trend, the research group proposed an approach to encourage the practice of Boccia, while promoting social inclusion and reducing the amount of time it takes for newcomers to the sport to become proficient by gaining knowledge of game tactics. The present work focus on the detection, in real-time, of Boccia gestures for the framework proposed in a previous work by using a wearable device to detect the gestures. To evaluate the correct functioning of the system, several types of tests were carried out. First, the developed machine learning model was evaluated in terms of accuracy, recall, among others. Then, the gesture detection system was tested with 15 participants that executed the different Boccia gestures while using the wearable placed on the wrist. Finally, tests were carried out to integrate the gesture detection module into the framework proposed in a previous work. The tests yielded positive results that allowed the validation of the use of the system in the Boccia game.

Keywords: Activity Monitoring · Boccia · Gesture Recognition · Machine Learning · Wearable

1 Introduction

The practice of sports, being fundamental, has been neglected. A large part of the national adult population is not physically active, which translates into a higher risk of non-communicable diseases and mortality [1]. The benefits of practicing sports are not only physical, but they are also a catalyst for personal development, which is even more expressive in people with disabilities [2]. The significance of social development in the life of individuals with disabilities becomes apparent considering the previous remarks. As a result, it is crucial to formulate and execute strategies aimed at mitigating the effects of disability-induced social exclusion and fostering inclusivity, which can be achieved through engagement in sports.

Boccia is a precision ball sport that has gained attention for its ability to promote inclusivity. The sport requires a combination of physical and mental skills, including

A. Burduk et al. (Eds.): ISPEM 2023, LNNS 790, pp. 268–280, 2023.
https://doi.org/10.1007/978-3-031-45021-1_20

hand-eye coordination, strategic planning, and decision-making. It can be played individually or in teams, and the goal is to throw or roll a set of colored balls as close as possible to a target ball, known as the jack. A system using augmented reality to enhance the Boccia game experience, aiding novice players in comprehending key tactical and physical aspects of the sport, ultimately resulting in a more enjoyable learning experience was proposed by the research team in [3]. One of the features of the proposed framework concerns the detection of the key gestures that can be executed by the players during a Boccia game using the data from a wearable device.

Several authors have proposed different approaches for the recognition of gestures using data from wearable devices [4–8]. Traditional approaches combine the use of hand crafted features with machine learning algorithms in order to detect the gestures [6, 7]. In recent times, deep learning methods have gained popularity and shown exceptional performance in several image recognition tasks. In fact, some studies have employed these techniques, specifically Convolution Neural Networks (CNNs), to address the gesture recognition problem, [9–11]. Several studies demonstrate that CNNs outperform traditional hand-crafted approaches as they can learn intricate motion features, [12–14]. Nonetheless, conventional approaches of gesture recognition using deep learning methods typically employ the use of full RGB images, which could result in an increase in the complexity of the model due to the large dimension of the input features. This results in more complex and slower models, less practical for solving real-time large-scale problems. In order to tackle the dimensionality problem and computation complexity, some authors proposed the use of skeleton based representation data obtained from a 2D camera in order to recognise gestures, [9, 10, 13]. This process consists in obtaining the 2D position of the skeleton joints from RGB images and then convert it into an image representation. This data is then fed as input to the deep learning model.

Following this trend, the present work focus on the developments concerning the gestures recognition module of the framework developed by the research team and presented in [3]. In order to fulfill this goal, a dataset was created and a deep learning model based on a Convolution Neural Network (CNN) was trained. The data from the wearable sensor is converted into an image representation. The final model runs in real-time on a CPU.

The following paper is organized in 5 sections. The proposed system and the developed approach to the gesture recognition process are described in Sect. 2. The results are presented in Sect. 3 and discussed in Sect. 4. Final remarks and future work are addressed in Sect. 5.

2 Proposed System

The proposed system consists of a wearable device M5StickC [15], a computer, and a camera placed directly above the Boccia court (Fig. 1). The view from the camera is used as input of a deep learning model in order to detect, in real-time, the Boccia balls within the camera Field-of-View (FOV) and sort them according to color (red, blue, or white), which was the focus of previous work [3, 16]. To compute the distance of each red and blue ball and the jack (white ball), the corresponding centroid coordinates of each ball within the camera FOV are used, allowing to map all balls and to compute the score for the current game situation.

For each move there is an associated throwing movement that may facilitate the task of placing the ball in the desired position. There are two main gestures that are usually executed by the players, the upper throwing and the under throwing gestures (Fig. 2). The present work focus on the development of a deep learning model to detect these two gestures in real-time.

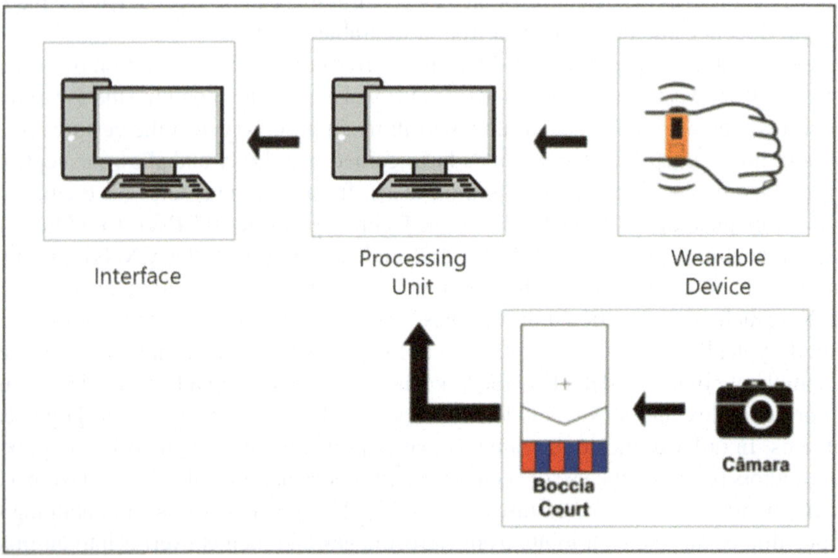

Fig. 1. Proposed system framework.

Fig. 2. Main gestures usually executed by Boccia players: on the left the upper throwing gesture and on the right the under throwing gesture.

2.1 Dataset

To detect movements, a set of data is collected from the M5StickC development board. These data are all coming from a 6-axis Inertial Measurement Unit (IMU) in the MPU6886. Three values are collected for each of the accelerometer and gyroscope axes, as well as the corresponding roll, pitch and yaw values.

In order to obtain data for training the deep learning model, the development board was attached to a bracelet. This bracelet was placed on the right wrist of the participants. Ten samples of data referring to each type of movement (Fig. 3) that is intended to be detected were captured. Therefore, each participant performed 10 throws over, 10 throws under and remained at rest for 10 times as well.

Fig. 3. Types of movements considered in the construction of the dataset.

Data capture was performed by twelve adult participants, which resulted in a total of 360 data files (12 participants, 3 types of recorded movements, 10 times each). The age group of the participants was between 18 and 80 years old and individuals of both sexes were invited. In each of those 360 files, the accelerometer and gyroscope values are arranged on the x, y and z axes, as well as the roll, pitch and yaw values during a time interval of two seconds of movement.

2.2 Data Processing

The first step is the normalization of the data values between 0 and 255 to integrate into color pixel values. Once normalized and in order to augment the amount of data to be provided to the neural network, extra handcrafted features are computed on the incoming data from the wearable device. The extra features computed are the cumulative moving average and the RMS values, both made for all the x, y and z components of the accelerometer and gyro, as well as the roll, pitch and yaw.

The wearable placed on the player's wrist provides the data frame, as seen in (1).

$$gyroX/gyroY/gyroZ/accX/accY/accZ/roll/pitch/yaw \tag{1}$$

The data obtained over several frames are used as input for the action recognition model to classify the action. All generated images have a 64 x 12 x 3 size, where 64 is the number of data samples per movement, 12 parameters from the wearable representing the height of the image, and 3 is the number of channels, Fig. 4.

The first three lines of the image are the gyroscope data, accelerometer data and RPY (roll-pitch-yaw) data. Each pixel has the data variable X as the red color component, the data variable Y as the green color component, and the data variable Z as the blue color component.

For the following three lines, the cumulative moving average values were used. The pixel assignment is very similar to the previous lines, but now the color components are matched to the MMA (cumulative moving average) for all the color components (x, y and z) of the received data.

For line number seven, eight and nine the reasoning is similar again. In this case, the assignment is made according to the RMS values of each data component.

In the last three lines, the pixel values are assigned according to the data components. Line ten was associated with all data components in X, line eleven with all data components in Y and finally, line twelve with all data components in Z. The detailed constitution of each pixel can be analyzed in Fig. 4. Figure 5 shows some samples of the generated images per class.

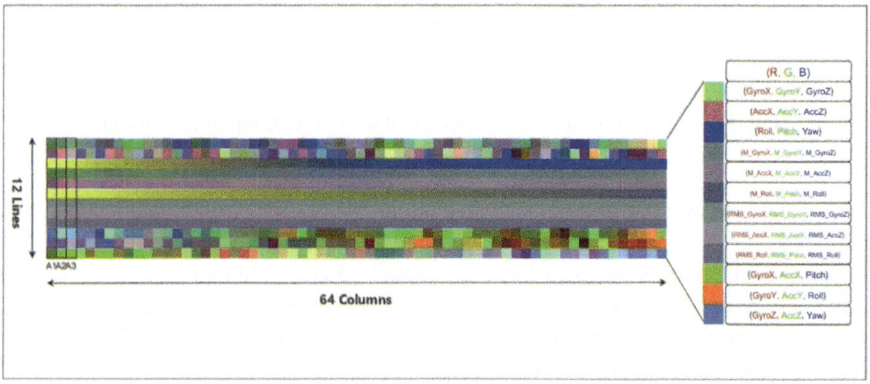

Fig. 4. Generated image-based representation from the incoming data from the wearable device.

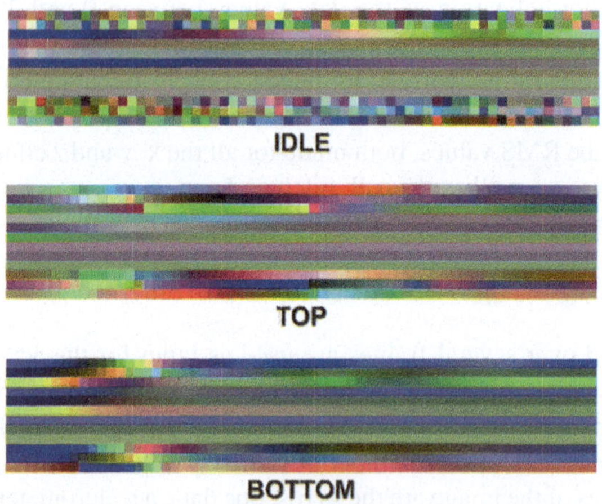

Fig. 5. Samples of the generated images per-class.

2.3 Developed Model

The model follows a CNN architecture and consists of a total of six convolutional layers each with a 3 x 3 kernel, a stride value of 1, see Fig. 6. It was decided to use groups of two 3x3 filters in a stack since these are less demanding from a computational point of view when compared to the use of only one filter of larger dimensions [9]. Thus, three groups of two stacked convolutional layers are left, where a batch normalization layer is added to each group. The ReLu function is used as an activation function.

A max polling layer with a size of 2 x 2 is added after the second and fourth convolutional layers. This layer's main function is to subsample the feature maps, maps generated by convolutions. As in this work the layer has a size of 2 x 2 the spatial resolution of the data will be halved. After the last convolutional layer, a global averaging pooling layer is used to replace the fully connected layers, since these layers can constitute an overfitting problem, [17]. To finalize the model, a fully connected layer was added to handle the classification part. The entire model was trained in TensorFlow and the processing capabilities of the google COLAB tool were used, since it allows the use of a dedicated GPU, in order to train the model more rapidly.

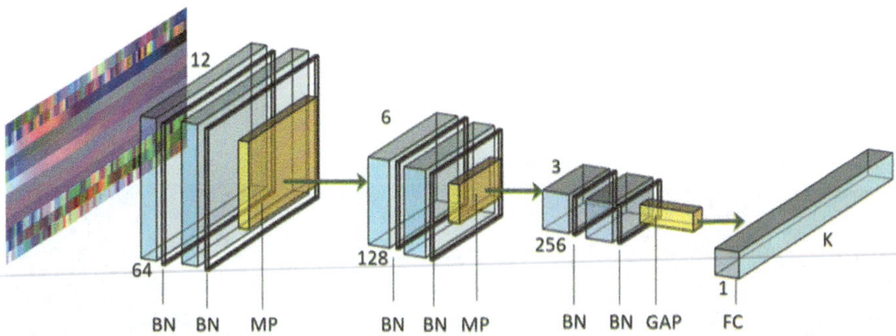

Fig. 6. The proposed deep learning model architecture.

The trained model was then imported to the developed Graphical User Interface (GUI).

2.4 GUI

In order to extract the data from the wearable device and to validate the trained deep learning model, a GUI was developed as seen in Fig. 7.

The GUI consists of five main areas. The data view area on the left side is where the user can watch the data flow coming from the wearable wristband. Then the generated image area, on the right side, where is possible to analyze the image that feeds the CNN model. In the centre, the activity area is used to store session data, and the communication area is responsible for establishing the connection between the wearable and the PC. Finally, the detection area located at the bottom of the centre area of the interface is where the predicted class of movement is displayed to the user via image.

As previously mentioned, this project follows the work done by the research team in [3], and so the GUI Fig. 8 was developed in the scope of the same project.

The GUI allows to monitor in real-time the Boccia game, by detecting the balls within the FOV of the camera. With the collected information it was possible to develop

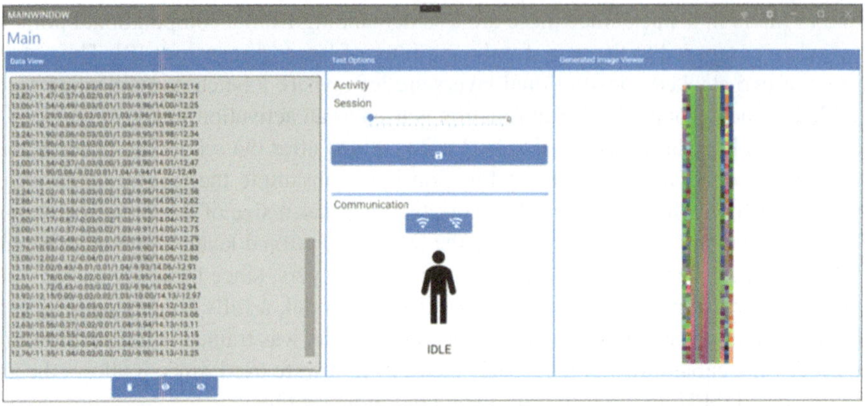

Fig. 7. Developed GUI used to collect the data and to validate the trained deep learning model.

a system that calculates the score of the game and that tells the player the best tactical play for that moment.

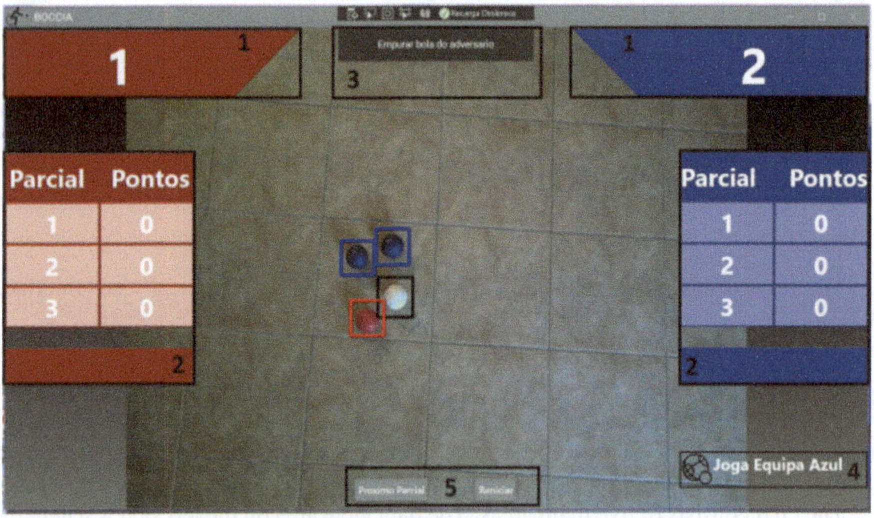

Fig. 8. The GUI developed by the research team in [3].

Figure 9 shows the update version of the GUI, where it was added an area that shows to the user the prediction of the deep neural network model regarding the type of gesture recognized.

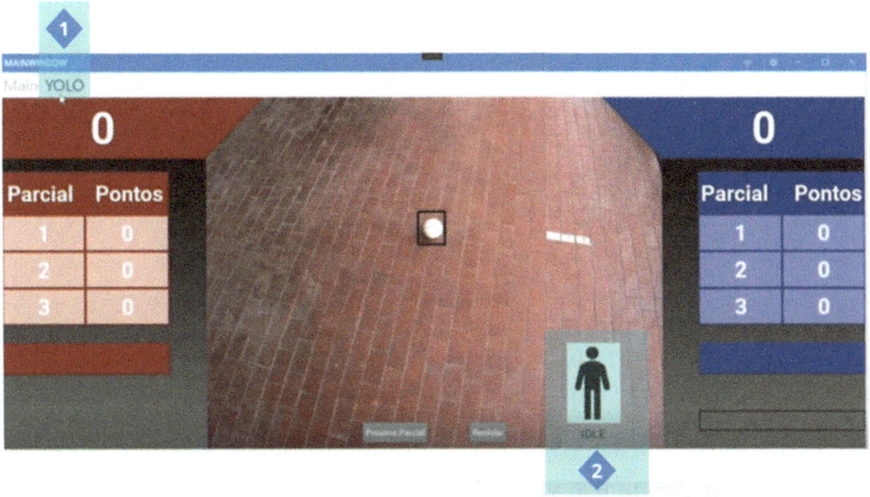

Fig. 9. The updated GUI with the prediction of the deep neural network model regarding the type of gesture recognized (2).

3 Results

This section presents the results obtained throughout the development of the proposed system. First, the performance of the deep learning model is evaluated through metrics. Afterwards, a pilot test was conducted in order to assess, the recognition accuracy of the deep leaning model in a real-world environment. Finally, the results of the integration of the gesture recognition subsystem in the system previously developed by the research group are addressed.

3.1 Model Results Train and Testing

The deep learning model was tested to evaluate its performance. The results of the full analysis – model accuracy, precision, recall, F1-Score, and confusion matrix are presented below.

Table 1 shows the overall performance of the gesture recognition model with an accuracy of 97.0%.

Table 2 presents the values of precision, recall and an F1-Score for each class of movements. For the under throw, the model achieved a precision of 100.0%, a recall and F1-Score of 95.0% and 98.0% respectively. Secondly, for the idle class the model achieved a precision, recall and, an F1-Score of 93.0%. Finally, for the upper throw the model achieved a precision of 96.0%, a recall value of 100.0% and an F1-Score of 98.0%.

The confusion matrix for the three classes of gestures can be observed in Fig. 10. More specifically, the under throw with a classification rate of 100%. The classification rate for the idle class and for the upper throw was 93.0% and 96.0%, respectively.

Table 1. Overall performance of the gesture recognition model.

Metric	Value
Accuracy	97%

Table 2. Performance of the model per-class.

	Precision	Recall	F1-Score
Bottom	1.00	0.95	0.98
Idle	0.93	0.93	0.93
Top	0.96	1.00	0.98

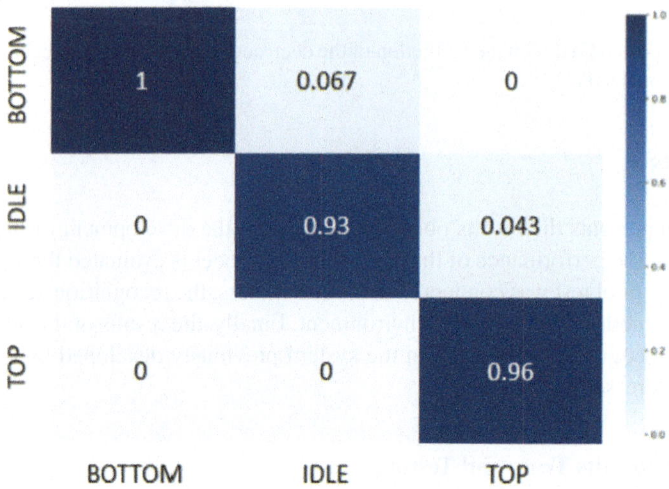

Fig. 10. Trained model confusion matrix.

3.2 Pilot Test Results

A pilot test was carried out to test the gesture detection system in real-time. The participants considered for the test were 15 adults aged 18 to 81 years old, with 26.7% of the participants being female and 73.3% being male. The experimental unit consisted of a processing unit (PC) and a wearable wristband equipped with the M5StickC. With the bracelet placed on the wrist of the subject, the researcher asked for the execution of the following sequence of gestures. (IDLE - TOP - IDLE - BOTTOM - IDLE - BOTTOM - IDLE - TOP – IDLE). The sequence was chosen so that both the upper throw and the under throw gestures were performed the same number of times (twice each). The sequence also ensured that there would always be a gap between each throw where the individuals would be idle between gestures. During the execution of the tests, all the results were noted down by a supervisor.

Of the data collected, the Fig. 11 was created. The cells represented in green are the set of gestures correctly predicted in real-time by the artificial intelligence model and the cells in red represent the gestures where the system failed to predict.

Table 3 shows the overall performance of the real-time test results with an accuracy of 98.0%.

Table 4 presents the values of precision, recall and an F1-Score for each class of movements. For the under throw, the model achieved a precision of 100.0%, a recall and F1-Score of 93.0% and 97.0% respectively. Secondly, for the idle class the model achieved a precision of 97.0%, recall of 99.0% and, an F1-Score of 98.0%. Finally, for the upper throw the model achieved a precision of 97.0%, a recall value of 100.0% and an F1-Score of 99.0%.

The confusion matrix for the three classes of gestures can be observed in Fig. 12. More specifically, the under throw with a classification rate of 100%. The classification rate for the idle class and for the upper throw was 97.4% and 96.8% respectively.

USER ID	IDLE	TOP	IDLE	BOTTOM	IDLE	BOTTOM	IDLE	TOP	IDLE
1	1	1	1	1	1	1	1	1	1
2	1	1	1	1	1	1	1	1	1
3	1	1	1	1	1	1	1	1	1
4	1	1	1	1	1	1	1	1	1
5	1	1	1	1	1	1	1	1	1
6	1	1	1	1	1	1	0	1	1
7	1	1	1	1	1	1	1	1	1
8	1	1	1	1	1	1	1	1	1
9	1	1	1	1	1	1	1	1	1
10	1	1	1	0	1	1	1	1	1
11	1	1	1	1	1	1	1	1	1
12	1	1	1	1	1	1	1	1	1
13	1	1	1	1	1	1	1	1	1
14	1	1	1	1	1	1	1	1	1
15	1	1	1	1	1	0	1	1	1

Fig. 11. Results of the experiment conducted

Table 3. Model test performance obtained from the tests conducted.

Metric	Value
Accuracy	98.0%

Table 4. Model test performance per-class obtained from the experiment conducted.

	Precision	Recall	F1-Score
Bottom	1.00	0.93	0.97
Idle	0.97	0.99	0.98
Top	0.97	1.00	0.99

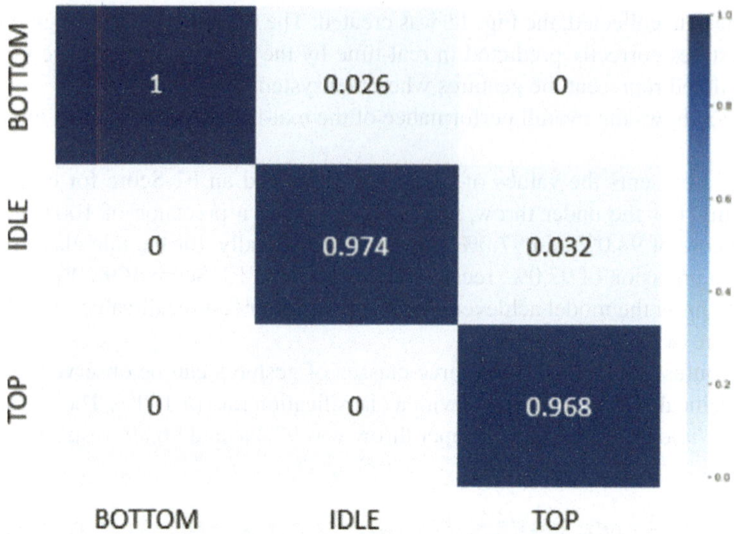

Fig. 12. Model test confusion matrix.

4 Discussion

Starting by analyzing the confusion matrix of the artificial intelligence model (Fig. 10), it is possible to conclude that all three classes had a precision over 90.0% (diagonal values). However, the model predicted 6.7% of cases as being idle when they were, in fact, an under throw movement. The model also predicted 4.3% of cases as being part of the upper throw class despite being, in fact, part of the idle class. In general, the results of the model performance-wise are positive since gesture recognition is performed correctly in most of the times.

Regarding the performance of the real-time results and analyzing the confusion matrix (Fig. 12), the under-throw class presents a 100.0% precision value. Concerning the upper throw class, there is a precision of 97.4% and finally, in the idle class a precision of 96.8%. However, the model also predicted 3.2% of cases as being upper throw despite being part of the idle class and 2.6% of the cases as being part of the idle class when they were, in fact, an under throw. The developed system had an accuracy value of 98.0%.

By comparing the confusion matrix of the model (Fig. 10) and the confusion matrix obtained experimentally in the real-time tests carried out (Fig. 12) is possible to get some conclusions. It's possible to observe that in the under-throw class, the results in terms of precision are the same in both situations, that the real-time tests and the idle class presented results slightly better than model results. Finally, in the upper throw class, the real-time tests also had a smaller error than the model prediction.

The computational time was computed considering the average execution time of the model. The tests ran on a notebook equipped with an Intel(R) Core (TM) i7–8650 quad-core CPU with 16 GB of RAM. The inference is done by using OpenCV integrated Deep Neural Network module that allows to import saved TensorFlow models [18]. Based on the performed tests, the average execution time of the model is 9 ms, which ensures the real-time capability of the developed approach.

5 Final Remarks and Future Work

Sports play a vital role in promoting physical and mental health, social inclusion, and overall well-being. Boccia has gained a significant attention, as it allows to promote the practice of sports and to foster inclusivity. In recent years, Boccia has been further enhanced by technological advancements with the goal to gamify and augment the user experience while promoting the practice of sports.

Following this trend, the research team proposed in [3] an augmented reality framework to facilitate learning and improve the player's experience while playing Boccia. One of the modules of the framework consists in detecting the user throwing gestures. The present work focus on the development of this module. In order to detect gestures, a deep learning model based on a CNN architecture was developed and trained. In order to train the model, data was collected from various throws executed by several participants. The data was collected through a wearable placed on the participants' wrist, integrated into a bracelet.

Regarding the results of the developed model, a precision value of 93% was obtained for the still class, 100% for the underhand launch class, and 96% for the overhand throw class. Concerning the real-time test, the following results were achieved: a precision value of 97.4% for the still class, 100% for the underhand throw class, and 96.8% for the overhand throw class. The developed approach was integrated in the final GUI. The model execution time is, on average, 9 ms.

Future work includes a continuous improvement of the presented approach by, for example, augmenting the dataset and implementing other deep learning architectures in order to improve the model performance. Additionally, experiments will be conducted in order to test the system proposed in [3] with the model developed in the present work.

Acknowledgments. The authors would like to thank the volunteers who helped test the developed system. This work has been supported by national funds through FCT – *Fundação para a Ciência e Tecnologia* within the R&D Units Project Scope: UIDB/00319/2020.

References

1. World Health Organization: WORLD REPORT ON DISABILITY (2011)
2. Burchell, A.: The Importance of Sport to the Disabled. The Commonwealth Health Minister's Book (2006)
3. Calado, A., Marcutti, S., Silva, V., Vercelli, G., Novais, P., Soares, F.: Towards a virtual coach for boccia: developing a virtual augmented interaction based on a boccia simulator. In: VISIGRAPP 2020 - Proceedings of the 15th International Joint Conference on Computer Vision, Imaging and Computer Graphics Theory and Applications (2020). https://doi.org/10.5220/0009142602170224
4. Weinland, D., Ronfard, R., Boyer, E.: A survey of vision-based methods for action representation, segmentation and recognition. Comput. Vis. Image Underst. **115**(2), 224–241 (2011). https://doi.org/10.1016/j.cviu.2010.10.002

5. Amendola, S., Bianchi, L., Marrocco, G.: Movement detection of human body segments: passive radio-frequency identification and machine-learning technologies. In: IEEE Antennas Propag Mag **57**(3), 23–37 (2015). https://doi.org/10.1109/MAP.2015.2437274
6. Wilson, D., Wilson, A.: Gesture Recognition Using The XWand (2004). http://www.ri.cmu.edu/pub_files/pub4/wilson_daniel_h_2004_1/wilson_daniel_h_2004_1.pdf%5Cn. http://citeseerx.ist.psu.edu/viewdoc/summary?doi=10.1.1.1.2780
7. Wu, J., Pan, G., Zhang, D., Qi, G.: Gesture recognition with a 3-d accelerometer. In: Proceedings of the 6th International Conference on Ubiquitous Intelligence and Computing, **5585**, pp. 25–38 (2009). https://doi.org/10.1007/978-3-642-02830-4_4
8. Chen, Y., Luo, B., Chen, Y., Liang, G., Wu, X.: A Real-time Dynamic Hand Gesture Recognition System Using Kinect Sensor, pp. 2026–2030 (2015)
9. Ludl, D., Gulde, T., Curio, C.: Simple yet efficient real-time pose-based action recognition. In: 2019 IEEE Intelligent Transportation Systems Conference, ITSC 2019 (2019). https://doi.org/10.1109/ITSC.2019.8917128
10. Pham, H.H., Khoudour, L., Crouzil, A., Zegers, P., Velastin, S.A.: Exploiting deep residual networks for human action recognition from skeletal data. Computer Vision and Image Understanding (2018). https://doi.org/10.1016/j.cviu.2018.03.003
11. Silva, V., Soares, F., Sena Esteves, J., Vercelli, G.: Human action recognition using an image-based temporal and spatial representation. In: 2020 12th International Congress on Ultra Modern Telecommunications and Control Systems and Workshops, ICUMT, pp. 1–6 (2021). https://doi.org/10.1109/ICUMT51630.2020.9222408
12. Du, Y., Wang, W., Wang L.: Hierarchical recurrent neural network for skeleton based action recognition. In: Proceedings of the IEEE Computer Society Conference on Computer Vision and Pattern Recognition (2015). https://doi.org/10.1109/CVPR.2015.7298714
13. Li, C., Hou, Y., Wang, P., Li, W.: Joint distance maps based action recognition with convolutional neural networks. In: IEEE Signal Process Lett. **24**(5), 624–628 (2017). https://doi.org/10.1109/LSP.2017.2678539
14. Song, S., Lan, C., Xing, J., Zeng, W., Liu, J.: An end-to-end spatio-temporal attention model for human action recognition from skeleton data. In: 31st AAAI Conference on Artificial Intelligence, AAAI 2017 (2017)
15. M5StickC ESP32-PICO Mini IoT Development Kit | m5stack-store. https://shop.m5stack.com/collections/m5-controllers/products/stick-c. Accessed 27 Feb 2023
16. Calado, A., Silva, V., Soares, F., Novais, P.: Ball detection for boccia game analysis. In: 2019 6th International Conference on Control, Decision and Information Technologies (CoDIT), pp. 1468–1473 (2019)
17. Lin, M., Chen, Q., Yan, S.: Network in network. In: 2nd International Conference on Learning Representations, ICLR 2014 - Conference Track Proceedings (2014)
18. OpenCV: Deep Neural Networks (dnn module) (2020). https://docs.opencv.org/master/d2/d58/tutorial_table_of_content_dnn.html. Accessed 8 Apr 2020

Ultrasound Tissue Differentiation (UTC) an Objective Diagnostic Tool in Prevention and Healing Monitoring Tendon Problems in Race Horses

Andrzej Golachowski(✉) and Barbara Golachowska

Muscat, Oman
veterinarysurg@gmail.com

Abstract. Tendon and ligament injuries in race horses are common and can contribute to the end of a sporting career. Tendon damage is caused by an imbalance between matrix synthesis by tenocytes and damage caused by excessive repetitive loading, overstrain and insufficient recovery time. Cumulative microtrauma are weakened collagenous, non-collagenous and vascular components of the tendon, finally resulting in tendon rupture. Treatment of tendon injuries is prolonged, scar formation replaces injured bundles which cause inferior strain and contribute to high injury recurrence rates. Since 1980 ultrasound examination is the primary diagnostic technique. However, one of its limitation is the difficulty to distinguish fibrotic tissue from normal tendon by grey level and detect early signs of overstrain to prevent injuries. Computed analysis of fiber patterns along the entire length of the tendon created from series of transverse images (UTC) is a new diagnostic tool in equine practice. It provides tomographic visualizations of the entire tendon, quantification of the architecture, integrity and accurate calculation with high repeatability. This technique provides strong evidence based results and contributes to prevention of injury as well as monitoring repair. The multimodal approach that shifts from curative to preventive medicine might have a significant impact to improve sport horses welfare.

Keywords: UTC · Tendonitis · injury · prevention · Racehorse

1 Introduction

Tendon and ligament injuries in race horses are common and can contribute to the end of a sporting career. Superficial Digital Flexor Tendon (SDFT) injury affects up to 30% of racehorses in training per year.

Affected horses, despite treatment require long rehabilitation which takes 9–18 months, are in high risk of tendon re-injury, which occurs in up to 80% of treated horses and preform on lower level than before injury [1, 2].

At gallop, a 500 kg horse landing on a single front limb around 150 times per minute and moving at speeds of over 60 km/hr generates forces that work close to the mechanical

A. Burduk et al. (Eds.): ISPEM 2023, LNNS 790, pp. 281–291, 2023.
https://doi.org/10.1007/978-3-031-45021-1_21

limits of tendons. Fast movement produces tendon strain (change in tendon length as a percentage of initial length) of up to 16%, where experimental tendon failure was observed at 12–20% tendon hyperthermia, core temperature reaches 43–45 C [1]. These tendons storing and releasing energy decreases the energetic cost of locomotion by 36% in the horse [3].

Extreme loading environments which repeat loading of the tendon without sufficient recovery time make it more susceptible to failure, even when submaximal loads are applied.

After maximal exercise, clinical and ultrasound examination do not reveal any abnormalities in healthy horses but an ultrastructural tendon response is detectable by the use of ultrasound tissue characterization (UTC) technique [4].

Biomechanical and biothermal effects of strain on the SDFT contribute to the accumulation of microdamage, which do not produce prominent clinical signs, until rupture is observed. Despite the morphologic information gained by ultrasonography it has been shown that it is impossible to predict injuries of the superficial digital flexor tendon [5, 6].

Tendons due to poor vascularization heal slowly. During the proliferative phase scar tissue consisting of type III collagen synthesized by fibroblasts, differs structurally and functionally from normal tendon structure [5, 6].

Over years, different treatments have been applied to improve tendon healing. For centuries firing (thermocautery) or blistering (applying causative ointments) of skin over injured tendons was advocated. Credible research in the 1980's demonstrated line (bar) firing of the skin overlying a tendon injury had neither a beneficial nor a detrimental effect on healing of the tendon, but pin firing into a damaged tendon was detrimental [7]. Nowadays the Royal College of Veterinary Surgeons considers all forms of firing and "blistering" to be ineffective and lacking justification as methods of treatment which should be discontinued, but they are still in use in many regions of the world.

In modern equine medicine, a combination of conservative, regenerative and surgical therapies are applied to improve the outcomes of tendon injuries. Controlled exercise program based on monitoring of tendon healing by ultrasound examination is the primary treatment.

Regenerative therapies including intralesional injection of platelet-rich plasma, stem cells, autologous conditioned serum, autologous protein solution, growth factors, amniotic fluid allograft, microvesicles are under clinical investigation. The latest research suggests that a combination of intralesional bone marrow aspirate, SDFT accessory ligament desmotomy and 12 months rehabilitation provides a better clinical outcome than conservative management (only rehabilitation) of SDFT injury in racehorses in terms of post-injury performance [8].

Despite the chosen treatment method healing being prolonged, failure to re-constitute normal structure contributes to high injury recurrence rates. Tendons sustain cumulative subclinical damage for undefined periods of time prior to rupture, implying that there is an opportunity for intervention before irreversible injury occurs, but there is a lack of evidence-based preventive programs reflecting equine training and detection of tendon microdamage [9, 10].

2 Pathophysiology of SDFT Injury and Repair

The SDFT acts to support the metacarpophalangeal and proximal interphalangeal joints during the stance phase and to actively and passively flex these joints and the carpus during the swing phase. The SDFT also acts as a biological spring, storing kinetic and potential energy as elastic energy during weight-bearing and returning it during propulsion, resulting in significant improvements in locomotor efficiency during high-speed gaits [1]. Tendons have a longitudinally oriented, hierarchical structure with the largest subunits termed fascicles (cross-sectional area of 0.8–1.2 mm^2) separated by the endotenon comprised of loose connective tissue carrying blood vessels, nerves and lymphatics from connective tissue sheaths. Tendons are composed predominantly of water, approximately 70%. The major constituents of the remaining 30% dry matter are collagen and a non-collagenous matrix [11].

Tenocytes are fibroblastic cells responsible for synthesis and turnover of the matrix. They are located between collagen fibril bundles within fascicles, where most lack direct contact with the vasculature located in the endotenon.

Excessive repetitive loading and overstrain combined with insufficient time between episodes cause direct damage to the matrix that cannot be repaired because it overwhelms the cellular capacity of tenocytes. Cumulative microtrauma weakens collagenous, non-collagenous and vascular components of the tendon which finally result in tendon rupture [9].

The majority of injuries are generally thought to occur as a combined result of continuous loading cycles above a certain intensity rather than due to a single overloading event. Degenerative changes in the matrix also occur induced indirectly by autocrine or paracrine modulation, hypoxia, ischemia with reperfusion injury, abnormal regulation of apoptosis and hyperthermia [1].

Diagnosis of matrix degeneration, before clinical signs of tendon injury are present, is made for research purposes by biochemical, molecular and cellular approaches on samples obtained from biopsy or research animal specimens.There are no biomarkers to monitor tendon condition in sport horses by blood test or other noninvasive methods.

Injured tendons undergo slow repair with fibrous scar tissue which is compromised in functional efficiency. This scar tissue is essentially the same as that forming in skin wounds and does not have the same matrix composition as the normal tendon. The high stiffness of scar tissue puts increased strain on adjacent areas of the tendon, resulting in a high incidence of recurrent injury [11].

Prognosis for tendon injuries is dependent on the severity of the initial injury, with little contribution from the therapeutic regime subsequently applied. Current treatment, at present, is not able to regenerate normal tendon tissue.

Following rupture there are overlapping phases of reactive inflammation, proliferation, remodeling and maturation.

Early signs of inflammation can be detected by careful clinical examination as focal swelling and pain at palpation before visible enlargement and lameness is observed. At this time tendon lesion is present. Non-prominent clinical signs of minor injuries including lack of lameness is often missed by caretakers which results in progressing injury and treatment delay.

Developing inflammation results in an increase of cross-sectional area (swelling) extending proximally and distally to the injury site with hemorrhage, fibrin clot formation, edematous swelling and infiltration of inflammatory cells into the defect. Enzymatic degradation by endogenous collagenases and proteases extends the lesion beyond the limits of the initial mechanical disruption. Monitoring surgically created lesion by computerized ultrasonography revealed 2- to 4-fold increase in length and an approximately 2-fold increase in width of lesions, which led to an increase in volume of up to 10 times. The large expansion of the lesion in a distal direction can be explained by progressive tearing along the lines of longitudinal arrangement of the fiber bundles under the influence of bearing weight. Enlargement of the lesion continued until 4 weeks after surgery [12].

Proliferation starts within the first week and can continue for 3 months. The inflammatory cells begin to be replaced by new small blood vessels and migrating fibroblasts which produce collagen type III to fill the defect.

The remodeling and 'maturation' phase takes up to 12 months with fibrous tissue becoming more prominent within the defect, with increasing synthesis of type I collagen and re-establishment of the crimp. Formation of scar tissue between tendon ends provides basic functional continuity at the site of disruption. Subsequent remodeling causes conversion of the provisional matrix to type I collagen (scar tissue) so that it is more capable of contracting as it matures, providing increased mechanical strength.

Collagenous tissue is randomly oriented and cellularity is increased. The original large-diameter collagen fibrils are not replaced following injury. The cross-sectional area of the tendon may remain increased, which may be a compensatory change due to lower elasticity of the scar tissue [11]. Repaired tendon tissue does not regain its original biochemical or mechanical properties. Persistent scar tissue significantly interferes with normal function. This failure to reconstitute normal structure explains the high recurrence rates, with re-injury most likely to occur at transitional areas between healthy and pathological tissue [11]. Proliferation of the scar between the tendon and adjacent tissues is also undesirable because these attachments later impede normal tendon gliding and function. Adhesion formation results in loss of movement, and consequently higher risk of functional disability [10].

Unlike in other tissues, such as bone or muscle, mature adult tendons appear to lack the inherent ability to remodel the structure of tendons to their pre-injury state. Instead, tendon healing involves the formation of fibrotic scars, which impairs mechanical strength of tissues, and subsequently, tendon injuries can result in significant dysfunction and disability. This may explain (in part) why 23%-67% of horses treated using conservative methods re-injure tendons within 2 years of the initial injury 10].

There is considerable interest in measures to detect and prevent SDFT injury.

Prompt detection of early stages of pathology is likely to have a greater outcome than advances in treatment, given that a tendon is a slowly healing and poorly regenerative tissue.

Conventional techniques for monitoring degenerative changes and screening tools in vivo are under constant development.

3 Diagnostic Approach to Tendon Injury

Ultrasonography since 1980's has been used to evaluate and monitor tendon and ligament injuries in horses. Despite of lot of advantages including affordability and portability of equipment, training, evidence based resources, ultrasonography diagnosis has some major limitations. This includes a restricted field of view; image acquisition, assessments that are subjectively dependent on the operator as well as different scoring systems that are focused on different parameters.

Objective measurements, including cross sectional areas, are repeatable, where echogenicity assessment which reflect to tendon structure has weak intra/inter rate agreement [15]. Diagnosis of tendons is made by evaluating their echogenicity which is reflected in the intensity (or brightness) and homogeneity. Machine settings are operator-dependent, where even minor changes in transducer angle, amplifier gain and displacement can affect the final assessment and diagnosis.

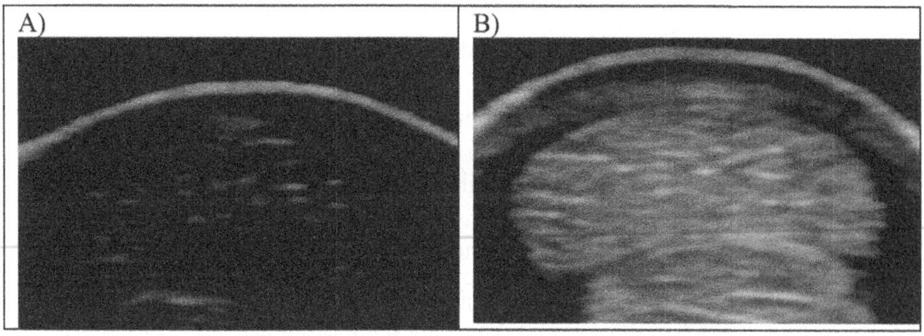

Fig. 1. Acute tendon lesion (A) and fibrotic tissue (B) 5 months after injury

Acute, early stages of tendon damage are easy to recognize due to obvious loss of echogenicity. Conversely, late granulation tissue and early fibrotic tissue cannot be distinguished from normal tendon tissue by use of the gray level (Fig. 1).

Low and high field MRI provide excellent diagnostic visualization of equine distal limb including objective detection of chronic scar tissue in tendons where were difficult to discern on ultrasound [16]. A major limitation of MRI as a screening tool in tendon pathologies is its cost and availability in a field practice.

Ultrasound Tissue Characterization (UTC) is computed processing and analysis of a series of ultrasound images at entire length of the tendon. UTC was researched by prof. Hans T.M. van Schie, and since 2007 is commercially available in equine practice.

Ultrasound tissue characterization captures contiguous transverse ultrasound images over the length of the tendon and semi-quantifies the stability of the echo texture over the length of the tendon into 4 echo types. Experimental research confirmed strong correlation between echo type and tissue composition, giving the opportunity to distinguish between normal, granulation, and fibrotic tissue within tendon, where basic grey level statistics could not. Computed processing series of scans provide standardized data-collection and analysis, a 3-dimensional ultrasound image of the tendon and high intra-

and inter-observer reliability. Even subtle changes over serial scans can be detected with high reproducibility [17]. Objective diagnostic modality provides excellent monitoring of the progression of tendon lesions and objective evaluation of repair processes in response to various treatments are one of the major advantages over conventional ultrasound. This new approach allows early detection of superficial digital flexor, deep digital flexor and musculus interosseous (suspensory ligament) tendinopathy, before clinical signs are present and can prevent serious tendon injury in sport horses [4].

The system consists of both hardware and software for quantitative evaluation of tendon structural integrity. Software quantifies the stability of the gray value of each corresponding pixel in contiguous transverse ultrasonographic images by means of mathematic routines. The correlation permits the exclusive discrimination between structure-related and non–structure-related echoes.

Standardized instrumental settings collect transverse ultrasound images at even intervals to create a 3-D ultrasound data-block. In this block, UTC-algorithms quantify the dynamics of echopatterns over contiguous images by intensity and distribution of relative grey levels of corresponding pixels and results are presented as 4 echo types:

- echo type I (green) represents intact and aligned fibres and fasciculi, with diameter over 0,35 mm
- echo type II (blue) represents discontinuous or wavy fibres and fasciculi, with diameter over 0,35 mm
- echo type III (red) represents smaller fibrils with diameter significantly lower than 0,35 mm
- echo type IV (black) represents amorphous tissue with cells and fluid

Based on experimental research comparing processed scans with histopathological findings it was established that normal flexor tendon is composed of (see Fig. 2):

- type I (green) intact bundles in 80–85%,
- type II (blue) discontinuous, wavy bundles in 10–15%,
- than 5% of the tendon

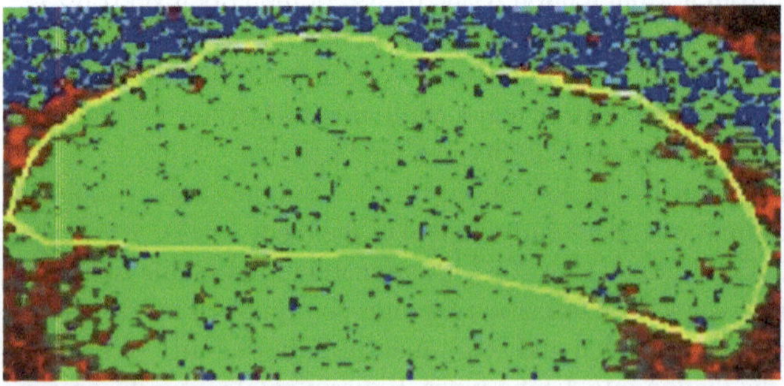

Fig. 2. UTC scan of normal tendon

UTC set consists of:
Linear ultrasound transducer 7–12 MHz.

- UTC tracker to drive ultrasound transducer in very accurate steps of 0.2 mm across the area of interest to collect successive 2-dimensional ultrasound slices for compounding into the 3-D data block.
- Stand off pad
- Computer with software to record and process acquired data

The exam is performed on a standing horse after hair is clipped with 0,1 mm blade and skin is cleaned with water and detergent. An ultrasonic gel is applied on both sides of the stand-off pad and the tracker device is placed against examined tendon. When proper placement is confirmed the automatic tracker is started. Scanning SDFT tendon takes around 45 s. It is very important that the horse bears weight on the examined leg and does not move when scanning is in progress. Even small movements produce artefacts regarding fiber pattern and can lead to misinterpretation. Nervous, fragile horses might require sedation to complete the exam.

Images are stored and processing provides tomographic visualization of the entire tendon, quantification of the architecture and integrity and accurate calculation of lesion size.

UTC imaging provides differentiation between short-duration exercise-effects from more serious changes like overstrain and degeneration, which are not possible by conventional ultrasound. Early recognition of overstrain injury and prevention of further damages is the major advantage of the system.

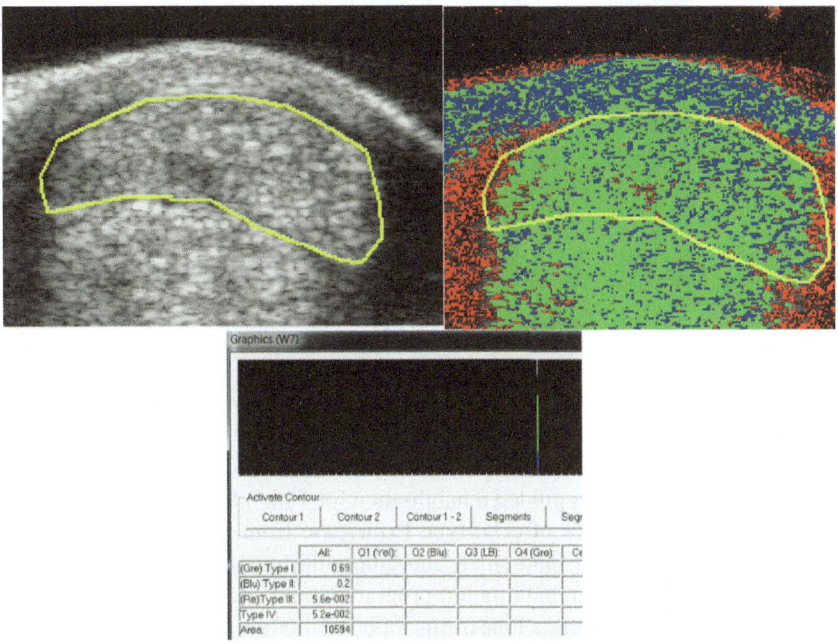

Fig. 3. Overstrain injury in a race horse

Ultrasound image presented in Fig. 3 suggest hypoechoic lesion which represent tendon damage.

Computed analyses revealed decrease of echo type I (from 85–90% in normal tendon to 69%) and an increase of echo type II (from 10–15% in normal tendon to 20%) with slight increase of echo type III and IV to 10% (in normal tendon below 5%). This change is characteristic for overstrains injury. Reduced speed exercise, physiotherapy and regular monitoring allowed preparation the horse for important race and prevent irreversible tendon injury.

A case presented on Fig. 4 and Fig. 5 is a patient without clinical signs of lameness and an unknown history of pervious injury and treatment. Tendon palpation results in mild pain reaction.

Ultrasound image would suggest diffuse tendonitis and hard to determine prognosis for further racing performance without follow up examinations.

UTC examination revealed lack of aligned intact bundles, poor organization of small bundles, a decrease of echo type I (from 85–90% in normal tendon to 6%), decrease of type II (from 10–15% to 2%). Type III (very small fibrils) and type IV (amorphic tissue, fluid) form more than 91% of cross-sectional area of the tendon. Diagnosis of severe tendonitis with poor prognosis for racing performance was made.

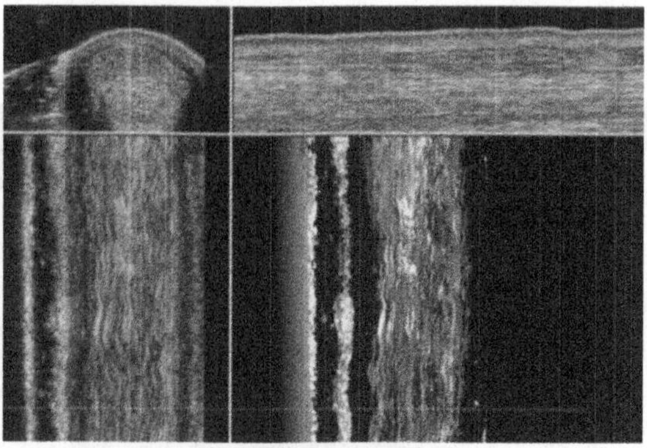

Fig. 4. Computed image of SDFT in greyscale

UTC imaging provides excellent monitoring of tendon healing for customized rehabilitation.

Figure 6 presents a patient 45 days after injury. Increase of echo type III indicates active fibrillogenesis. Further guided rehabilitation for 7 months results in improvement (Fig. 7) of organized bundles of echo type I (increase from 28% to 69%), echo type II (from 15% of the tendon) and small and amorphic fibrils decrease (from 51% to less than 15%). A standard rehabilitation program would suggest canter exercise, but based on injury healing patterns a longer time of rehabilitation was recommended. After 18 months from injury the horse returned to races.

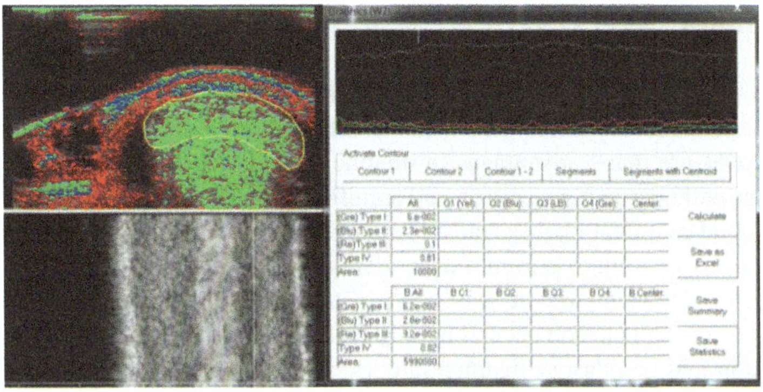

Fig. 5. Processed image of SDFT, echotypes are presented and calculated

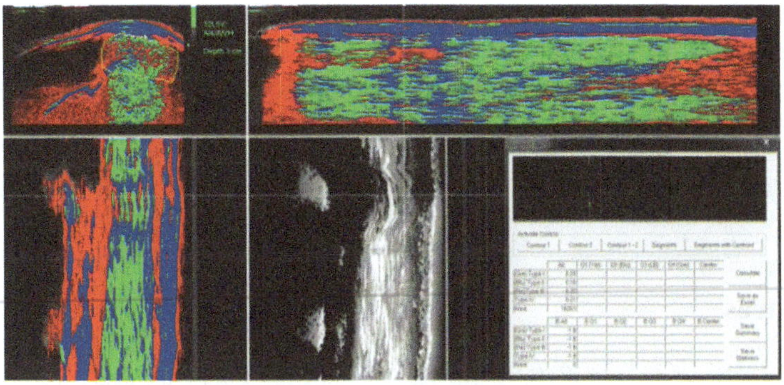

Fig. 6. Tendon injury 45 days after diagnosis

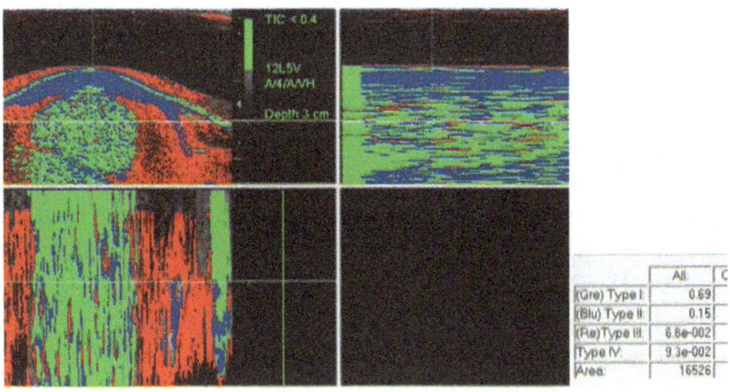

Fig. 7. Same tendon after 7 months of rehabilitation

4 Summary

Tendon injuries are the most prevalent in race horses and result from a cumulation of microdamage, where repeated stress without sufficient recovery time make it susceptible to failure. Despite many advantages in treatment, to date none of the evidence-based therapies provide tissue regeneration and significantly reduce re-injury risk.

Ultrasound examination is the most common diagnostic modality but it cannot detect early signs of overstrain to prevent further damage.

Computed ultrasonographic tissue characterization provides early tendinopathy detection before acute lesion occur, but only pathologies within superficial, deep digital flexor tendon and suspensory ligament are accessible for examination. UTC and MRI provide objective detection of chronic scar tissue in tendons where they were difficult to discern on only ultrasound examination.

The multimodal approach to prevent injuries in equine athletes needs continuous development. The horse racing industry is missing evidence-based training programs to reduce incidence of injury as well as reliable and easy to use aids for early detection pathologies. A shift from curative to preventive medicine will have a significant impact to improve sport horses' welfare.

Author Contributions. All authors have significantly contributed to the first draft of the manuscript, to its revisions, and they all approved its final form.

References

1. Patterson-Kane, J.C., Firth, E.C.: The pathobiology of exercise-induced superficial digital flexor tendon injury in thoroughbred racehorses. Vet. J. **181**, 79–89 (2009)
2. Ülke, Ç.G., Deniz, S.I., Nureddin, Ç.: Evaluation of return rates to races in racehorses after tendon injuries: lesion-related parameters. J Equine Vet Sci. **87**, 102931 (2020). https://doi.org/10.1016/j.jevs.2020.102931. Epub 2020 Jan 23 PMID: 32172919
3. Biewener, A.A.: Muscle-tendon stresses and elastic energy storage during locomotion in the horse. Comp. Biochem. Physiol. Part B Biochem. Mol. Biol. **120**, 73–87 (1998). http://www.sciencedirect.com/science/article/pii/S0305049198000248
4. Docking, S.I., Daffy, J., van Schie, H.T., Cook, J.L.: Tendon structure changes after maximal exercise in the thoroughbred horse: use of ultrasound tissue characterisation to detect in vivo tendon response. Vet. J. **194**(3), 338–342 (2012). https://doi.org/10.1016/j.tvjl.2012.04.024. Epub 2012 Jun 1 PMID: 22658820
5. Avella, C.S., Ely, E.R., Verheyen, K.L.P., Price, J.S., Wood, J.L.N., Smith, R.K.W.: Ultrasonographic assessment of the superficial digital flexor tendons of national hunt racehorses in training over two racing seasons. Equine Vet. J. **41**, 449–454 (2009)
6. Williams, I.F., Heaton, A., McCullagh, K.G.: Cell morphology and collagen types in equine tendon scar. Res Vet Sci. **28**, 302–310 (1980). https://www.sciencedirect.com/science/article/abs/pii/S0034528818327139
7. Silver, I.A., Rossdale, P.D.: A clinical and experimental study of tendon injury, healing and treatment in the horse. Equine Vet. J. **15**, 1–42 (1983)
8. Murphy, D.J., Kö-Peternelj, V., Aleri, J.W.: Intralesional bone marrow and superior check desmotomy is superior to conservative treatment of equine superficial digital flexor tendonitis. Equine Vet. J. **54**, 1047–1054 (2022). https://doi.org/10.1111/evj.13553

9. O'Brien, C., Marr, N., Thorpe, C.: Microdamage in the equine superficial digital flexor tendon. Equine Vet. J. **53**(3), 417–430 (2021). https://doi.org/10.1111/evj.13331. Epub 2020 Sep 2 PMID: 32772396
10. Patterson-Kane, J.C., Becker, D.L., Rich, T.: The pathogenesis of tendon microdamage in athletes: the horse as a natural model for basic cellular research. J. Comp. Pathol. **147**(2–3), 227–47 (2012). https://doi.org/10.1016/j.jcpa.2012.05.010. Epub 2012 Jul 11. PMID: 22789861
11. Dowling, B.A., Dart, A.J., Hodgson, D.R., Smith, R.K.: Superficial digital flexor tendonitis in the horse. Equine Vet. J. **32**(5), 369–378 (2000). https://doi.org/10.2746/042516400777591138. PMID: 11037257
12. Geburek, F., Gaus, M., van Schie, H.T., Rohn, K., Stadler, P.M.: Effect of intralesional platelet-rich plasma (PRP) treatment on clinical and ultrasonographic parameters in equine naturally occurring superficial digital flexor tendinopathies - a randomized prospective controlled clinical trial. BMC Vet Res. **12**(1), 191 (2016). https://doi.org/10.1186/s12917-016-0826-1. PMID: 27604193; PMCID: PMC5015224.
13. Alzola, R., Easter, C., Riggs, C.M., Gardner, D.S., Freeman, S.L.: Ultrasonographic-based predictive factors influencing successful return to racing after superficial digital flexor tendon injuries in flat racehorses: A retrospective cohort study in 469 Thoroughbred racehorses in Hong Kong. Equine Vet J. **50**(5), 602–608 (2018). https://doi.org/10.1111/evj.12810. Epub 2018 Feb 23. PMID: 29352495; PMCID: PMC6099230
14. Smith, R.K., Birch, H.L., Goodman, S., Heinegård, D., Goodship, A.E.: The influence of ageing and exercise on tendon growth and degeneration–hypotheses for the initiation and prevention of strain-induced tendinopathies. Comp. Biochem. Physiol. A. Mol. Integr. Physiol. **133**(4), 1039–1050 (2002). https://doi.org/10.1016/s1095-6433(02)00148-4. PMID: 12485691
15. Alzola Domingo, R., Riggs, C.M., Gardner, D.S., Freeman, S.L.: Ultrasonographic scoring system for superficial digital flexor tendon injuries in horses: intra- and inter-rater variability. Vet Rec. **181**(24), 655 (2017). https://doi.org/10.1136/vr.104233. Epub 2017 Dec 7. PMID: 29217766
16. Magnetic Resonance Imaging Application to Live Horse for Diagnosis of Tendinitis
17. van Schie, H.T., Bakker, E.M., Jonker, A.M., van Weeren, P.R.: Computerized ultrasonographic tissue characterization of equine superficial digital flexor tendons by means of stability quantification of echo patterns in contiguous transverse ultrasonographic images. Am. J. Vet. Res. **64**(3), 366–375 (2003). https://doi.org/10.2460/ajvr.2003.64.366. PMID: 12661879

Location of Obstacles on Transport Routes in a Centralized Multi-agent System

Grzegorz Iskierka(✉), Bartosz Poskart, and Kamil Krot

Wrocław University of Science and Technology, Wrocław, Poland
grzegorz.iskierka@pwr.edu.pl

Abstract. The commercially available autonomous mobile robots navigate the environment with the previously prepared map and the obstacles currently detected in the immediate vicinity of the robot. The purpose of this paper is to create a solution for early detection of potential obstacles to eliminate the already blocked paths in advance. The centralized multi-agent system could be implemented in a production enterprise with mixed transportation methods, i.e. mobile robots and personnel-operated carts and forklifts. The authors present a Node-RED based system to manage the robot fleet based on the input of other agents.

Keywords: AMR · RTLS · Multi-Agent System

1 Introduction

Until recently, in commercially available logistics solutions using AMR (Autonomous Mobile Robots), all units were treated as completely separate agents. Due to the autonomous navigation of mobile robots, with less complexity of the envisaged transport operations, there is no need to implement a Fleet Management System (FMS), however, when organizing internal transport for larger production, its implementation should be considered. Commercially available solutions are characterized by an approach dedicated to robots of a specific manufacturer, where the most commonly available solutions are characterized by a centralized architecture. Examples include manufacturers such as Mobile Industrial Robots, KUKA, Agilox. In addition to the solutions offered directly by the manufacturers of autonomous mobile robots, there are also fleet management systems that enable the management of a heterogeneous fleet of mobile robots. They are offered by companies such as Freedom Robotics, Formant, InOrbit or Meili Robots. Due to the access to an increasing number of mobile robots from various manufacturers and in order to facilitate their integration in multi-agent systems, in 2021 the first interoperability standard for autonomous mobile robots was published [1], which is compatible with the communication of many commercially available autonomous mobile robots.

In the literature, there are more and more solutions based on decentralized architecture, where system agents exchange information with each other in the immediate vicinity, instead of with the central supervisory unit. When the environment is not well researched, has an unknown topology from the point of view of the system, or does not

A. Burduk et al. (Eds.): ISPEM 2023, LNNS 790, pp. 292–299, 2023.
https://doi.org/10.1007/978-3-031-45021-1_22

have a properly prepared infrastructure, as in the case of search missions [2], most often you can find systems with decentralized architecture, where communication takes place between neighboring agents, thus creating a network. The problem with this approach is the stability of such a network, which can be determined by examining the vertex stability of the graph of a given multi-agent network [3]. However, a system based on centralized architecture provides greater stability and often greater communication possibilities between agents [4]. In some cases a mixed architecture may be observed, where some planning is done in a centralized and others in decentralized architecture, like in the multi-agent load towing application presented by Chenyu Yang et al. in [5], where the authors struggled with detecting obstacles on the fly, which could be solved by applying an external obstacle detection system as presented in this paper. The typical approach offered commercially and being focused on in the literature is one where obstacle avoidance is done by individual robots on the fly and not in advance, like in the papers presented by Sukkpranhachai Gatesichapakorn et al. in [6] and Anh-Tu Nguyen and Cong-Thanh Vu in [7].

Many small and medium-sized enterprises are willing to automate transport operations, but due to high implementation costs, they are forced to gradually expand the infrastructure, as shown by the analysis by B. Pinto et al. in the publication [8]. A gradual approach to the implementation of autonomous transport solutions may cause in-house transport to be carried out in a mixed way, i.e. with the use of autonomous mobile robots and personnel. Most of the presented systems, however, tend towards systems using only autonomous mobile robots. It should be borne in mind that each implementation of autonomous mobile robots in an enterprise, whether total or partial, may have negative effects, which in the context of safety was presented by A. Markis et al. [9].

Considering the above observations, unforeseen obstacles may appear on the transport roads of autonomous mobile robots, hindering and slowing down or, in some cases, preventing autonomous logistic operations completely. For this reason, the natural direction of development of fleet management systems for autonomous mobile robots should be their extension to monitor the location of potential obstacles on transport roads or in storage areas. Such action will allow the mobile robot to change its route earlier, before it encounters an obstacle on the transport road, thus reducing the risk of its complete blocking.

There are many solutions on the market that allow the location of objects in the production space. Many of them are solutions based on Ultra-Wide Band (UWB) radio communication, such as those offered by Eliko or Kathrein Solutions, or using the Bluetooth standard, such as those offered by Blueiot. Due to the lack of solutions dedicated to fleet management systems for mobile robots offered directly by AMR manufacturers, it is necessary to develop a superior system whose task will be to manage and optimize the assigned transport tasks and their routes depending on the position of both mobile robots and personnel along with the manual transport equipment participating in internal logistics operations. The system proposed by the authors can become a universal basis for further development of the route planning and task scheduling algorithms for the scientific community to test and implement the working solutions in the industrial applications. The following sections present the architecture and algorithm of the proposed operation.

2 System Architecture

The architecture of multi-agent systems can be centralized or decentralized. The developed solution uses a centralized architecture due to the reliability of the connection between the agents and the central unit using the existing network infrastructure at the target implementation site of the solution. The developed centralized system for monitoring and managing agent tasks comes down to software monitoring transport equipment (e.g. transport carts, pallet trolleys, forklifts) and taking it into account when planning tasks and routes of mobile robots included in the system. The system architecture is shown in Fig. 1.

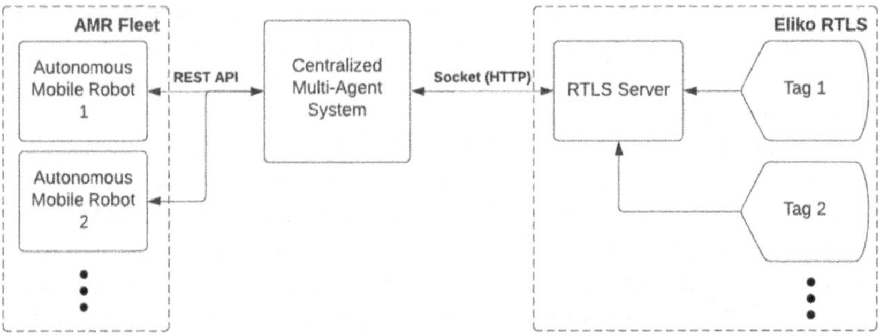

Fig. 1. Architecture of the centralized multi-agent system.

The presented structure of the system consists of two subsystems - a fleet of autonomous mobile robots and a tag location system in the transport space in the production hall.

It should be noted that the fleet of robots could be managed by a fleet management system offered by the manufacturer of the robots, or the individual agents of the system could be directly connected to the developed system, as was implemented in the presented solution. In the case of the MiR100 mobile robots used, communication via the REST API interface in the local wireless network in version 2.7.1 of the Mobile Industrial Robots [10] software was used. Using the API programming interface provided by the manufacturer, it is possible to download the current position of the mobile robot in the workspace and other parameters of the current status of the robot.

The real-time locating system Eliko RTLS, consisting of a server, a network of antennas and tags, was used to locate the transport equipment. In this system, the tags generate an ultra-wide band radio signal, which, after being intercepted by the antennas, allows the location of the tag to be calculated in three-dimensional space. The technology of ultra-wide band radio communication used in the solution is characterized by low energy consumption, which allows the tags to be loaded on average once every three months with a one-second refresh rate of their position. The architecture of the Eliko RTLS system is shown in Fig. 2. More details about the location system can be found in the documentation [11].

The positioning of tags takes thanks to the previous calibration of the antennas positioned in the production hall. In order to facilitate the configuration and subsequent

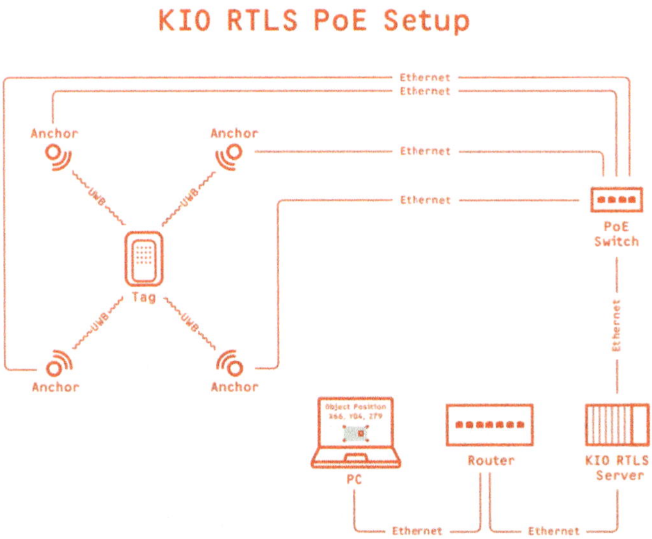

Fig. 2. Architecture of the Eliko RTLS [11].

transfer of data from both systems to a single coordinate system, a map of the hall downloaded from the MiR100 mobile robot was introduced to the system and the origin

Fig. 3. Eliko RTLS configuration panel.

of the Eliko RTLS coordinate system was placed in the south-western corner of the hall (see Fig. 3).

The proposed centralized multi-agent system was developed based on the Node-RED platform, the advantage of which is the support for many industrial communication protocols, including the HTTP protocol used for communication with both systems. It should be borne in mind that the position data of mobile robots and tags from both systems are expressed in relation to other coordinate systems, so their positions had to be brought to a common, superior coordinate system in order to present the agents of both subsystems on a common map and to detect potential collisions along the route.

3 Algorithm for the Early Detection of an Obstacle

The developed system allows to take action to prevent delays in logistics activities inside the production plant and to notify the maintenance department of the threat of disruption to internal transport. Figure 4 shows the algorithm of the system's operation, which is divided into two decision paths - monitoring of designated sensitive transport zones and automatic interference in the trajectory of the mobile robot in order to prevent detected potential difficulties in internal transport. The first of the decision paths offered by the system allows for continuous monitoring of high-risk zones and notification of the maintenance department in the event of a blockage of a transport road caused by plant personnel, which allows them to remove the obstacle, while registering the occurrence in order to prevent it from happening in the future. The second path of the algorithm allows for premature detection of potential obstacles near the determined trajectory of the autonomous mobile robot and its correction so as to prevent blocking the transport

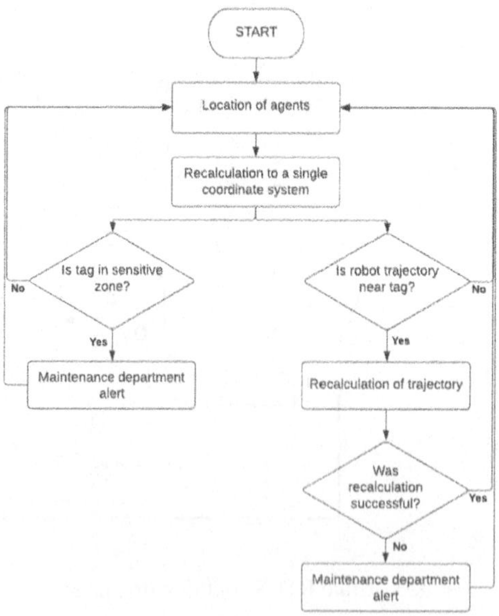

Fig. 4. System operation algorithm.

road. If the modification of the trajectory by the multi-agent system is not possible, the maintenance department is notified.

Figure 5 shows a map made available in the system for the maintenance department, which includes designated sensitive zones, Eliko RTLS system tags and autonomous mobile robots.

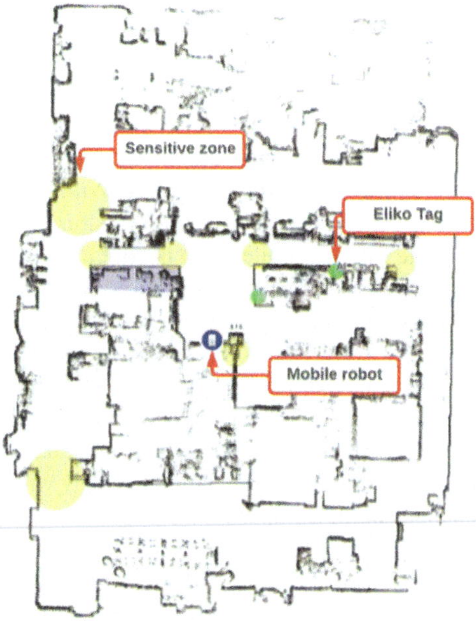

Fig. 5. A map of the production hall displayed in the developed system.

Sensitive zones (see Fig. 5) can be freely defined by the maintenance department depending on the needs. In addition, the system allows for continuous monitoring of the location of mobile robots and tags, and thus transport equipment, as well as the recording of events which occurred in the past, in order to improve internal transport.

Fig. 6. Sensitive zone violation message.

Figure 6 shows an exemplary message for the maintenance department when a sensitive zone is violated by one of the tracked pallet trolleys, which allows taking action to prevent further difficulties in internal transport.

4 Conclusion

The issue taken up in this publication was to solve the problem of the appearance of obstacles on the transport routes of autonomous mobile robots in small and medium-sized enterprises, where, in particular, there may be mixed in-house transport and the movements of mobile robots are in no way coordinated with personnel-based transport. The developed multi-agent system allows for the monitoring of the location of agents of both subsystems, i.e. mobile robots and personnel, taking into account potential obstacles on transport roads when planning the trajectory of mobile robots.

The developed system, due to its architecture, allows the use of sub-systems from various manufacturers and is not closed to solutions dedicated to specific autonomous mobile robots. Bearing this in mind, however, one should remember about the limitations resulting from the solutions used in the demonstrator. With the current version of the REST API software of the MiR100 mobile robot (version 2.7.1), the impact on the determination of a new route for the mobile robot is difficult due to the inability to enter a "forbidden zone" (as referred to in the documentation) from the API level. This issue has been solved with the use of "direction zones", although the manufacturer ensures that the possibility of editing "forbidden zones" will be implemented in version 3.0 of the software. Another limitation resulting from the Eliko RTLS tag localization system is the need to load tags on average once every three months. Nevertheless, the developed system only illustrates the concept of a working environment, where manual transport and the use of autonomous robots can operate simultaneously within a common working space without major changes in the company's infrastructure.

As directions for further development, the system can be integrated with the company's ERP system, extended with additional functions related to storage or implementation not only in the indoor environment, but also in the outdoor environment. It should be emphasized that the prepared solution is fully scalable and allows for extension with further subsystems.

References

1. MassRobotics AMR Interoperability Working Group. "Mass Robotics Interoperability Standard." (2021)
2. FDrew, Daniel S.: Multi-agent systems for search and rescue applications. Current Robotics Reports **2**(2), 189–200 (2021). https://doi.org/10.1007/s43154-021-00048-3
3. Nikodem, M.: "Wierzchołkowa Stabilność Grafów." Akademia Górniczo-Hutnicza im. Stanisława Staszica w Krakowie (2014)
4. Iskierka, G., Kamil, K., Bartosz Oskar P.: Grafowa Interpretacja Systemów Wielorobotowych I Ich Implementacja W Stosie Automatyzacji Przedsiębiorstwa. In: Inżynieria Zarządzania: Cyfryzacja Produkcji. Aktualności Badawcze 4, edited by Ryszard Knosala, pp. 81–91. Warszawa: Polskie Wydawnictwo Ekonomiczne (2022)
5. Yang, C., et al.: Collaborative navigation and manipulation of a cable-towed load by multiple quadrupedal robots. IEEE Robotics and Automation Letters **7**(4), 1–8 (2022). https://doi.org/10.1109/LRA.2022.3191170
6. Gatesichapakorn, S., Takamatsu, J., Ruchanurucks, M.: ROS based autonomous mobile robot navigation using 2D LiDAR and RGB-D Camera. First International Symposium on Instrumentation, Control, Artificial Intelligence, and Robotics (ICA-SYMP), pp. 151154. IEEE (2019). https://doi.org/10.1109/ICA-SYMP.2019.8645984

7. Nguyen, A.-T., Vu, C.-T.: Obstacle avoidance for autonomous mobile robots based on mapping method. In: Proceedings of the International Conference on Advanced Mechanical Engineering, Automation, and Sustainable Development 2021 (AMAS2021), pp. 810816 (2022). https://doi.org/10.1007/978-3-030-99666-6_118
8. Pinto, B., Silva, F.J.G., Costa, T., Campilho, R.D.S., Pereira, M.T.: A strategic model to take the first step towards industry 4.0 in SMEs. Procedia Manufacturing **38**, 637645 (2019). https://doi.org/10.1016/j.promfg.2020.01.082
9. Markis, A., Papa, M., Kaselautzke, D., Rathmair, M., Sattinger, V., Brandstötter, M.: Safety of mobile robot systems in industrial applications. In: Proceedings of the ARW & OAGM Workshop, pp. 26–31 (2019)
10. Mobile Industrial Robots. 2.7.1 MIR100 REST API. Accessed 6 Jan 2023 (n.d.). https://www.jugard-kuenstner.de/fileadmin/daten/Downloads/Intralogistik/MiR_Transportsystem/MiR100_MiR200/MiR_Rest-API.pdf
11. Eliko. "Eliko RTLS Installation Guide." Accessed 7 Jan 2023. https://eliko.tech/wp-content/uploads/2021/10/eliko_rtlss_installation_guide.pdf

Converging Data Mesh and Microservice Principles into a Unified Logical Architecture

Francisco Morais[1], Nuno Soares[1,2], João Bessa[1], José Vicente[3], Pedro Ribeiro[2], José Machado[4](✉), and Ricardo J. Machado[2]

[1] CCG/ZGDV Institute, Campus de Azurém, Edifício 14, Guimarães, Portugal
{francisco.morais,nuno.soares,joao.bessa}@ccg.pt
[2] ALGORITMI Research Centre, University of Minho, Guimarães, Portugal
{pmgar,rmac}@dsi.uminho.pt
[3] InsideLimits, Vila Nova de Gaia, Portugal
[4] MEtRICs Research Centre, University of Minho, Guimarães, Portugal
jmachado@dem.uminho.pt

Abstract. Companies have invested in Domain-Driven Design to support data-driven business, but this strategy has led to bottlenecks, demanding the decentralisation of analytical data by domain. This paper proposes a converged strategy for a domain-specific decentralisation of analytical data. Sequentially applied, the elicitation of cloud computing microservices, their identification uses using functionally decomposed UML use cases as a result of the domain-driven design, the application of a logical architecture derivation method followed by the application of microservices patterns, and, finally, the application of data mesh principles to organize packets or reorganize objects in new packet boundaries, result in the derivation of a converged architecture containing both microservices quantum and data mesh quantum as a distributed architecture mesh.

Keywords: Data Mesh · Microservices · Cloud Computing · Domain-Driven Design · Logical Architecture

1 Introduction

With the advent of the convergence of operating technologies (OT) and information technologies (IT), working in a real time integration, led to the use of the term Smart Manufacturing [1]. While smart factories have always relied on skilled operators with years of experience with specific equipment's or on sensors monitoring equipment's, the unprecedented computational capabilities and ubiquity have provided access to data that can be used to enhance efficiency, sustainability, quality, and time-to-market. The fact is that for large or small manufacturers, its suppliers to stay valued part of the supply chain need to face

A. Burduk et al. (Eds.): ISPEM 2023, LNNS 790, pp. 300–314, 2023.
https://doi.org/10.1007/978-3-031-45021-1_23

this integration of OT and IT. And this integration is leading to a more self-managed manufactured systems, tying a more limited human contact but more sophisticated to program and operate.

While NIST provides a more elaborated definition for smart manufacturing as a vision of a next-generation manufacturing with enhanced capabilities [2], also advices for a review of the body of pertinent standards to help technology adoption and development, introducing new paradigms based on distributed manufacturing services. At the same time, it promotes test beds with key components such as Computer-Aided technologies (CAx), the use of fabrication machine tools (CNC), and inspection equipment (CMM) to extend existing production-focused concepts that enable smart manufacturing research and development across the product lifecycle[3]. The challenges and requirements for introducing new digital technologies are brought on by the introduction of smart devices accessible as services on a network, embedded intelligence at different levels, predictive analytics, and cloud computing technology [2]. Cloud-hosted computing resources boost the high demands of the digital technologies in use and are a key technology to achieve smart manufacturing.

The electronics production market has manufacturing process support areas, such as ICT testing systems, that are characterised by enormous technological challenges due to the increase in variety and decrease in size, requiring a greater capacity to produce highly efficient and highly accurate test systems. This leads to a large number of test points and the need to decrease oscillations, errors, and tolerances arising from the manual operations of the fixture production process. Automated conversion of data received from customers into test systems is therefore one of the core steps to overcome those technological challenges.

As referred to in [4], while many organisations have invested in central data housing with the expectation to drive business based on data and at the same time invested in Domain-Driven Design (DDD) to own their domain, including the information needs of the business, such an approach has led to bottlenecks. On the one hand, data teams cannot handle all the analytical questions of management and product owners quickly enough because they spend too much time fixing broken data and too little time discovering and understanding the necessary domain data. On the other hand, the domain teams have to get the necessary data-driven insights from the overloaded data team. So, how to handle such a bottleneck? A Domain-oriented decentralisation for analytical data is the core idea behind the data mesh concept. Enabling domain teams to perform cross-domain data analysis on their own domain and interconnect data as APIs in a microservice architecture [4]. We explore its concepts to build an analytical microservice, but using microservices principles and architecture. Further, we intend to use this experiment towards a microservice-based application implementing a distributed data mesh, applied in a case study of the electronics production market in its manufacturing process support areas, namely the one responsible for the ICT testing system.

InsideLimits is a Portuguese company whose core business is the development and provision of PCB (Printed Circuit Board) testing solutions for the electronics production market. These test systems include the fixture bed, the respective test programs, and the test report. Under the iFixturing project, one research goal is to develop an ICT Test Support Software to reduce production time and cost. This software tool will allow the conversion of the data received from the customer to test systems as well as generate a report of the components tested by a given test table. We framed this objective within the larger research goal of conceiving a domain-specific cloud platform with decentralization of analytical data capable of handling all the management analytical questions and producing the necessary data-driven insights. A platform that explores the converged concept of microservices with analytical roles, along with other roles serving other business capabilities.

This paper describes these efforts and is structured as follows: Sect. 2 starts by establishing some theoretical background and summarising the state of the art concerning Data Mesh Principles and Microservices Principles. Section 3 presents the work performed, depicting and discussing the strategies undertaken for the componentization of the data product as a convergence concept, and the definition of the microservice structure. Section 4 describes the experiment and the various activities taken up to the implementation and testing of the microservice. Finally, Sect. 5 concludes with remarks about the undertaken project and some considerations for future work.

2 Background

2.1 Data Mesh Principles

The conversion of data has to do with transforming and manipulating the data so it can be converted into a format that is compatible and functional in its new location, usually one that a new system will recognize. The process of conversion consists of multiple steps, with its core being the need to extract and reformat data from a source system for use in a target system. While the conversion is simple, we may hand-develop our own data converter, but soon we will look for ETL tools that improve the coding of the data transformation logic. Thinking in a wider application of data storing and conversion as a data platform, we recently assisted in a paradigm shift, moving from dominant architectural patterns in data and analytics from data lakes and their predecessors, data warehouses, to a concept of domain-focused data products, enabling us to move beyond a monolithic data lake to a distributed data mesh.

As stated in [5], the growing complexity of the labyrinth of data pipelines connecting operational data and analytical data is a familiar sight to anyone who tries to connect these two planes. And so, data mesh attempts to connect these two planes under a different structure, as an inverted model and topology based on domains and not technology stacks, with a focus on the analytical data plane. Moreover, [5] refers to the fact that operational and transactional data is driven largely by microservices architecture, with data hidden inside of each microservice and accessed through the microservice API. However, analytical data remains a point of friction at scale, with constant change of the data landscape, proliferations of both sources and consumers of data, diversity of transformation and processing that use cases require, and speed of response to change. So, it suggests keeping those planes concerns separate while enumerating the core principles guiding the development of the analytical plane as a data mesh.

The first principle "Domain ownership" tells that data mesh should make the business domain's bounded context a good candidate for distribution of data ownership because that is where the impact of continuous change and evolution lies. Ownership and serving of the analytical data should respect the business domain. The domain's interface to the rest of the organisation not only includes the operational capabilities but also access to the analytical data that the domain serves. And so, the architecture should arrange the analytical data by domains.

The second principle "Data as a product" should overcome the consequences of the increasing number of domains from the first principle of decentralization, which are exacerbated by the cost of discovering, understanding, trusting, and using quality analytical data. Thus, analytical data provided by domains must be treated as a product, the consumers as customers, and be treated as the smallest unit of architecture that can be independently deployed with high functional cohesion including all the structural elements required for its function: code, data, and infrastructure, at the granularity of a domain's bounded context. Such a data contract needs to have six basic qualities: discoverability (have a registry or data catalogue), addressability (have a unique address), trustworthiness (have data lineage and provenance and provide acceptable service level objectives), self-describing semantics and syntax (data schemas), interoperability (polyglot domain data sets), security (access control policies, SSO, and RBAC policies). Also, this principle's implementation promotes organisation by introducing roles such as domain data product owner and data product developer. The former having objective measures that ensure data is delivered as a product (data quality, decreased lead time of data consumption, and data user satisfaction through a net promoter score). And the latter is responsible for building, maintaining, and serving the domain's data products. This way, accountability for data quality shifts upstream, as close to the source of the data as possible.

The third principle, "Self-serve data platform", promotes convergence of operational data and data infrastructure. Usually, big data technology stacks diverge from operational platforms. But nothing prevents running Spark on the same orchestration system docker or kubernetes. So, such a platform must include a long list of capabilities, such as scalable polyglot data storage, data product schema, data pipeline declaration and orchestration, data product lineage, etc. This leads to automation, where using cloud infrastructure as a substrate reduces cost and effort and provides on-demand access to the data infrastructure. In a logical view, those capabilities form a "triple-plane" data platform, where a plane is a level of existence and serves different profiles of users: a data infrastructure provisioning plane (advanced data product developers use this interface directly to low-level data infrastructure), a data product developer experience plane (the main interface a data product developer uses), and a data mesh supervision plane (searching or browsing the mesh of data products as a graph of connected data products through execution of a data semantic query).

The fourth principle "Federated computational governance" is needed to enable data products to interoperate in order to set operations on them at scale. A set of global rules, applied to all data products and their interfaces, to ensure a healthy and interoperable ecosystem and equilibrium between centralised domain decisions and decentralised decisions in each domain, in a way that embraces change and multiple interpretative contexts. For example, the semantics and syntax of a domain data model definition must be left to the local domain, while the identification of the domain or its use is a global concern. The unified identification allows correlation information about users or service orchestration.

With these principles, a logical architectural model emerges, joining analytical and operational data together under the same domain while maintaining technical differences, such as hosting, compute technologies, or ways of querying and accessing data. It establishes a common language and a blueprint of the components of the mesh: data product, platform, and the required standardizations. Data mesh introduces data product as its architectural quantum [6], the smallest unit of architecture that can be independently deployed with high cohesion, including all structural elements required for its function [5] (Fig. 1).

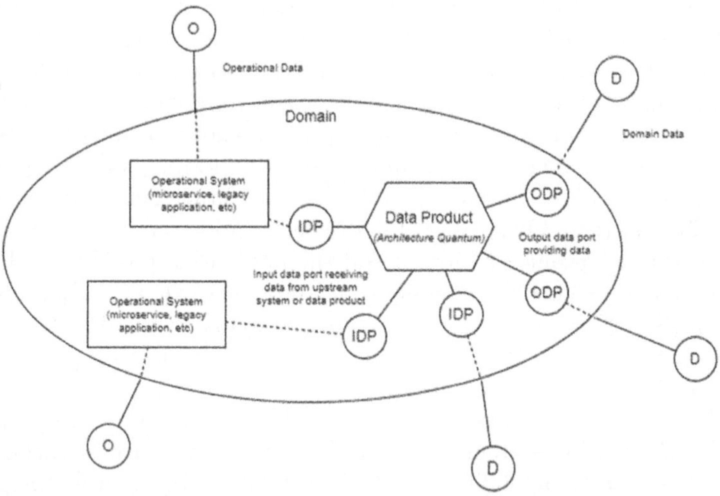

Fig. 1. Data mesh notation: domain, its (analytical) data product and operational system [5].

2.2 Microservice Principles

In a microservices architecture, the bounded context serves as the quantum boundary and may include dependent components such as database servers, search engines, and reporting tools that contribute to delivered functionality [6]. Microservices derive from SoA, as when one thinks about structuring an application by decomposing it into multiple services that can be classified as different types, such as subsystems or tiers. But while the SoA concerns large central brokers, central orchestrators, and the enterprise service bus (ESB), where the preferred share of information between bounded contexts is to unify around a single representation, in microservices they do it via messaging, requiring less coordination for each change in architecture [6].

Microservices are an approach to building server applications as a set of small services, where each service runs in its own process and, if it needs to share information, communicates with other services using distributed protocols, most commonly HTTP request-response or lightweight messaging. So, microservices implement a domain, business capability, or workflow within a certain context boundary. Should own its related domain data model and domain logic, having at the same time sovereignty over its data but sharing its management. While there isn't a formal definition of microservices architectural style, common characteristics contribute to describing architectures that fit the microservices designation.

Although there is no globally accepted definition, nine qualities that most microservices design should display are agreed upon and established. [7,8].

The first one is the Componentization via services, turning them into independently deployable units and making them easier to evolve, as well as providing a more explicit component interface. The authors [7] define a component as a unit of software that is independently replaceable and upgradeable. And services as out-of-process components that communicate with a mechanism such as a web service request or a remote procedure call. The componentization contributes to a smooth evolving application as long as the decomposition into multiple services results in cohesive service boundaries and evolution mechanisms.

The second characteristic requires services to be organised around Business Capabilities. Diminishing the number of contexts and having more explicit separation required by service components make it easier to keep team boundaries clear. So, to avoid silos application architectures, use cross-functional teams to ensure that applications are organised based on capabilities. (because of Conway's Law [7]).

The third characteristic advocates Products over Projects, preferring the notion that a team owns a product over its full lifetime to the aim of delivering a piece of software, which is then considered to be completed and the project team that built it is disbanded. This way, an on-going relationship is built where the question is, "How can software assist its users to enhance business capability?" This relationship is promoted by the smaller granularity of services.

The fourth characteristic is to have smart endpoints and dumb pipes, favouring the use of simple choreographed protocols because, being decoupled and cohesive, they act more as filters, receiving a request, applying logic, and producing a response. So, using the principles and protocols that the world wide web is built on, it favours the use of HTTP request-response with resource APIs or messaging using a light message bus in a simple implementation (a message router only) just to provide reliable asynchronous fabric, leaving the smarts in the end points that are producing and consuming messages.

A fifth characteristic is having a Decentralized Governance, avoiding single technology platforms. Partitioning a system into services means having more options when building each of them. On the other hand, decentralising responsibility to the development teams for all aspects of the software they build (including operating the software 24/7) leads to a strong incentive to focus on quality when writing code. At the same time, open source practises allow for the sharing of useful tools that other developers can use to solve similar problems.

A sixth characteristic is a Decentralized Data Management, where responsibility for data across microservices has implications for managing updates. Distributed transactions are difficult to implement because consistency may only be eventual, and problems are handled with compensating operations. So, the business practise is to handle a degree of inconsistency to respond quickly to demand while having a process to deal with mistakes. This trade-off is pursued as long as fixing mistakes costs less than losing business under great consistency.

The seventh characteristic is to have an Automated Infrastructure, which has been favored recently by the reduced operational complexity in building, deploying, and operating services. The key features pursued are automated tests and automated deployment, in order to have confidence in working software, which also means working software "up" in the pipeline for each new environment.

The eighth characteristic is to Design for failure so that services as components can tolerate the failure of services, due to unavailability, for example, and respond gracefully, introducing a reflection on how service failures affect the user experience. Introducing patterns like circuit breaker, bulkhead, and timeout is important to achieve communicating applications, as is the introduction of automated testing and real-time monitoring of the application. Because microservices emphasize choreography and collaborative events, these can lead to emergent behavior, so monitoring architectural elements or business metrics can detect anomalous behavior and quickly resolve it.

The ninth characteristic is to apply an Evolutionary Design to favor frequent, fast, and controlled changes to software. The key property of a component is the notion of independent replacement and upgradeability [7]. Replaceability is based on modular design, which intends to keep things that change at the same time in the same module. Upgradeability is based on simplicity and speed of release because one only needs to redeploy the modified service. And the key tool to achieve those properties is the service decomposition, which allows you to control changes without slowing them down.

With these principles in mind, a logical architectural model emerges around business capabilities, producing strong module boundaries that are reinforced by a modular structure under the same domain, maintaining independent deployment and technology diversity. It establishes a common language and a blueprint of the components of a microservice: componentization via services, with smart endpoints and dumb pipes, in an infrastructure automation platform. Microservices introduce boundary context as their architectural quantum, the smallest unit of architecture that can be independently deployed with high functional cohesion, including all structural elements required for their function.

A very similar quantum in data mesh, where it considers the business domain's bound context a good candidate for distribution of data ownership. It considers data as a product with all the structural elements required for its function: code, data, and infrastructure, at the granularity of a domain's bounded context.

3 Development

3.1 Componentization of the Data Product as a Convergence Concept

Domain-Driven Design (DDD) is about designing software based on models of the underlying domain [9]. It divides a complex domain up into multiple bounded contexts and maps out the relationships between them [7]. DDD also acts as a conceptual foundation to decompose the design into objects or functions, resulting in a natural correlation between service and boundary contexts. As per Eric Evans's book, DDD defines the bounded context as the place where everything related to the domain is visible internally but opaque to other bounded contexts. It's a logical frame inside which a model evolves [10], having only one concern and operating with one concept. If a domain's bounded context is a common quantum for both data mesh and microservices principles, can we converge those principles into a unified logical architecture?

While data mesh considers bounded context as domain-oriented analytical data, imputing operational data and outputting analytical data, microservices don't distinguish between serving operational or analytical data and consider bounded context as a broad concept, as a domain-oriented capability containing behavior, information, and presentation [11]. So, the convergence should consider the componentization of the data product's quantum via service. Resulting in a data product that is independently deployable, is built as a separate process for each bound context, and uses distributed protocols to be a distributed service. Moreover, this componentization led to reducing computational governance efforts due to using microservices planes instead of data mesh planes, like the microservices monitoring plane instead of the mesh supervision plane or infrastructure automation instead of the data infrastructure plane.

By componentization, we consider a unit of software that is independently replaceable and upgradeable, as per [7] definition. So, reasoning a component via service, means a single service changes to only require that service to be redeployed. A service may consist of multiple processes that will be developed and deployed together, such as an application process and a database that's only used by that service. This componentization should encapsulate the three structural components of code, data, and metadata, as well as the infrastructure coming from the data product. The code component should include not only data pipelines for Data Flow when reasoning about data availability, transformation, latency, and analyzing data between processes, but also include Control Flow when reasoning about order of computation and how locus of control moves inside a process, considering Data and Control flow as referred to in [12,13].

Such a service could use data flow to ingest operational data and handle its persistence. And at the same time, use data flow to receive domain data, aggregate it with its operational data using control flow, do more transformation, and deliver analytical data.

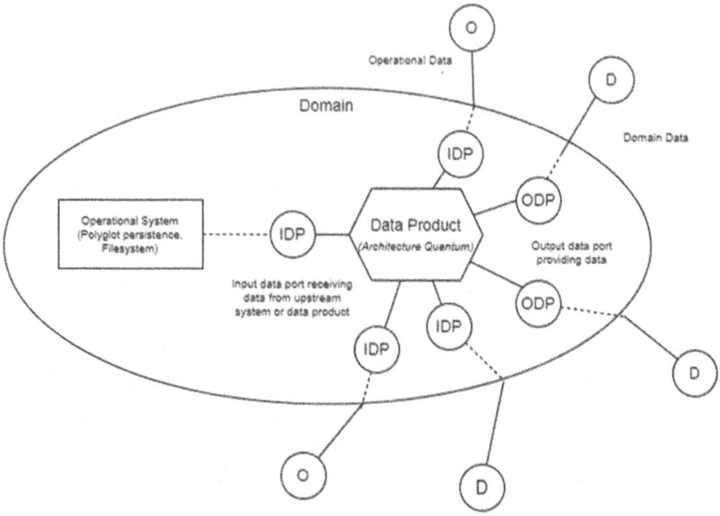

Fig. 2. Componentization of the data product quantum via service. Notation: domain bounded context, its data product and operational system.

This convergence aggregates in a microservice the transactional nature of the operational data, keeping state, and serving business capabilities to the application. And at the same time, aggregate views of the facts of the business over time as analytical data, modeled to provide retrospective or future-perspective insights in a single-bound context. It should aim to be decoupled and cohesive, receiving a request, applying its own domain logic, and producing a response, all choreographed using simple RESTish protocols. Moreover, it should manage its own operational data, either using different instances of the same database technology or entirely different database systems.

3.2 Reasoning About Structuring a Microservice

The Requirements Model aims to define the scope of the system and specify its behaviour, using use case diagrams. Once finished, one may start structuring the system independently of the actual implementation, thus focusing on the logical structure of the system. Since we look for a bound context, that is decoupled and cohesive, deployable and upgradeable, we follow an object-oriented analysis method that roots in [10], which uses three objects (entity objects, interface objects, and control objects) to model each of the analysis dimensions of Presentation, Behaviour and Information. One of the main advantages of this approach is that it is more adaptable to changes since one object may contain, if required, all dimensions, information, behaviour and presentation. Modifying information, for example, often requires modifying some behaviour. So, each of these object types captures at least two of the three dimensions, with a certain inclination towards one of the dimensions. So changeability is local since it affects preferably one object in the system. And this gives us some cohesion and loose coupling between objects. A method, like the 4SRS [14], based on the analysis model of [10], allows us to do the transition from the requirements model to the logic model, doing object packaging and aggregation around bounded contexts, and in this way derive our services. From this perspective, we get microservices as a package with a set of object types expressing information, behaviour and presentation dimensions. Applying a pattern language for microservices [15] as the infrastructure deployment pattern allows us to do the componentization of the package and make it independently replaceable and upgradeable.

4 Experiment

4.1 Requirements Model

The scope of the work determines the boundaries of the business area to be studied and outlines how it fits into its environment. We conducted an analysis of the existing business processes, including automated and manual processes that will be replaced or changed by the new software application. We used the Business Process Model Notation for building process models to study the current situation, providing the basis for understanding the effects of proposed changes.

The context of the work as a diagram identifies the boundaries of the work, as understood by [16]. It includes the work context and its adjacent systems (other domains to be understood). And the interfaces between them indicate why adjacent systems are important. This diagram describes the inputs and outputs that flow between the work and the adjacent systems. The names used in the inputs and outputs define the data dictionary. This is understood as a business event and represents a discrete piece of work that contributes to the overall functionality, helping to discover detailed requirements. They also form the first data model that, when crossed with events, rises and checks the existence of operations (e.g. CRUD).

In inspecting each business use case, it is determined with stakeholders the product boundary, which parts should be automated and which parts should be done by others. The use case diagram identifies the boundaries between actors and the product, which are the first boundary contexts of the system.

It is the main diagram used in the discussion with the stakeholders in order to validate and discover requirements. The requirements are organized in levels, and these levels are directly related to the refinement level and range from the most abstract to the most specific.

The Use-Case refinement is represented as a tree structure. In the root levels, it is represented the most abstract requirements that are related to a lower level of refinement. As leafs start to appear, so does the most refined requirements (leaf requirements). With a high level of refinement, these requirements are the ones that are translated into the system's functions and are the ones in which the operational flow is described.

In order to build our experiment, we refined the use-case diagram to two levels. And selected one analytical use case to be used in our experiment: U.C.11.1 pcf conversion management to nets, as described in the diagram in Fig. 3.

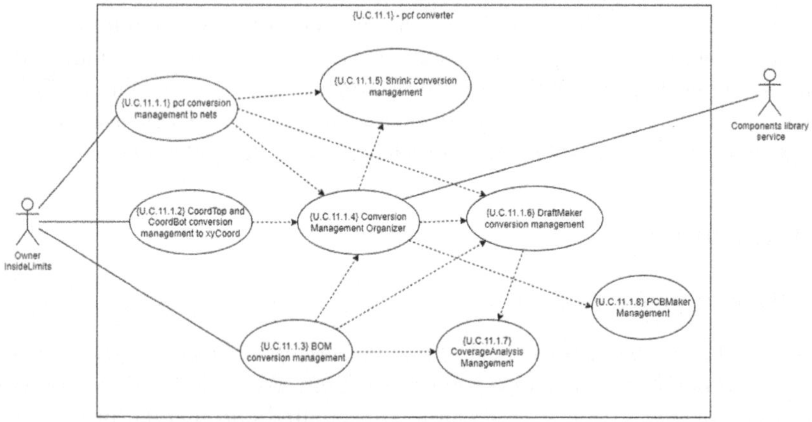

Fig. 3. UML Use Case Diagram level2.

4.2 Logic Model

To do the transition from the requirements model to the logic model, we applied the 4SRS method, supporting our object-oriented analysis[10], to model each of the analysis dimensions of the use case, resulting in a package (microservice) with two cohesive objects. One interface object to handle code for APIs that provide access to data, semantic and syntax schema, observability metrics, and other metadata. And one control object to handle code for data pipelines responsible for consuming, transforming, and serving data received in our case from other upstream microservices. Based on these object descriptions taken from 4SRS, the class diagrams were built for the data flow interface (i). Moreover, we applied design decisions to build the logic for the control object (c).

In Fig. 4, the logical model of our microservice is depicted. And Fig. 5, describes the code for data pipelines responsible for consuming, transforming, and serving data received from an upstream microservice. The circles in the figure represent the ETL operations: in blue, extraction; in yellow, transformation; and in green, load.

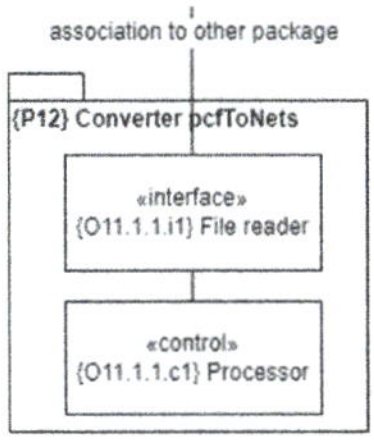

Fig. 4. UML Package Diagram.

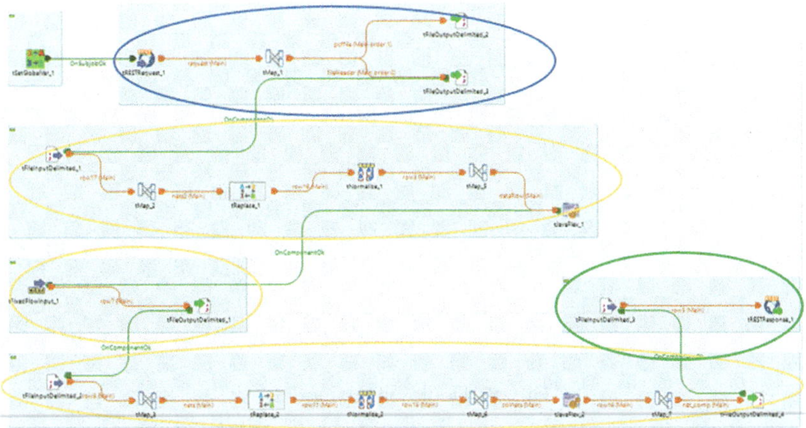

Fig. 5. Package converter with the API and Data pipeline logic on Talend Studio.

4.3 Deployment Model

By architecting the system as a set of services, each one is deployed as a set of service instances for throughput and availability. Design decisions follow Microservices Architecture (MSA) Patterns for infrastructure and applications. [15]. At the infrastructure level, it was applied the service per container to the package, to meet the non-functional requirements of being installable (deployable) and scaleable (upgradable) independently, allowing them to be programmed in several programming languages and different framework versions. To let a client or upstream service access our service, we have applied an API Gateway pattern to handle requests proxied/routed to the service. For MSA communications, a remote procedure invocation pattern was used Remote Procedure Invocation pattern, resulting in tight runtime coupling for the duration of the request and emphasizing the scalability of interaction between components and the independent deployment of components. We also need to handle cross-cutting concerns (observability, testing, and discovery) or non-functional requirements (NFR) security using microservice patterns, but this was not considered in our current experiment.

The functionality is provided by customized microservices implemented with open-source technology, which at least in the prototype phase proved to be a more cost-effective solution. The components were virtualized in docker images and installed in a virtual machine in a private cloud, but with public IP access, orchestrated by a docker server (for horizontal scalability, docker swarm can be used or kubernetes), along with the custom microservice "converter pcf to nets" (pcfToNets) and API gateway service (agw).

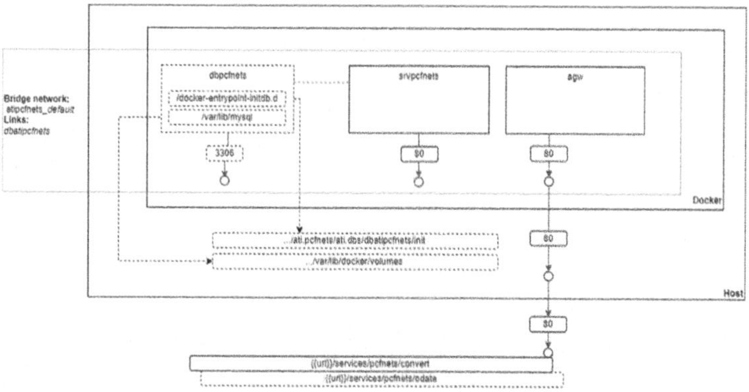

Fig. 6. Deployment model.

The "pcfnets" microservice was developed in Talend Open Studio, with synchronous REST-based inter-process communication. The service developed in Talend uses the Postman testing tool. The Application Gateway component was also implemented with open-source technology, as a containerized microservice using a NGINX API gateway/reverse proxy. Figure 6 illustrates the container model, corresponding to the componentization of the data product as a microservice, along with the Application Infrastructure Pattern API gateway. The package could benefit from the database per service pattern, having in this case a database schema that is private to the service, acting as an operational data retention database, that could be fed by an interface in the package. This design is represented in dashes in Fig. 6, just as an example to describe the conceptual model in Fig. 2, not making part of our experiment package. The result of the implementation of the microservice is shown in the following figures. Figure 7 describes the domain data is served as a batch file in an unstructured text format (figure on the left) and producing intermediate analytical data in an unstructured text format.

In Fig. 8, we show the result of testing the microservice using a scalable API testing tool (postman) to send an HTTP/s POST request with a form-data body and domain data served as a file in text format to the microservice. And getting its response, with the analytical data in text format, corresponding to the expected file depicted in Fig. 7, on the right.

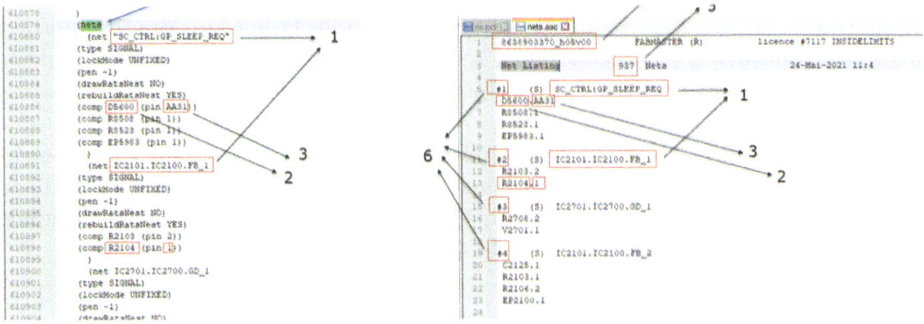

Fig. 7. Extract File on the left, and after Transformations the Load File on the right.

Fig. 8. Testing the microservice converter pcfnets with postman.

5 Conclusions

Organisations have invested in central data housing and Domain-Driven Design to drive business based on data, but this approach has led to bottlenecks. Data teams cannot handle all the analytical questions of management and product owners quickly enough, while domain teams have to get the necessary data-driven insights from the overloaded data team. The answer to the bottleneck could be a domain-specific decentralisation of analytical data, which is the basic principle of the data mesh concept.

Undergoing work explores the converged concept of a microservice with an analytical role, along with other microservice roles serving other business capabilities, to build a cloud platform. We use some methods to elicitate cloud requirements and derive a logic architecture, pursuing at the same time Data Mesh Principles, towards a microservice-based application implementing a cloud platform. Moreover, it is applied in a case study of the electronics production market in its manufacturing process support areas, namely the one responsible for the ICT testing system. The results will be published in further work.

Acknowledgments. This work is co-funded by the European Regional Development Fund (ERDF) through the North Regional Operational Program (NORTE 2020) of the Portugal 2020 Program [Project No. 43922, with acronym "iFixturing"; Funding Reference NORTE-01-0247-FEDER-043922]

References

1. National Institute of Standards and Technology. What is Smart Manufacturing?
2. NIST. Current standards landscape for smart manufacturing systems. Technical report, U.S. Department of Commerce, National Institute of Standards and Technology (2016)
3. National Institute of Standards and Technology. Smart Manufacturing Systems (SMS) Test Bed
4. Dehghani, Z.: Data Mesh Paradigm Shift in Data Platform Architecture (2020)
5. Dehghani, Z.: Data mesh principles and logical architecture (2020). https://martinfowler.com/articles/data-mesh-principles.html. Accessed 20 Feb 2023
6. Ford, N., Parsons, R., Kua, P., Sadalage, P.: Building Evolutionary Architectures. O'Reilly Media Inc., Sebastopol (2022)
7. Lewis, J., Fowler, M.: Microservices: a definition of this new architectural term (2014). https://www.martinfowler.com/articles/microservices.html. Accessed 20 Feb 2023
8. Newman, S.: Building Microservices: Designing Fine-Grained Systems. O'Reilly Media, Inc., Sebastopol (2015)
9. Fowler, M.: Boundedcontext (2014). https://www.martinfowler.com/bliki/BoundedContext.html. Accessed 20 Feb 2023
10. Evans, E., Evans, E.J.: Domain-Driven Design: Tackling Complexity in the Heart of Software. Addison-Wesley Professional, Boston (2004)
11. Jacobson, I.: Object-Oriented Software Engineering: A Use Case Driven Approach. Pearson Education India, Delhi (1993)
12. Shaw, M., Garlan, D.: Software Architecture: Perspectives on an Emerging Discipline. Prentice-Hall Inc., Upper Saddle River (1996)
13. Kazman, R., Bass, L., Clements, P.: Software Architecture in Practice. Addison-Wesley Professional, Boston (2012)
14. Machado, R.J., Fernandes, J.M., Monteiro, P., Rodrigues, H.: Transformation of uml models for service-oriented software architectures. In: 12th IEEE International Conference and Workshops on the Engineering of Computer-Based Systems (ECBS 2005), pp. 173–182. IEEE (2005)
15. Richardson, C.: Microservices Patterns: With examples in Java. Manning (2018)
16. Robertson, S., Robertson, J.: Mastering the Requirements Process. ACM Press Books, Addison-Wesley (1999)

Development and Comparative Analysis of IT Tools Supporting Data Processing in Custom Manufacturing of Orthoses

Anna Dudkowiak(✉), Ewa Dostatni, Filip Górski, Aleksander Tomczak, and Karolina Krzyżanek

Poznan University of Technology, Pl. Marii Skłodowskiej-Curie 5, 60-965 Poznan, Poland
anna.dudkowiak@put.poznan.pl

Abstract. In order to gain a market advantage these days, companies must adapt to a market situation in which it is the customer who makes demands on manufacturers. With this in mind, product customization, i.e. offering products according to individual specifications or customer preferences, is becoming increasingly important. This means that more and more frequently manufacturers not only offer products of different types, but also a particular product in different variants, and even allow the customer to interfere in the appearance or design of the product. This variety of offerings means that companies are struggling to collect and manage data sets related to specific products and customers that are not fully standardized. This raises issues such as increased production time or the time to find specific information for their own use or in response to customer requests. An example of a customized product is limb orthoses. They are customized for a specific customer based on the patient's specified anthropometric dimensions. The article attempts to streamline the process of managing patient data collected during the orthosis design and manufacturing process using 3D printing technology. The process is supported by the AutoMedPrint system, which is a proprietary solution of researchers from Poznan University of Technology for standardizing the process of saving patient data in a nested folder structure. The purpose of this article is to support the data management process by developing two IT solutions to support the process and comparing them.

Keywords: Data Quality · Product Customization · Data Processing

1 Introduction

1.1 Data and Information Quality

Data processing is a key aspect of enterprise management. Converting data into information enables the control of processes in an enterprise and making informed decisions at various management levels, based on up-to-date and reliable information. Data availability and quality, as well as the speed of acquisition of data integrated into enterprise management systems, are real game changers [1].

A. Burduk et al. (Eds.): ISPEM 2023, LNNS 790, pp. 315–327, 2023.
https://doi.org/10.1007/978-3-031-45021-1_24

The quality of decisions made by the managing personnel is a function of the quality of information comprising data and facts collected and sorted in a particular purpose [2]. Collected from various sources, data may be of different quality [3]. The topic of data quality has been discussed by many authors. According to Maślankowski [4], data quality is determined based on the criteria of accuracy, usefulness and up-to-dateness. According to Hernes and Bytniewski [3], data quality is measured using the function of usefulness, and the most important data features are readiness and up-to-dateness, as well as incontestability, clarity, and reliability. They hail the importance of relatively quick acquisition of up-to-date data from reliable sources in response to user query. Lee et al. [5] define data quality as a hierarchy of attributes, as shown in Fig. 1. Each attribute is a logical product of the subordinate attributes, i.e., it is completed when all its components are completed.

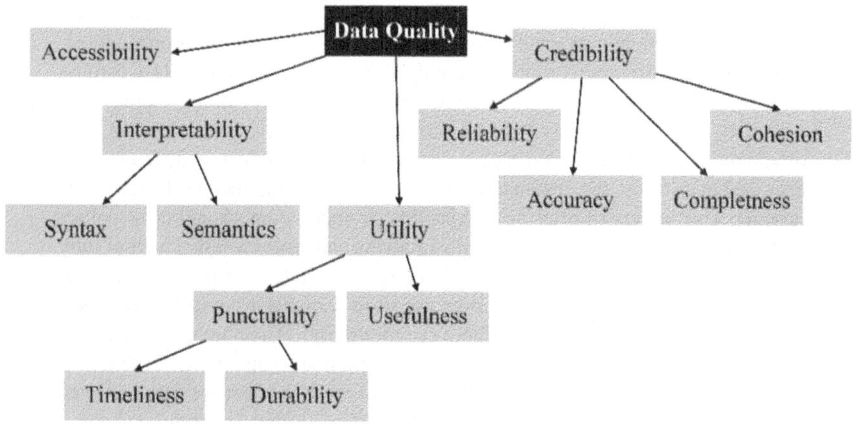

Fig. 1. Hierarchy of data quality attributes (source: [5]).

The term data quality is often used interchangeably with information quality. However, we often speak of data quality with reference to a technical matter, and of information quality with reference to a non-technical matter [6]. Information quality assessment is then considered as a comprehensive assessment of a set of related attributes. Numerous attributes of information quality can be found in the literature of the subject, yet most of them are one and the same attribute, just named differently. The most common ones are up-to-dateness, availability, and purposefulness.

Too many attributes of information quality can make it problematic to unequivocally determine their values, especially of those, whose measures are not specified in detail [7, 8]. For this reason, the authors have selected the most important, in their opinion, attributes, which are listed in Table 1, together with a brief description thereof.

Definitions of information quality often include the notion of subjectivity or relativity, as they are the preferences of the recipient that determine values assigned to attributes. In infological terms, quality of information is determined through the assessment of attributes assigned to information, where the assessment is influenced by various factors, such as psychological (human personality), social (directions and forces of interaction

Table 1. Attributes of information quality (own elaboration based on: [7, 9]).

Attribute	Description
Up-to-dateness	Information corresponds to the actual state of things
Availability/Readiness	Information is readily available
Precision	Information is precise and consistent with the actual state of things
Security	Inadvertent manipulation of information in order to change the result and influence a limited group of recipients is impossible
Clarity	Information is understandable to the recipient
Application	Information is readily applicable and meaningful to the recipient
Coherence	Information is coherent with the purpose
Conciseness	Information does not include any irrelevant pieces

among people), linguistic (the structure of the language in which the information is conveyed) and semantic (sense and relevance of information) [10]. It implies that quality of information is the degree to which requirements of a particular recipient are met [10]. The recipient is the one who assesses the quality of information received. Information is considered of high quality if it meets expectations, as assessed by the consumer of information [11]. What is considered high-quality information by one group of recipients can be considered low-quality by others. The recipient is typically a system user in an enterprise. However, the ultimate recipient is often a customer, waiting for a quick response to their query.

Today, meeting the requirements posed by recipients of information and useful management of data would never be possible without the implementation of IT solutions. Universal systems, offered by software manufacturers, are increasingly often replaced with dedicated, tailor-made solutions—a trend driven by specific requirements of system end users who are recipients of information, as well as by changes on the market. To gain the competitive edge, modern enterprises customize their products to specific customer requirements and transition to tailor-made manufacturing, what involves the processing of datasets for each customer individually.

1.2 Data Management in Customized Manufacturing

Nowadays, enterprises customize their offering to meet customer needs. Changes on the market and the introduction of new technologies within the Industry 4.0 paradigm shed a new light on customization [12]. Owing to new technologies, customization and personalization of products, construed as modifying or creating products to meet specific customer requirements or preferences [13, 14], has become possible without extending lead times or increasing the costs of manufacturing. Not to mention the profitability of such offering on the modern competitive market, in some cases, customization and personalization is not as much a desired option, as it is a required necessity. It often entails the need to collect individual customer data.

However, customized manufacturing calls for a more complex enterprise data management system. A manufacturing process which requires the storage of customer data and data on customized products raises challenges concerning efficient data collection and processing. What is more, the storage of files in various formats, e.g., digital product models, requires standardization. On many occasions, inappropriate data storage or saving creates an informational chaos and delays data processing.

An example of customized product are orthoses. An orthosis is a medical product which supports and secures a certain body part during treatment or convalescence, or forces it to remain in a desired position [15]. Orthoses may range from universal, relatively inexpensive to custom ones, made to order, based on anatomic measurements [16]. The manufacturing of custom orthoses (which are much more effective and comfortable in use than universal ones) involves a lot of manual work, such as measuring the patient (creation of the so-called negative based on a plaster cast) or creating a positive model by putting multiple layers of synthetic and other materials one over the other [17]. The process is lengthy, and the costs surge. In a modern approach, new technologies, such as 3D scanning and printing, are utilized. With the 3D scanning technology, contactless anthropometric measurements of the patient are made, and with the 3D printing technology, the product can be manufactured quickly based on a digital model. This way, a custom orthosis can be created within a mere few hours [18].

Scientists of the Poznań University of Technology have developed the AutoMedPrint system for systematizing and standardizing the storage of data for the manufacturing of custom orthoses. The system supports the designing and manufacturing of custom orthoses with the use of the 3D scanning and printing technologies [18]; however, it does not support the processing of collected data or storing of the data in a single database. In this study, the authors attempt to develop an IT solution supporting patient data analyses based on the requirements posed by recipients of information.

2 Practice

2.1 How AutoMedPrint Works

AutoMedPrint is an original solution developed by scientists of the Poznań University of Technology under the project entitled "Automation of design and rapid manufacturing of individualized orthopedic and prosthetic supplies on the basis of data of anthropometric measurement". The system supports quick, automated designing of custom prostheses and orthoses and creation of the same using the 3D printing technology [18]. The system can help reduce the duration of measurements to less than an hour, and using 3D printing, the product can be made and verified for correctness within several dozen hours [19].

The process is initialized by entering essential patient data. In step two, measurements are made using a 3D scanner at a specially designed measurement station. In this way, a cloud of points is created, which represents a digital model of the object (i.e., the limb). Based on the cloud of points, the system develops 3D models which, possibly improved and approved by the designers, are sent to a 3D printer. The final steps include manufacturing (in parts), finishing, and verification of the manufacturing accuracy and comfort of use on the patient. Upon approval, the product is assembled. The operation of the system is shown in Fig. 2.

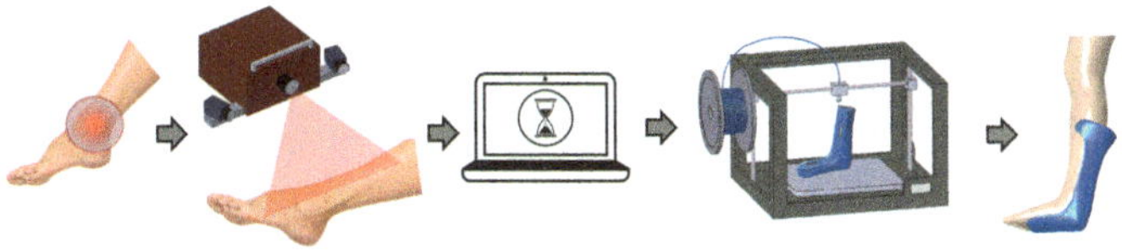

Fig. 2. Operation of AutoMedPrint (source: [17–19]).

The storage of patient data in AutoMedPrint is automated. Patients are assigned ID numbers (customer numbers). The root folder, being the first data storage level, is named after that ID number. The root folder contains data on the patient and the products developed for them, saved in a hierarchical order (see Fig. 3). The root folder contains a text file named *PatientData.txt* and folders corresponding to particular products developed for the patient, named *Product_x* (where x = patient product number). The *PatientData.txt* file contains essential patient data, such as name and surname, gender, year of birth, and email address. Attributes within the text file are separated with the character ';'.

The *Product_x* folder contains another text file named *ProductData.txt*, where basic product attributes, such as type, body side, patient's height and dimensions are saved, separated with the character ';'. In this folder, subfolders named *CAD*, *CAM*, *Scanning*, and *STL* are created, which constitute a set of product models in different formats. Scans performed with a 3D scanner are saved in the.stl format in the *Scanning* folder. Subsequently, a file named *Merged.stl* is created automatically, which combines all the scans performed, as well as a file named *Reconstructed.stl*, where information on cleaning and reconstruction is saved.

The *Reconstructed.stl* file is subjected to an automated data extraction algorithm. As a result of this operation, an.xls spreadsheet is generated containing data which feed into an autogenerated 3D model (saved in the.ipt or.iam format) in the *CAD* folder. In the subsequent step, the 3D model files are saved automatically in the *STL* folder, in the.stl format. The files are loaded automatically and on their basis, programs for 3D printers are generated. The program files are saved in the *CAM* folder.

2.2 User Requirements

AutoMedPrint supports the designing and manufacturing of custom orthoses and prostheses. The system standardizes data saving and eliminates possible human error from the process. However, considering the growing number of customers, a dialogue between the system users and system designers has revealed a need for streamlining data processing—a functionality which AutoMedPrint does not currently offer. The system supports data entering and saving in folders, but it does not create an affiliated database which would store all the data. The data is saved in various formats, which means that when looking for particular information, the user needs to browse manually, while the administering of automatically created folders is time-consuming and onerous. With the current data saving method, searching for a folder of a particular patient is done using external folders named with numbers, what creates confusion (it would be easier to search by name and surname). Responding to a customer query or acquiring information on the

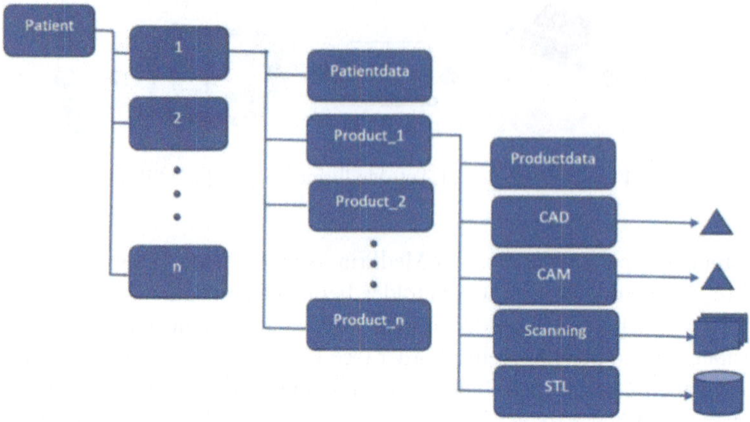

Fig. 3. Patient data saving scheme on the server, performed by AutoMedPrint (own elaboration).

progress of works related to a particular patient is challenging even to experienced staff. The current data saving method does not support data analyzing or creating clear lists or reports. An efficient analysis of the data sorted by patients meeting certain criteria is currently not possible either.

To meet the needs, a decision was made to develop and implement a system supporting the processing of data saved in AutoMedPrint. The functional and non-functional system requirements have been defined, as follows:

- implementation of a software program to store patient data and orders placed in a single location,
- intuitive data availability and quick retrieval,
- creation of patient lists based on order status,
- searching for patient data by inputting first and last name,
- clear and logical presentation of data (lists/reports),
- quick implementation of the system at a low cost.

An attempt was made to develop two software tools supporting the processing of data saved automatically by AutoMedPrint. Taking into account the requirements of users related mainly to quick implementation at low cost, due to the accessibility of software and licenses by the University, Microsoft software, Excel and Access, were selected. These solutions are also characterized by simplicity and intuitiveness in use, while maintaining the possibility of advanced data processing. In addition, working in the software did not require expanding the developers' knowledge, which accelerated the process of developing tools for processing the collected data.

2.3 Application Working in the MS Excel Environment

One of the solutions is an application based on MS Excel spreadsheets. In step one, data was imported using Power Query. The *Patients* folder was set as the source file, containing all patient folders saved as described in chapter *How AutoMedPrint works.*

As a result of importing data from a folder which contained numerous files in various formats, a randomly ordered dataset was obtained, which required initial preparation. The data were converted, filtered, sorted and linked until the desired effect was achieved. The resultant spreadsheet is shown in Fig. 4.

patientID	patientName	patientSex	patientBirthDate	patientEmail	productModificationDate	productType	productSide	patientHeight
12		K	1997	x	28.05.2021	WHO	P	175
12		K	1997	x	28.05.2021	WHO	P	175
13		K	1997	x	21.05.2021	WHO	P	170
13		K	1997	x	28.05.2021	WHO	P	170
13		K	1997	x	28.05.2021	WHO	P	170
10		M	1997	x	21.05.2021	WHO	P	188
14		M	1996	x	21.05.2021	WHO	P	178
15		M	1995	x	21.05.2021	WHO	P	192
16		M	2001	x	28.05.2021	WHO	P	182
16		M	2001	x	28.05.2021	WHO	P	182
10		M	1997	x	26.05.2021	WHO	P	188
16		M	2001	x	21.05.2021	WHO	P	183
17		M	2001	x	21.05.2021	WHO	P	182

Fig. 4. Sorted data in a table (own elaboration).

In the next step, a spreadsheet named *UI* was developed in order to filter the data sorted in the table with two inserted slicers. The first slicer filters data by name and surname of the patient, the second one - by a number of products. The two slicers combined make it possible to use the second slicer only on the data filtered by the first slicer. This is shown in Fig. 5. The first slicer, *product_3,* has filtered out the patients to whom at least three products are assigned; the second slicer can be used only to the patients filtered out by the first slicer.

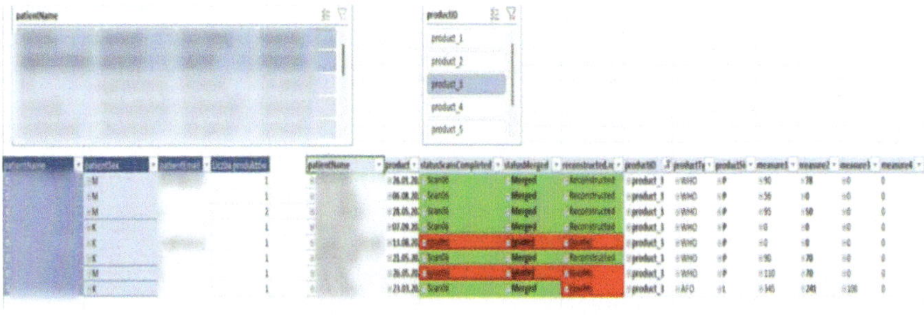

Fig. 5. Slicers in MS Excel (own elaboration).

The filtered data displayed in the spreadsheet includes patient personal data, dimensions, and information of the assigned product. Additionally, conditional formatting has been used to change the appearance (color) of cells. In this way, empty cells can be easily identified, which indicate that the product design is incomplete.

Users interested in checking the product status can use *UIpoStatusie* (Fig. 6). Three slicers inserted into the spreadsheet make it possible to check the product status, i.e., whether all the scans have been made and the *merged.stl* and *reconstructed.stl* files generated. By checking the *(empty)* field with a selected slicer, a list of designs whose status is unfinished will be displayed.

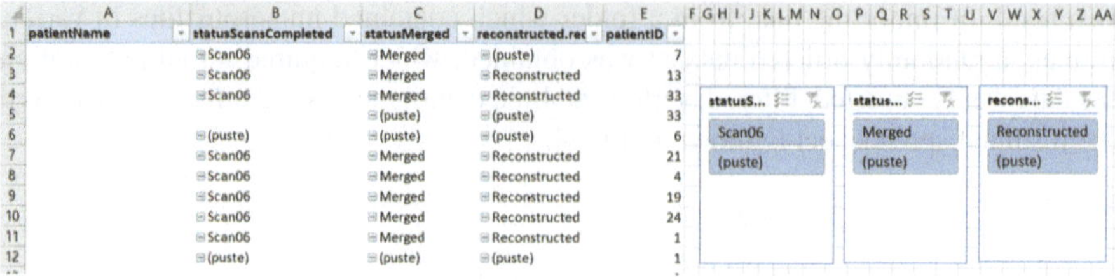

Fig. 6. User interface when using status slicers (own elaboration).

2.4 Application Working in the MS Access Environment

Another solution is an app based on the MS Access database software. For the purpose of the application, entities and their attributes, as well as relations between objects, were identified. On this basis, an implementation model was designed, as shown in Fig. 7. A relational database in MS Access was designed according to the model.

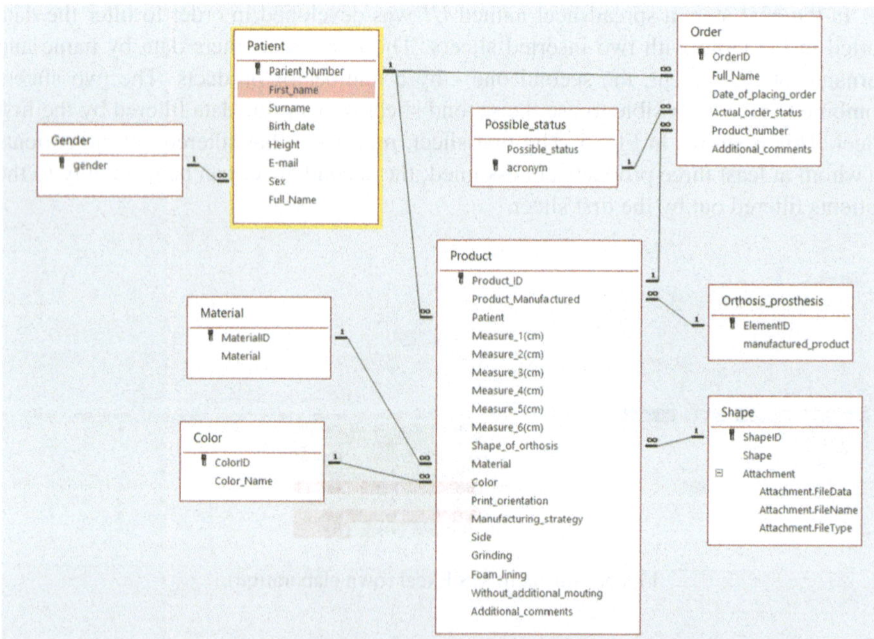

Fig. 7. Implementation model of an MS Access database (own elaboration).

The user never gets direct access and does not perform any operations the tables where the necessary data are stored. All data processing-related operations are performed through forms, which protect the datasets against direct user intervention and thus against inadvertent data manipulation.

The database application comprises 11 forms. The start form is shown in Fig. 8. It consists of four tabs for easy navigation.

The *patients* tab has two buttons to open the forms. With the first button, a new patient can be added (the *add new patient* button), the second one enables viewing previously added patients (the *view existing patients* button). Similarly, the *products* tab has two buttons which enable adding a new product and viewing the existing products by means of dedicated forms. The third tab in the start form enables adding a patient order for a new product (the *add new order* button), viewing the order database (the *view orders* button) and updating the order status (the *modify order status* button). The *view orders by patient* button enables searching the database by all the existing wordings, phrases and names. This function facilitates the searching of the database by a sequence entered (e.g., patient's surname), which was one of the functional requirements defined by end users.

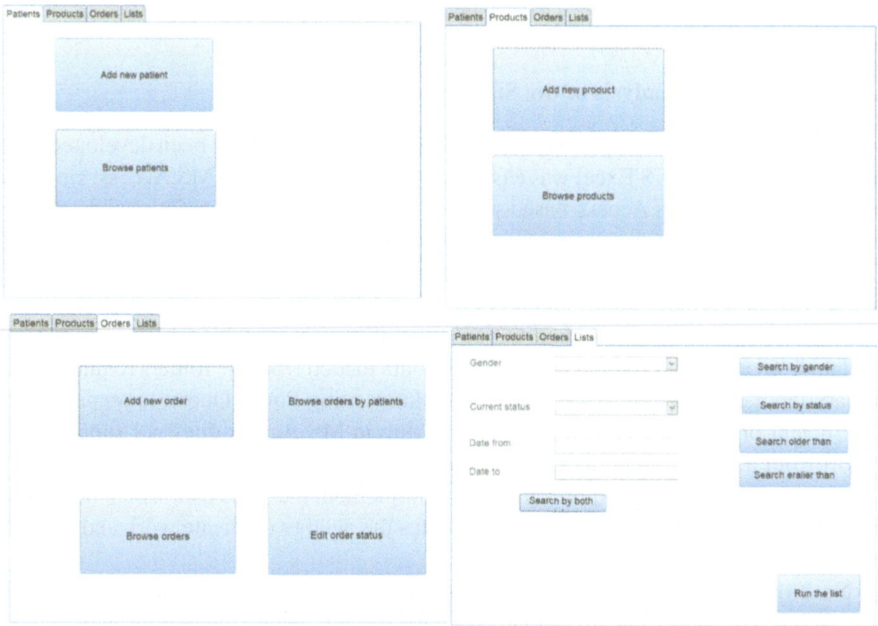

Fig. 8. Tabs in the start form of the MS Access-based application (own elaboration).

The key tab for data processing and drawing up of lists/reports is the *lists* tab. Unlike in other tabs, where the pre-programmed buttons are used for navigation through forms which handle data processing operations, the buttons in the *lists* tab enable queries. Queries generate selective search results (lists) based on predefined criteria, such as gender (the *search by gender* button), order status (the *search by order status* button), and the order date (the *search for older than*, *search for newer than*, and *search by both dates* buttons). Figures 9 and 10 show the database search results to queries about females entered into the database (Fig. 9) and queries concerning the 'patient added' status (Fig. 10).

Limitation by gender Kwerenda

Patient_Num	First_name	Gender	Surname	OrderID	Product_manufactured	Actual_order_status	ProductID
1	Ewa	W	Dostatni	3	WHO orthosis	Scanning done	3
2	Ewelina	W	S	23	Cosmetic hand prosthetic	Entered patient	13
3	Marcelina	W	B	24	AFO orthosis	Entered patient	15
4	Maria	W	Skłodowska-Cu	27	AFO Orthosis	Entered patient	16

Fig. 9. Searching the database by gender (own elaboration).

Limitation by status Kwerenda

Patient_Num	First_name	Surname	OrderID	Product_manufactured	Date_of_pla	Actual_order_status	ProductID
[illegible]	Maria	Skłodowska-Cu	27	AFO orthosis	10.08.2022	Entered patient	16
5	Filip	Górski	2	AFO orthosis	24.05.2022	Entered patient	2
10	Aleksander	Smyczkowski	10	WHO orthosis	23.05.2022	Entered patient	8
12	Ewelina	S	23	Cosmetic hand prosthetic	19.07.2022	Entered patient	13
13	Marcelina	B	24	AFO orthosis	08.08.2022	Entered patient	15

Fig. 10. Searching the database by order status (own elaboration).

2.5 Comparative Analysis of the Solutions

Both software tools draw on the software based on which they have been developed. The designing process in MS Excel was easier and quicker than that in MS Access, since the development of the MS Access-based solution required expertise in relational databases.

The key difference between the solutions, in terms of their utility, lies in the data source. The MS Excel-based application can import data from a folder. It means that with the current patient data saving method by AutoMedPrint, in a nested folder structure, choosing the root folder as the data source path will generate a table which systematizes the data from all the subfolders. However, the data loaded from the folders require initial preparation, i.e., conversion, sortation, and linking, using Power Query, to prepare them for efficient processing. The application working in MS Access does not support data import from a folder due to its database nature. Therefore, in its current state, the MS Access-based application requires the user to implement the same data which have been previously saved in AutoMedPrint, what creates the need to duplicate work and is a real hindrance.

Nevertheless, in the authors' opinion, the MS Access-based application is more intuitive, as it is operated by means of interactive forms with navigation buttons which make completion of the process easy. Its use does not require expertise in the operation of MS Access. Neither has the end user access to relations or table designs, thus the relational database is protected against data manipulation. The MS Excel-based application is, on the other hand, a tool which requires from the end user some knowledge of data import and linking of data during the initial preparation, so that the data can be loaded correctly into the appropriate location in the MS Excel environment. Due to the fact that the end user has access to all spreadsheets with imported data, there is a possibility that the data may be inadvertently manipulated, e.g., unintentionally deleted or modified. It may result in lower information quality in the opinion of the information recipient, as, on each occasion, reliability of the processed data can be questioned.

Each of the solutions meets the requirement of quick data retrieval based on predefined criteria. Both the MS Excel-based and the MA Access-based applications support filtering by status. Owing to the two slicers, the MS Excel-based application facilitates search for particular patients and search for patients with a specified number of products. The MS Access-based application additionally supports data filtering by date and gender. Moreover, the entire database can be searched by a user-predefined sequence, what makes the process of searching for patients considerably easier.

Considering the subjectivity of information quality assessment, referred to in the theoretical part of this paper, the tools were presented for comparison by three end users—information recipients. The data import part was skipped and focus put on data processing features, on the assumption that data had been properly imported. For the comparison, the data quality attributes shown in chapter 1.1, Table 1, were used. Each attribute was assigned a grade on a scale of 1–5, where 5 was the highest grade. Table 2 shows the mean grades for each of the data quality attributes, based on the assessment by three end users. The users jointly agreed on the weights of the attributes, as shown in Table 2.

Table 2. Mean grades for information quality attributes assigned by end users (own elaboration).

Attribute	Weight (0–1)	MS Excel-based application (0–5)	MS Access-based application (0–5)
Up-to-dateness	1	4.67	4.67
Availability/Readiness	1	4.67	4.33
Precision	1	4.67	4.67
Security	0.75	1.67	4.33
Clarity	0,5	4.33	3.33
Application	1	5	5
Coherence	1	4	3.67
Conciseness	0.5	3	3.33
	Final result	27.93	28.92

The final result was obtained based on the mean grades and weights. According to the three end users, the MS access-based application is a slightly better solution, mainly owing to the security of information. Additionally, the response time to a query about order status was measured. Each of the tools reduced the response time to less than a minute (without considering the import/data entering process at the current state of development of the applications). The response time to the same query without the support of an IT data processing solution could be even up to an hour.

3 Summary

It can be concluded unequivocally that each of the solutions has streamlined patient data processing. Searching for patients and the related data has become more intuitive and less time-consuming. The collecting of data in datasets makes it possible to perform advanced analyses and produce lists based on selected orders. The MS Access-based solution has turned out to be slightly more advantageous, considering the fact that it ensures better data security, and its database nature (relation-based) eliminates data redundancy.

The end users, who are also information recipients, have assessed both applications as useful. In their opinion, further work should focus on enabling automated data import from automatically saved folders, so that at each start-up, the tool loads up-to-date data and saves them in appropriate locations. This task, however, requires some expertise in the SQL or VBA programming language. Nevertheless, the programming of the database to load data automatically is a key factor minimizing the duration of preparatory operations when processing data.

References

1. Haug, A., Zachariassen, F., van Liempd, D.: The costs of poor data quality. J. Industrial Eng. Manage. **4**(2), 168–193 (2011)
2. Howard, G.R., Lubbe, S., Klopper, R.: The impact of information quality on information research. Alternation Special Ed. **4**, 288–305 (2011)
3. Hernes, M., Bytniewski, A.: Data quality in integrated management information systems. Scientific papers of the University of Economics in Wroclaw, **471** (2017)
4. Maślankowski, J.: Analyzing the quality of data extracted from websites using big data solutions. Annals of the College of Economic Analysis **38**, 167–177 (2015)
5. Lee, Y.W., Pipino, L., Funk, J.D., Wang, R.Y.: Journey to Data Quality. MIT Press (2006)
6. Madnick, S.E., Wang, R.Y., Lee, Y.W., Zhu, H.: Overview and framework for data and information quality research. J. Data Inf. Quality **1**(1), 1–22 (2009)
7. Grudzień, Ł.: Impact of process characteristics on information requirements of process document users. Mechanical Engineering, pp. 127–139 (2013)
8. Górski, F.: Rapid design and manufacture of orthoses and prostheses. Lectures, Poznan University of Technology
9. Krzesaj, M.: Using M.J. Eppler's quality framework to evaluate the information quality of PWSZ websites. Overview of the Organization, **9**, pp. 47–56 (2019)
10. Unold, J.: Information system vs. qualitative view of information. Scientific Papers / Academy of Economics in Katowice, str., pp. 163–170 (2004)
11. Laudon, K.C., Laudon, J.P.: Essentials of Management Information Systems. 4th edn. Prentice-Hall (2001)
12. Szozda, N.: Product customization in modern supply chains - trends, strategies, application examples. Scientific Journals. Organization and Management / Silesian University of Technology, pp. 537–550 (2018)
13. Newell, F.: The New Rules of Marketing: How to use One-to-One Relationship Marketing to be the Leader in Your Industry. McGraw-Hill, Nowy Jork (1997)
14. Gilmore, J.H., Pine, B.J.: The four faces of mass customization. Harv. Bus. Rev. **75**(1), 91–101 (1997)
15. Baronio, G., Volonghi, P., Signoroni, A.: Concept and design of a 3D printed support to assist hand scanning for the realization of customized orthosis. Appl. Bionics Biomech, **2017** (2017)
16. Andringa, A., van de Port, I., Meijer, J.W.: Long-term use of a static hand-wrist orthosis in chronic stroke patients: a pilot study. Stroke Res. Treat., **2013** (2013)
17. Górski, R., et al.: Design and additive manufacturing of an individualized specialized leg orthosis. In: Górski, F., Rychlik, M., Pacurar, R. (eds.) Advances in Manufacturing III, **5**, pp 31–44, Springer (2022)

18. Górski, F., Wichniarek, R., Kuczko, W., Żukowska, M., Lulkiewicz, M., Zawadzki, P.: Experimental Studies on 3D Printing of Automatically Designed Customized Wrist-Hand Orthoses. Materials **13**(18) (2020)
19. Górski, F., Wichniarek, R., Kuczko, W., Żukowska, M., Rybarczyk, J., Lulkiewicz, M.: Evaluation of a prototype system of automated design and rapid manufacturing of orthopaedic supplies. In: Górski, F., Rychlik, M., Pacurar, R. (eds.) Advances in Manufacturing III, **5**, pp 1–15, Springer (2022)

Design of Experiments: An Overview and Future Paths

Hugo Silva[1], André S. Santos[2], Leonilde R. Varela[3], Justyna Trojanowska[4(✉)], and Kristina Berladir[5]

[1] Institute of Engineering, Polytechnic Institute of Porto - ISEP/IPP, Porto, Portugal
1181214@isep.ipp.pt
[2] Interdisciplinary Studies Research Center (ISRC), Institute of Engineering, Polytechnic Institute of Porto - ISEP/IPP, Porto, Portugal
abg@isep.ipp.pt
[3] Algoritmi Research Centre/ LASI, University of Minho, Guimarães, Portugal
leonilde@dps.uminho.pt
[4] Faculty of Mechanical Engineering, Poznan University of Technology, Poznan, Poland
justyna.trojanowska@put.poznan.pl
[5] Faculty of Technical Systems and Energy Efficient Technologies, Sumy State University, Sumy, Ukraine
kr.berladir@pmtkm.sumdu.edu.ua

Abstract. Process optimization is increasingly important as competition deepens. As such, the optimization of process parameters allows the manufacturer to improve its competitive advantage. In that context, planning experiences to test the limits of the system is progressively more important. In that context, Design of Experiments is a group of techniques that allow a better understanding of every process, when correctly applied. During the last two centuries, several methods were developed to approach Design of Experiments. In this paper, the history and methods, focusing on the most important, are examined in detail. The last chapter describes the future paths of planned experiences, addressing its connection with digital simulation, making its use a lot more accessible. The algorithmic and mathematical evolution is also described, relating it to the software that can be used to execute planned experiences.

Keywords: Design of Experiments · Process Optimization · RSM · Taguchi Method

1 Introduction

In this paper, the different methods of Design of Experiments are described, along with their applications, advantages, and disadvantages. Its aims to indicate when one method is more appropriate than any other. This form of planned experiences relies heavily on statistical knowledge and process control, to guarantee the correct application of the changes introduced in the system.

A. Burduk et al. (Eds.): ISPEM 2023, LNNS 790, pp. 328–341, 2023.
https://doi.org/10.1007/978-3-031-45021-1_25

This paper focuses first on explaining the statistic tools generally applied around DOE (Design of Experiments), followed by the main concepts of planned experiences.

Then, each method is described, from OFAT (One-Factor-At-a-Time) to RSM (Response Surface Methodology). This is continued by a bibliometric research on the topic, the future paths of DOE, and, finally, the conclusions learned on this paper.

2 Literature Review

2.1 Hypothesis Testing and ANOVA

DOE resorts to planning, data collection and subsequent data treatment and analysis, dealing with both independent and dependent variables [1]. This methodology is based on the statistical knowledge of tools, such as: ANOVA (Analysis of Variance), t-tests, among others [2]. For the correct operation of this type of statistical tool, a previous analysis is recommended to distinguish between relevant factors from the irrelevant ones. This fits perfectly with the DMAIC (Define, Measure, Analyse, Improve, Control) methodology, where, prior to the implementation of DOE in the improve phase, a statistical analysis is performed on the data collected in the analyse phase.

The application of any iteration of DOE should be followed by some statistical approval tool, to confirm conclusions drawn in the experimental phase. One of the most used is ANOVA. To be applied, ANOVA one must assure normality, but there are cases where that is not established. To overcome this obstacle, transformations such as logarithmic are often used [2].

ANOVA and t-tests, use hypothesis testing as their starting point. The assumption of this type of test arises in the study of two hypotheses, the null hypothesis, H_0, and the alternative, H_1. The first suggests the non-existence of a relationship between cause and effect, and it is initially assumed to be true. The second, the hypothesis that seeks to reject the null, must be strongly substantiated with valid arguments.The procedure usually follows the process shown on the Fig. 1.

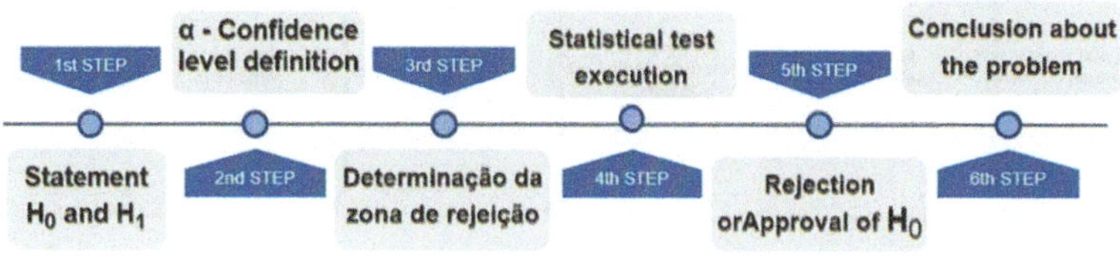

Fig. 1. Hypothesis Testing Steps

However, despite the effectiveness of hypothesis testing, there is a possibility that the wrong conclusion could be drawn. This possibility is defined as the region outside the safety level. There are two types of errors [3]:

Type I error occurs when rejecting H0, even though it is true. This type of error is represented by the significance level, α, which is the probability of the error occurring [4].

Type II error, on the other hand, is represented by the value of β, in which case H0 is incorrect, yet it was not rejected [5]. The complementary value of β, that is, 1-β, is the power of the hypothesis test.

ANOVA emerges as an important tool within statistical study by comparison method, applied on independent samples.

F-test, created by Ronald Fisher, is a way to define ANOVA, because it follows an f distribution, which evaluates the ratio between variances [6]. Equation 1 defines it:

$$F = \frac{Between - group\ Variability}{Within - group\ variability} = \frac{\sum_{i=1}^{k} n_i\left(\overline{Y}_i - \overline{Y}\right)/(k-1)}{\Sigma_{ij=1}^{n}\left(Y_{ij} - \overline{Y_i}\right)^2/(N-k)} \tag{1}$$

2.2 Design of Experiments

Historically, this methodology is basically a process of planning and handling experiences. The beginning of the development and implementation of this tool is not known. It is, therefore, based on the concept of experiment, which can itself be defined as the alteration of parameters to promote the discovery of new information. Its foundations were built by Ronald Fisher during the 1920s and 1930s, which had significant development after the end of the Second World War [7].

The complexity of a DOE is determined by the methodology used, as well as the number of factors (variables or inputs to be studied) and the number of levels for each of the factors. The levels mean the extreme points of the planned amplitude, usually ordered in increasing order, for the study in question. The larger the range, the easier it is to identify the effect of one of the factors or interaction of factors [2].

The steps in the development and application of a DOE, which may vary slightly depending on the author, are shown in Fig. 2.

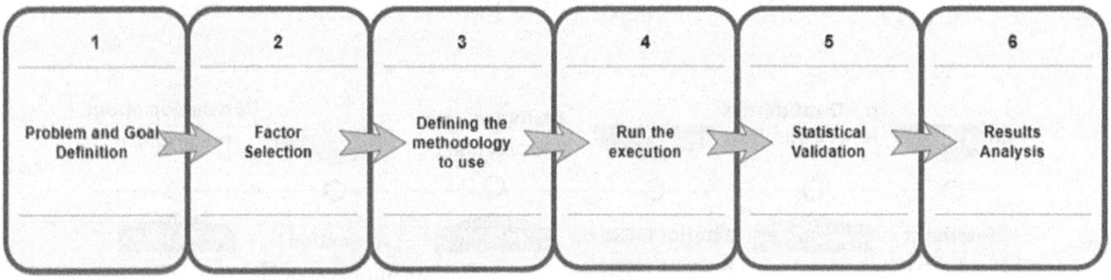

Fig. 2. DOE implementation usual steps

In DOE there are two key concepts, randomization, and replication. The former is the opposite of the event of performing tests with some logical order, that is, they are performed randomly to avoid biased conclusions. Replication can be explained as the repetition of the same conditions in two different tests, giving consistency to the information collected in the equivalent runs [8]. Associated to these two concepts, there is another complementary one, named blocking. It tries to guarantee impermeability to the experiment, making it less sensible to external factors, designated noise factors that may affect the process. These factors are generally uncontrollable. A golden rule is: block everything that is possible, randomize the rest [2].

A distinction can be made between two types of experiments: comparative and screening. The first is the simplest iteration, comparing alternatives using statistical tools to find the best alternative. The second seeks to understand which factors influence the process [2]. Examples of comparative experiments arise in randomized designs that use blocking, that seek to study one factor, always considering others that may affect the process. On the other hand, full or fractional factorial experiments are examples of screening experiments.

Inputs will be the factors that affect the process, being these controllable factors or not. They can be considered normal factors, that is, controllable variables, which affect the process, or they can be noise factors, variables that affect the process but are not under the control of the user [2]. The latter type of factors is undesired, but should also be considered, and the robustness of the experiment reveals the sensitivity of the process to these types of factors [9], something that is addressed in a later subpoint.

Outputs are the result of the process, the inputs, and the interactions between the inputs. Both outputs and inputs must be measurable.

The classic DOE model uses Eq. 2 as an assumption, and it assumes that all measured responses have an associated error value. ym is the measured response, yt is the theoretical response, and ε is the term referring to the assumed value [10].

$$y_m(x) = y_t(x) + \varepsilon \tag{2}$$

2.3 OFAT

OFAT (One-Factor-At-a-Time) is the simplest iteration of the DOE, since, as the name suggests, only one factor is changed at a time [11].

For a small number of factors, OFAT is an interesting solution. However, when the number of factors increases, it becomes very labor intensive and, with other techniques, irrelevant. This is because OFAT requires replications to achieve the same statistical power as other DOE variations [1].

For the same number of factors, OFAT presents a larger number of experiments than factorial DOE. Figure 3 demonstrates this phenomenon. In the left representation is the factorial method for factors A and B, and on the right are the same factors arranged in an OFAT. The factorial version requires four trials, while the OFAT requires six, illustrated by the dots. This demonstrates the argument explained in this paragraph.

Taking into consideration the explanation given in this sub-point, some advantages of using other DOE methods, such as factorials, against OFAT can be highlighted [11]:

- Identification of effects related to interaction between factors.
- Use of less time and resources for the same amount of information.
- Reduction of variability, due to the use of all observations for each of the effects.

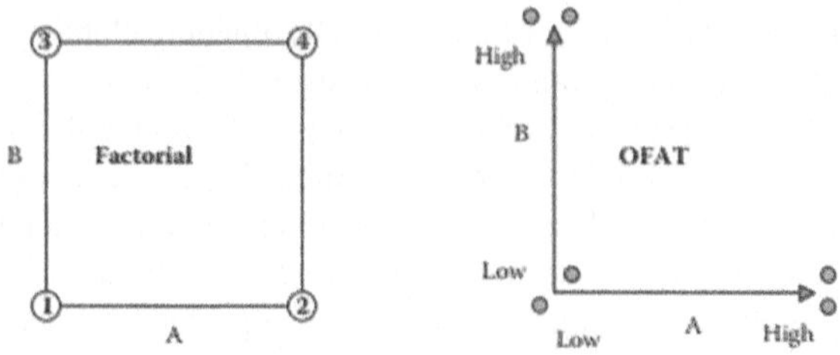

Fig. 3. Two level comparison between a factorial experience and OFAT [2]

2.4 Full and Fractional Factorial

Full factorial experiments utilize the totality of factors and their respective levels. As such, all possible effects and interactions between factors are addressed, and interactions tend to be the catalyst for process improvement optimization [12]. The expression used to calculate the number of tests, N, in full factorial experiments is presented in Eq. 3 [1]:

$$N = L^k \tag{3}$$

where L is the number of levels of the test, and k is the number of factors.

The number of tests grows rapidly as the number of levels and the number of factors grow, so this type of DOE is usually preferable for tests with two or three levels [1].

In tests with two levels, coding with positive and negative signs is recommended. In the following example, Table 1, considering outputs Y1 and Y2 and factors A and B with two levels, a typical table for a two-level coded DOE is presented, with purely representative values.

Table 1. Example of a two-level coded factorial experiment [2]

Index	Run	A	B	Y_1	Y_2
1	3	-	-	74	3.1
2	1	+	-	75	3.5
3	2	-	+	71	1.6
4	4	+	+	80	1.2

The derivation of factorial experiments in which not all tests are used are called fractional factorial experiments. These contain a fraction of a full factorial, accompanied by a complementary trial with the overall mean of the factors evaluated [2]. This type of planned experiment has shown to have a lot of use in the industry [13].

A pertinent concept to carry out the study of fractional factorials is the concept of blocking. This can be summarized as the joining of similar data into blocks, with the goal of reducing noise. This concept is most often applied when the sample lacks homogeneity [13].

To calculate the number of trials to be performed in a fractional factorial test, in this case having two levels per factor, the following Eq. 4 is used, where N is the number of trials, n the number of factors and k a value defined by the desired partial fraction [14]:

$$N = 2^{n-k} \tag{4}$$

One of the concepts most associated with fractional factorial experiments is the concept of resolution. This is a basic DOE procedure, and its index measures the greater or lesser possibility of effect distortion occurring. The lower the resolution, the more likely the distortion will occur [2]. This can be observed, in a properly coded experiment, when factors exhibit the same sequence of signals that effects between factors exhibit. To calculate the appropriate resolution, a value is added to the number of factors.

As such, for two-level factors, there are three predefined resolutions, and their index appears in the value of n, in the previous equation, present in Table 2.

Table 2. Fractional Factorial Resolutions

Index	Description
III	Leads to the estimation of the main effects, although its conclusions generally are suspicious, due to the lack of information caused by the distortion of effects
IV	Allows clear distinction of the interaction between two factors
V	Allows clear distinction of the interaction between three factors

2.5 Taguchi Method and Orthogonal Matrix

An important concept for the study of DOE is the orthogonal matrix. These matrixes are a result of the balanced Eq. 4. Their conception resulted from the work of several authors throughout history, such as Bush (1952) and Rao (1946–1949). This type of matrices can be defined as [15]: "Matrices of multifactorial experiments where the columns correspond to the factors, the entries in the columns correspond to their test levels, and the rows correspond to the test runs". The identified factors form a so-called balanced matrix since these factors are considered equally [9].

Genichi Taguchi introduces to the scientific community a new way of approaching the planning of experiments, this becoming known as Taguchi's Method. It belongs to the classification of fractional factorial experiments since it partially addresses a universe of possible tests for the respective factors and levels. The method allows a change of perspective in the interpretation of the factors. With one of its two goals being to reduce the number of trials, it uses a special version of orthogonal matrix [16]. With this new design, the experiments performed are less sensitive to process "noise" than full factorial experiments, which is the other goal of the Taguchi method [17].

Taguchi's method seeks to provide robustness to DOE by controlling the inputs and examining their effect on the outputs [18]. This method therefore starts at the control factors.

Genichi Taguchi suggests that quality is measured between the expected value of an output and its actual value. With this, a new quadratic function on quality loss is introduced [19].

Once L, the consequent loss value of the effect, is obtained, the value of the loss function, three types of SN (Signal-to-Noise) ratio quotients are accomplished. This is the indicator that varies with the iteration being applied. Table 3 presents the three solutions. In this μ represents a mean, σ the standard deviation, Yi the measurement taken at trial i and n the number of trials performed.

Table 3. The three Solutions for the SN ratio [16]

Iteration	Equation
The larger, the better	$\frac{s}{N} = -10\log\left(\frac{1}{n}\sum_{i=1}^{n}\frac{1}{Y_i^2}\right)$ (5)
The smaller, the better	$\frac{s}{N} = -10\log\left(\frac{1}{n}\sum_{i=1}^{n}Y_i^2\right)$ (6)
The closer to the nominal value, the better	$\frac{s}{N} = 10log\left(\frac{\mu^2}{\sigma^2}\right)$ (7)

However, there are three main arguments in the scientific community that criticize the use of the Taguchi Method [9]:

- SN ratios do not reveal exactly how control factors affect the trial response.
- Interactions between factors, something of high importance, are mostly not considered when applying this method.
- It is not possible to understand the behavior of noise factors, although identified, and their interactions with control factors.

2.6 RSM and Split-Plot

RSM (Response Surface Methodology) pioneered by Box and Wilson, due to their works published around 1951. These provided the basis for this methodology. Later, authors such as Jerome Myers evolved the methodology between the 1970s and 1990s, making the concepts sub adjacent to the RSM methodology more accessible to the scientific and industrial community [20]. It is mainly used on industries such as chemical and biomedical.

This methodology can be defined as the set of techniques applied to numerical information, presented on a surface that demonstrates all possible combinations, as well as the optimal solution, if it exists. These are a different type of models, the optimization one, differing from the ones previously described, which would fall under the category of screening. RSM can be presented with a linear or quadratic model, and a linear solution is preferable since it is more accessible to the user. Both can generate misunderstood regions, named as residual [7].

Following the previous subpoint, referring to the Taguchi Method, RSM can address the same problems, these being robust design problems, having better performance on the matter [20].

In a two-phase approach, Vining and Myers sought to develop a better solution to this type of challenge. In a first phase, the authors assume that the control factors undergo replication, combining with the technique developed by Myers and Carter decades earlier, to solve problems associated with the replication phenomenon. In the second phase, the combination of noise and control factors is made in a single design. As a result, Eq. 8 emerges that allows the graphic representation in two dimensions [20].

$$y(x, y) = \beta_0 + xt\beta + xtBx + yt\gamma + xt\Delta y + \varepsilon \tag{8}$$

where xt represents the set of control factors, regulated by the β vector that contains the regression of this type of factors, and yt the set of noise factors, having the vector γ the same effect as β, the difference being only in the type of factors. B represents the matrix resulting from the control factors, and its diagonal reveals details about the interactions between them. ε represents the residual value associated with the equation [20].

To complement the study, it is necessary to understand the use of center points. These seek to provide a measure of stability and variability, as well as facilitate the verification of curvatures. They should initiate or terminate the attempts, as well as these should not be random, they should provide more control to the user.

For a better understanding of RSM, a case study is presented that addresses the optimization of cellulose extraction from banana stalk [21]. The following example seeks to demonstrate the application of the methodology in a recent article, reaffirming the importance of RSM in both the industry and academic context. Three factors of three levels are introduced, with the respective encoding, as represented in Table 4.

Table 4. Case Study considered factors [21].

Parameters	Relation of code and original variables		
	-1	0	1
Concentration of NaOH, C_A (%)	1	2	3
Concentration of H_2O_2, C_P (%)	13	15	17
Microwave radiation time, t (min)	10	20	30

Next, the trials are presented, with the respective measured responses, considering the assumed center points, Table 5.

Based on Table 5, graphical representations are performed for all responses and the RSM method is applied to each one. For the purposes of literature review, only one of the representations performed in the case study is presented, Fig. 4.

Split-plot experiments were born within the agricultural context in the 1930s, where the core of the method emerged in the division of plots of land [22]. Being an old method, this has undergone a revival in the last 20 years [23].

For the industrial application that this type of experiments can have, it is important to define the difference between the two types of factors: HTC (Hard-to-Change), also called whole-plot, are the factors where there is greater difficulty in controlling their variation and randomization; ETC (Easy-to-Change) are parameters where the user can

Table 5. Case Study Runs and Outputs [21].

No. of experiments	X_1	X_2	X_3	C_A (%)	t (mins)	C_P (%)	Y_C (%)	Y_H (%)	Y_L (%)
1	-1	-1	-1	1	10	13	74.91	25.05	3.14
2	1	-1	-1	3	10	13	75.65	24.35	2.63
3	-1	1	-1	1	30	13	76.99	23.05	2.53
4	1	1	-1	3	30	13	79.16	20.85	2.46
5	-1	-1	1	1	10	17	78.15	21.84	2.52
6	1	-1	1	3	10	17	78.45	21.58	2.42
7	-1	1	1	1	30	17	77.42	22.56	3.35
8	1	1	1	3	30	17	79.43	20.52	3.53
9	$-\alpha$	0	0	0.32	20	15	76.54	23.45	3.15
10	$+\alpha$	0	0	3.68	20	15	79.02	20.99	2.85
11	0	$-\alpha$	0	2	3.20	15	77.75	22.24	1.94
12	0	$+\alpha$	0	2	36.80	15	80.21	19.82	2.42
13	0	0	$-\alpha$	2	20	11.64	75.86	24.1	2.92
14	0	0	$+\alpha$	2	20	18.36	78.99	21.03	3.35
C1	0	0	0	2	20	15	82.15	17.89	2.16
C2	0	0	0	2	20	15	81.17	18.86	2.19
C3	0	0	0	2	20	15	82.21	17.81	2.16
C4	0	0	0	2	20	15	82.17	17.84	2.23

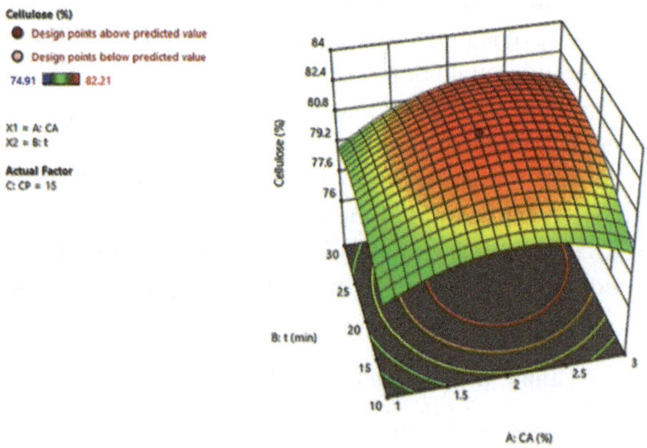

Fig. 4. Case Study RSM 3D representation [21]

have greater control, giving the model its name, split-plot. The same consideration for experiment cost is also applicable [22].

Previously, split-plot designs were more used in the presence of categorical variables, and currently these have more varied applications, due to the optimization models and algorithms [23]. All the tools, besides the basic concept of the methodology, make split-plots designs more robust than methodologies like Taguchi's method, explored earlier, thus presenting itself as the preferable model when robustness is sought.

There is also a derivation of split-plot designs, called incomplete designs. These arise when the initial model cannot be applied due to the constraints of the universe under study. This solution was designed by Robinson (1967) [24].

3 Bibliometric Research

To understand the relevance of DOE to the scientific community, a bibliometric research was conducted. For that purpose, Biblioshiny, from Bibliometrix, and the app Dimensions were used.

As such, the research focused on the theme Design of Experiments, between the years of 1995 and 2022, including. As shown on the Fig. 5, the main area to develop this term was Engineering, followed by Biomedical Sciences, Biology, etc.

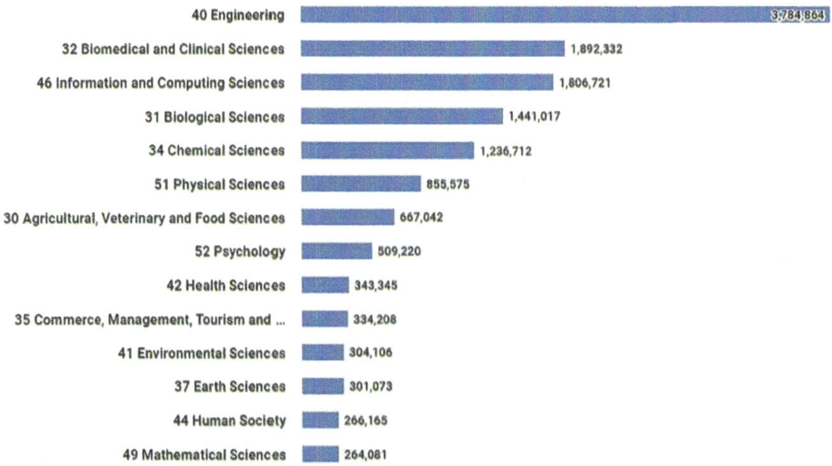

Fig. 5. DOE distribution by area of application

The words most used in DOE articles, present in their abstract, are shown in Fig. 6. The larger the word size, the more often it was used in the analyzed articles. Again, the results reveal words such as "model", "factors", "parameters" and "method" as very important words in this context.

Fig. 6. Most used words in the abstract of DOE-related papers

This bibliometric research reflects the growth of DOE in the scientific community, and its use and research quadrupled in two decades, reflected in Fig. 7. This demonstrates the growth of experience planning, resulting in the optimization of parameters and, consequently, a better knowledge of the production processes of various industries.

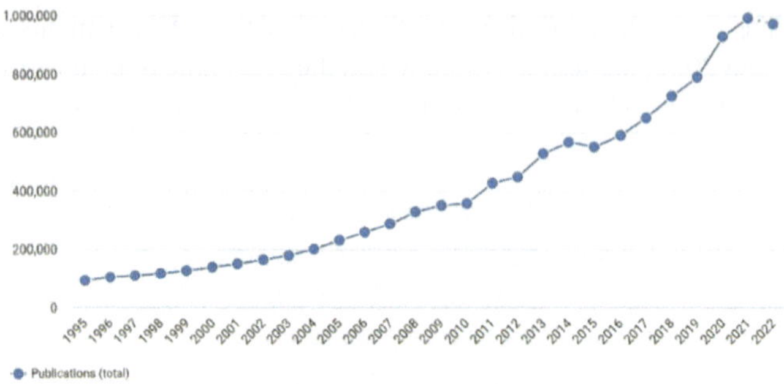

Fig. 7. Growth of DOE in the scientific community since 1995

4 Future Paths

The methods presented in this work date mostly to past centuries, and many of them are still applied today. In this section are presented two subtopics referring to different modern iterations of DOE.

With the development of technology and taking into consideration the increased importance of parameter and process optimization, the Design of Experiments methodology is associated with new forms of improvement, highlighting its use in simulations. The use of both concepts at the same time, DOE and simulation, allows for a viable prediction model, and the cost and time of testing is substantially reduced. It is also the best alternative when analytical tests cannot be performed [25]. Another advantage associated with DOE in simulations is the fact that it is less sensible to replication or order placement, making the experimentation process more accessible and faster [25–27].

Another subtopic under development today is ODOE (Optimal Design of Experiments). It is found primarily in research related to chemical engineering. This is associated with the concept of uncertainty, present in all planned experiments. In this context there are two techniques that ensure the optimization of a DOE when the uncertainty is mentioned, these being: stochastic programming and robust optimization, with the difference between the two being in the replication of tests. These techniques seek to reduce the time and resources spent on poorly planned experiments, where the degree of parameter uncertainty invalidates the tests performed [28–31].

Within the context of robust optimization, split-plots design, previously mentioned before, are one of the best solutions. In face of the new phase of development of this method, it is associated with optimization models, as well as programmed algorithms. Authors such as Trinca and Gilmour (2001), Jones and Goos (2012) are responsible for some of these current algorithms, bringing new ways to approach experiments, resorting to ratios, tools or using other methodologies for that purpose [23].

In recent years, software such as JMP and Design Expert have been developed to help users who wish to carry out DOE. These can facilitate the demonstration of results, or even demonstrate to the user, once the factors under study have been placed, which attempts they should perform.

Considering the previous paragraphs, it is possible to conclude that techniques of optimisation, programming, and mathematical development, using algorithms, dominate the future of the development of planned experiments. DOE thus becomes easier and more accessible to be applied. With the help of the mentioned software, the user practically only needs to define the factors that wishes to study and execute them, saving time with their planning, as well as also avoiding errors due to lack of knowledge of the methods. It can be predicted that the growth of planned experiments will confirm the trend of the last decade and will be increasingly applied in an industrial context.

5 Conclusions

This paper addresses the various methods of DOE, allowing for the reader to understand which method is appropriated to each situation. If there are resources and the goal is to extensively understand the process, a full factorial may be the best method. On the other hand, if one tries to plan a study where noise factors may be highly influential to the process, the Taguchi method, or a robust one, makes more sense.

Although most of these methods are decades old, they can still be applied effectively on most cases. Following the concepts related to the future paths, it can be understood that DOE is evolving in a way that its cheaper and easier to run. Now intertwined with the technological evolution, methods like RSM allow the user to know the range of the parameters of the study, if the total optimization of the goal. This paper also describes the use that new DOE *software* can have. Algorithm-based solution appear to be an attractive option, allowing the user to have a customized planned experiment.

In conclusion, the bibliometric research affirms the growing tendency of planned experiences, and the prediction of DOE gaining even more relevance in the next years can easily be made.

Acknowledgement. The research was partially supported by the Polish National Agency for Academic Exchange within the project "Strengthening the scientific cooperation of the Poznan University of Technology and Sumy State University in the field of mechanical engineering" (agreement no. BPI/UE/2022/8–00). This work was also supported by national funds through the FCT-Fundação para a Ciência e Tecnologia through the R&D Units Project Scopes: UIDB/00319/2020, and EXPL/EME-SIS/1224/2021.

References

1. Román-Ramírez, L.A., Marco, J.: Design of experiments applied to lithium-ion batteries: a literature review. Appl. Energy **320**, 119305 (2022)
2. Anderson, M.J., Whitcomb, P.J.: DOE Simplified: Practical Tools for Effective Experimentation. CRC press (2017)
3. Vickers, A.J.: Parametric versus non-parametric statistics in the analysis of randomized trials with non-normally distributed data. BMC Med. Res. Methodol. **5**(1), 1–12 (2005)
4. Berger, R.W., Benbow, D.W., Elshennawy, A.K., Walker, H.F.: The Certified Quality Engineer Handbook. ASQ Quality Press (2006)
5. Byrne, G.: A statistical primer: understanding descriptive and inferential statistics. Evid Based Libr Inf Pract **2**(1), 32–47 (2007)

6. Rouder, J.N., Engelhardt, C.R., McCabe, S., Morey, R.D.: Model comparison in ANOVA. Psychon. Bull. Rev. **23**(6), 1779–1786 (2016)
7. Draper, N.R., Pukelsheim, F.: An overview of design of experiments. Stat. Pap. **37**(1), 1–32 (1996)
8. Vanaja, K., Shobha Rani, R.H.: Design of experiments: concept and applications of Plackett Burman design. Clin Res Regul Aff, **24**(1), 1–23 (2007)
9. Hedayat, A.S., Sloane, N.J.A., Stufken, J.: Orthogonal Arrays: Theory and Applications. Springer Science & Business Media (1999)
10. Giunta, A.A., Wojtkiewicz, S.F., Eldred, M.S.: Overview of modern design of experiments methods for computational simulations. In: 41st Aerospace Sciences Meeting and Exhibit (2003). https://doi.org/10.2514/6.2003-649
11. Czitrom, V.: One-factor-at-a-time versus designed experiments. Am. Stat. **53**(2), 126–131 (1999)
12. Yahiaoui, I., Aissani-Benissad, F.: Experimental design for copper cementation process in fixed bed reactor using two-level factorial design. Arab. J. Chem. **3**(3), 187–190 (2010)
13. Grömping, U.: An algorithm for blocking regular fractional factorial 2-level designs with clear two-factor interactions. Comput. Stat. Data Anal. **153**, 107059 (2021)
14. Han, X., Liu, M.-Q., Yang, J.-F., Zhao, S.: Mixed 2-and 2r-level fractional factorial split-plot designs with clear effects. J Stat Plan Inference **204**, 206–216 (2020)
15. Kacker, R.N., Lagergren, E.S., Filliben, J.J.: Taguchi's Orthogonal Arrays are Classical Designs of Experiments (1960)
16. Yuce, B.E., Nielsen, P.V., Wargocki, P.: The use of Taguchi, ANOVA, and GRA methods to optimize CFD analyses of ventilation performance in buildings. Build. Environ. **225**, 109587 (2022)
17. Hasanzadeh, R., Mojaver, P., Chitsaz, A., Mojaver, M., Jalili, M., Rosen, M.A.: Biomass and low-density polyethylene waste composites gasification: Orthogonal array design of Taguchi technique for analysis and optimization. Int. J. Hydrogen Energy **47**(67), 28819–28832 (2022)
18. Munawar, M.A., et al.: Investigation of functional, physical, mechanical and thermal properties of TiO2 embedded polyester hybrid composites: a design of experiment (DoE) study. Progress in Natural Science: Materials Int. **28**(3), 266–274 (2018)
19. Collin, L.R.D., Pamplona, E.O.: A utilização da função Perda de Taguchi na prática do Controle Estatístico de Processo. Escola Federal de Engenharia de Itajubá-IEM/DPR (1997)
20. Myers, R.H., Montgomery, D.C., Vining, G.G., Borror, C.M., Kowalski, S.M.: Response surface methodology: a retrospective and literature survey. J. Qual. Technol. **36**(1), 53–77 (2004)
21. Van, N.T.T., et al.: Cellulose from the banana stem: optimization of extraction by response surface methodology (RSM) and characterization. Heliyon, p. e11845 (2022)
22. Borrotti, M., Sambo, F., Mylona, K.: Multi-objective optimisation of split-plot designs. Econom Stat (2022)
23. Smucker, B.J., del Castillo, E., Rosenberger, J.L.: Model-robust designs for split-plot experiments. Comput. Stat. Data Anal. **56**(12), 4111–4121 (2012)
24. Mandal, B.N., Parsad, R., Dash, S.: Incomplete split-plot designs: construction and analysis. Stat Probab Lett **166**, 108869 (2020)
25. Jankovic, A., Chaudhary, G., Goia, F.: Designing the design of experiments (DOE)–an investigation on the influence of different factorial designs on the characterization of complex systems. Energy Build **250**, 111298 (2021)
26. Trojanowski, P., Filina-Dawidowicz, L.: Diagnostic and repair centers locating methodology for vehicles carrying sensitive cargo. Transportation Research Procedia **55**, 410–417 (2021). https://doi.org/10.1016/j.trpro.2021.07.004

27. Sousa, R.A., Varela, M.L.R., Alves, C., Machado, J.: Job shop schedules analysis in the context of industry 4.0. In: 2017 International Conference on Engineering, Technology and Innovation: Engineering, Technology and Innovation Management Beyond 2020: New Challenges, New Approaches, ICE/ITMC 2017, pp. 711–717 (2018). https://doi.org/10.1109/ICE.2017.8279955
28. Kaiser, S., Engell, S.: An integrated approach to fast model-based process design: Integrating superstructure optimization under uncertainties and optimal design of experiments. Chem Eng Sci. 118453 (2023)
29. Husar, J., Knapcikowa, L.: Possibilities of using augmented reality in warehouse management: a study. Acta Logistica **8**(2), 133–139. https://doi.org/10.22306/al.v8i2.212
30. Vieira, G.G., Varela, M.L.R., Putnik, G.D., Machado, J.M., Trojanowska, J.: Integrated platform for real-time control and production and productivity monitoring and analysis. Romanian Review Precision Mechanics, Optics and Mechatronics **50**, 119–127 (2016)
31. Arrais-Castro, A., Varela, M.L.R., Putnik, G.D., Ribeiro, R.A., Machado, J., Ferreira, L.: Collaborative framework for virtual organisation synthesis based on a dynamic multi-criteria decision model. Int. J. Computer Integrated Manuf. **31**(9), 857868. https://doi.org/10.1080/0951192X.2018.1447146

SOC Evaluation of Lithium-ion Battery Based on MFC Sensors

Lyu Yan[1](✉), Geng Liang[2], and Gao Jie[1]

[1] Faculty of Information Technology, Beijing University of Technology, Beijing 100124, China
lvyan@bjut.edu.cn

[2] Faculty of Materials and Manufacturing, Beijing University of Technology, Beijing 100124, China

Abstract. The impact of different dimensional parameters of the macro fiber composite (MFC) sensor on its response performance was investigated using finite element simulation software. The frequency domain analysis of performance parameters, such as the electromechanical coupling coefficient, center frequency, and resonant frequency characteristic relationship, was carried out using different simulation models. Based on the simulation results, a MFC sensor suitable for acoustic detection was designed and customized. Then an ultrasonic guided wave detection system for lithium-ion batteries using the optimized MFC sensor was established. And the intrinsic relationship between the state of charge (SOC) and guided wave propagation properties was experimentally investigated. The results showed that the time of flight, signal amplitude, and SOC exhibited an approximately linear distribution, indicating the feasibility and reliability of characterizing the SOC of lithium-ion batteries based on MFC sensors. The mapping relationship between the state of charge and the propagation characteristics provides a foundation for the nondestructive evaluation and quantitative estimation of the state characteristics of lithium-ion batteries.

Keywords: Lithium-ion batteries · Ultrasonic guided wave · State of charge · MFC sensors

1 Introduction

Lithium-ion batteries have gained widespread adoption in various industries such as new energy vehicles and aerospace due to their high specific energy, cycle life, operating voltage, and unique performance advantages [1, 2]. As technological advancements continue to propel the use of these batteries, ensuring their monitoring and safety has become increasingly critical. In recent years, several incidents of battery-related accidents have sparked concerns in society, making it crucial to develop a reliable and efficient approach to monitor the performance status of these batteries [3]. Therefore, the development of quick and precise detection methods for assessing the state of lithium-ion batteries is a key challenge that must be addressed within the battery industry and the new energy sector [4, 5].

A. Burduk et al. (Eds.): ISPEM 2023, LNNS 790, pp. 342–355, 2023.
https://doi.org/10.1007/978-3-031-45021-1_26

The accurate estimation of important state parameters, such as State of Charge (SOC) and State of Health (SOH), is crucial in ensuring the safety and performance of lithium-ion batteries. Despite several commonly used SOC measurement methods, such as ampere-time integration [6], open circuit voltage [7], Kalman filter [8], and neural network methods [9], these methods suffer from various uncertainties, resulting in large estimation errors and complex calculation processes. Similarly, the commonly used SOH estimation methods can be classified into three categories: model methods, data-driven methods, and fusion-type methods [10]. Among these, the fusion-type method has recently gained attention due to its high estimation accuracy and stable prediction results.

As the battery ages, the repetitive formation of a solid electrolyte interface film (SEI) internally leads to a reduction in charging capacity due to the consumption of Li-ions. During the charging and discharging process of lithium-ion batteries, the potential difference between the positive and negative electrodes drives the movement of lithium ions through the separator, resulting in changes in mechanical parameters such as density and elastic modulus of the electrodes [11]. The ultrasonic detection technology has emerged as a promising non-destructive testing method in the battery field.

The Macro Fiber Composite (MFC) is a highly performing actuator and sensor with a thin and flexible structure. It comprises piezoelectric fibers, epoxy resin, and forked finger electrodes. Specifically, the MFC is composed of a rectangular piezoelectric ceramic rod that is sandwiched between an adhesive, electrodes, and a polyimide film layer. The crossed electrode pattern is applied to add or output the voltage from the ribbon piezoelectric rod, making it highly sensitive. The MFC is hermetically sealed, making it durable and polarizable, thus making it suitable for use on flat surfaces as a sensing device. The epoxy resin covering the MFC surface serves a dual purpose: to protect the internal piezoelectric ceramic rods and provide flexibility. This flexibility is an essential feature of the MFC, as shown in Fig. 1. The MFC's thin and flexible nature makes it adaptable to various working conditions, such as curved surfaces and narrow spaces [12]. In this regard, MFC are selected as the ultrasonic excitation and reception sensor, which can be tightly arranged within individual cells of the battery module, expanding the application of ultrasonic detection method in lithium-ion battery testing.

Meanwhile, MFC sensors have proven to be a versatile actuator in various research applications, including cantilever beam damping, radar antennas, and vibration generation. Additionally, MFCs can be utilized as transducers to excite and receive acoustic waves. The combination of MFCs with non-destructive testing technology presents a promising solution to the limitations of traditional probes in complex working conditions. Many scholars have made significant contributions in the field of non-destructive testing (NDT) through the application of MFCs. For example, Gang et al. [13] attached two MFCs of type M2814-P1 to a 2 mm aluminum plate as ultrasonic signal excitation and receive probe, respectively, with a 200 mm distance between the two MFCs. A high voltage pulsed square wave signal was excited using the RAM5000. The results of the received signal indicate the visibility of the S0 mode of the guided wave in the range of 0.25 MHz to 0.45 MHz, and the A0 mode of the guided wave in the range of 0.2 MHz to 0.35 MHz. The MFC successfully excites and receives the guided wave signal, as confirmed by the dispersion curves derived from theoretical calculations for a 2 mm thick aluminum plate. Cui et al. [14] have introduced a novel method for measuring and detecting axial cracks in cylindrical structures, using the Macro Fiber Composite (MFC) sensor in pipe inspection for the first time. The MFC serves as both an excitation sensor to generate a torsional wave for crack detection and a receiver for the sensor, enabling the detection and measurement of crack size. Numerical simulations and experimental studies have verified the feasibility of this approach. In another study, Tiwari et al. [15] investigated the applicability of MFC in structural health monitoring and NDT. By measuring the resonant and anti-resonant frequencies of the M2814-P1 MFC, the team obtained reference values for frequency selection when using the MFC for ultrasonic applications in NDT. Furthermore, the team used a scanning laser vibration to measure the three-dimensional spatial displacement values generated by the MFC, confirming its ability to generate and receive S0/A0 guided wave modes in the low-frequency ultrasonic band range. Although domestic scholars are relatively new to research on MFC-based guided waves, they have also made significant progress in this field.

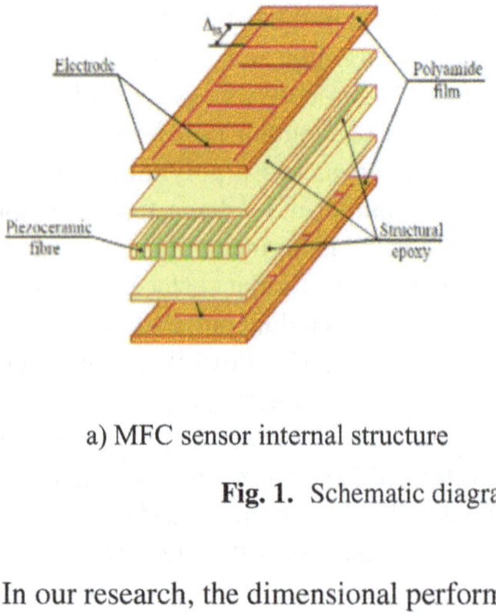

a) MFC sensor internal structure

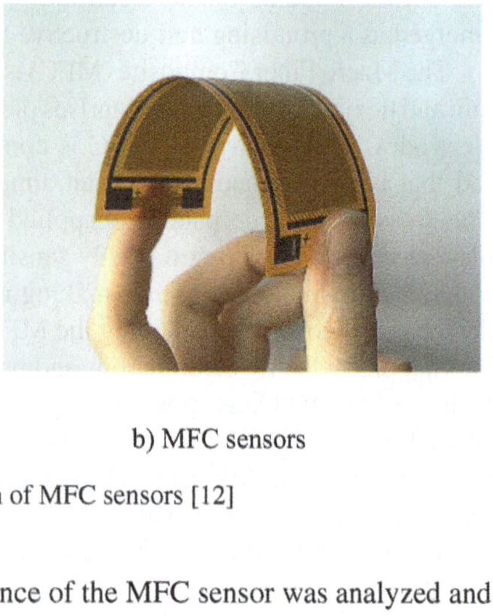

b) MFC sensors

Fig. 1. Schematic diagram of MFC sensors [12]

In our research, the dimensional performance of the MFC sensor was analyzed and optimized using simulate method. Subsequently, an experimental system is constructed using the optimized MFC sensor for ultrasonic guided wave detection of lithium-ion batteries. Moreover, the guided wave propagation characteristics of Li-ion batteries are investigated under different SOC. The mapping relationship between SOC and acoustic signals was analyzed in detailed.

2 MFC Sensor Simulation Optimization

2.1 Simulation Model Building and Reliability Analysis

The finite element simulation model of piezoelectric fiber structure was established using COMSOL software, which involved the actuating layer comprising intermediate piezoelectric fibers and an epoxy resin composite. And the electrostatic part and solid mechanics part of the simulation software for multi-physics field coupling were used. The excitation was provided by applying an electrical potential to the corresponding position of the forked finger electrodes through the electrostatic field. The inverse piezoelectric effect of the piezoelectric material converted the electrical signal input into a mechanical energy output, resulting in the sensor producing an actuating effect. The flow chart of model establishment is shown in Fig. 2.

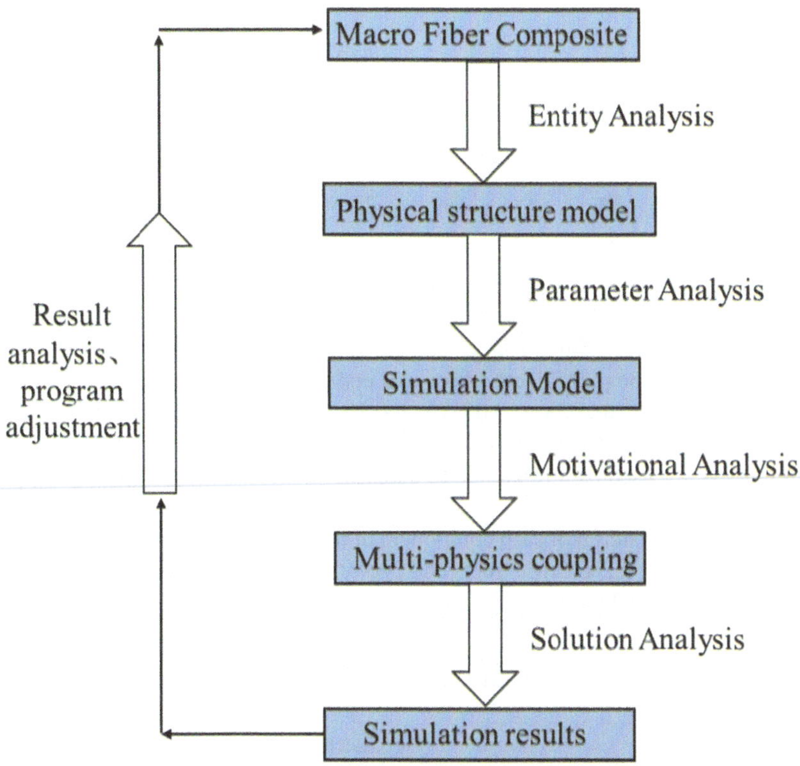

Fig. 2. MFC sensors simulation model flow chart

A 3D finite element simulation model of the MFC-2814-P2 piezoelectric fiber actuator layer was constructed in accordance with the structural parameters of the existing MFC model, as illustrated in Fig. 3. For clarity, a local area of the model was enlarged and presented in Fig. 3b). The simulation model consists of various components, including a geometric model, material properties, solid mechanics, electrostatic field, and multi-physics field coupling of piezoelectric effects, as well as meshing, study type, and analysis of results, in order to provide a comprehensive understanding of the structural relationship of the MFC.

The actuation area of the MFC-2814-P2 simulation model is approximately $28 \times 14\ mm^2$, and comprises of 35 piezoelectric fibers. The gaps between the adjacent piezoelectric fibers are filled with epoxy resin, which provides bonding and flexibility to

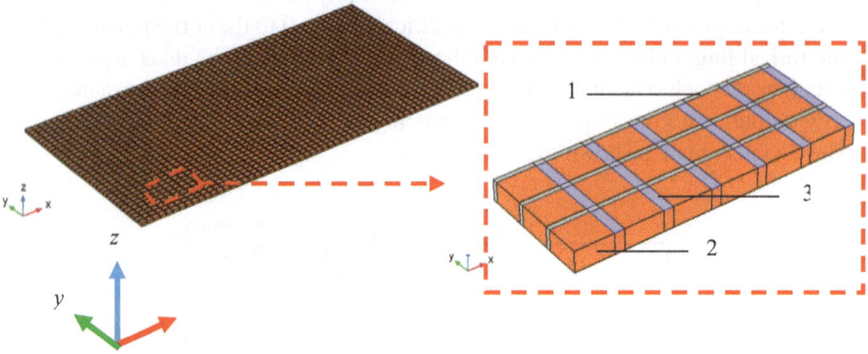

1. Epoxy 2. Piezoelectric fibers 3. Negative electrode

Fig. 3. 3D simulation model of MFC-2814-P2

the structure. The electrodes are arranged perpendicular to the direction of the fibers and epoxy resin, and they are parallel to the upper and lower surfaces of the actuator layer. Each side of the actuator layer has the same electrode, as illustrated in Fig. 3. The dimensions of the piezoelectric fiber and epoxy resin are reported to be around 28 $\times$ 0.355 $\times$ 0.18 mm^3 and 28 $\times$ 0.05125 $\times$ 0.18 mm^3, respectively [16]. The interval between two adjacent electrodes is 0.5 mm, and the electrode width is 0.1 mm. These measurements are used as a reference to construct a three-dimensional MFC-2814-P2 simulation model.

In defining the material of each component of the MFC piezoelectric fiber composite, we considered PZT piezoelectric fibers and epoxy resin as the primary components. The corresponding material parameters outlined in Tables 1 and 2.

Table 1. Material parameters of PZT-5A

Elastic constants, GPa					Piezoelectric constant,C/m^2			Dielectric constant, $\varepsilon/\varepsilon_0$	
C_{11}	C_{12}	C_{13}	C_{33}	C_{44}	e_{15}	e_{31}	e_{33}	ε_{11}	ε_{11}
120.3	75.2	75.1	110.9	21.1	12.29	-5.35	15.78	919.10	826.60

Table 2. Material parameters of polymer phase

Materials	Density ρ/kg/m^3	Poisson's ratio ν	Young's modulus E/Pa
Piezoelectric fibers	1960	0.36	3.5×10^9

To demonstrate the validity of the developed finite element simulation model, simulations were conducted to investigate the impedance characteristics of the model in the frequency domain. In order to set up the model, the calculation equation type of each parent module was modified to the frequency domain. The fixed constraints within the solid mechanics module were removed and the terminal voltage condition was set to 1V as the excitation. The solver was set to the frequency domain solution, with frequency sweeps ranging from 10 kHz to 100 kHz in 0.1 kHz increments. Using the "1/abs(es.Y11)" command in the COMSOL finite element simulation software, the impedance value of the model was calculated. The simulated results were compared with the impedance analyzer results in the sweep range to evaluate the consistency between the characteristic frequencies. The comparison between the simulated calculations and the impedance analyzer results is presented in Fig. 4. It depicts the impedance characteristics of the MFC-2814-P2 in the frequency domain, where the red line represents the simulation calculation results, and the black line represents the actual impedance analyzer measurement data. The frequency sweep range was set from 10 kHz100 kHz, during which the MFC displayed four primary vibration modes named T1, T2, T3 and T4. The resonant frequencies corresponding to the simulation and actual measurement results in each mode are denoted as fti and Fti (i = 1,2,3,4), respectively. Table 3 presents the statistical results of the resonant frequencies for each mode. The comparison between the simulated and measured results reveals a good agreement in the resonant frequency values for each mode, with a maximum relative error of 6.8%, which is controlled within 7%. The error can be attributed to the fact that the three-dimensional entity of the copper electrode was not included in the modeling of the COMSOL finite element simulation software, which affected the uniformity of the applied electric field.

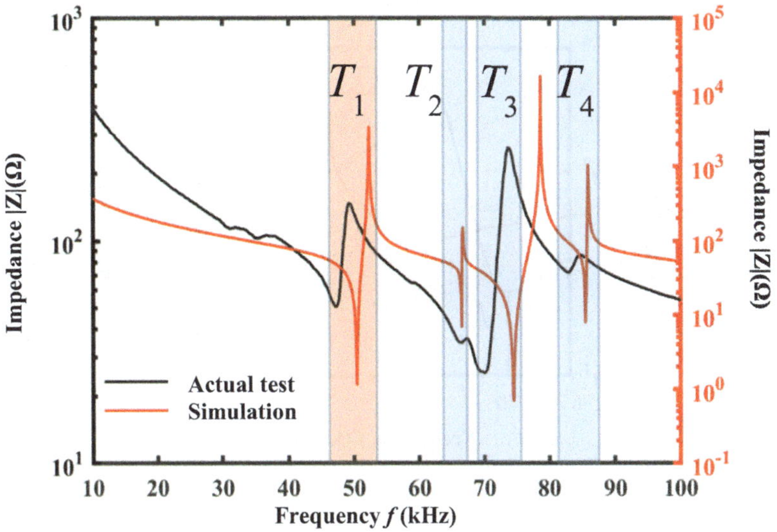

a) Simulated and measured impedance curves

Fig. 4. Simulation and experimental impedance curves of MFC sensor

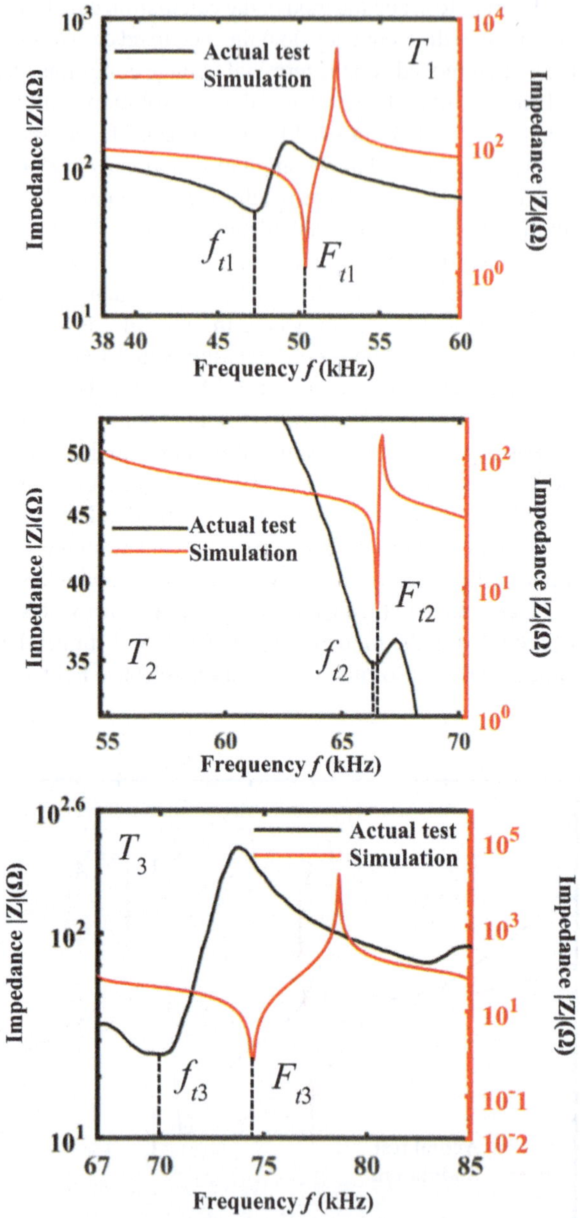

Fig. 4. *(continued)*

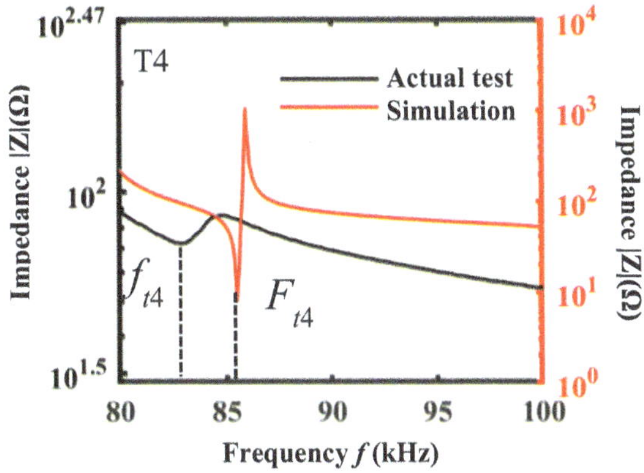

Fig. 4. (*continued*)

Table 3. Comparison of the resonant frequencies T_1-T_4 vibration modes

	T_1	T_2	T_3	T_4
Simulation	50.4 kHz	66.5 kHz	74.5 kHz	85.5 kHz
Experimental	47.2 kHz	66.4 kHz	70.0 kHz	82.9 kHz

2.2 Effect of Dimensional Parameters on MFC Performance

2.2.1 Length Variation

To investigate the effect of sensor length on response frequency, we constructed frequency domain simulation models for a range of single piezoelectric fibers with varying lengths. The lengths of the fibers were adjusted in 2 mm increments within the range of 2 mm to 28 mm, and each model contained both piezoelectric and epoxy phases. To streamline the simulation process, we utilized the parametric scanning module in COMSOL to build and calculate MFC frequency domain simulation models with the same structure but different lengths in a seamless manner. Meanwhile, the fundamental frequency vibration modes of each different length was selected. The corresponding admittance curves are depicted in Fig. 5. As we can show, the resonant frequency of the MFC gradually decreases when the fiber length increasing. And the response frequency eventually converges to 50 kHz. The resonant and anti-resonant frequencies of the f1 at different lengths were extracted, and the corresponding electromechanical coupling coefficients were calculated. Figure 6 presents these characteristic coefficients for varying MFC lengths, indicating that the electromechanical conversion efficiency of the sensor remains constant beyond a fiber length of 6 mm.

To evaluate the energy values generated by the sensor at varying lengths, we extracted and analyzed the displacements that can be produced by actuation along the x-fiber direction for different models (l = 2 to l = 32, 2 mm steps). The results are presented in Fig. 6. As illustrated in the Fig. 6, the absolute value of displacement gradually increases

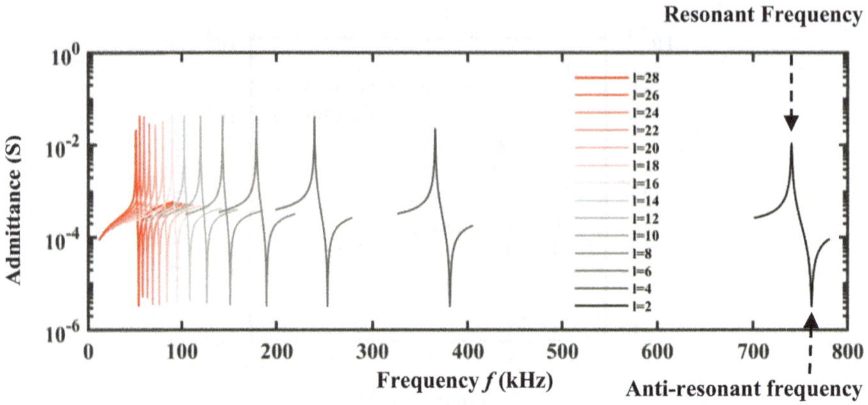

Fig. 5. Admittance curves of different lengths MFC sensors

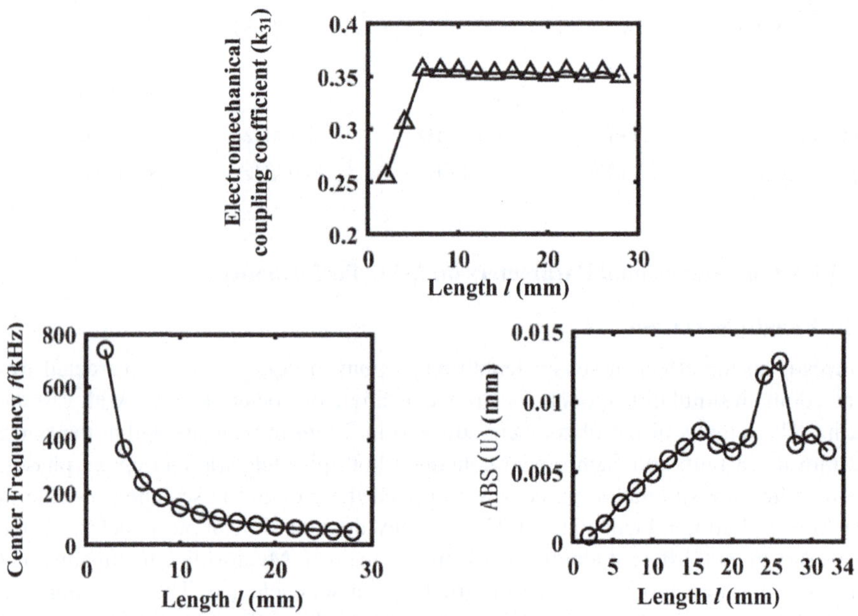

Fig. 6. Variation of MFC performance parameters for different lengths

with length, with the largest displacement occurring at 26 mm within the simulated range. This indicates that the MFC at a length of 26 mm has a stronger vibration capability in the fiber direction compared to other sizes, and the larger displacement can reflect the vibration energy of the MFC sensor to a certain extent, thereby enhancing the amplitude of the excitation signal.

2.2.2 Width Change

Based on the above simulation, the constituent unit was kept at a fixed length of 26 mm, the effect of MFC width on response frequency was investigated. And the width b of the MFC was varied by adjusting the value of the number of constituent units. The b was varied from 4 to 32 in increments of 4. To study the effect of MFC width on sensor performance, the frequency sweep range was adjusted to 20 kHz 90 kHz, increasing simulation efficiency.

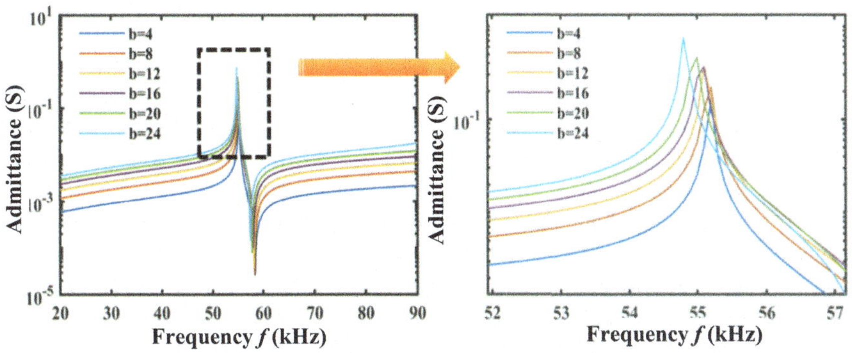

a) Admittance curve of MFC sensors for different *b*

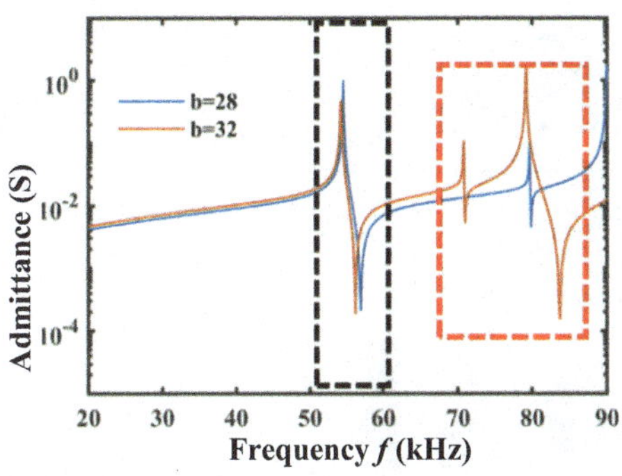

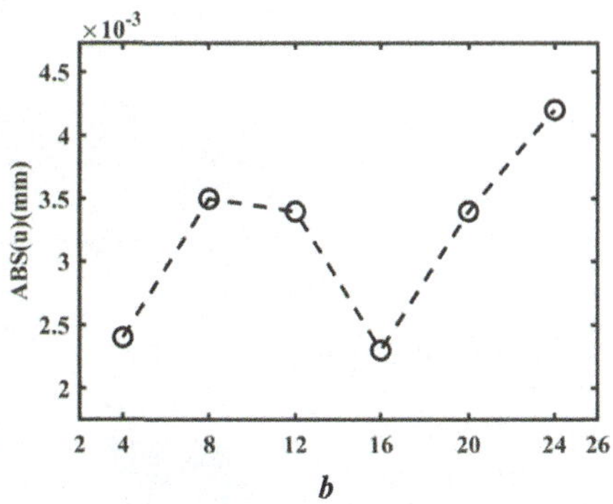

b) Admittance curve of MFC with *b*=28-32 c) ABS(u) of MFC with *b*=4-24

Fig. 7. Admittance properties of MFC sensors with different widths b

As we can show, the resonant frequency of the MFC sensor decreases with the increase in width when the b is range from 4 to 24. Although the width impact on the resonant frequency of the MFC sensor, the overall effect is minor, and the resonant frequencies of the MFC simulation models with varying widths fall within an acceptable range. However, when the number of constituent cells b in the simulation model is 28 or 32, other modes show obvious peaks in the frequency domain, as depicted in the red area in Fig. 7b). The occurrence of mixed resonance peaks in the low-frequency range reduces the sensor's consistency and tends to affect its normal performance. Therefore, the sensor size with only single vibration mode in the low-frequency range is chosen to ensure the sensor's consistency as much as possible. When the value of b is between 4 and 24, the elongation of the sensor along the length (ABS(u))is highest when b = 24.

3 Experimental Results and Comparison

To evaluate the MFC sensor's feasibility for detecting lithium-ion battery state parameters, we selected a commercial lithium-ion soft pack battery (model: TS-MT-IP-mini4), which is commonly used in Ipads and manufactured by Guangdong Pinsent Technologies Co. The optimized MFC sensor was employed for ultrasonic guided wave signal excitation and reception. An experimental system consisting of function/signal generator, oscilloscope, Neware CT-4008T-5V6A was built, as shown in Fig. 8. The commercial Ipad Li-ion battery size is 145 × 100 × 2.2 mm with a capacity of around 5124 mAh. The batteries were charged and discharged in nominal capacity between 3.0 V-4.35 V. By using the ultrasonic guided wave detection system, we conducted ultrasonic testing experiments on commercial lithium-ion batteries during discharge. Specifically, the acoustic signals were collected during the discharge process, and the acquired time domain signals are processed by band-pass filter and Hilbert transform.

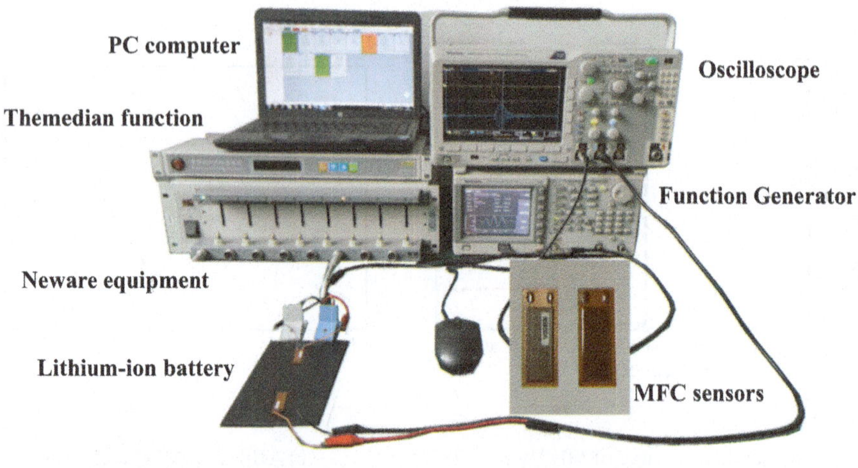

Fig. 8. Ultrasonic guided wave detection system for Lithium-ion battery

The collected acoustic signals during the discharge process are plotted after band-pass filtering, as shown in Fig. 9. The reference signal at SOC = 0% marked in green. The color of remaining signals changes from red to black, which representing the discharge process with gradually decreasing SOC. Meanwhile, we process the acquired acoustic signals and the original excitation signals using the inter-correlation algorithm, to obtain the TOF of acoustic signals at different SOC. The mapping relationship between SOC and acoustic characteristic parameters (signal amplitude and TOF) during the discharge

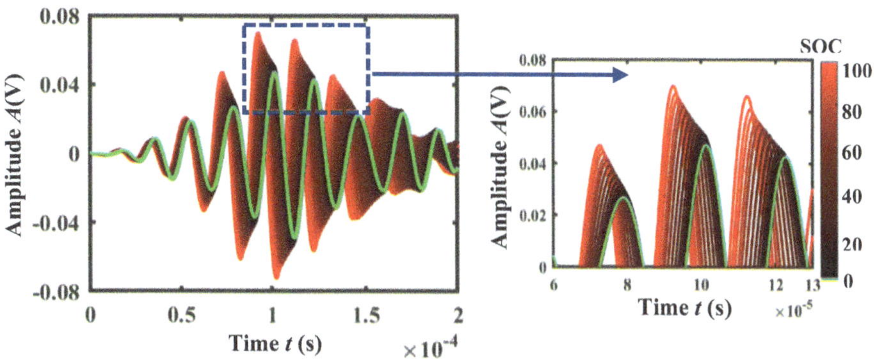

Fig. 9. Guided wave signal characteristics during discharge

process are established, as shown in Fig. 10. In addition, as the SOC of the lithium-ion battery decreases during the discharge process, the amplitude of the ultrasonic guided wave signal decreases gradually. Moreover, as the SOC increases, the time of flight (TOF) shows a decreasing trend. These observations are mainly attributed to the changes in the mechanical properties of the electrodes. Furthermore, it proved that the optimized MFC sensor in the acoustic characterize the state parameters of Li-ion batteries is feasible.

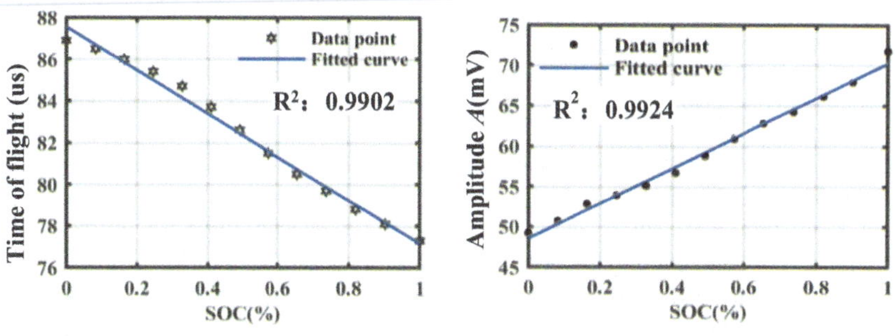

a) Time of flight

b) Signal Amplitude

Fig. 10. Guided wave behavior of the cell during discharge

4 Conclusion

This research investigates the effect of the dimensional parameters of the MFC sensor on response performance through finite element simulation software. Furthermore, an ultrasonic guided wave detection system for lithium-ion batteries is established using the optimized MFC sensor, the intrinsic connection between SOC and acoustic propagation characteristics was revealed experimentally. The main findings of this study can be summarized as follows:

(1) The variation of the MFC sensor resonant frequency at different lengths was explored, as well as the influence of the length on performance parameters (such as electromechanical coupling coefficient, center frequency, and ABS(u)). The results indicate that the MFC sensor has better performance with 26 mm length.
(2) The effect of MFC sensor width variation on its performance was analyzed. The results show that when the width cells b is 24 (9.75 mm), the sensor has only f1 fundamental frequency vibration mode in the low frequency range (20 kHz-90 kHz). And there is no other mixed resonance peaks, and the ABS(u) value is relatively maximum. Meanwhile, we performed impedance matching analysis on the customized MFC sensor, and the measured results are consistent with the simulation conclusions.
(3) The experimental results show that, there is a strong mapping relationship between the acoustic propagation characteristics and SOC of the Li-ion battery. Specifically, as the SOC decreases, the TOF will be gradually increase, and the signal amplitude becomes smaller. It has been proved the ability of the ultrasonic guided waves detection technology using MFC sensors to characterize the SOC of Li-ion batteries. Furthermore, the intrinsic relationship between acoustic parameters and the state of charge can help the non-destructive quantification of the state parameters of the lithium-ion battery.

References

1. Yan, J.D.: Current status and development analysis of lithium-ion batteries. Acta Aeronautica et Astronautica Sinica, **35**(10), 2767–2775 (2014)
2. Yang, B., Cui, N., Wang, M.: Internal short circuit fault diagnosis for lithiumion battery based on voltage and temperature. In: 2019 3rd Conference on Vehicle Control and Intelligence (CVCI) (2019)
3. Li, Z., Huang, J., Liaw, B.Y., et al.: On state-of-charge determination for lithium-ion batteries. J. Power. Sources **348**, 281–301 (2017)
4. Nam, K.W., Bak, S.M., Hu, E., et al.: Combining in situ synchrotron X-ray diffraction and absorption techniques with transmission electron microscopy to study the origin of thermal instability in overcharged cathode materials for lithium-ion batteries. Adv. Func. Mater. **23**(8), 1047–1063 (2013)
5. Waag, W., Käbitz, S., Sauer, D.U.: Experimental investigation of the lithium-ion battery impedance characteristic at various conditions and aging states and its influence on the application. Appl. Energy **102**, 885–897 (2013)
6. Lei, Z.L., Liu, T., Sun, X.M., et al.: Extended state observer assisted coulomb counting method for battery state of charge estimation. Int. J. Energy Res. **45**(7), 3157–3169 (2020)
7. Roscher, M.A., Sauer, D.U.: Dynamic electric behavior and open-circuit-voltage modeling of LiFePO4-based lithium ion secondary batteries. J. Power. Sources **196**(1), 331–336 (2011)
8. Zhang, S., Zhang, C., Jiang, S., et al.: A comparative study of different adaptive extended/unscented Kalman filters for lithium-ion battery state-of-charge estimation. Energy **246**, 123423 (2022)
9. Rivera-Barrera, J.P., Muñoz-Galeano, N., Sarmiento-Maldonado, H.O.: SoC estimation for lithium-ion batteries: review and future challenges. Electronics **6**, 102 (2017)
10. Jiang, J., Gao, Y., Zhang, C., Wang, Y., Zhang, W., Liu, S.: Online diagnostic method for health status of lithium-ion battery in electric vehicle. J. Mechanical Eng. **55**(20), 60–72 (2019)

11. Hsieh, A.G., et al.: Electrochemical-acoustic time of flight: in operando correlation of physical dynamics with battery charge and health. Energy Environ. Sci. **8**(5), 1569–1577 (2015)
12. Wilkie, W.K., Bryant, R.G., High, J.W., et al.: Low-cost piezocomposite actuator for structural control applications. Smart structures and materials 2000: industrial and commercial applications of smart structures technologies. International Society for Optics and Photonics 3991, 323–334 (2000)
13. Ren, G., Jhang, K.Y.: Application of macrofiber composite for smart transducer of lamb wave inspection. Adv. Mater. Sci. Eng. **2013**, 281575 (2013)
14. Cui, L., Liu, Y., Soh, C.K.: Health monitoring of cylindrical structures using torsional wave generated by piezoelectric macro-fiber composite. Health Monitoring of Structural & Biological Systems 7984(10), (2011)
15. Tiwari, K.A.: Investigation of 3D displacement characteristics of macro fiber composite transducer (MFC-P1). Materiali in Tehnologije **52**(2), 235 (2018)
16. Bilgen, O.: Aerodynamic and Electromechanical Design, Modeling and Implementation Of Piezocomposite Airfoils. Virginia Tech (2010)

Indicators of Hierarchical Structure Model of Supporting the Production Management Process – A Framework

Artur Meller[1](✉), Mariusz Piechowski[2], and Arkadiusz Gola[3]

[1] Faculty of Mechanical Engineering, Poznan University of Technology, Plac Marii Skłodowskiej-Curie 5, 60-965 Poznan, Poland
artur.meller@doctorate.put.poznan.pl
[2] IQ-Workshops, ul. Koronna 3/31A, 60-652 Poznan, Poland
mariusz.piechowski@iq-workshops.pl
[3] Department of Production Computerisation and Robotisation, Lublin University of Technology, ul. Nadbystrzycka 36, 20-619 Lublin, Poland
a.gola@pollub.pl

Abstract. Increasing customer requirements oblige producers to improve methods of managing the production process. One of the directions of improvement is the digitization of production management processes. The digitization process significantly affects the way of monitoring and controlling the production process by optimizing its resources thanks to the use of effective methods of collecting and processing information. One of the most important tools for monitoring the course of the process is the analysis of Key Performance Indicators (KPI). They are used to measure and evaluate the strategically, operational and tactical parameters that characterize the functioning of the enterprise. This article presents an novel model of the hierarchical structure of indicators, the task of which is to support the managerial staff in the decision-making process. The model allows to optimize and control of the production process at various decision levels based on reliable information resulting from its parameters.

Keywords: Key Performance Indicators · KPI · Hierarchical Model · Decision Support · Production Management

1 Introduction

The progressing global economic crisis related to the pandemic and the war in Ukraine has resulted in radical changes in the long-term policy of companies related to ensuring competitiveness on the market [1]. As a result, manufacturing companies with complex business processes are subject to even more stringent restrictions than ever before [2]. Many affected companies are changing their strategic goals to ensure shorter delivery times and increased product volatility, and expand their expansion into new markets [3]. New economic and political conditions as well as changing customer requirements force the improvement of production process management methods and adaptation of production systems to greater reliability, flexibility and productivity [4].

A. Burduk et al. (Eds.): ISPEM 2023, LNNS 790, pp. 356–367, 2023.
https://doi.org/10.1007/978-3-031-45021-1_27

The article presents the concept of a hierarchical model supporting decision-making processes based on key performance indicators. The first part of the article describes the assumptions of such a system and the most important features of KPIs. The hierarchical model of indicators for individual levels and areas of management as well as the course of action for the selection, implementation and improvement of a system based on this type of indicators are also described. Based on the developed model, the conducted analysis was described, which enabled the selection of key performance indicators of the company. The developed guidelines and design assumptions of the IT system supporting decision-making processes based on KPIs enable: automation of data processing procedures, optimization of the goal supervision function at individual decision-making levels, improvement of the reporting system and calculation algorithms.

The main assumption is to develop a tool for effective decision making that enables optimization and control of the production process at various decision levels based on reliable information resulting from its parameters.

2 The Importance of KPI Indicators in Production Processes

Modern production systems integrate advanced mechanical, electronic and IT solutions [5, 6]. The complexity of these systems means that in order to fully understand the control process at all decision-making levels and to enable the proper assessment of its condition, knowledge from all engineering disciplines should be consolidated [7]. When making decisions, it is crucial to obtain useful information about the state of the process, on the basis of which we can effectively eliminate or reduce the causes of disturbances [8]. Obtaining such information can be done by monitoring the state of the process and its parameters, i.e. quality, efficiency, effectiveness, time consumption [9].

One of the tools supporting decision-making processes are KPIs. Indicators enable monitoring and assessment of the process status in many production and auxiliary areas, as well as at many decision-making levels [10]. The inference model based on KPIs should contain key information for the company's decision-making processes. Information provided to managers, engineers and operators must be provided in a legible, transparent and understandable manner [11]. Indicators, wherever possible and reasonable, should present both absolute and relative values [12]. The absolute values themselves do not indicate the process state assessment scale, and the percentages cannot be aggregated. Indicators at the level of strategic goals should be result indicators, i.e. they should reflect the intended effects [13]. On the other hand, at the operational level, in addition to result indicators, output indicators should also be defined, which will indicate activities leading to the achievement of goals [14]. For such defined goals, it is necessary to identify appropriate indicators supporting the decision-making process.

3 Features of Performance Indicators

Efficiency indicators are widely used in enterprises, but they are usually used in a limited way, i.e. they either relate to selected aspects of management, e.g. maintenance, or only selected indicators are used, e.g. OEE [15]. From among dozens of KPIs, it is possible to choose those that will allow to design a system supporting the decision-making process in the entire enterprise [16]. Thanks to such a system, it is possible to verify and evaluate: assumptions, goals, activities [17]. Effective decision-making depends on the features that should characterize the indicators. The effectiveness of this tool in the decision-making process depends on these features.

One of such features is the scope of the indicator's impact, i.e. the indicator's significance. Each management model has its key (strategic) goals that affect its effectiveness, so the team must carefully analyze the current state of the company, its limitations and possibilities in order to define a clearly defined main goal to which the entire company will be aiming at all decision-making levels [18]. In each model, the most important will be other indicators related to:

a. eliminating waste: identification of losses, cost analysis, efficiency of production equipment,
b. implementation of the organization's goals, personnel management, and work efficiency,
c. product and service quality, employee improvement process, process efficiency, risk assessment.

The second feature is the precision and accuracy of mapping the process state, i.e. the reliability of the indicator. The indicators should make it possible to assess the state of the process at different decision-making levels:

a) strategic, enabling key decisions to be made for the entire enterprise, e.g. PMP (Percent Maintenance Planned) indicator - reflecting the effectiveness of preventive actions,
b) tactical, where decisions made are important for departments (branches) or production lines, e.g. FPY (First Past Yeld) indicator - reflecting the effectiveness of the production process,
c) operational, where decisions are made by the operator on a given position, e.g. the effective time utilization index (ECP) - reflecting the normative working time at a production position.

Another feature is the completeness of the index components. Often the indicator is representative at many decision-making levels, e.g. OEE indicator, the value of which may reflect the condition of: machines, groups of machines, as well as the entire production process. In this case, one has to face many problems, e.g.:

a) How to aggregate (calculate) the indicator value: sum or average the value?
b) How to interpret the value of the indicator and its constituent parameters, ie: should the technological and maintenance losses be taken into account together or separately in the performance parameter?
c) How to choose key indicators for a given area? Is the list of indicators in the machining department the same as the list of indicators in the assembly department?

Another characteristic feature is the influence on the moment of making a decision, i.e. the timeliness of the indicator value. This feature is also important in the decision-making process, i.e. the assessment of the process status on the basis of historical indicator values (from previous weeks, days and even hours) becomes useless. Monitoring the value of an indicator in real time requires its digitization - a multidimensional combination of the real and virtual world, i.e. the physical parameters of the indicator (e.g. count, temperature, time, availability) and the IT system (devices, meters, algorithms, software).

4 Design Assumptions

The proper selection of indicators requires proper preparation. In the first step, a thorough analysis of the current state of the process should be carried out in relation to the defined goals in line with the adopted management strategy. To conduct such an analysis, a team should be appointed, whose members will represent all areas of the company. The analysis should identify:

- areas significantly influencing strategic goals (e.g. efficiency of the production line),
- key area characteristics (e.g. machine availability, changeover times),
- features for the indicated characteristics (e.g. working time, number of shortages),
- components of features (e.g. failure time, tool calibration time),
- physical values of features (e.g. time to diagnose the cause of a failure, time to regenerate/purchase a tool) along with historical data that will enable the determination of the limit values of the indicator.

After the analysis, its scope should be narrowed down to the most important aspects: areas, characteristics and features, e.g. by using the Pareto-Lorenz diagram. The conducted analysis will enable the effective selection of indicators. The transfer of the company's mission and strategy into a set of data and values, using KPIs, provides an accessible and transparent opportunity to create and provide a framework for the strategy measurement and management system. In an era where success in running a business is often determined by access to information, the ability to obtain, process, interpret and draw appropriate conclusions based on it, resulting in the right decision made by managers, methods of selecting and implementing performance indicators come with help. There are many methods for selecting and implementing performance indicators. The choice of method should also be adjusted to the strategy and goals of the enterprise, taking into account its current problems. The most effective methods of implementing and maintaining the system of selecting indicators include:

a. The method of "winning KPIs" - associated with critical success factors (CSFs) that determine the competitive advantage of a given enterprise. In principle, each KPI is non-financial and subject to cyclical measurement [19].
b. Balanced Scorecard (BSC) method - characterized by maintaining a balance between measures of effectiveness relating to the description of past events and measures used to forecast future events, as well as between measures relating to internal processes implemented in the organization, and those related to the external measurement of customer satisfaction [20].
c. The Hoshin Kanri method - allows to build relationships between the established strategy and business execution, based on the identified goals. It includes vertical and horizontal adaptation of communication, activities and functions in the organization to its strictly defined and defined vision, through the involvement of all employees of

the company, at every level, from top management to direct production employees. A systematic approach to implementing the company's vision uses the Plan-Do-Check-Act Cycle (PDCA) for Continuous Improvement as well as the "catchball" process [21, 22].

5 Model and Concept of the System Implementation

The developed structure of the indicator management model reflects the hierarchical structure of management levels. Each node of the structure can be described in an object-oriented way, assigning it parameters and attributes relevant to each management level as well as features related to the areas of the enterprise. Feature attributes will allow to create points that aggregate information at individual levels and areas of management. Such a structure makes it possible to link indicators with object attributes and verify the impact of indicators on the defined management objectives. An additional advantage of such a solution is the detection of correlations between individual parameters of indicators. The developed model is shown in Fig. 1.

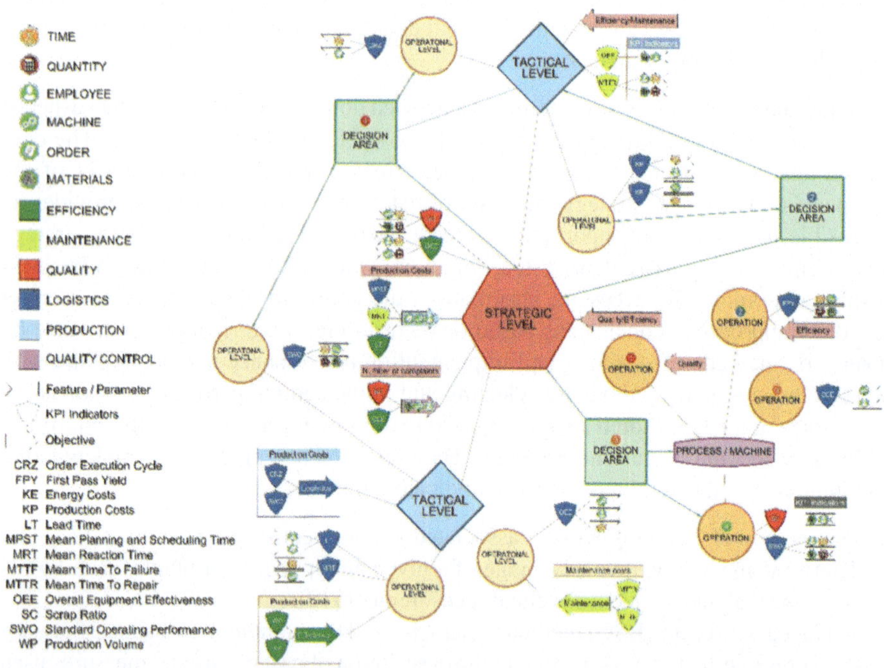

Fig. 1. Structure of the hierarchical model of indicator management (Own study).

The implementation of a system based on indicators is a long process and therefore requires the development of a plan. The plan should be divided into stages with defined goals and objectives. Each stage should have a set time frame and milestones. Figure 2 shows the system implementation procedure.

The criteria for selecting indicators should also take into account methodological correctness, i.e.:

a. the number of indicators should not be too large (a system based on a smaller number of indicators is considered more effective),
b. indicators should not be directly dependent on each other, i.e. a high value of one indicator is the result of a low value of another, e.g. an increase in production efficiency in relation to the quality of products,
c. indicators should have a direct impact on the defined goals, e.g. an increase in the value of the OEE indicator translates into an increase in sales.

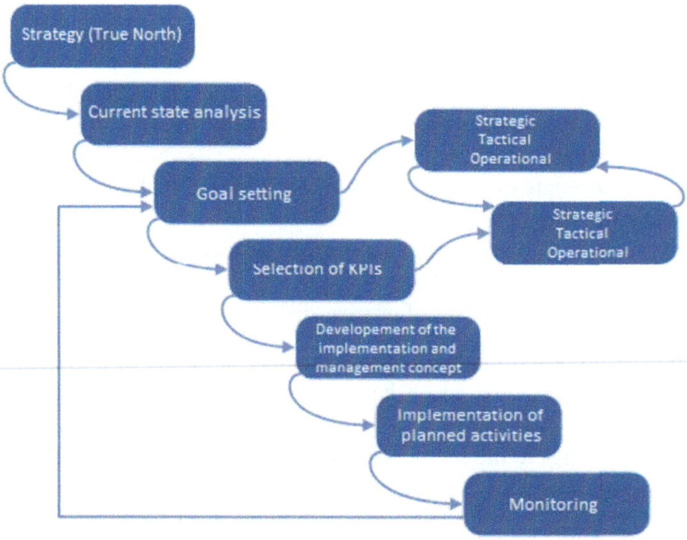

Fig. 2. Stages of system's implementation (Own study).

Each indicator should be assigned an engineer (manager, operator) responsible for its verification, optimization, periodic reviews, historical data control and correction of reporting algorithms. For such selected indicators, metrics should be developed including a description of:

a. the scope and level of impact of the indicator on the achievement of the defined strategic, tactical and operational goals,
b. accuracy and precision of mapping the state of the process at individual decision-making levels and areas of influence,
c. description of the individual components of the indicator influencing its effectiveness by defining the assessment scale as well as base and target values,
d. precise description of the indicator type: its name, measurement unit and data sources at individual indicator monitoring points.

Such a record is the basis for the development of a position manual, especially for operators making decisions at the operational level.

If the analysis of the current state of the process has been performed and appropriate indicators have been selected on its basis, it is still necessary to build an IT system that enables the management and improvement of the entire structure of indicators covering selected decision-making positions. Managing indicators without an IT system, as experience shows, is in 90% doomed to failure. The IT system should contain algorithms for calculating indicators (including methods of aggregation and propagation of values to higher decision levels), a notification system (as well as warning against critical values), clear and precise visualization of the description and indicator indicator and be available at every decision-maker position. It should also include procedures for verifying the compliance of the indicator with the design assumptions and ensuring the validation of the indicator structure with the results of the current state analysis. The decision model based on KPIs enables:

a. making decisions based on real and up-to-date information about the state of the process, without the need to interpret raw data and use advanced inference methods,
b. assessment of the state of the process based on the processing of data from various sources in combination with expert knowledge,
c. assessment of the state of the process at many decision-making levels and in many areas of the company's operation,
d. easy evaluation of the process status, especially by operators at the lowest decision-making level,
e. easy identification of the causes of process disturbances.

Based on the proposed methodology of conduct, a model of a system supporting decision-making processes based on key performance indicators was designed. The research was carried out at Fabryka Armatur "Swarzędz" sp. z o.o (FAS) in Rabowice (Poland), which belongs to the sector of small and medium-sized enterprises operating in the metal industry. As a leader in the consumption of brass in Poland, the company has modern technologies in the field of hot forming, heat treatment and machining. The extensive machinery park includes, among others hydraulic presses for waste-free die forging, or specialized transfer machines for the subtractive processing of tee-type products. This allows for the fulfillment of orders for the demanding automotive market (steel rings of the gearbox synchronizer) and for water and heat metering (water meters, heat meters), based on the stringent requirements for products in contact with drinking water.

The vision of digital development, but most of all the adoption of an appropriate business strategy, has become the starting point for selecting key performance indicators. For this reason, it was important to establish and verify the strategic direction in which the analyzed industrial plant is going to identify, designate and define its strategic areas. For this purpose, an in-depth SWOT analysis (S - Strengths, W - Weaknesses, O - Opportunities, T - Threats) of the enterprise was used, where individual process managers in the company, mostly managers of individual organizational units, identified strengths and weaknesses, as well as opportunities and threats to the organization. Subsequently, the surveyed group was asked to evaluate each of them individually in each group of factors on a scale of 1–4, where 1 is an irrelevant factor, and 4 is the most important factor, which allowed us to draw an average for each factor classifying the company's capabilities. Thus, the following results were obtained: S, 3.30; W - 2.88; O - 3.16; T -

3.06. Their interpretation indicates the advantage of strengths and opportunities, which in turn predisposes to define and adopt the company's strategy as aggressive (maxi-maxi). It consists in the maximum use of the synergy effect related to the strengths of the organization, as well as the opportunities arising from its environment.

Having established the direction in which the company is going, the next step was to define the areas essential for the production department, which was undertaken by a team of managers and specialists representing individual departments of the company, and where during the work carried out, the organization, performance, maintenance and quality were specified. For the areas selected in this way, the main context was described, goals were set, and key activities or stages in achieving the goal were defined. Quantitative and qualitative requirements were created to define the target state and actions - what exactly is to be carried out to achieve the goal. Target performance indicators were proposed, and a RASCI (Responsible, Accountable, Support, Consulted, Informed) matrix was created - defining the scope of responsibility and the role that individual people are to play in achieving the goal. Verification whether the goal was correctly set was performed using the SMARTER method (Specific, Measurable, Attainable, Relevant, Time-bound, Exciting, Recorded).

In the next step, the work of the implementation team was based on the use of elements of the Hoshin Kanri concept, i.e. a method of strategic management based on running a company in a strictly defined and agreed direction, the so-called True North, which was previously defined after performing the SWOT analysis. One of its elements is the organizational improvement template, the so-called X-matrix. On the basis of defining the strength of the influence of individual matrix components on each other (directly, indirectly, no interaction), it was possible to determine the correlation between them. The obtained image allowed for a precise verification and determination of irrelevant relationships, and thus omitting them at a later stage of the work, and, above all, for determining those critically important, resulting in the need to subject them to in-depth analysis, aimed at determining potential KPIs.

The effect of using such an approach was the determination of a set of indicators, closely integrated with the surveyed organization, corresponding to its needs and in line with the digital transformation that the company undergoes, encompassing four main areas of production. In such a statement, seven indicators will be monitored, which are part of the multi-area support for the organization of the production process:

- MRT – Mean Reaction Time
- MTTR - Mean Time To Repair
- FPY - First Pass Yield
- SC - Scrap Ratio
- MPST - Mean planning and scheduling time
- LT - Lead Time
- OEE - Overall equipment effectiveness (OEE = Availability * Effectiveness * Quality ratio).

The designated indicators are a response to the current challenges faced by the analyzed company, and their implementation and monitoring is to serve to assess the state of the production process at defined decision-making levels: strategic, tactical and operational. The individual KPIs correspond to the problems actually diagnosed by the appointed implementation team, as shown in Table 1.

The proposed comprehensive approach, which covers important parts of the business activity, is aimed at introducing modifications and improvements to the management processes of the production organization. Hence, the correct determination of indicators itself, although it is an absolutely key task, will not bring the desired results without a reliable definition of the indicator description standard. Such a metric should contain a range of information relating both to the precisely defined KPI itself, but also to the conditions related to the environment in which it will function. For this reason, a comprehensive sheet was created, divided into three main parts, and consisting of: basic characteristics (name of the indicator, what it relates to, definition, area of influence, owner, calculation formula), detailed characteristics (measurement method, source of the measured value, unit of measure, range of values, trend, frequency of measurement) and reporting (reporting frequency, place of reporting, alarm levels, corrective actions, method of presentation).

Table 1. List of KPIs and the corresponding production problems.

No	KPI	Problem description
1	MRT	Too long inactivity from reporting a failure to taking corrective action
2	MTTR	Too long time from reporting a failure, through diagnosis to problem removal
3	FPY	Occurring destabilization of the process resulting in corrections and corrections at its later stages
4	SC	Occasional periodic increases in the level of defective products
5	MPST	Production supervision that uses a significant part of its working time to plan and schedule production
6	LT	Occurring mismatch between the production schedule and the customer's order
7	OEE	Lack of information on the result of the correlation between availability, use and quality obtained for selected production stations

The above study was the basis for defining the assumptions for the designed IT system enabling support of the decision-making process monitoring the goals set in the indicated areas of the company's operation and at individual decision-making levels:

a. Remodeling of production task registration forms: registration of additional information (parameters for indicators), automation of data entry (RFID, QR codes), verification of data entered by the operator, additional data access terminals (POS computers, tablets and mobile phones), monitors for data visualization,
b. Extending the database with additional objects (tables and views): registration of KPI parameters, automatic generation of reports and summaries, generation of notifications and alerts,
c. Additional modules for data collection and working time recording devices: counting systems for products at workplaces, machine working time recording readers (Golem, Historian) and operators (RFID cards),

d. Organizational change related to:
   - maintenance: failure service and maintenance,
   - registration of tasks performed by machine operators,
   - logistics processes: production logistics, maintenance logistics,
   - production planning: scheduling production orders, - project management: new launches,
   - quality control: first item inspection, registration of deficiencies.

6 Conclusions

A modern production company that wants to compete in a dynamically changing market is obliged not only to meet the challenges related to its external environment, e.g. socio-cultural (work life balance), economic and legal (frequent changes in legislation), geopolitical (unpredictable events), technological (digitization), but the more it has to efficiently adapt to the current situation, appropriately shaping its internal policy relating to its own organization and management methods. The rapid development of information technologies and the accompanying widespread digitization, prompts entrepreneurs to look for their own path to control production processes, based on modern business methods using quick access to large data sets, often possible to obtain in real time.

The development of the system required multiple analyzes and processing of historical data to find out the current reference point of the organization. This is a considerable amount of work, the execution of which, by itself, will not bring any benefits, but it is necessary for further work. The appointment of a multifunctional team also had a great impact on the shape of the system, as it has often turned out that by going deep into the details of individual issues (goals, key activities), despite many years of work in the enterprise, these people did not realize what daily work in production looks like. Explaining these issues and describing production from the position of managers managing this process allowed for understanding the mechanisms used by other employees.

The next step of the authors is the implementation of the developed tool in the enterprise and the verification of its assumptions and goals. The results achieved so far indicate the right direction of research, the result of which is the ability to effectively influence the control and optimization of the production process.

As shown in the article, KPIs are a useful tool supporting the decision-making processes of managers, production engineers and operators at all levels of the company's operation to verify the set goals. They also enable multi-area monitoring of the process and dynamic response to any irregularities. In addition, in a simple and understandable way, they signal the state of the process and enable its precise correction, as well as affect the effectiveness of decisions made.

Acknowledgments. This work was created as part of the "Implementation Doctorate" program, Contract no. DWD/4/22/2020 of 06.11.2020.

References

1. Mykhnenko, V., Delahaye, E., Mehdi, N.: Understandning forced internal displacement in Ukraine: insights and lessons for today's crises. Oxf. Rev. Econ. Policy **38**(3), 699–716 (2022)

2. Wang, K., et al.: A novel strategy of regional microstructure design in manufacturing. Mater. Lett. **333**, 133644 (2023)
3. Huawei, T.: Does gross domestic product, inflation, total investment, and exchanges rate matter in natural resources process volatility?. Resourc. Policy **79**, 103013
4. Pizoń, J., Gola, A.: The meaning and directions of development of personalized production in the era of Industry 4.0 and Industry 5.0. In: Machado, J., et al. (eds.) Innovations in Industrial Engineering II. icieng 2022. LNME. Springer, Cham (2023). https://doi.org/10.1007/978-3-031-09360-9_1
5. Antosz, K., Paśko, Ł, Gola, A.: The use of artificial intelligence methods to assess the effectiveness of Lean Maintenance concept implementation in manufacturing enterprises. Appl. Sci. **10**(21), 7922 (2020)
6. Bocewicz, G., Wójcik, R., Sitek, P., Banaszak, Z.: Towards Digital-Twin-driven performance Evaluation methodology of FMS. Appl. Comput. Sci. **18**(3), 5–18 (2022)
7. Vidal, G.H., Coronado-Hernandez, J.R., Minnard, C.: Measuring manufacturing system complexity: a literature review. J. Intell. Manufac. early access. https://doi.org/10.1007/s10845-022-01974-5
8. Piechowski, M., Szafer, P., Wyczolkowski, R., Gladysiak, V.: Concept of the FMEA method-based model supporting proactive and preventive maintenance activities. IOP Conf. Seri. Mater. Sci. Eng. **400**(6), 141170 (2018)
9. Paśko, Ł., Antosz, K.: Neural model of manufacturing process as a way to improve predictability of manufacturing. In: Gapiński, B., Ciszak, O., Ivanov, V. (eds.) Advances in Manufacturing III. MANUFACTURING 2022. LNME. Springer, Cham (2022). https://doi.org/10.1007/978-3-031-00805-4_3
10. Jahangirian, M., Taylor, S.J.E., Young, T., Robinson, S.: Key performance indicatiors for succesful simulation projects. J. Oper. Res. Soc. **68**(7), 747–765 (2017)
11. Morella, P., Lamban, M.P., Royo, J., Sanchez, J.C., Latapia, J.: Development of a new KPI for the economic quantifications of six big losses and its implementation in a cyber physical system. Appl. Sci. **10**(24), 9154 (2021)
12. Wittbrodt, P., Łapuńka, I., Baytikenova, G., Gola, A., Zakimova, A.: Identification of the impact of the availability factor on the efficiency of production processes using the AHP and Fuzzy AHP methods. Appl. Comput. Sci. **18**(4), 116–129 (2022)
13. Jasiulewicz-Kaczmarek, M., Antosz, K., Zhang, C, Ivanov, V.: Industry 4.0 technologies for sustainable asset life cycle management. Sustainability **15**(7), 5833 (2023)
14. Burduk, A., Lapczynska, J., Popiel, P.: Simulation modeling in production effectiveness improvement – case study. Manage. Product. Eng. Rev. **12**(2), 75–85 (2021)
15. Gola, A., Nieoczym, A.: Application of OEE coefficient for manufacturing lines reliability improvement. AEBMR-Adv. Econ. Bus. Manage. Res. **31**, 189–194 (2017)
16. Zarate, P.: Decision making process: a collaborative perspective. Stud. Inform. Control **17**(2), 225–230 (2008)
17. Grzelak, M., Borucka, A., Guzanek, P.: Application of linear regression for evaluation to production processes effectivenes. Lect. Notes Mech. Eng. **261309**, 36–47 (2021)
18. Kuryło, P., et al.: Selected determinants of machines and devices standarization in designing automated production Processes in Industry 4.0, Materials **16**(1), 312, 2023
19. Parmenter, D.: Key Performance Indicators – Developing, Implementing and Using Winning KPIs, 4th edn. Wiley, Toronto (2019)
20. Kaplan, R.S., Norton, D.P.: Strategiczna karta wyników. Jak przełożyć strategię na działanie. Wydawnictwo Naukowe PWN, Warszawa (2022)
21. Melander, A., Andersson, D., Elgh, F., Fjellstedt, F., Löfving, M.: Implementing Hoshin Kanri. How to manage strategy through policy deployment and continuous improvement. Routledge, Taylor & Francis Group, Abingdon (2021)

22. Thürer, M., Maschek, T., Fredendall, L., Gianiodis, P., Stevenson, M., Deuse, J.: On the integration of manufacturing strategy: deconstructing Hoshin Kanri. Manag. Res. Rev. **42**(3), 412–426 (2019)

Gamification-Based Crowdsourcing as a Tool for New Product Development in Manufacturing Companies

Mateusz Molasy(✉), Maria Rosienkiewicz, Joanna Helman, and Mariusz Cholewa

Department of Laser Technologies, Automation and Production Organization, Wroclaw University of Science and Technology, Wyspiańskiego 27, 50-370 Wrocław, Poland
mateusz.molasy@pwr.edu.pl

Abstract. The paper discusses new phenomena of crowdsourcing, as an element of open innovation. New approach of the use of gamification elements in crowdsourcing is presented and the possibility of use of gamification-based crowdsourcing in supporting manufacturing companies. The term gamification is defined and its elements are discussed in theoretical review. Several areas of supporting manufacturing companies by gamification-based crowdsourcing are identified, however the paper and the empirical experiment that has been performed and presented focuses on new product development. The experiment of designing new product by crowd is discussed and the use of gamification elements of challenges, quests, teams and leaderboards to attract the participants of the experiment and make it successful are presented. It has been shown that manufacturing companies can involve external individuals in new product development process with attracting them by gamification-based crowdsourcing in order to more efficiently manage this process.

Keywords: production · efficiency · new product development · open innovation · crowdsourcing · digitalization · gamification

1 Introduction

Manufacturing companies, in order to remain competitive while facing challenges regarding 4th industrial revolution, continuously need to improve their products and processes, using various types of tools in an innovative way [1]. One of the recently booming concepts in this area has been the concept of open innovation (OI), which enables external entities (including individuals) to participate in the innovation processes in organizations [2]. Several empirical studies have noted that higher degrees of openness to external knowledge improves organizations' performance in the manufacturing industry. By implementing OI tools and methods manufacturing companies can improve their collaboration process throughout the whole supply chain, improve new product development process, minimize production waste, ensure better working conditions, increase general efficiency in manufacturing. One of the examples include P&G

A. Burduk et al. (Eds.): ISPEM 2023, LNNS 790, pp. 368–379, 2023.
https://doi.org/10.1007/978-3-031-45021-1_28

that recently adopted an OI approach and developed a new strategy leading to include ideas from outside for their new products. Other examples of organizations that opened up their boundaries and now exchange technology and knowledge with other external partners include GE, Samsung, Lego, NASA [3]. Research on innovation development trends clearly shows that number of organizations implementing elements of OI is regularly growing [4] and will keep on growing due to the fact that OI has been recognized as an essential part of one of the United Nations Sustainable Development Goals for 2030 [3].

Moreover, an ongoing digital disruption is transforming the manufacturing industry. A large part of the production challenges have been recently related to the digitization of various organizational aspects [5]. An important aspect that is inextricably linked to OI and digitalization is platformization defined as the transition from individual products/services to platforms as intermediaries in transactions and organizing value creation processes [6]. It has been shown that OI approach, through utilization of digital innovation platforms, offers significant improvements in companies' efficiency and almost all services supporting OI are used via platforms [7].

Due to constantly increasing level of digitalization more and more interest is being focused on gamification, defined in general as the implementation of the most common and enjoyable mechanics of video games, in non-video game contexts [8]. Studies on gamification confirm that it leads to a higher degree of engagement and involvement [9] which can be crucial for the successful use of tools and methods of OI approach. As quite often it is challenging to involve individuals to participate in innovation processes and make their tacit knowledge visible, games seem excellent solutions for involving and increasing understanding of the concept and likeliness of using its tools and methods [10].

As gamification has been providing benefits such as motivation, increased happiness, reduced stress, and improved productivity, it can be applied to the world of manufacturing. Increased employee motivation leads to increased production output and lowers the rate of accidents. Increased happiness and less stress increases employee retention rates and reduce burnouts. Improved productivity obviously impacts the manufacturing company's efficiency [11].

The aim of this paper is to present the possibility of use gamification-based crowdsourcing in order to more efficiently support manufacturing companies in new product development process.

2 Literature Review

2.1 Gamification

The term "gamification" was first coined in 2002, however the concept itself is older. Loyalty cards or frequent flyer programs of airlines might be considered early examples of gamification the way it is understood today [12]. Several analyzed definitions of gamification, presented in the Table 1, can be summarized as *implementation of the most common and enjoyable mechanics of video games, in non-video game contexts* [13].

Table 1. Popular definitions of gamification. Source: own elaboration on the basis of [12].

Author	Definition
Pelling (2011)	*"Applying game-like accelerated user interface design to make electronic transactions both enjoyable and fast"*
Zicherman and Cunningham (2011)	*"The use of game design elements in non-game contexts"*
Burke (2012)	*"The use of game mechanics and game design techniques in non-game contexts to design behaviours, develop skills or to engage people in innovation"*
Werbach and Hunter (2012)	*"The use of game elements and game-design techniques in non-game contexts"*
Huotari and Hamari (2012)	*"A process of enhancing a service with affordances for gameful experiences in order to support user's overall value creation"*
Werbach (2014)	*"The process of making activities more game-like"*

The most commonly implemented modern models of gamification include (1) badges, (2) leaderboards, (3) points and levels and (4) challenges and quests [14]. **Badges** can be attained by meeting nonrelative goals, i.e. achievement benchmarks or relative, that are limited achievements rewarded to users relative to their peers and they can be considered as similar to Olympic medals [15]. As Maslow proves in his hierarchy of needs, presented in 1943, people have the need for high evaluation of themselves, for self-esteem, self-respect and the esteem of others [16]. Therefore, earning badges and points and being among the best performers on the leaderboard is confirmation that one has provided valuable content and is recognized and acknowledged for it [17]. **Leaderboards** dynamically rank the progress of individual users and their achievements as compared to their peers, providing a sense of how well they are performing as compared to their peers [15]. Although high-rank positions on leaderboards positively affect motivation, low-rank positions may produce damaging effects on motivation [18]. **Points and leveling systems** are intended to inform users of the level of familiarity and reward continued expertise in using the system, that can be achieved through assessing the users on the basis of predetermined values points and depending on assessment results granting icons or titles that reflect their levels. **Progress bars** are common features, because users can see how many points they have attained along a continuum [14, 15]. Through continued **challenges and quests** users are motivated to continue using a system. The impact is even higher if challenges validate understanding of the goals of the application by the users. They have meaning to the users because of an underlying narrative, comprising checkpoints or benchmarks, indicating that the user is, indeed, using and progressing through the system as it was meant to be used [15]. Examples of gamification elements and their possible use in manufacturing organizations is presented in the Table 2.

One of the types of games is a design game – rarely competitive with the winner of the game, but rather based on open tasks that allow the participants to make their own

Table 2. Examples of possible use of elements of gamification in manufacturing. Source: [11].

Game element	Example of use in manufacturing
Badges	Distinguishing workers' who have and have not achieved a particular training level or a set of experience
Leaderboard	A visual scoreboard that facilitates healthy competition among shopfloor workers or teams that helps increase progress towards a specific production-related goal, showing identifiers and current scores of the competitors
Points, levels	Workers who successfully completed a set of objectives or reached certain number of points reaches the next level
Progress bars	A graphical element that visualizes the progression along the completion of the task that motivates workers to continue working toward the completion of a task by highlighting how much they have already completed

interpretations and find meaningful focus; engaging and creating a relaxed and informal atmosphere that increases creativity, utilize different senses and include ambiguous and open-ended props that force the players to be explicit in describing how they understand and interpret them. Therefore, the aim of design games is to inspire design and to help facilitate a co-design process proposed by users and other stakeholders [10]. This type of the game seems highly valuable and useful in product development process.

Following the success of the gamification-based social media services (like *Foursquare* that has popularized awarding badges as an approach of engaging users after completing a specific activity, for instance, checking in in the social network context) the idea of using elements of design of games in non-game contexts to motivate and increase user activity and retention has rapidly gained traction in organizations [13]; including becoming booming trend in the marketing field because due to delivery of customers' behaviour and motivating them in task performance [8]. Currently, gamification becomes even one of the strategies to improve organizational processes that increases motivation, loyalty, participation and commitment of the members of an organization and supporting knowledge workers. Therefore, it is starting to be considered as a tool for practicing managers, specialized consultants or academia's researchers [12].

In the Gartner report [19] it has been shown that gamification has now been adopted by more than 70% of businesses belonging to the Global 2000 list. This trend is estimated to continue to grow as more organizations gamify their business operations. Moreover, these companies are likely to keep on investing in more innovative gamified solutions. Companies such as Oracle, Hewlett Packard, IBM, and Google will help drive this trend. Small businesses will also gamify their recruiting and sales activities in response to increasing competition in the SME space [20]. Gamification has been shown to improve employee engagement by over 60 percent [21]. It is estimated that gamification will be even more popular with millennial workers, as this generation raised on computers and video games. Today, millennials account for 35 percent of the global workforce. By 2025, they will account for 75 percent of the global population. This is another reason why organizations are gamifying their processes to cater to this generation's distinct workplace needs. Gamification satisfies millennials' appetite for collaborative and engaging interactions [20].

2.2 Open Innovation

Henry Chesbrough, the author of the concept of open innovation states that the innovation process is currently facing a "paradigm shift". The way how companies develop new ideas and bring them to the practical use is undergoing a fundamental change – the transition from a closed innovation model to an increasingly open one [2]. Open Innovation helps to master the innovation game and to stay competitive in fast changing environments and rapid development of digital technologies (i.e., robotics, artificial intelligence and high-performance computing), along with the transformation of the role of individuals and communities in the socioeconomic context [4]. Several challenges on the way to success include finding the right knowledge or partner for exchange, securing know-how against the leakage during the cooperation or the possibility of knowledge absorption [22]. In recent years open innovation 2.0 (OI2.0) has been replacing open innovation; defined as "a new paradigm based on a Quadruple Helix Model where government, industry, academia and civil participants work together to co-create the future and drive structural changes far beyond the scope of what anyone organization or person could do alone" [4]. Tools and methods of OI2.0 include crowdsourcing, crowdfunding, microworking, Social Product Development and sharing economy [23]. Crowdsourcing is defined as "an online, distributed problem-solving and production model that leverages the collective intelligence of online communities to serve specific organizational goals" [24]. Any of the stakeholders, including individuals might be integrated to work on a specific development via on-line platform.

Most popular element of crowdsourcing seems to be crowdfunding that is a kind of funding model, where individuals "use the Internet to contribute relatively small amounts of money to support the creation of a specific product or the investment in a specific business idea" [4]. An analysis of the crowdfunding phenomenon is important from both scientific and practical point of view due to a number of reasons: crowdfunding has influenced significantly a process of new product development (NPD). Through presenting the product idea to the public, crowdfunding has created an opportunity for engineers and designers to receive an immediate market feedback. It can be quick and with little effort to verify if potential customers are interested in the new product. What is more, the new product is therefore easily linked to potential investors. As it has been underlined by H. Forbes and D. Schaefer "a crowdfunding campaign can be launched with minimal cost, no proof of sales and, in three of the models, no release of equity" [25]. It's much more difficult to engage backers. Crucial for the success of crowdfunding campaign is an active crowd of participants and their will and high motivation to participate in the challenge for free or for a small monetary compensation. In general, active crowd of participants is crucial for successfulness of crowdsourcing and therefore the motivation of the crowdsourcees is of great importance [26]. Organizations attempt to make the work activity more like playing a game to provide other motives for working than just the monetary compensation [26]. It has been found that success of crowdsourcing also depends in high extend on social interactions, community engagement and the popularity of the channel. Crowdsourcing tasks, due to their nature, can be considered as gamified. Crowdsourcing websites and channels use game mechanics to achieve the desired output – namely, gamification has been widely applied into OI platforms [9]. Social interactions among donors positively lead to the success of crowdfunding campaigns and are mediated by third-party endorsements [27].

Open innovations bring many benefits to both big and small enterprises. They include reducing R&D costs, modernizing production processes, reducing the risk of implementing innovations [22]. Nowadays different organizations use crowdsourcing and engage a mass of people via the Internet. Although growing on popularity, there are still barriers in using these tools. 99% of students use social media and seems ready to enter OI2.0 era, lack of their knowledge about the open innovation paradigm, its elements, platforms, and opportunities that OI brings is still high. Moreover, lack of trust to OI tools and methods have been identified – over 35% of surveyed students do not trust OI [23]. On organizational level OI processes require changes in several aspects in organizations, both 'hard' and 'soft' such as culture [28], including the shift to an open culture and the creation of new routines [29].

Several theories explain the gamification mechanisms as a booster in behavior towards innovation. They explain gamification effectiveness in terms of "increased levels of motivation, engagement and positive reinforcement generated by the gamification design in individuals and organizations to obtain desired attitudinal, cognitive or behavioral change" [30]. Gamification approaches has been found to be beneficial in supporting the innovation process of companies especially during the early stages of innovation, in particular in co-creation with customers or in crowdsourcing (i.e. idea generation, either directly through crowdsourcing platforms or through the generation of individual engagement to participate in the ideation process). Similarly, gamification can be a useful approach in co-creating and enhancing collaboration among organizations. Gamification has been proven to be efficient in the education field, especially in targeting problems of students' evasion, disengagement, and lack of motivation in educational environments [31]. Game design elements, such as simulation and fun, and technology endowment contribute to innovativeness [30].

Research concludes, that gamification in the crowdsourcing context is still at an early stage of development [10, 26, 30], however features like a progress bar, leader-boards or avatars encourage individuals to participate in challenges. It has been proven that gamification approaches can support manufacturing companies on the workshop-level production management (see Table 2) and in the innovation process. Some research proves that crowdsourcing is an efficient tool for new product development. Several studies focuses on the use of gamification for idea generation, including directly through crowdsourcing platforms, helping mostly in enhancing creativity, other finds gamification as a useful approach in co-creating and enhancing collaboration among organizations [30], but no research has been found that directly refers to use gamification-based crowdsourcing to new product development. In order to find how certain elements of gamification can support crowdsourcing in new product development in manufacturing companies, the experiment have been performed and analyzed by authors.

3 Methodology

Authors of this paper being leaders of Interreg CE funded project SYNERGY [32], developed the Synergic Crowd Innovation Platform (SCIP; "Synergy Platform"; https://synergy.pwr.edu.pl). Within the platform, different types of newly designed services enhancing crowd innovation initiatives has been designed to be available, to help manufacturing companies facing industrial and innovation challenges. Services include crowdsourcing, simulated crowdfunding and infrastructure sharing. In the crowdsourcing functionality the platform users (i.e. manufacturing companies, individuals, researchers…) can upload problems to be solved (e.g. technological, research) and other users solve these problems and present their solutions. On the basis of results of research on best practices and success stories of crowd innovation approaches [7] and workshops dedicated to defining needs of potential users of the platform it was agreed that gamification-based crowdsourcing will be implemented in order to effectively support the functionality of the platform and its utilization in manufacturing industry. Practical use of gamification-based crowdsourcing in new product development process has been tested and proven to be promising.

4 Designing Gamification-Based Crowdsourcing Functionality on OI Platform and Testing It on New Product Development

In the Synergy Platform a user that is logged-in can add a new challenge by clicking the button "Add challenge"'. Then a pop-up window will appear with a form where the details about the challenge should be filled in. The finished version of the challenge's information form shall be submitted as "Released version" where all required data have to be filled in. After administrator's verification and approval, it is visible on the challenge list in the "Crowdsourcing challenges for innovation" functionality of the Synergy Platform. In order to support the challenge, it is possible to send a public comment visible to everyone and/or a private comment visible only to the author of the campaign. After clicking the tab "Teams" platform users can add a team that would like to solve the challenge. Non-team users can send a request to be added to the team. After clicking the tab "Solutions" the list of submitted solutions of the challenge can be seen and users can add a new solution challenge. If the user would like to submit a new solution, the button "Add solution" within "Solutions" section should be pressed. Description of the proposed solution is here required with possible files attached. The author of the challenge can evaluate the solution and change its status to make the solution visible to everyone (by default each submitted solution is visible only to the submitter and challenge giver). Figure 1 shows an exemplary application screen.

From the gamification point of view, it was agreed that the rating system would be the crucial functionality – to give the possibility of assigning "stars" (1/5, 2/5,…, 5/5) or the number of points for both the proposed ideas and platform users. This functionality is to allow i.e. a member of the team to assess other team member as a credible person as well as the challenge giver to evaluate the applicant. Users' assessment is based on their activity on the platform: number of applications, ideas and solutions submitted. Projects and solutions are evaluated by the applicant and optionally by the platform community as well. The system automatically ranks users. To allow a user to search for challenges, projects and/or users a browser needed to be created. Users are able to hide their profile if they don't like to be visible to others. The second gamification-based functionality, that refers to leaderboards, is a possibility of setting up teams.

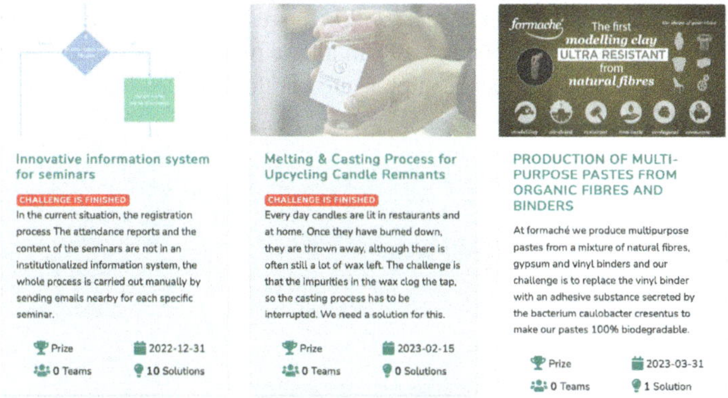

Fig. 1. A screenshot of simulated crowdsourcing functionality of the platform.

Ready-to-use OI platform has been developed and a gamification-based crowdsourcing-related experiment "*Design and prototype model*" was performed. It was intended as an opportunity for micro-working and Social Product Development as a revolutionary process to design, create and manufacture new innovative products, favoring collaboration and the exchange of competences among platform's users that are manufacturing companies. As an example of possible challenge that a manufacturing company can develop to support their product development process and the agreed one was "*House of the future*". The platform users were asked to develop a house ornament, object or piece of furniture with possibility of integration of advanced technologies (sensors, AI) that can help people with everyday activities and improve their quality of life. The requirement included presentation of a 3D CAD model (that could be 3D printed) and a document explaining the problem they intend to solve. Platform users were asked to explore the challenge on the platform and elaborate a first draft of a solution or associate to solutions already presented on the platform.

Solution providers (or challenge takers) were no limited to a specific target group: anyone from the general public (SMEs, BSOs, students, etc.) could be involved. They had to form research teams of at least 2 platform users being international or belonging to the same country. They had to work together to a research plan explaining how the solution envisaged solved the problem and met the requirements presented in the challenge. Overall, more than 3000 possible solution providers were intercepted by project partners via direct contacts (mails, invitations, etc.) or online (social media, website, newsletter, etc.). Most of them were individuals from the general public, secondary school or university students and researchers, representatives of BSO, companies (incl. Manufacturing SMEs) and start-ups.

The research plan had to contain also a timeline and a budget proposal for the implementation of the solution. Once uploaded on the Synergy Platform, platform users had look for support from other platform's users and they had to promote it online and collect as many "likes" as possible. At the end, the 3 best solutions which got highest number of likes were awarded with a voucher based on a ranking. Vouchers could be used to pay for services or buy consumables, such as stationary or electronic devices. They were assigned to one person (team leader) within the research team, who was responsible of ensuring that the voucher is benefited from all the members of the research team.

The experiment workflow has been presented in the Fig. 2.

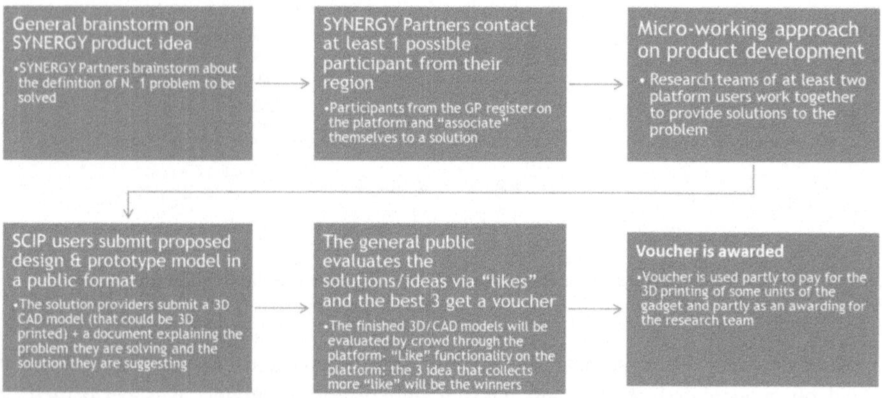

Fig. 2. Workflow of the experiment "*Design and prototype model*"

Overall, 13 research teams were created. 12 of them presented the solutions. Number of "likes" and prizes for each of the solutions are presented in the Table 3.

Table 3. Number of "likes" granted to each of the solutions in "*Design and prototype model*" gamification-based crowdsourcing experiment.

Name of the solution	Number of "likes"	Granted voucher
Eg's	163	1.250€
The intelligent clothespin	162	1.000€
Smart lock	149	750€
Flaura	136	–
Travel Dish	57	–
Torque Wrench	40	–
Automatic insect repellent	34	–
Post Box	26	–

(continued)

Table 3. (*continued*)

Name of the solution	Number of "likes"	Granted voucher
Can holder	24	–
Automatic hedge-trimmer	23	–
Playstation CD holder	7	–
ToastR	4	–

5 Conclusions

OI platform with the gamification-based crowdsourcing functionality has been developed by authors of this paper. One of the aims of development was to test innovative OI solutions with the use of elements of gamification to support manufacturing companies and test the use of these gamification tools and methods for smoothening the process and making it more attractive for the users, including manufacturing companies that would be able to more efficiently manage their processes. Identified elements of gamifications have been implemented and tested in the experiment "*Design and prototype model*". According to the feedback received from the participants (regions that performed the experiment are: Poland, Austria, Croatia, Germany, Italy, Slovenia) the awareness about open innovation definitely raised among the participants and enthusiasm for development of creative solutions, in particular with regard to working with 3D printers and the development of CAD models. It has enabled the existing knowledge on 3D printing to be expanded on an individual level, mainly to high school students and young researchers. The experiment offered a perfect opportunity to introduce high school students to the topic of crowd innovation and open innovation and thus raise awareness for this type of collaborative work. Gamification elements included in the experiment made it more attractive and resulted in successful implementation.

The aim of this paper has been achieved. The possibility of use gamification-based crowdsourcing in order to more efficiently support manufacturing companies in new product development process has been presented and tested. It is assumed that gamification elements (especially challenges, quests, teams and leaderboards) involving employees or external stakeholders (incl. Individuals) in products development and/or improvements in products or processes can be the promising option for manufacturing companies in developing new products, however detailed functionalities and efficiency of each of them need further research. Due to limitations regarding project implementation so far only few gamification elements could have been tested on the Synergy Platform.

References

1. Kotowska, J., Markowski, M., Burduk, A.: Optimization of the supply of components for mass production with the use of the ant colony algorithm. In: Burduk, A., Mazurkiewicz, D. (eds.) Intelligent Systems in Production Engineering and Maintenance – ISPEM 2017. pp. 347–357. Springer International Publishing, Cham (2018). https://doi.org/10.1007/978-3-319-64465-3_34

2. Chesbrough, H.W.: Open Innovation: The New Imperative for Creating and Profiting From Technology. Harvard Business School Press, Boston, Mass (2011)
3. Obradović, T., Vlačić, B., Dabić, M.: Open innovation in the manufacturing industry: a review and research agenda. Technovation **102**, 102221 (2021). https://doi.org/10.1016/j.technovation.2021.102221
4. European Commission. Directorate General for Communications Networks, Content and Technology.: Open innovation 2.0 yearbook 2017–2018. Publications Office, LU (2018)
5. Burduk, A., Musiał, K.: Genetic algorithm adoption to transport task optimization. In: Graña, M., López-Guede, J.M., Etxaniz, O., Herrero, Á., Quintián, H., Corchado, E. (eds.) International Joint Conference SOCO'16-CISIS'16-ICEUTE'16. pp. 366–375. Springer, Cham (2017). https://doi.org/10.1007/978-3-319-47364-2_35
6. Bygstad, B., Hanseth, O.: Transforming Digial Infrastructures Through Platformization (2018)
7. Cholewa, M., Molasy, M., Rosienkiewicz, M., Helman, J.: Analysis of platforms supporting open innovation approach. In: Batako, A., Burduk, A., Karyono, K., Chen, X., Wyczółkowski, R. (eds.) Advances in Manufacturing Processes, Intelligent Methods and Systems in Production Engineering. pp. 475–490. Springer, Cham (2022). https://doi.org/10.1007/978-3-030-90532-3_36
8. Huotari, K., Hamari, J.: A definition for gamification: anchoring gamification in the service marketing literature. Electron. Mark. **27**, 21–31 (2017). https://doi.org/10.1007/s12525-015-0212-z
9. Behl, A., Dutta, P.: Engaging donors on crowdfunding platform in Disaster Relief Operations (DRO) using gamification: a Civic Voluntary Model (CVM) approach. Int. J. Inf. Manag. **54**, 102140 (2020). https://doi.org/10.1016/j.ijinfomgt.2020.102140
10. Kauppinen, S., Luojus, S., Lahti, J.: Involving citizens in open innovation process by means of gamification: the case of WeLive. In: Proceedings of the 9th Nordic Conference on Human-Computer Interaction. pp. 1–4. ACM, Gothenburg Sweden (2016). https://doi.org/10.1145/2971485.2971526
11. Keepers, M., Nesbit, I., Romero, D., Wuest, T.: Current state of research & outlook of gamification for manufacturing. J. Manuf. Syst. **64**, 303–315 (2022). https://doi.org/10.1016/j.jmsy.2022.07.001
12. Shpakova, A., Dörfler, V., MacBryde, J.: Changing the game: a case for gamifying knowledge management. World J. Sci. Technol. Sustain. Dev. **14**, 143–154 (2017). https://doi.org/10.1108/WJSTSD-01-2017-0002
13. Deterding, S., Dixon, D., Khaled, R., Nacke, L.: From game design elements to gamefulness: defining "gamification." In: Proceedings of the 15th International Academic MindTrek Conference on Envisioning Future Media Environments - MindTrek 2011, p. 9. ACM Press, Tampere, Finland (2011). https://doi.org/10.1145/2181037.2181040
14. Zichermann, G., Cunningham, C.: Gamification by Design: Implementing Game Mechanics in Web and Mobile Apps. O'Reilly Media, Sebastopol, Calif (2011)
15. Miller, A.S., Cafazzo, J.A., Seto, E.: A game plan: Gamification design principles in mHealth applications for chronic disease management. Health Inform. J. **22**, 184–193 (2016). https://doi.org/10.1177/1460458214537511
16. Maslow, A.H.: A theory of human motivation. Psychol. Rev. **50**, 370–396 (1943). https://doi.org/10.1037/h0054346
17. Ďurinik, M.: Gamification in knowledge management systems. Cent. Eur. J. Manag. **1** (2015). https://doi.org/10.5817/CEJM2014-2-3
18. Na, K., Han, K.: How leaderboard positions shape our motivation: the impact of competence satisfaction and competence frustration on motivation in a gamified crowdsourcing task. Internet Res. **33**, 1–18 (2023). https://doi.org/10.1108/INTR-12-2021-0897

19. Gamification 2020: What Is the Future of Gamification? https://www.gartner.com/en/documents/2226015. Accessed 8 Nov 2022
20. Rodriguez, P.: 19 Gamification Trends for 2021–2025: Top Stats, Facts & Examples. https://www.growthengineering.co.uk/19-gamification-trends-for-2022-2025-top-stats-facts-examples/. Accessed 8 Nov 2022
21. Chang, J.: 54 Gamification Statistics You Must Know: 2021/2022 Market Share Analysis & Data. https://financesonline.com/gamification-statistics/. Accessed 8 Nov 2022
22. Walecka-Jankowska, K., Zimmer, J.: Open innovation in the context of organisational strategy. Eng. Manag. Prod. Serv. **11**, 86–95 (2019). https://doi.org/10.2478/emj-2019-0023
23. Rosienkiewicz, M., Helman, J., Cholewa, M., Molasy, M.: Open innovation readiness assessment within students in poland: investigating state-of-the-art and challenges. Sustainability **14**, 1213 (2022). https://doi.org/10.3390/su14031213
24. Brabham, D.C.: Crowdsourcing. The MIT Press, Massachusetts; London, England (2013)
25. McAdam, M., Debackere, K.: Beyond 'triple helix' toward 'quadruple helix' models in regional innovation systems: implications for theory and practice: Beyond 'triple helix' toward 'quadruple helix' models. RD Manag. **48**, 3–6 (2018). https://doi.org/10.1111/radm.12309
26. Morschheuser, B., Hamari, J., Koivisto, J.: Gamification in crowdsourcing: a review. In: 2016 49th Hawaii International Conference on System Sciences (HICSS). pp. 4375–4384. IEEE, Koloa, HI, USA (2016). https://doi.org/10.1109/HICSS.2016.543
27. Calic, G., Mosakowski, E.: Kicking off social entrepreneurship: how a sustainability orientation influences crowdfunding success: kicking off social entrepreneurship. J. Manag. Stud. **53**, 738–767 (2016). https://doi.org/10.1111/joms.12201
28. Lazzarotti, V., Manzini, R.: Different modes of open innovation: a theoretical framework and an empirical study. Int. J. Innov. Manag. **13**, 615–636 (2009). https://doi.org/10.1142/S1363919609002443
29. Harland, C., Nassimbeni, G., Schneller, E.: The SAGE Handbook of Strategic Supply Management. SAGE Publications Ltd, 1 Oliver's Yard, 55 City Road, London EC1Y 1SP United Kingdom (2013). https://doi.org/10.4135/9781446269886
30. Gimenez-Fernandez, E., Abril, C., Breuer, H., Gudiksen, S.: Gamification approaches for open innovation implementation: a conceptual framework. Creat. Innov. Manag. **30**, 455–474 (2021). https://doi.org/10.1111/caim.12452
31. Oliveira, W., et al.: Tailored gamification in education: a literature review and future agenda. Educ. Inf. Technol. **28**, 373–406 (2023). https://doi.org/10.1007/s10639-022-11122-4
32. Rosienkiewicz, M., Helman, J., Cholewa, M., Molasy, M.: SYNERGY project: open innovation platform for advanced manufacturing in central Europe. In: Burduk, A., Chlebus, E., Nowakowski, T., Tubis, A. (eds.) Intelligent Systems in Production Engineering and Maintenance, AISC, vol.835, pp. 306–315. Springer, Cham (2019). https://doi.org/10.1007/978-3-319-97490-3_30

Analysis Safety and Quality of Products for Internal Supply Chains

Julia Giera(✉), Ewa Kulińska, and Małgorzata Dendera – Gruszka

Faculty of Production Engineering and Logistics, Opole University of Technology, Opole, Poland

{j.giera,e.kulinska}@po.edu.pl

Abstract. The constant pursuit of effective operations, improvement of customer service and effective implementation of logistics tasks requires the fulfillment of the basic element, which is to maintain the safety of goods, so that the company can achieve specific goals. High turnover of goods in the supply chain, both deliveries and shipments, increases the likelihood of disruptions. The challenge in supply chain management is to ensure safety and maintain the required quality of goods in the FMCG industry. The aim of the article is to assess existing risk factors in internal supply chains in terms of maintaining an appropriate level, quality and safety of goods. A qualitative risk management method was used, on the basis of which the risk factors with the greatest impact on the safety of goods were identified. Based on the obtained results, preventive measures to be implemented in logistics companies were developed. The implementation of new solutions to the links of the supply chain will make it possible to improve the risk management system with elements of safer distribution of stocks and to create a flexible and resilient risk management system for the internal supply chain.

Keywords: Safety of products · FMCG industry · internal supply chain

1 The Safety of Products in Logistics Process

1.1 Stages and Functioning of the Supply Chain

The constant pursuit of effectively performed activities, improvement of customer service and effective implementation of logistics tasks requires organizations to take measures to improve the functioning of the supply chain. Elements of supply chains are movable goods is their key component [2]. The stream of movable goods comprises the following links [15, 16]:

- Supply - obtaining raw materials and semi-finished products needed for the production of products,
- Production - transformation of goods to obtain a finished product,
- Manipulation of materials - activities of physical movement and storage of products,
- Distribution and after-sales customer service - activities that provide consumers with access to products.

A. Burduk et al. (Eds.): ISPEM 2023, LNNS 790, pp. 380–392, 2023.
https://doi.org/10.1007/978-3-031-45021-1_29

The security strategy for movable goods provides the ability to manage product risk at all stages of the supply chain, including the warehouse process. It is a set of technological activities that must occur in the correct order [1, 13]. The phases of the process in warehouse enterprises consist of the following elements: the process of acceptance, storage, completion and release of goods [10]. All are subject to registration and control. The overall functioning of the warehouse and production companies consists of individual activities that are carried out during the flow of goods through the warehouse.

1.2 Supply Chain of Fast Moving Consumer Goods

The striving for no occurrence of risk factors in the process of storage is key to the entire supply chain. In order to be able to counteract the occurrence of risk factors, it is necessary to understand the concept of the word risk. According to E. Kulińska, this risk is the possibility of an event that will affect the achievement of the organization's goals. It has measurable effect and probability. It is seen as a negative phenomenon that should be avoided. However, it also provides some perspectives, such as the possibility of innovation and beneficial change resulting from the investment of funds [3]. The goal of risk management is to maximize the chances of achieving the organisation's goals [4, 7], and to apply an appropriate risk management method in warehouse management to prevent interference.

In the area of storage of goods in the warehouse process, inventories can be classified in terms of the following factors: physical characteristics of products, type of recipients and affiliation of articles to a given industry. One of the most common sectors in warehouse management is the FMCG (Fast Moving Consumer Goods) industry [11]. The FMCG sector is one of the largest sectors on the world market [17]. The Fast Moving Consumer Goods industry consists of products:

- Foods,
- Cosmetics,
- Detergents,
- Cigarettes,
- Hygiene items,
- Pet foods,
- Over-the-counter drugs,
- Alcohols.

A large stream of products between individual logistics processes and then between individual links of the supply chain is characteristic of the FMCG industry, where the demand for warehouse services is constantly at a high level [11]. Articles from the FMCG industry usually have a short product life cycle, which is why supply chain fluidity is important in logistics, where proper storage of goods from this industry is an important element [11].

Supply chains of the FMCG industry are dynamic. The rapid marketability of products results in the need for constant replenishment of stocks. FMCG is an industry of fast-moving articles, which is why products belonging to this industry are bought by consumers every day [6]. The supply chain with the flow of articles from the FMCG industry is difficult and requires high discipline [11]. This is due to the constant high flow of articles, which is why it is so important to prevent occurrence of risk factors.

2 Research Facilities

2.1 Enterprises Providing Warehousing and Production

The research was conducted on 11 companies providing storage and production services in Poland. All facilities manage items belonging to the FMCG industry. Of these enterprises, 7 entities provide warehouse outsourcing services (marked with the letters A - G), and 4 entities are manufacturing companies (marked with numbers 1 - 4).

Table 1. Enterprises providing warehousing and production

Sign	Products
Warehouse company	
A	Snacks
B	Loose products, for example flour
C	Over-the-counter drugs
D	Dried spices
E	Chocolate products
F	Ready meals for people
G	Cosmetics
Manufacturing company	
1	Food for children and babies
2	Toilet paper, handkerchiefs, napkins
3	Pet food
4	Fruit and vegetable mousses

Source: Own study

A logistic audit was carried out in the research facilities regarding the characteristics of risk factors for FMCG supply chains. The research period is the first and second quarter of 2022. The data used for the calculations are statistical data from the companies' computer systems.

2.2 Risk Factors for FMCG Products in Supply Chains

Identification of risk factors in the supply chains of the FMCG industry is possible thanks to the audit carried out in research entities. In this article, the analysis concerns internal supply chains, where production and storage facilities are examined. One of the most important elements of the proper functioning of supply chains is the safety and quality of movable goods. This is one of the most important challenges of modern supply chains. The figure (Fig. 1) below presents a diagram of the analysis of elements affecting the possibility of risk factors in supply chains.

Risk factors in supply chain apply to each stage of the warehouse and production process. The table (Table 2) presents the risk factors for supply chain of goods from the FMCG industry (Table 1).

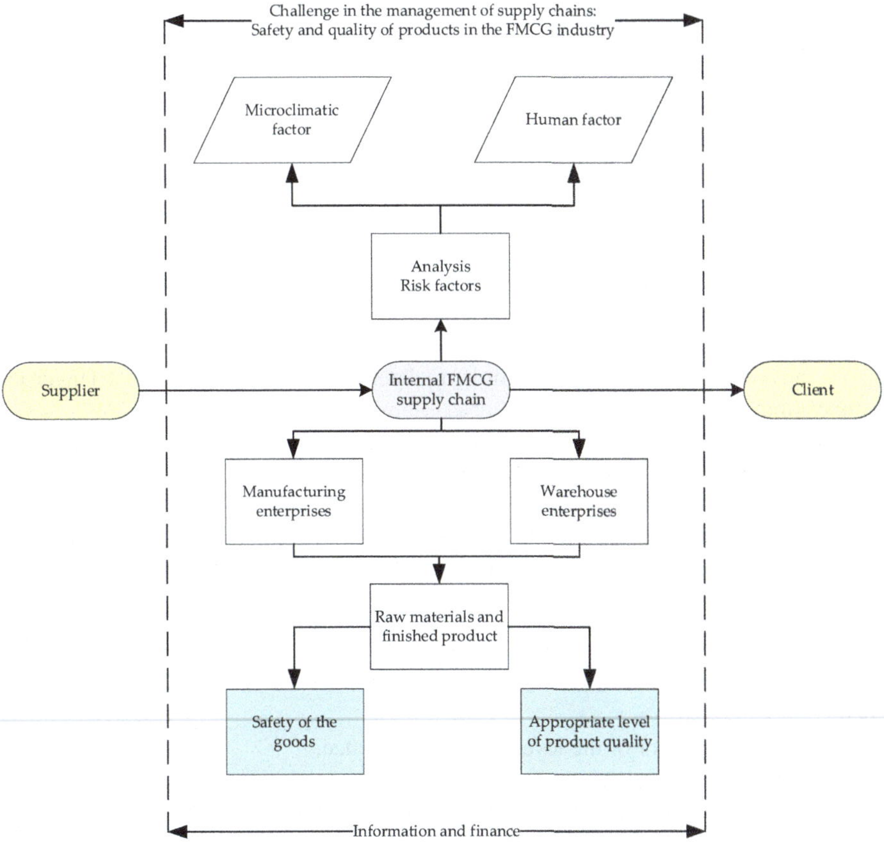

Fig. 1. Analysis how to make a safety products for internal supply chain. Source: Own study.

Table 2. Risk factors for supply chain of goods from the FMCG industry

Risk factors for supply chains
1. Occurrence of a microbial agent,
2. Contamination of the product with glass,
3. Contamination of the product with dust,
4. Bioterrorism,
5. Damage to products by employees,
6. Permeation of odors between products,
7. Incorrect climatic factors,
8. Careless driving of drivers,
9. Incorrectly monitored conditions of transportation goods in supply chains

Source: Own study

High rotation of goods in the supply chain, both in terms of the number of deliveries and dispatch of inventories, increases the likelihood of disturbances that affect directly FMCG products.

3 Qualitative Method for Risk Management

3.1 FMEA Method for Supply Chains

The FMEA method (failure modes and effects analysis) is an analysis of the types and effects of possible risk factors. It is a method of counteracting and eliminating the negative consequences of a risk factor. The method involves the identification of risk factors and their assessment [9, 14] in order to determine the risk factors most dangerous to the examined entities.

The essence of the FMEA method is to analyze possible product failures, their causes and effects, as early as possible to eliminate the failure. The FMEA method enables systematic product and process improvement through further analysis [5, 8].

FMEA analysis involves assigning values to three indicators: risk factor detection, significance of error, magnitude of damage effects, failures and probability of error occurrence, followed by the application of the formula:

$$R = D * S * P \tag{1}$$

where:

R - General indicator of the level of risk occurrence,

D - Risk factor detection,

S - The significance (severity) of the error, the magnitude of the effects of damage/failure,

P - Probability of error (damage, failure).

The following table (Table 3) defines the components of the FMEA method and determines the factors affecting product quality. The characterized variables are intended to give the risk factors a uniform form and to define how many elements affect the possibility of risk factors occurring.

Table describes the potential types of defects and effects of the defect and preventive measures are also highlighted. The individual risk determination values for the surveyed business entities are presented in the tables below. Below are tables with markings for risk level indicators. The following table (Table 4) specifies the probability of error (P).

The next table (Table 5) characterizes the error detection and product damage indicator.

The next table (Table 6) defines the error severity (significance) indicator, which is the extent of damage and failure effects.

The next table presents the results of statistical data analysis after applying the FMEA method. Ratings have been assigned for individual determinations for the risk level indicator. The table below presents the detection of risk factors in the entities surveyed.

The table (Table 7) specifies the risk factor level indicator for FMCG articles. The given result is the product of three indicators for a particular examined entity:

Table 3. Defines the components of the FMEA method

Potential type of defect	Potential effect of the defect	Potential causes of the defect	Factors affecting product quality
Occurrence of a microbial agent	Withdrawal of a batch of goods, financial losses, food poisoning of recipients, decrease in product rank, loss of customers.	Incorrect storage of goods, air conditioning failure.	A correct flow of information is necessary between the links of the supply chain in order to obtain reliable data about the conditions of storage of goods.
Contamination of the product with glass	Withdrawal of a damaged batch of goods, financial losses, withdrawal from the contract.	Lack of attention from employees, distraction, monotonous work mode, poor protection of goods, poor maintenance of machines and devices.	Thorough maintenance of machines and devices, care for cleanliness.
Contamination of the product with dust	Withdrawal of lots, financial losses.	Lack of cleanliness in the production plant or warehouse.	Systematic cleaning of the storage facility, in particular the floor of the storage hall. Cleaning should not only apply to the surface of the warehouse and productions facility, but also to devices supporting storage of inventories, devices for maintaining security in the warehouse and devices for transporting goods.
Bioterrorism	Poisoning of people, fatalities.	Exacerbation of conflicts between criminal groups, deterioration of public mood, increase in support from political leaders.	Control of people entering the warehouse and productions facility. The verification should include persons from outside and working in a given warehouse. People from outside should always move around the building assisted by a person employed in the institution.

(*continued*)

Table 3. (*continued*)

Damage to products by employees	Loss of trust in employees, dismissal of employees, staff shortages, financial losses.	Lack of involvement of employees in the proper functioning of the company. Lack of training and experience of employees.	Employee control, inventory control, high penalties for theft, possibility of job loss.
Permeation of odors between products	Withdrawal of product lots.	Air conditioning failure, lack of adequate ventilation, poor placement	Appropriate placement of goods in storage places. Segregation of goods based on odors. In the case of highly odor-emitting goods, isolate them from the rest of stocks. If it is not possible to isolate the goods, research entities should invest in additional walls dividing the warehouse into zones.
Incorrect climatic factors	Product spoilage.	Storage in inappropri-ate conditions, faulty ventilation and air conditioning, too much sunlight, no protection against.	Constant monitoring of warehouse conditions, air conditioning and ventilation control, storage of goods in dark places, covering of windows/skylights.
Careless driving of drivers	Damage to products, loss of products.	Inattention of the driver, traffic accident, transporting goods in inappropriate condi-tions.	Insurance of goods.
Incorrectly monitored conditions of transportation goods in supply chains	Shortening the shelf life of products, spoiling the goods.	Lack of control over employees, distraction of employees, carelessness of employees, failure of IT systems, failure of air conditioning and wrong humidity management.	Introduction of precise procedures regarding the storage of goods.

Source: Own study

Table 4. Designation of probability of error (damage, failure) indicator (P)

Probability of ocurance [%]	Rate	Description
0–5	1	Very low
6–20	2	Low
21–30	3	
31–40	4	Average
41–50	5	
51–60	6	
61–70	7	High
71–80	8	
81–90	9	Very high
91–100	10	

Source: Own study

Table 5. Designaton of risk factor detection indicator (D)

Error/damage detection (D)	Rate	Description
91–100	1	Very high
71–90	2	
61–70	3	High
51–60	4	
41–50	5	Average
31–40	6	
26–30	7	
16–25	8	Low
6–15	9	
0–5	10	Improbable

Source: Own study

- indicator of the significance of the error, the magnitude of the effects of damage, failure,
- error probability indicator,
- detection of risk factor.

Then, the sum of the level of occurrence of a particular risk factor was calculated. The numbers from 1 to 9 there are risk factors.

The calculated value of the R indicator allows to determine which risk factors for supply chains of goods from the FMCG industry are the most dangerous for the examined

Table 6. Designation of error severity indicator (S)

Action	Rate	Severity	Characteristics
The process of accepting goods into the companies	1	Very low	The error does not cause a break in the warehouse and production process of FMCG articles. Requires minor corrective actions.
	2		
	3	Low	The error requires corrective action. Without the involvement of additional people.
	4		
	5	Average	The disruption requires the implementation of an additional number of people.
	6		
	7	High	Disruption forces reorganization of the warehouse process. There are additional costs for the logistic process.
	8		
	9	Very high	The risk factor prevents the proper functioning of the warehouse and production process. The disruption affects the next links in the supply chain. There are high additional storage costs.
	10		

Source: Own study

entities. The determined value of R = 250 means that risk factors below this value do not pose the greatest threat to the entities surveyed. In turn, risk factors with an R-value greater than 250 have a direct probability of occurrence of a risk factor, and the effect of a given threat will have a negative impact on the entities surveyed. Preventive mechanisms have been developed for risk factors whose value exceeds R = 250, which are presented in the table below (Table 8).

The obtained results enabled the development of preventive mechanisms for the following risk factors:

- occurrence of a microbial agent on the stored goods, e.g. mold,
- dirt or dust on the product,
- bioterrorism,
- permeation of odors between products.

The proposed preventive mechanisms allow for changes to be made without investing large financial resources by research entities. The most dangerous risk factor is bioterrorism, where the effect of the risk factor affects the subjects most negatively, but the likelihood of a given risk factor being unlikely.

The applied method allowed to identify of the most dangerous risk factors for supply chains of goods from the FMCG industry. However, all risk factors should be monitored regularly, as new risk factors may appear over time, and the likelihood and effect of a given risk factor also changes.

Table 7. Risk factor level indicator for FMCG products in supply chains

Risk factor level indicator for FMCG products in supply chains												
Enteprise	**A**	**B**	**C**	**D**	**E**	**F**	**G**	**1**	**2**	**3**	**4**	**Amount**
1	**45**	**21**	**48**	**32**	**18**	**18**	**96**	**27**	**54**	**32**	**48**	**439**
D	5	1	3	4	2	6	4	3	3	2	4	
P	1	3	2	1	1	2	3	1	2	2	2	
S	9	7	8	8	9	10	8	9	9	8	6	
2	**30**	**10**	**14**	**32**	**9**	**9**	**18**	**36**	**20**	**20**	**20**	**218**
D	3	1	2	2	1	4	1	2	2	2	1	
P	1	1	1	2	1	1	2	2	1	1	2	
S	10	10	7	8	9	10	9	9	10	10	10	
3	**48**	**30**	**48**	**12**	**21**	**21**	**30**	**105**	**16**	**18**	**84**	**433**
D	3	2	2	1	1	2	2	3	2	1	3	
P	2	3	4	3	3	2	3	5	1	3	4	
S	8	5	6	4	7	4	5	7	8	6	7	
4	**90**	**50**	**45**	**96**	**128**	**128**	**126**	**90**	**63**	**48**	**70**	**934**
D	9	5	5	6	8	8	7	9	7	6	7	
P	1	1	1	2	2	1	2	1	1	1	1	
S	10	10	9	8	8	8	9	10	9	8	10	
5	**4**	**10**	**6**	**9**	**8**	**8**	**6**	**5**	**32**	**9**	**12**	**109**
D	1	2	1	1	1	1	1	1	2	1	2	
P	1	1	1	1	1	2	1	1	2	1	1	
S	4	5	6	9	8	6	6	5	8	9	6	
6	**48**	**120**	**48**	**10**	**32**	**32**	**6**	**72**	**24**	**12**	**28**	**432**
D	8	5	2	1	4	6	3	2	3	2	2	
P	2	3	3	2	2	3	1	4	2	1	2	
S	3	8	8	5	4	4	2	9	4	6	7	
7	**6**	**16**	**16**	**30**	**18**	**18**	**6**	**40**	**32**	**18**	**18**	**218**
D	2	2	1	2	3	2	1	1	2	1	1	
P	1	1	2	3	1	1	2	4	2	3	2	
S	3	8	8	5	6	7	3	10	8	6	9	
8	**8**	**12**	**28**	**54**	**32**	**32**	**12**	**4**	**12**	**16**	**12**	**222**
D	1	2	2	3	2	2	1	1	1	2	1	
P	2	1	2	2	2	1	2	1	2	1	2	
S	4	6	7	9	8	5	6	4	6	8	6	
9	**3**	**27**	**14**	**16**	**21**	**21**	**24**	**18**	**48**	**24**	**16**	**232**
D	1	3	2	2	1	4	2	1	2	2	2	
P	1	1	1	2	3	1	3	2	3	2	1	
S	3	9	7	4	7	7	4	9	8	6	8	

Source: Own study

Table 8. Preventive mechanisms affecting the safety and quality of FMCG

Preventive mechanisms affecting the safety and quality of FMCG	
Occurrence of a microbial agent	A correct flow of information is necessary between the links of the supply chain in order to obtain reliable data about the conditions of storage of goods. Control of climatic factors in the storage hall, in particular the level of humidity. The tightness of the ramps is very important in order to have a constant air temperature in the warehouse. In periods with a higher concentration of humidity in the air, investments in moisture absorbers are needed.
Contamination of the product with glass	Thorough maintenance of machines and devices, care for cleanliness.
Contamination of the product with dust	Systematic cleaning of the storage facility, in particular the floor of the storage hall. Cleaning should not only apply to the surface of the warehouse facility, but also to devices supporting storage of inventories, devices for maintaining security in the warehouse and devices for transporting goods.
Bioterrorism.	Control of people entering the warehouse facility. The verification should include persons from outside and working in a given warehouse. People from outside should always move around the building assisted by a person employed in the institution.

Source: Own study

4 Conclusions

Maintaining an appropriate level of security in the distribution of goods from the FMCG industry in logistics companies is one of the key elements of internal supply chains. In order to provide companies with adequate protection against the occurrence of risk factors, it is necessary to properly identify the possibility of a given risk factor and perform a disruption analysis in order to create a flexible and resilient system. The aim of the article was to assess the existing risk factors in internal supply chains in terms of maintaining an appropriate level of quality and safety of goods.

The applied qualitative risk management method allowed for the classification of disruptions in logistics processes using three indicators: probability, effect and detection of a given risk factor. On this basis, the four most dangerous risk factors were identified along with preventive activities, the advantage of which is low financial outlay:

- Occurrence of a microbial agent - control of climatic factors, in particular the level of humidity, with an indication of elements facilitating the maintenance of proper storage conditions for goods,
- Contamination of the product with glass - attention to cleanliness and the use of appropriate devices for handling goods,
- Contamination of the product with dust - systematic cleaning of facilities and devices,
- Bioterrorism - verification of employees and intermediate persons in contact with the goods in the supply chain.

The assessment of risk factors made it possible to identify the most dangerous disruptions in internal supply chains. Implementable preventive actions to the links of the supply chain make it possible to implement changes immediately and thus maintain an appropriate level of quality and safety of goods in logistics companies.

References

1. Baras, N., Chatzisavvas, A., Ziouzios, S., Dasygenis, M.: Improving automatic warehouse throughput by optimizing task allocation and validating the algorithm in a developed simulation tool. Autom. MDPI **2**(3) (2021)
2. Burduk, A.: Assessment of risk in a production system with the use of the FMEA analysis and linguistic variables. In: Corchado, E., Snášel, V., Abraham, A., Woźniak, M., Graña, M., Cho, S.-B. (eds.) HAIS 2012. LNCS (LNAI), vol. 7209, pp. 250–258. Springer, Heidelberg (2012). https://doi.org/10.1007/978-3-642-28931-6_24
3. Burduk, A.: Evaluation of the risk in production systems with a parallel reliability structure taking into account its acceptance level. In: Corchado, E., Kurzyński, M., Woźniak, M. (eds.) HAIS 2011. LNCS (LNAI), vol. 6679, pp. 389–396. Springer, Heidelberg (2011). https://doi.org/10.1007/978-3-642-21222-2_47
4. Kleindorfer, P.R., Saad G.H.: Managing disruption risks in supply chains. Production and Operations Management (2005)
5. Kulińska, E.: Aksjologiczny wymiar zarządzania ryzykiem procesów logistycznych. Modele i eksperymenty ekonomiczne, (Axiological dimension of risk management in logistics processes. Economic models and experiments), Oficyna Wydawnicza Politechniki Opolskiej, Opole 2011 (in Polish)
6. Kulińska, E.: Podstawy logistyki i zarządzania łańcuchem dostaw, (Basics of logistics and supply chain management), Oficyna Wydawnicza Politechniki Opolskiej Akademicki Inkubator Przedsiębiorczości, Opole 2009 (in Polish)
7. Kulińska, E., Odlanicka – Poczobutt, M., Kulińska, K.: Diagnosis of operational malfunctions and failures by the FMEA method in the industrial robots sector. Res. Logist. Product. **8**(1) (2018)
8. Liczmańska, K.: Strategie marek produktowych na rynku FMCG, (Product brand strategies on the FMCG market), Wyd. Naukowe Uniwersytetu Mikołaja Kopernika, Toruń 2016 (in Polish)
9. Liu, X.: The role of enterprise risk management in sustainable decision-making: a cross-cultural comparison. Sustain. MDPI **11**(10) (2019)
10. Mahmood, S., Dinmohammadi, F.: An FMEA-based risk assessment approach for wind turbine systems: a comparative study of onshore and offshore. Energies MDPI **7**(2) (2014)
11. Mahmood, S., Enjema, E., Kolios, A.: An integrated fta-fmea model for risk analysis of engineering systems: a case study of subsea blowout preventers. Appl. Sci. MDPI **9**(6) (2019)

12. Michlowicz. E.: Zarys logistyki przedsiębiorstwa, (Outline of the company's logistics), Wyd. Akademia Górniczo – Hutnicza, Kraków 2012 (in Polish)
13. Miller, P.: Analiza wpływu innowacyjnego narzędzia na wspomaganie działalności firm z branży FMCG na rynku Polskim, (Analysis of the impact of an innovative tool on supporting the activities of FMCG companies on the Polish market), Zeszyty Naukowe, Ekonomiczne Problemy Usług nr 122, Szczecin 2016 (in Polish)
14. Nagaraju, D.B., Thejaswini, H.D.: Consumers perception analysis-market awareness towards ecofriendly fmcg products-a case study of mysore district. IOSR J. Bus. Manage. **16**(4). Ver. V (2014)
15. PN-84/N-01800: Gospodarka magazynowa - Terminologia podstawowa, (Warehouse management - Basic terminology), NKP 162 ds. Logistyki. 1984, Kodów Kreskowych i Gospodarki Magazynowej (in Polish)
16. Rusecki, A.: Praktyczne zastosowanie metody FMEA na przykładzie produkcji koła pasowego w wybranym przedsiębiorstwie, (Practical application of the FMEA method on the example of a pulley production in a selected enterprise). Qual. Product. Improve. Nr **1**(8) (2018) (in Polish)
17. Saderova, J., Rosova, A., Sofranko, M., Kacmary, P.: Example of warehouse system design based on the principle of logistics. Sustain. MDPI **13**(8) (2021)
18. Sharma, N.: Marketing strategy on different stages PLC and it's marketing implications on FMCG products. Int. J. Mark. Finan. Serv. Manage. Res. **2**(3) (2013)
19. Ślusarczyk, Z., Janczewski, J. [red.]: Zarządzanie innowacyjne w gospodarce i biznesie, (Innovative management in economy and business), Wyd. Akademii Humanistyczno – Ekonomicznej w Łodzi **2**(23) (2016) (in Polish)

Development of Acoustic Beam Deflection Electromagnetic Acoustic Transducer for Surface Crack Detection of Wind Turbine Main Shaft

Gao Jie[1], Dong Baoyu[2], Lyu Yan[1](✉), and He Cunfu[1]

[1] Faculty of Information Technology, Beijing University of Technology, Beijing 100124, China
lvyan@bjut.edu.cn

[2] Faculty of Materials and Manufacturing, Beijing University of Technology, Beijing 100124, China

Abstract. The main shaft serves as a critical component in the wind turbine transmission system, and carries the torque transmitted by the turbine hub. However, the surface of the main shaft is vulnerable to fatigue cracks under the external harsh environmental factors. And the cracks are a serious threat to the safe operation of the turbine. To detect these defects, an electromagnetic acoustic transducer designed to fit the central hole of the spindle was introduced. Based on the Lorentz force effect and the magnetostriction, the finite element simulation model of the EMAT transducer was established, to analyze the acoustic beam deflection angle and acoustic field intensity change range. Also, the details analyze the law of influencing factors on transducer efficiency, such as magnet size, number, layer number, coil splitting and so on. The optimized transducer detects effectively surface defects on the spindle sample in the crack detection experiments. Moreover, the sector scanning detection of surface cracks can be demonstrate by adjusting the frequency, and the defect positioning error is less than 3 mm.

Keywords: wind turbine main shaft · center hole · electromagnetic acoustic transducer · acoustic beam deflection

1 Introduction

As a green and clean energy, Wind energy possesses the advantages of short construction periods, low operating costs, and minimal environmental pollution [1]. Wind power development has become a critical component of the country's energy revolution and structure adjustment, which has been strongly supported by national policies [2]. Larger wind turbines are more cost effective and are grouped together into wind farms, which provide bulk power to the electrical grid [3]. With the urgent demand for electricity, the structural health of wind turbines seems to be more significant.

The main shaft of the wind turbine serves as a key component in the transmission system. Along with the growing rotor diameters and hub heights, the loads on the main

A. Burduk et al. (Eds.): ISPEM 2023, LNNS 790, pp. 393–405, 2023.
https://doi.org/10.1007/978-3-031-45021-1_30

shaft and the main bearing are continuously increasing. The fatigue cracks, from wind turbine main shaft surface scratch, would further extend through alternating load and impact load, which finally led to the main shaft fracture. This kind of intolerable cracks seriously threaten the performance of the main shaft. It is necessary to perform health monitoring during the service period. Ultrasonic non-destructive testing presents an effective way of detecting such cracks, and their position and extension depth on the spindle surface can be determined through the ultrasonic echo signal [4].

Due to the limitations of wind turbine packaging process, arranging transducers in the spindle body position for ultrasonic non-destructive testing will be proves challenging. Numerous studies have been conducted by domestic and international scholars. For hollow axles, REVA Company [5] employed ultrasonic phased array technology to realize axle defect detection on the outer surface of the axle body. The phased array probe is used to carry out helical scanning along the inner surface of the central hole, and defect imaging detection is realized. Ogi et al. [6, 7] achieved unidirectional oblique incidence of acoustic waves using unequally spaced zigzag coils EMAT, and the focused characteristic of the excited acoustic beam was superior to that of bidirectional pointing EMAT.

Meanwhlie, the related scholars employ ultrasonic oblique probes or ultrasonic phased array probes at the central hole to detect on the outer circumference defects of the main shaft [8–10]. Due to the piezoelectric transducers require the couplant, the layout of the center hole is limited, and the existing detection technology requires downtime for testing [11–13]. Therefore, it is necessary to develop an electromagnetic acoustic transducer suitable for the layout of the central hole, and spindle outer surface defect detection without couplant assistance.

In our research, a transducer suitable for the layout of the center hole for detecting transverse cracks on the outer circumferential surface of the wind turbine main shaft was developed. A simulation model of the electromagnetic acoustic transducer with double-turnback coils was established by using finite element simulation software, and the distribution law of acoustic directivity of different frequency excitation signals was analyzed in detailed. Based on the detect experiments on the spindle specimen, It have been verified that the developed transducer can realize to detect the surface defects of the spindle.

2 EMAT Transduction Mechanism

The oblique incident shear wave electromagnetic acoustic transducer is composed of permanent magnet, return coil. When the current is passed through the return coil, the surface of the sample will induce the pulsed eddy current of the same frequency, and the Lorentz force will be generated under the action of the biased static magnetic field generated by the permanent magnet. Under the action of the Lorentz force, the specimen will be vibrate periodically and propagate as an ultrasonic wave, the excitation of obliquely incident shear vertical (SV) waves can be realized.

The obliquely incident θ of SV wave is related to the wire spacing d of the reflex coil, the wave speed v, and the excitation frequency f. When the Eq. (1) is satisfied, the obliquely incident SV wave can be excited, where the λ is the wavelength.

$$\sin\theta = \frac{v}{2df} = \frac{\lambda/2}{d} \tag{1}$$

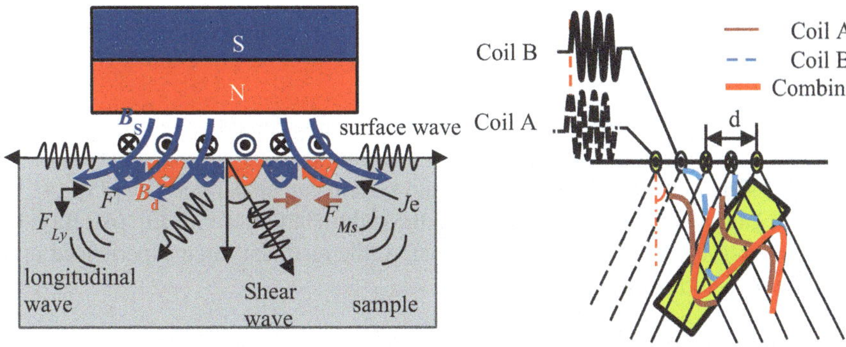

a) Schematic diagram of the return coil EMAT

b) Transduction mechanism of unidirectional oblique incidence

Fig. 1. Transduction mechanism of obliquely incident shear wave EMAT

To achieve unidirectional radiation of ultrasonic waves, two sets of the combined excitation of return coils are employed. Both sets of coils are supplied with the same excitation current amplitude and frequency. Meanwhile, the phase difference is T/4, and the spacing between adjacent meandering coils is d/2. This configuration results in the enhancement of ultrasonic waves propagating at an angle of θ on one side, and weakening on the other side, as shown in Fig. 1.

3 Finite Element Modeling for EMAT Deflection Analysis

Based on the principle of double coil excitation of unidirectional radiation transverse wave, a finite element simulation model based on the joint effect of Lorentz effect and magnetostriction is established by using COMSOL software. The simulation takes into account the requirement that the ultrasonic wavelength needs to be smaller than the defect to be detected when using ultrasonic bulk waves to detect small defects. In addition, the transducer coil spacing and excitation signal frequency are set to satisfy Eq. (1). Four groups of transducer coil spacing, including 1 mm, 1.2 mm, 1.4 mm, and 1.6 mm were set up separately. And the deflection range of the acoustic beam is analyzed by changing the coil excitation frequency. The built model simulates the acoustic wave generation process by using the force loading form, as shown in Fig. 2.

The sample is a semicircle with the radius of 50mm. The finite element model uses the triangular element with a maximum mesh size of 0.2mm, where the mesh size is less than one-eighth of the ultrasonic length. Since the sample is made of ferromagnetic material [13], the transduction process is subject to a combination of Lorentz and magnetostriction effects, as shown in Eq. (2). Where the fMs is the magnetostrictive force, fL is Lorentz

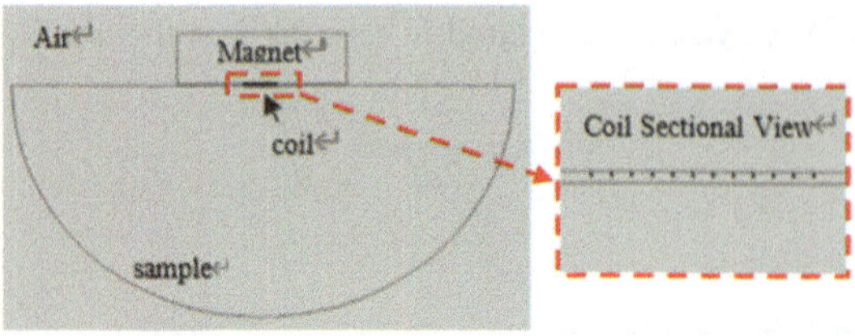

Fig. 2. EMAT simulation model of obliquely incident shear wave

force. In ferromagnetic materials, the fMs can be represented by Eq. (3), where H is the dynamic magnetic field strength, and e is the inverse pressure magnetic coefficient matrix. The fMs in the rectangular coordinate system can be derived from Eq. (4), and the Eq. (5) can be obtained by multiplying the right-hand matrix of Eq. (4). Based on the inverse pressure magnetic coefficient matrix, the resultant force experienced in the model can be determined.

$$f = f_L + f_{Ms} \tag{2}$$

$$f_{Ms} = -\nabla_t\left(e^T H\right) \tag{3}$$

$$f_{Ms} = \begin{bmatrix} f_{Msx} \\ f_{Msy} \\ f_{Msz} \end{bmatrix} = -\begin{bmatrix} \frac{\partial}{\partial x} & 0 & 0 & 0 & 0 & \frac{\partial}{\partial y} \\ 0 & \frac{\partial}{\partial y} & 0 & 0 & 0 & \frac{\partial}{\partial x} \\ 0 & 0 & 0 & \frac{\partial}{\partial y} & \frac{\partial}{\partial x} & 0 \end{bmatrix} \begin{bmatrix} e_{11} & e_{21} & e_{31} \\ e_{12} & e_{22} & e_{32} \\ e_{13} & e_{23} & e_{33} \\ e_{14} & e_{24} & e_{34} \\ e_{15} & e_{25} & e_{35} \\ e_{16} & e_{26} & e_{36} \end{bmatrix} \begin{bmatrix} H_x \\ H_y \\ 0 \end{bmatrix} \tag{4}$$

$$f_{Ms} = \begin{bmatrix} f_{Msx} \\ f_{Msy} \\ f_{Msz} \end{bmatrix} = -\begin{bmatrix} e_{11} & e_{21} & e_{16} & e_{26} \\ e_{16} & e_{26} & e_{12} & e_{22} \\ e_{15} & e_{25} & e_{14} & e_{24} \end{bmatrix} \begin{bmatrix} \frac{\partial H_x}{\partial x} \\ \frac{\partial H_y}{\partial x} \\ \frac{\partial H_x}{\partial y} \\ \frac{\partial H_y}{\partial y} \end{bmatrix} \tag{5}$$

When the excitation frequency of double-folded coil EMAT is 1.5 MHz, the stress distribution cloud map can be obtained by the proposed model, as shown in Fig. 3. It can be observed that the double-folded coil EMAT can effectively excite the unidirectional oblique incident ultrasonic shear waves.

The acoustic wave amplitude was extracted and normalized at the position of the 45mm radius semicircular arc, and the acoustic wave amplitude at different frequencies for the dual foldback coil EMAT sensor with various coil spacing can be acquired, as illustrated in Fig. 4. The simulation explores the effect of different frequencies on the variation of acoustic wave directivity and amplitude. The directivity results were normalized, and the -6dB method was used to calculate the acoustic wave pointing range. The findings reveal that for the 1.0mm coil spacing EMAT, the acoustic beam pointing angle can be deflected within a range of 26°–62°, when the frequency is adjusted

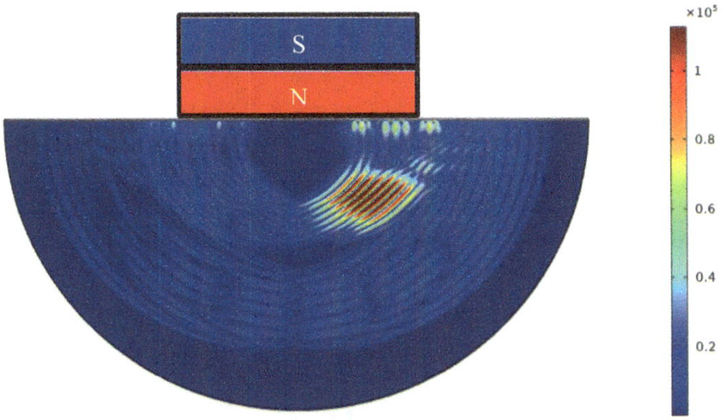

Fig. 3. Stress distribution cloud map in the sample

between 1.6 MHz–3.5 MHz. Where the acoustic wave will be covered approximately at 59°. Similarly, for the 1.2mm coil spacing EMAT, the acoustic beam pointing angle can be deflected within the range of 26°–62° when the frequency is adjusted between 1.4 MHz–3 MHz. Where the acoustic wave will be covered at around 58°. In comparison, the acoustic beam pointing angle for the 1.4mm coil spacing EMAT can be deflected within the range of 24°–60° when the frequency is adjusted between 1.2 MHz–2.8 MHz, with the acoustic wave coverage of about 60°. For the 1.6mm coil spacing EMAT, the acoustic beam pointing angle can be deflected within the range of 24°–62° when the frequency is adjusted between 1 MHz–2.5 MHz, with the acoustic wave coverage of approximately 60°. The results indicate that the deflection of the acoustic beam pointing angle of 25°–60° can be achieved for EMAT transducers with different spacings. And the deflection range of the acoustic wave is approximately 60°, which satisfies the requirements of scanning and detection.

Furthermore, the variation trend of the acoustic wave amplitude of the double-fold EMAT transducers with different spacings at different angles were analyzed, and the results are presented in Fig. 5. As the deflection angle increases, the acoustic wave amplitude exhibits an initial increase followed and then decrease. The maximum value is attained at around 45°, with the deflection angle ranging between 35° and 55°. And the acoustic wave amplitude was optimal when the transducer with a coil spacing of 1.6 mm.

4 Quantitative Detection of Spindle Surface Cracks

4.1 Diffraction Wave Detection Method

To measure the propagation time of the diffracted wave, the time of flight (TOF) from the excitation transducer to the receiving transducer can be used. Based on the TOF of the wave and the position of the transducer, the diffraction point can be located on the ellipse [14]. This is because the two foci of the ellipse are the positions of the transducer pairs, and the product of the TOF and the wave speed is equal to the sum of their

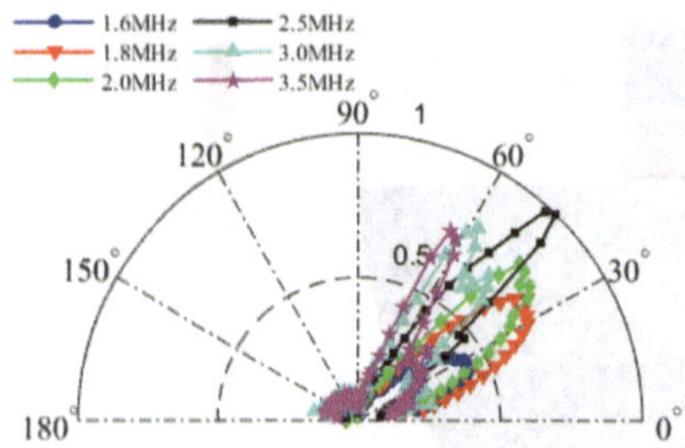

a) The trend of EMAT acoustic wave amplitude at a spacing of 1 mm.

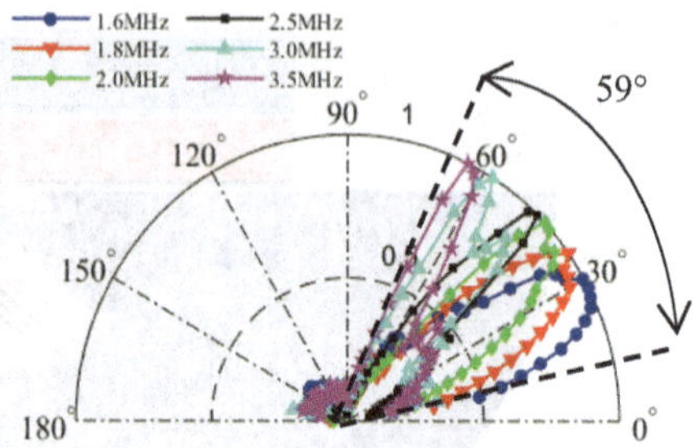

b) The range of EMAT acoustic wave angle at a spacing of 1 mm.

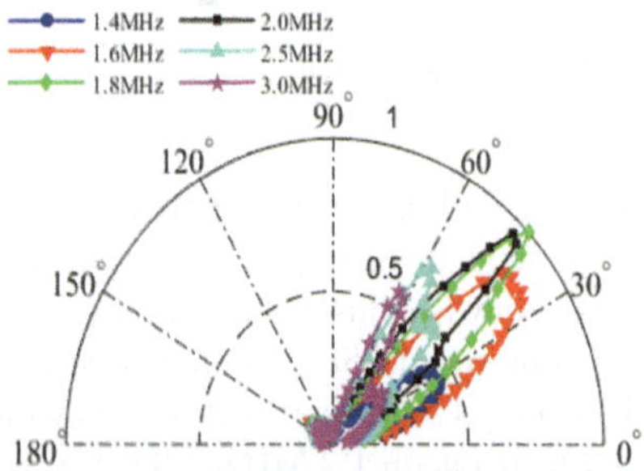

c) The trend of EMAT acoustic wave amplitude at a spacing of 1.2 mm.

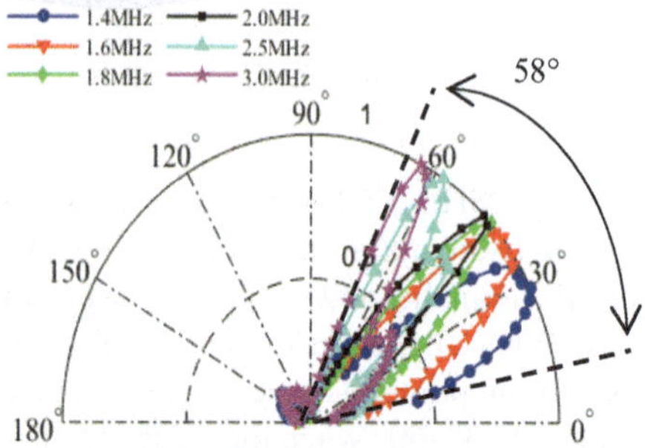

d) The range of EMAT acoustic wave angle at a spacing of 1.2 mm.

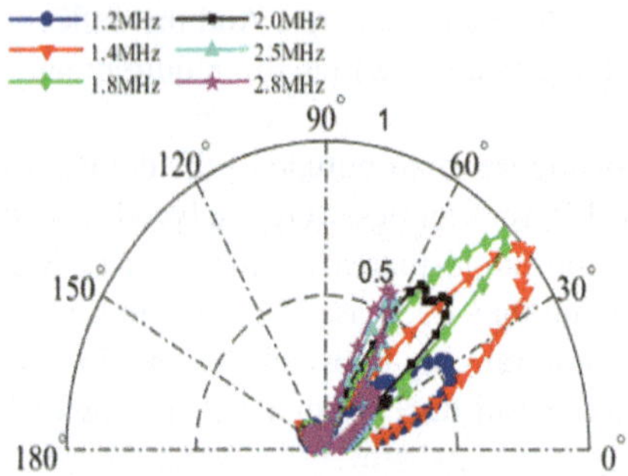

e) The trend of EMAT acoustic wave amplitude at a spacing of 1.4 mm.

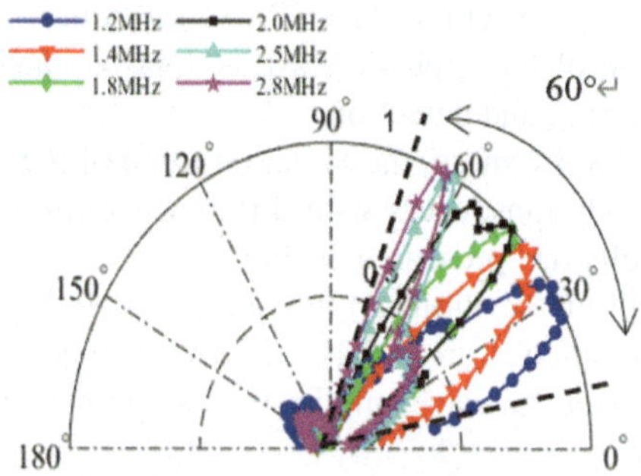

f) The range of EMAT acoustic wave angle at a spacing of 1.4 mm.

Fig. 4. Acoustic deflection range of different spacing EMAT at different frequencies

distances from the diffraction point to the excitation-receive transducer pair. Therefore, when the abscissa of the diffraction point is determined, the ordinate can be determined by solving the elliptic equation. Meanwhile, the precise position of the diffraction point can be obtained effectively. The characteristic process is illustrated in Fig. 6.

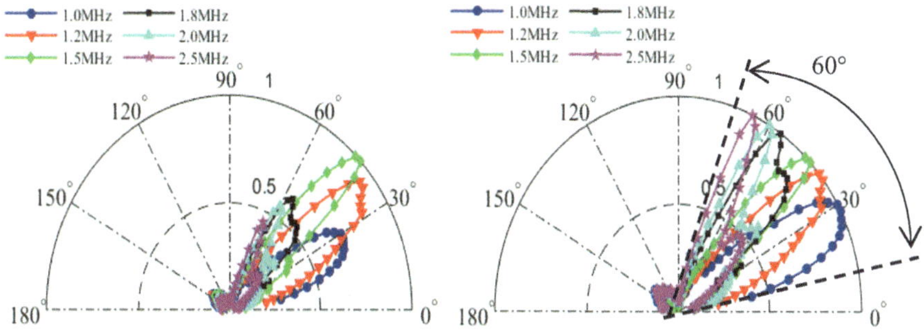

g) The trend of EMAT acoustic wave amplitude at a spacing of 1.6 mm.

h) The range of EMAT acoustic wave angle at a spacing of 1.6 mm.

Fig. 4. (*continued*)

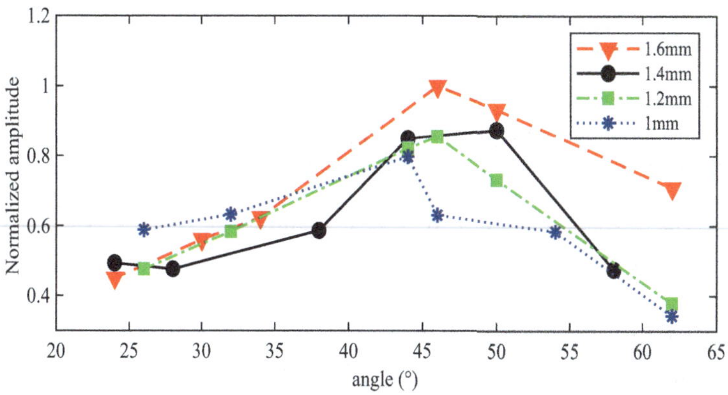

Fig. 5. Amplitude variation of EMAT with different spacing at different deflection angles

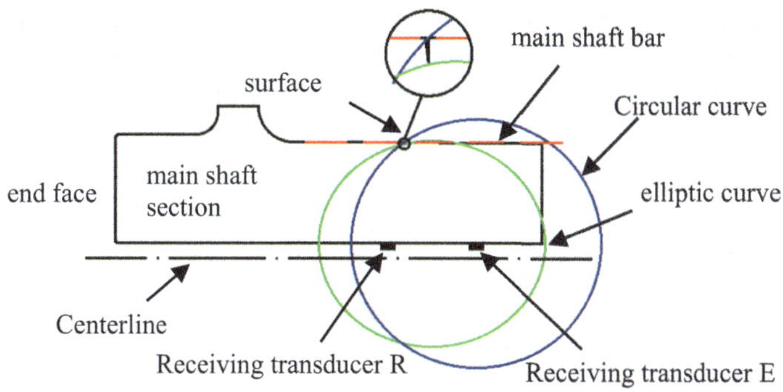

Fig. 6. Acoustic characterization of spindle surface cracks

To locate the position of spindle surface crack, an excitation transducer is used to generate ultrasonic waves in the specific direction towards the central hole of the spindle. The end angle reflection echoes from the opening of the surface crack are received, and the distance L from the crack position to the excitation transducer is determined based on the TOF t_1 and the wave velocity c_s. The circle can be obtained by using the excitation point and distance as the center and the radius, and the intersection of the circle with the high radius of the spindle circumference is the opening position. The axial coordinate of the lower point is recorded.

Another receiving transducer in the central hole is used to receive the diffraction echo signal of the crack tip. The diffracted wave TOF ts of the and cs are used to determine the distance of the acoustic wave from the excitation point to the defect tip and then to the receiving point, based on the diffraction time difference method. The elliptical trajectory of the acoustic wave determines the elliptical mathematical model with the excitation point and the receiving point as the focal points. The long axis a, short axis b, and focal length c of the ellipse model are numerically calculated using the Eq. (6).

$$\begin{cases} \mathrm{a} = (t_s \times c_s)/2 \\ c = L_E - L_O \\ b = \sqrt{c^2 - a^2} \end{cases} \tag{6}$$

The axial distance coordinates can be determined by using the reflected echo from the surface crack opening, and the elliptical trajectory parameters can be obtained by the diffraction echo from the crack tip. That is to say, the position coordinates of the crack tip can be acquired, and thus the crack depth information can be quantitatively characterized.

4.2 Spindle Crack Evaluation Model

The longitudinal section of the main shaft was obtained by cutting the shaft body along its cylindrical axis X. Then, the origin O was defined as the center position of the end face, and the excitation transducer and receiver transducer were located at $E(\mathrm{x_E}, \mathrm{y_E})$ and $R(x_R, y_R)$, respectively. The radius of the cross section of the shaft section where the crack was located at m. The TOF of reflected wave was designated as t_0, and the TOF of diffracted waves by the receiving transducer was denoted as t_1, as shown in Fig. 7.

The circular trajectory of the blue line can be represent as:

$$\begin{cases} |\mathrm{QE}| = r \\ r = vt_0 \end{cases} \tag{7}$$

The green elliptical trajectory can be represent as:

$$\begin{cases} |\mathrm{PE}| + |\mathrm{PR}| = 2a \\ 2a = vt_1 \end{cases} \tag{8}$$

According to the parameter relationship, the circular and the ellipse equations can be determined. The coordinates of crack tip can be determined by combining the Eqs. (9–10), yield:

$$(x - x_E)^2 + (y - y_E)^2 = r^2 \tag{9}$$

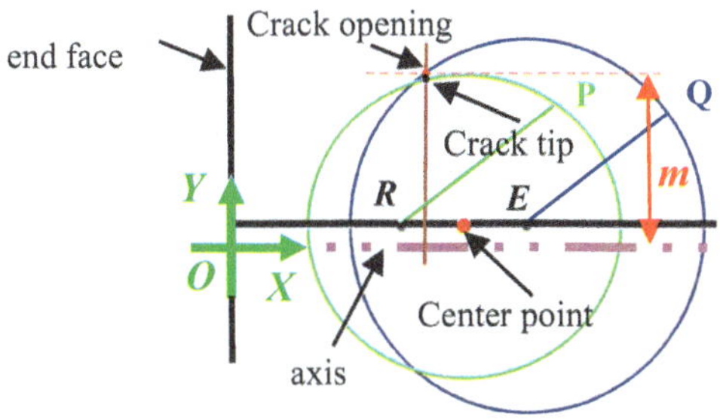

Fig. 7. Quantitative mathematical model of spindle surface crack

$$\frac{(x-x_o)^2}{a^2}+\frac{(y-y_o)^2}{b^2}=1 \tag{10}$$

$$\begin{cases} x_j=\sqrt{r^2-(m-y_E)^2}+x_E \\ y_j=\sqrt{\left[1-\frac{(x_j-x_o)^2}{a^2}\right]\times b^2}+y_o \end{cases} \tag{11}$$

The crack opening position can be determined using the horizontal coordinate, while the longitudinal coordinate information helps in identifying the location of the crack tip. Furthermore, the radius information of the shaft end position enables the calculation of the extension depth d of the main shaft crack, as demonstrated in Eq. (12).

$$d=m-y_j \tag{12}$$

5 Experimental Analysis

5.1 Transducer Directivity Experiment

The detection system has been established by comprising high-power pulse generator/receiver, oscilloscope, EMAT sensors and so on, as illustrated in Fig. 8. The semicircular test block is 100 mm in radius and thickness, and the EMAT transducer positioned locates at its center. Meanwhile, the received piezoelectric transducer fixed by clamping tool, and move along the outer circumference of the test block. The pitch-catch experiments use a standard five cycle Hanning-windowed sine signal with the excitation signals of the two sets of coils differing in phase by T/4. The piezoelectric transducer was moved at intervals of 2°, ranging from 20° to 160°.

The direct shear wave amplitudes can be required by the piezoelectric transducer at each angle in turn. It is also normalized and the polar coordinate receiver characteristic curve is plotted as shown in Fig. 9.

The double-folded coil EMAT transducer exhibits unidirectional directivity in acoustic wave excitation. The acoustic waves located in the superposition region can cause the

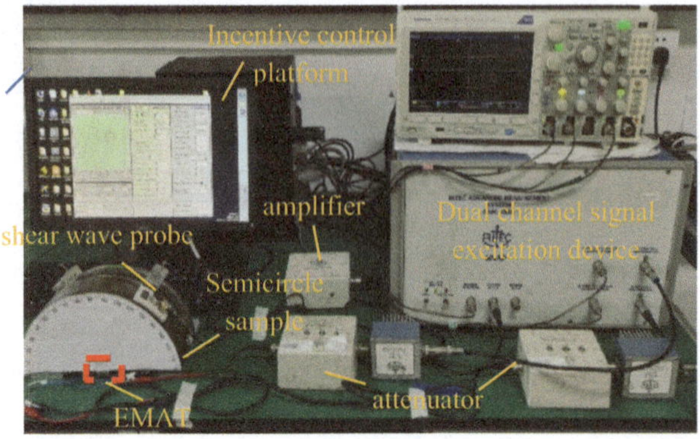

Fig. 8. EMAT performance verification platform for double-folded coil

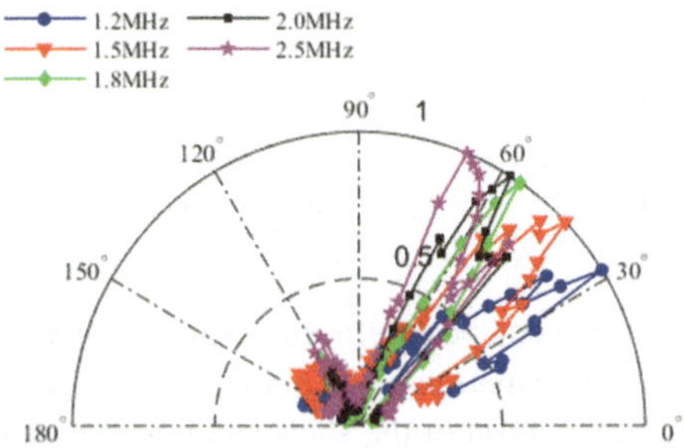

Fig. 9. Beam directivity of double-folded coil EMAT

pressure amplitude to increase, while the acoustic waves in the offset region will cause the pressure amplitude to weaken, and eventually form a unidirectional sound beam. By adjusting the frequency of the excitation signal, the wavelength of the acoustic wave is altered. Based on this, the direction of the strongest acoustic beam in the superposition process can be changed and achieve acoustic wave excitation with different pointing angles. For the double-return EMAT transducer with a coil spacing of 1.6mm, the frequency range of 1.2 MHz–2 MHz can change the deflection angle of the acoustic beam from 24° to 58°. And the wave directivity at different frequencies is normalized, determining the overall coverage of the sound wave to be about 50°, as indicated by the amplitude drop of 6dB. Since the transverse wave sensor needs to be in contact with the surface of the sample, the non-uniform coupling of the contact surface will lead to partial errors between the experimental results and the simulation data.

5.2 Spindle Detection Experiment

The developed transducer is conducted on detecting the cracks of the wind turbine spindles, the experimental setup is illustrated in Fig. 10, where the excitation transducer is positioned at the central hole of the main shaft. The piezoelectric transducer at the central hole can acquire the diffraction shear wave, and the time-domain waveforms were obtained. The experimental results are depicted in Fig. 11.

Then, submitting the TOF into the above crack evaluation model in Sect. 4, the crack propagation depth can be obtained effectively. Where the TOFs including the reflected wave and diffracted wave, respectively. Based on this, the evaluation results are shown in Fig. 12.

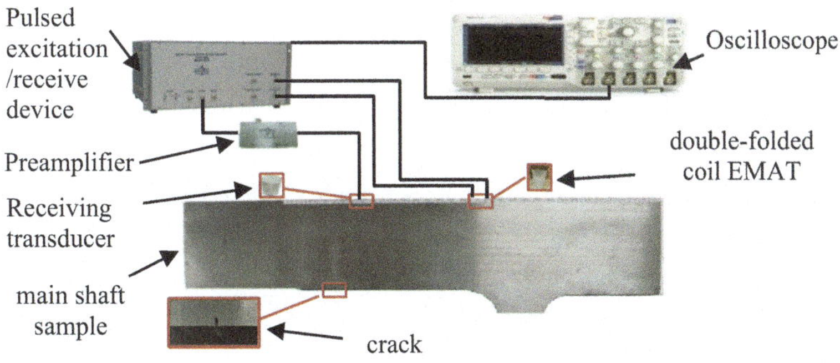

Fig. 10. Experimental system for Surface Cracks of Spindle

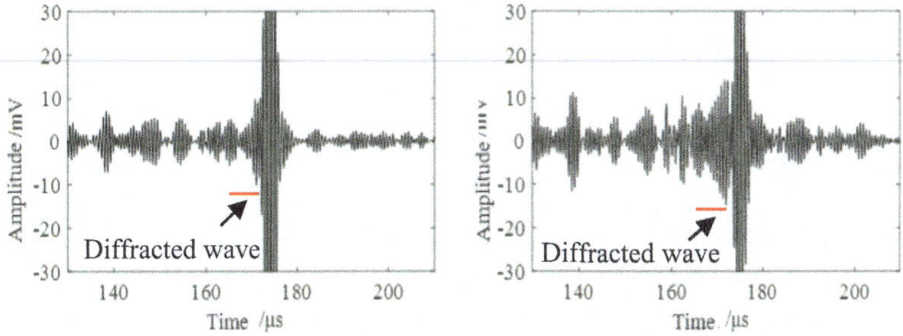

a) 710mm receiving point time domain waveform

b) 720mm receiving point time domain waveform

Fig. 11. Time domain waveforms of diffracted waves received at different positions

Table 1 presents the TOF, quantification results, and quantification errors of diffracted waves at different positions. For the 5 mm depth crack, the average quantification error is 1.15 mm, and the maximum quantification error is less than 3mm. However, there were variations in the crack quantification error of the receiving points at different positions, which is associated with the accuracy of the crack diffraction wave TOF.

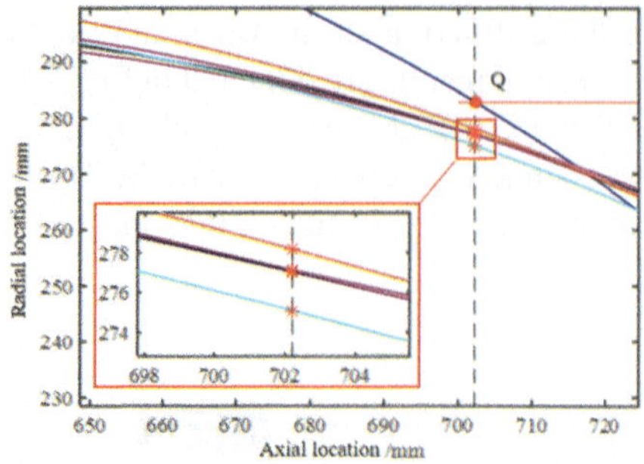

Fig. 12. Spindle surface crack localization results

Table 1. Quantification results (QR) of crack depth at different receiving positions

Position (mm)	TOF (μs)	QR (mm)	depth (mm)	Error (mm)
690	169.676	5.83	5	0.83
680	169.836	5.94		0.94
670	170.232	5.86		0.86
660	169.600	7.89		2.89
650	171.892	4.78		0.22

6 Conclusion

In this research, the EMAT simulation model of a double-bending coil based on the combined effect of magnetostriction and Lorentz force is established for the wind turbine spindle. And the effect of coil spacing and excitation frequency on the deflection of the obliquely incident transverse wave acoustic beam is analyzed. The results indicate that, the transducer achieves the around 60° beam deflection for oblique incident shear waves, and the strongest acoustic field intensity distributes along the pointing angle of around 45°. Based on the corresponding detection experiments, the results also show that the transducer can effectively detect surface cracks on the main shaft, with an average quantification error of 1.15mm for a 5mm depth crack. Furthermore, the quantification error can be reduced by increasing the receiving positions of diffracted wave. By adjusting the frequency of the excitation signal, the scanning detection of cracks at different positions of the spindle can be realized. Moreover, the location and depth of the crack can be determined quickly by using the proposed EMAT transducer.

References

1. Li, D., Ho, S., Song, G., et al.: A review of damage detection methods for wind turbine blades. Smart Mater. Struct. **24**(3), 033001 (2015)
2. Cobb, A.C., Fisher, J.L., Shiokawa, N., et al.: Development of a pseudo phased array technique using EMATs for DM weld testing. American Institute of Physics Conference Series. American Institute of Physics, 1015–1022 (2015)
3. Kawashima, K.: Electromagnetic acoustic wave source and measurement and calculation of vertical and horizontal displacements of surface waves. IEEE Trans. Sonics Ultrasonics **32**(4), 514–522 (1985)
4. Wang, S., Li, Z., Kang, L., et al.: Modeling and comparison of three bulk wave EMATs 2645–2650 (2011)
5. Marty, P.N., Engl, G., Krafft, S., et al.: Latest development in the UT inspection of train wheels and axles. In: Proceedings of the 18th World Conference on Nondestructive Testing, vol. 1620. Durban, South Africa (2012)
6. Ogi, H., Hirao, M., Ohtani, T.: Line-focusing of ultrasonic SV wave by electromagnetic acoustic transducer. J. Acoust. Soc. Am. **103**(5), 2411–2415 (1998)
7. Ogi, H., Hirao, M.: Line-focusing electromagnetic acoustic transducers for the detection of slitdefects. IEEE Trans. Ultrason. Ferroelectr. Freq. Control **46**(2), 341 (1999)
8. Zhai, G., Wang, K., Wang, Y., et al.: Modeling of Lorentz forces and radiated wave fields for bulk wave electromagnetic acoustic transducers. J. Appl. Phys. **114**(5), 244106–244157 (2013)
9. Jia, X., Ouyang, Q., Zhang, X.: An improved design of the spiral-coil EMAT for enhancing the signal amplitude. Transducers **17**(5), 1106 (2017)
10. Ribichini, R., Nagy, P.B., Ogi, H.: The impact of magnetostriction on the transduction of normal bias field EMATs. NDT E Int. **51**(10), 8–15 (2012)
11. Uchimoto, T., Takagi, T., Ichihara, T., et al.: Evaluation of fatigue cracks by an angle beam EMAT–ET dual probe. NDT E Int. **72**, 10–16 (2015)
12. Petcher, P.A., Dixon, S.: Weld defect detection using PPM EMAT generated shear horizontal ultrasound. NDT E Int. **74**, 58–65 (2015)
13. Hill, S., Dixon, S.: Frequency dependent directivity of periodic permanent magnet electromagnetic acoustic transducers. NDT E Int. **62**, 137–143 (2014)
14. Nath, S.K., Balasubramaniam, K., Krishnamurthy, C.V., et al.: Sizing of surface-breaking cracks in complex geometry components by ultrasonic Time-of-Flight Diffraction (TOFD) technique. Insight-Non-Destruct. Test. Condit. Monit. **49**(4), 200–206 (2007)

Analysis of Human Errors in the Traditional and Automated Order-Picking System

Agnieszka A. Tubis(✉), Juni Rohman, Anna Smok, and Dominika Dopart

Faculty of Mechanical Engineering, Wroclaw University of Science and Technology, Wyspiańskiego Street 27, 50-370 Wroclaw, Poland
agnieszka.tubis@pwr.edu.pl

Abstract. Processes related to the automation and digitization of warehouse systems intensified after 2020. The reasons for this can be found in two areas: (1) the development of the Logistics 4.0 concept supported by the popularization of tools provided by Industry 4.0 and (2) the effects of the COVID-19 pandemic, particularly disruptions and delays in logistics flows caused by employee absenteeism. Managers see automation as a solution to eliminating human errors that reduce the company's logistics service level. However, automation does not eliminate humans from logistics processes but only changes their roles. This means that instead of errors made by a single contractor, systemic errors are made by the person planning and monitoring the entire process. For this reason, it is crucial to identify potential adverse events and assess the risk of their occurrence to implement preventive solutions. The article aims to present the research results on analyzing adverse events in the picking process at a selected manufacturer. The tests were conducted in a real system - in a warehouse of finished products. The research aimed to identify adverse events caused by human errors or the functioning of the automatic load-handling system. For this reason, the measurements were carried out for the traditional picking process (person-to-goods method) and after introducing an automatic storage system (goods-to-person system).

Keywords: Human Error · Order-Picking · AS/RS system

1 Introduction

For over a decade, we have been observing the constant development of the Industry 4.0 (I4.0) concept, which introduces significant technological and organizational changes in the entire business environment. The advantage of the fourth industrial revolution is built on providing solutions based on a high level of digitization of processes, automation, virtualization, and centralization [1]. The implemented solutions aim to increase the efficiency and effectiveness of operations and reduce human errors. Initially, the changes introduced concerned mainly the digitization of production systems and the creation of the so-called *Smart Factories* [2]. However, the transformation of production systems also forced changes in supporting processes. For this reason, the area of research interest is currently the continuous development of concepts such as Logistics 4.0 and Maintenance 4.0. In particular, operations related to internal logistics are an area of implementing many solutions that improve the picking and internal transport processes.

A. Burduk et al. (Eds.): ISPEM 2023, LNNS 790, pp. 406–419, 2023.
https://doi.org/10.1007/978-3-031-45021-1_31

The phenomenon of automation and digitalization of cargo handling processes intensified in particular after 2020. The reasons for this can be found in two areas: (1) the development of the Logistics 4.0 concept supported by the popularization of tools provided by Industry 4.0 and (2) the effects of the COVID-19 pandemic, particularly disruptions and delays in logistic flows caused by employee absenteeism. The COVID-19 pandemic revealed a strong dependence on warehouse processes on the availability and psycho-physical condition of employees involved in them. On the other hand, the development of Logistics 4.0 provided managers with tools to reduce this dependence.

Automation and digitalization of logistics processes are also an attempt to reduce human errors, particularly in simple and repetitive operations. The phenomenon of human errors intensifies with the increased employee turnover rates in logistics systems. New employees require not only training and acquiring specific skills but also gaining experience and skill in performing duties. This means that in the initial period of their employment, the risk of making mistakes by them is high. Taking into account the fact that the employee turnover rate in Poland has been growing in recent years (in 2022, it amounted to 22%) and taking into account the approaching demographic decline, many enterprises are investing in automated solutions that, in their opinion, will allow them to protect themselves against these phenomena.

Automation and digitization of logistics processes are often considered a remedy for human errors in simple warehouse operations. However, it should be noted that the implemented Industry 4.0 solutions may cause other types of interference, which should also be analyzed. The implemented cyber-physical systems require higher competencies from the people operating them, and human errors made in the process of programming and maintaining these systems have higher consequences than those committed by operators in traditional logistics service systems. For this reason, the risk assessment carried out as part of the ex-ante analysis is particularly important. The results are the basis for designing solutions protecting against adverse events.

The article aims to present the research results on identifying adverse events occurring in the picking process at a selected manufacturer. The tests were carried out in a real system - in the warehouse of finished products. The research aimed to identify adverse events caused by human errors and the possibilities of reducing them by implementing Industry 4.0 solutions. For this reason, measurements were carried out for the traditional picking process (man-to-goods method) and after the introduction of automated storage and retrieval systems (goods-to-man system). Additionally, an ex-ante analysis was carried out for the implemented goods-to-man system, the purpose of which was to identify potential adverse events. The causes of their occurrence should be limited, and the risk of their occurrence should be constantly monitored.

The article's structure includes 5 sections, including the introduction and conclusion. Section 2 provides a basic overview of the literature in the area studied. Section 3 presents and characterizes the successive stages of the conducted research. The research results obtained are described in Sect. 4. The conclusions are included in Sect. 5.

2 Literature Review

The human factor is the subject of numerous studies on production engineering [3, 4] and transport [5, 6]. In both of these technical systems, the human factor is analyzed primarily from the point of view of possible errors that may interfere with the correct implementation of the supported processes. This is mainly because the human factor is mentioned in numerous studies as the cause of most registered accidents [7]. It is worth noting that the problem of human error is analyzed in the literature in two aspects [8]: (1) The person approach focuses on the errors of individuals, blaming them for forgetfulness, inattention, or moral weakness; (2) The system approach concentrates on the conditions under which individuals work and tries to build defenses to avert errors or mitigate their effects. However, as Read et al. [9] note, both approaches illustrate the evolution of security management practices in an organization.

Most human factors research focuses on optimizing the design of human-system interactions to improve system performance and operator well-being [10]. For this reason, the basic assumption for research in this area is the systematic use of knowledge concerning relevant human characteristics to achieve compatibility in designing interactive systems of people, machines, environments, and devices to ensure specific goals [11]. As part of these studies, the mental and physical requirements for operators operating a given system are often determined. Their identification is used to determine the demand for training, the required knowledge, and competencies of employees, but also analyzes the causes and effects of fatigue, discomfort, and injuries among employees [12]. Failure to take these requirements into account often translates into operator errors, a decrease in the efficiency of the entire process, and failure to achieve the goals assumed for a given system correctly. Such conclusions are consistent with the theory of social engineering systems, which emphasizes the importance of a good match between system technology and human operators who work together to achieve a common goal [12].

The development of the Industry 4.0 concept, observed for several years, and the accompanying digitization and automation of enterprise processes mean that human errors have changed their nature. Industry 4.0 solutions have introduced evident changes in how human work is organized and performed [13]. Therefore, the critical issue is the awareness that implementing these solutions is not only a process of adapting new technologies but, above all, is a fundamental change taking place in organizational strategy, business processes, organizational knowledge, and the whole socio-technical organizational system [14]. This is confirmed by the research presented in [15], according to which digitization introduces changes regarding: (1) technical capabilities and technological infrastructure in the organization, (2) organization strategy and development directions, (3) human factor potential, (4) integration all stakeholders around the organization. This means that technological and organizational changes also significantly impact the human factor involved in the implemented processes. One example of this impact is the fact that digital transformation forces the development of specialized capabilities in an organization's employees and workers [16, 17]. [13] The authors of numerous publications indicate that I4.0 introduced changes to the rules of competition, value delivery functions, labor market, socioenvironmental norms, and even educational priorities (among others [18, 19]).

Industry 4.0 connects physical and digital systems in real-time using new technologies. Thanks to this, it represents a new level of organization and control of the entire value chain of the product life cycle [20]. These technologies are changing the system of work but also the way that work is done. Thanks to this, I4.0 changes the traditional way of doing business, introducing agile, flexible, and higher-quality solutions [21]. The dynamic nature of Industry 4.0, the constant paradigm shift in the dynamics of digital technologies, and radical changes in the business environment have made the concept of Industry 4.0 undergo revolutionary changes in recent years [22]. In the first phase of the development of this concept, the primary attention of researchers was focused primarily on aspects related to the digitization of production processes. For this reason, one of the key concepts within Industry 4.0 is the so-called Smart Factory, also termed digital or intelligent factory [2]. Today, more and more researchers are also dealing with the use of Industry 4.0 technology in production-related processes, including the development of Logistics 4.0 [23]. The development of the fourth revolution in this way means that when identifying the economic benefits of implementing I4.0 solutions, researchers divided them into two clusters [22]:

1. Opportunities offered for manufacturing productivity, np. Improved production monitoring and control, labor productivity, and reduced inventory.
2. Opportunities for overall business competitiveness, such as organizational agility, supply chain integration, increased profitability, and innovation opportunities

Logistics 4.0 is a combination of intelligent technologies whose applications are in inventory management, warehousing, distribution, and transportation [24]. It is some new technological direction that combines technologies intending to make the entire supply chain more efficient and effective while changing the focus of enterprises on value chains, maximizing value for consumers and customers by increasing competitiveness through digitalization [25, 26]. Logistics 4.0 uses new technologies, such as cyber-physical systems, to support the functioning of traditional logistics systems [27]. New solutions improve networking and automation of storage system transportation and decentralized software control [28]. Among the most frequently discussed solutions in this area are the AGV systems - Automated Guided Vehicle (e.g. [29, 30]) and UAV - Unmanned Aerial Vehicle (e.g. [31, 32]). However, other automatic solutions that improve warehouse processes related to the transport and handling of loads are equally important.

The most important benefits for companies implementing Logistics 4.0 solutions include [33]: simplified monitoring of logistics systems, increased environmental considerateness, increased logistics awareness, minimized waste of costs, time, and energy, creation of new business models, design of flexible logistics processes that respond promptly to consumer demands. These benefits are noted primarily in resource planning, transportation management systems, warehouse management systems, and intelligent transportation systems superstructure [34]. For this reason, the development of Logistics 4.0 is becoming increasingly important from the point of view of the development of the enterprise, but also of entire supply chains.. This is confirmed by the number of publications relating to these solutions, which has steadily increased since 2017 [26].

Digital solutions of Industry 4.0 have been used primarily in internal logistics processes and enabled the design of intelligent warehouses, also known as Warehouses 4.0.

These warehouses are designed and operate following the basic principles of I4.0, which include interoperability, virtualization, decentralization, real-time, service orientation, modularity, and reconfigurability [35]. These warehouses respond to the current need to effectively manage all processes in the warehouse and design a dynamic warehouse facility that will be easily accessible [36]. The order-picking process should be distinguished among the most frequently automated warehouse operations. This is confirmed by the number of publications on this issue and the fact that in the literature, one can even find research on the concept of Order Picking 4.0 [37].

3 Characteristics of the Research Subject and Research Procedure

3.1 Characteristics of the Research Subject

The research was conducted in a selected logistics company dealing with the picking and distributing goods from various sectors. Until 2022, all orders were picked traditionally, i.e., person-to-goods. In 2022, the company's management decided to automate the picking process in response to emerging disruptions. Two categories of products have been separated for the automation process, the so-called Beauty and FMCG groups, but only to small-size individual orders. These processes have been indicated for modification due to high repeatability, simplicity of the procedure, and high time consumption.

A warehouse handling system based on automatic stacker cranes was introduced to the picking process of both product categories. The automatic storage system has two main functions. The first is the storage of products in containers designated for this purpose. In this part, several stacker cranes move, whose task is to insert and remove containers with products in the appropriate place on the rack and then transport them to the feeder, which is the second part in which the order-picking process takes place. Containers move along paths defined by the system. The employee is delivered products and cartons into which subsequent order items are inserted. The detailed implementation of the process in a traditional and automated system is described in Sect. 4.

3.2 Research Method

The analyzes were carried out in two research periods. The first research on the traditional picking system was carried out in 2021. The second research phase related to the automated system and used observations and measurements carried out in 2022/2023. The research procedure is shown in Fig. 1.

The first stage of the research concerned analyzing the picking process of selected product categories. The internal documentation of the company (applicable procedures) and observations in the picking zone were used to carry out the process analysis. On this basis, the processes for the traditional picking system were mapped (Fig. 2).

The next analysis stage identified human errors occurring in the picking process in the "man to product" system. For this purpose, accompanying observations were carried out in the picking area. In addition, based on the "What-If" analysis, undesirable events were identified that were not registered during the observations and may occur in the analyzed process in the event of non-compliance with the applicable procedures. Then, these events were assessed by managers in terms of their frequency of occurrence in the analyzed process. Potential causes of their occurrence were also indicated for the identified human errors.

The third stage of the research concerned the mapping and analysis of the picking process carried out in the "goods-to-man" system. This process is also shown in Fig. 2.

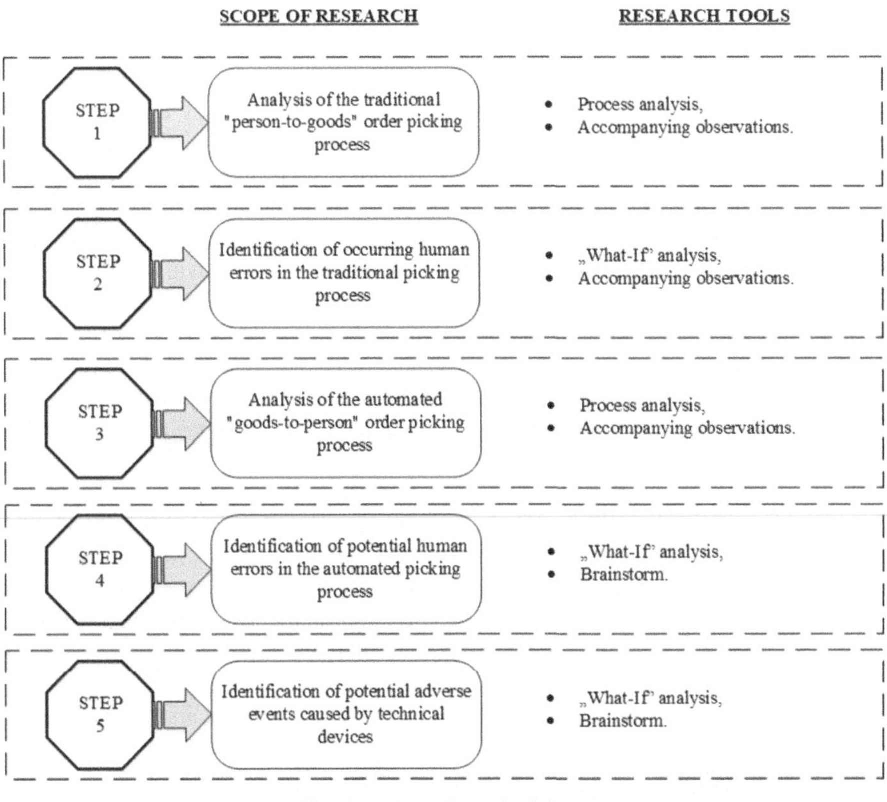

Fig. 1. Research methodology.

In the fourth step, the research team identified possible human errors that could occur in the new picking process. These errors were investigated based on the accompanying observations and the "What-If" analysis. However, it should be emphasized that while in stage 2, the identification of errors used historical data and knowledge of past events, in stage 4, the identification of adverse events is predictive. The purpose of the analysis is to prevent the occurrence of possible human errors, hence the need for their "pre-emptive" identification and implementation of preventive measures/solutions. For this reason, it has become essential to identify the possible causes of the described errors that employees may commit.

Due to the purpose of the analyzes presented in stage 4 defined in this way, a similar conclusion was made in the next stage regarding possible disturbances in the picking process, but caused by the operation of the technical system that supports handling processes. It should be noted that the implementation of automated solutions reduces the number of possible adverse events caused by human error but increases the possibility of interference from the technical system.

4 Results

4.1 Characteristics of the Traditional and Automated Order-Picking Process

The analysis covered two categories of products presented in Sect. 3. Until 2022, employees picked orders in the traditional "person-to-goods" system. They used hand pallet trucks for movement, while in the picking process, they were supported by hand scanners. The preparation of many orders consisting of a small number of products was very time-consuming. During the picking process, most of the time, employees moved between the racks in search of the following items from the handled order. At the same time, changes in the labor market caused more and more problems with the availability of employees. The large share of new employees constantly appearing increased the number of human errors in the picking process, which resulted in the need for additional control procedures and the risk of lowering the level of logistics service provided. These premises caused the company's management to decide to automate the picking process of both analyzed commodity groups. Figure 2 shows both person-to-goods and goods-to-person picking processes.

In the automatic warehouse, each container is marked with a code allowing direct identification and product filling. The containers are placed in individual warehouse locations, waiting for picking. The system determines the order of releases based on information on submitted orders. After the container leaves the warehouse, it moves along the path on designated tracks. Initially, they wait in the queue, and then when the product is taken from the previous container and sent to the warehouse, the container is moved from the waiting area to the place of picking up the order. If products remain in the container, it is sent back to the warehouse, where it is stored again. If the container remains empty after collection, it is sent to a place where employees manually decode the container markings and are reused to pack products.

Comparing both processes presented in Fig. 2, it should be noted that the automation of the picking process changed the scope of tasks performed by people preparing orders for shipment. First of all, the excessive movement of employees in the storage area has been eliminated. Employees no longer have to go from rack to rack to pick up the necessary products. Based on orders, the automatic system determines the order in which product containers are issued to the employee and then displays the number and type of product on the screen. The employee's task is limited to picking the product from the container and placing it in the appropriate cardboard box without moving between the racks. The process of correctly placing the product in the appropriate carton is additionally supported by a monitoring system that assesses the correctness of the packed product based on barcodes.

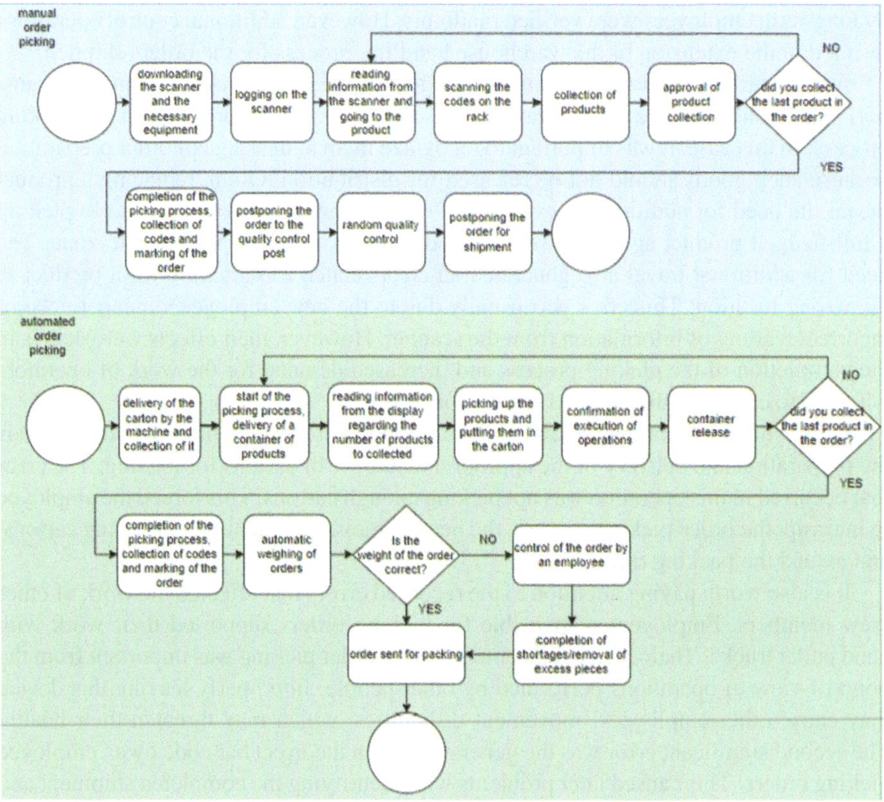

Fig. 2. Manual and automated order-picking process

4.2 Identification of Human Errors in the Picking Process in the "Person-to-Goods" and "Goods-to-Person" Systems

The research methods described in Sect. 3 were used to identify adverse events. In a traditional picking system, repetitive human errors were recorded. Due to the effects of their occurrence, they can be divided into two categories: (1) errors affecting the level of logistics service related to the completed order and (2) errors affecting the work of other crew members. These errors are shown in Fig. 3.

Among the most frequently recorded adverse events caused by man in the manual picking process, there are: (1) missing products in packed orders, (2) packed incorrect products, (3) excessive amounts of products. These problems persisted despite the preventive measure of using a scanner with a user-friendly interface. The scanner provided information on the type of order, the number of items in the order, and the location of the item on the storage rack. Its use supported operations performed during picking but did not eliminate the occurrence of the above errors. For this reason, the company introduced the need for additional verification of prepared orders. Orders prepared by newly employed persons were verified in their entirety. On the other hand, orders completed by long-term employees were verified randomly. However, additional control operations resulted in the extension of the warehouse handling process for the orders shipped.

Another group of adverse events was cargo damage in the process of internal transport. The result of these events was the presence of defective products in the picking process. In this case, it was important to recognize them at the stage of order preparation because such goods should not be released for distribution. Often, replacing a product meant the need for additional movement of the picking employee, who had to pick up a full-fledged product again and return the damaged one to the appropriate zone. The need for additional travel also generated an error related to searching for a product in the wrong location. This error was usually due to the new employee's inexperience or incorrect reading of information from the scanner. However, their effects were delays in the completion of the picking process and increased demand for the work of operators who had to complete the planned number of orders.

As part of the preparatory activities for the picking process, an important operation is the preparation and delivery of the appropriate number of cartons for loading. The error that occurred in this operation was not picking enough cartons. This forced the employee to interrupt the order-picking process, the need to move to replenish the missing cartons, and extend the packing time.

It is also worth paying attention to the recorded errors that affected the work of other crew members. Employees responsible for picking orders supported their work with hand pallet trucks. Their correct placement during order picking was important from the point of view of operations performed by other people. Improperly leaving this device may cause other employees' movement difficulties, which may threaten their health. The second significant error was the generation of an incorrect bar code by an employee picking orders. This caused later problems with identifying the completed shipment and delayed in its delivery to the customer.

The automatic warehouse system eliminated human errors related to transport and searching for a product in the wrong location - the share of orders whose picking time was extended due to the wrong location was reduced to almost zero. In addition, errors associated with downloading the wrong product have been reduced, as shown in Table 1.

However, it is worth noting that the indicator of the number of orders not completed by the quantity specified in the order refers to those events when the error was identified in the control process. In the case of the "goods-to-person" system, error detection is very high; therefore, the calculated indicator is reliable. On the other hand, in the "person-to-goods" system, the control was often random, which means that the number of errors made could be much higher than the recorded values. Therefore, it can be assumed that the scale of the benefits achieved is much greater than presented in the above results.

The above benefits have been achieved by introducing appropriate labeling and providing hints to the person picking orders. On the screen in the packaging area, the system displays the number, type, and even the product's location in the transport container. In addition, a security system has been introduced to monitor the correctness of the completed order:

1) Light confirmation of the compliance of the completed product with the order. If a container with the wrong product appears at the picking station or an employee picks up the product and places it in the wrong container, the system reports an error. By

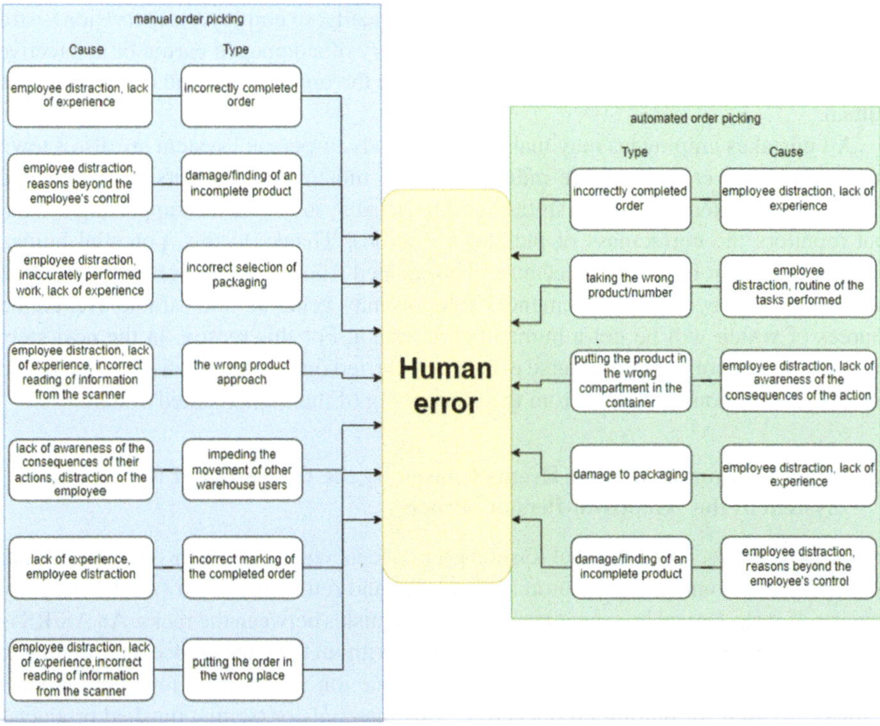

Fig. 3. Human error in the manual and automated order-picking process.

Table 1. Comparison of occurring adverse events in the "person-to-goods" and "goods-to-person" system.

Adverse event	Person-to-goods	Goods-to-person
Incorrectly completed orders (per year)	260	120
Share of orders with too few products	1,2%	0,5%
Share of orders with too many products	1%	0,3%
Share of damaged products in the handling process	0,8%	0,1%

reading the barcode while placing the product in the container, the system verifies its compliance with the order and confirms its correctness or lack thereof.

2) Automatic verification of the weight of the completed order. If an employee picks an incorrect amount of products for picking, the load is blocked and redirected to the picking zone for detailed checking. The system is equipped with a scale that verifies the shipment's actual weight with the estimated size. This makes it possible to supplement or limit the number of products in a completed order before it leaves the warehouse area.

However, not all errors are monitored systematically, so employee supervision is still necessary. An example of such errors is the delivery of a damaged carton or a defective product. These errors still need to be identified by the order picker and eliminated by a human.

All mistakes employees may make in the "goods-to-person" system are also shown in Fig. 3. However, as we have indicated above, most of these errors are eliminated by the implemented automated storage and retrieval system and the supporting system that monitors the correctness of picking operations. Thanks to this, potential human errors indicated in Fig. 3 are detected and eliminated from the order before it leaves the warehouse. However, the implemented solution may generate undesirable events, the sources of which will be not a human but a system. For this reason, in the next step, the identification of ex-ante adverse events was carried out, which allowed to determine potential disruptions resulting from the functioning of the implemented system.

4.3 Identification of Adverse Events Caused by the Operation of the Technical System in the "Goods-to-Person" Process

Automated storage and retrieval systems have been widely used in distribution and production environments. An automated storage and retrieval system (AS/RS) usually consists of racks served by cranes running through aisles between the racks. An AS/RS is capable of handling containers, cartons, and trays without the interference of an operator; thus, the system is fully automated. Both in production and distribution environments AS/RSs are used for putting products (e.g., raw materials or (semi-) finished products) in storage and for retrieving those products from storage to fulfill an order [38].

One of the goals of introducing an automatic storage system was to eliminate human errors in the traditional picking system. However, the replacement of manual pallet trucks with a complex technical system resulted in the possibility of a new type of adverse event in the analyzed process.

The causes of errors in an automatic warehouse system may be: (1) parts or mechanical components of the system and (2) software or system operation that supports the working devices. In general, the mechanical part of the system consists of a motor as the driving force, the rack itself as a storage area, containers, cartons, and trays as place-to-place goods, conveyers as transportation routes from or to shelves, and sensors as readers in the system. Each of these elements can fail and disrupt or stop the operation of the entire automatic storage system. In such a situation, the entire process is suspended despite the readiness of other devices and warehouse employees.

In the initial implementation phase in the examined company, the operation of the automatic warehouse system was stopped. These stoppages were caused by wedging shipping containers or jamming bearings in the roller. On the other hand, the result of system errors (software and control system) was most often caused by picking the wrong container, delivering the container to the wrong picking station, or returning the container to the wrong location. The operating team eliminates these errors by identifying the causes of their occurrence and removing the occurring disturbances.

However, it should be noted that the structure of the storage products is not static. We observe numerous changes related to the type of stored assortment, rotation of individual products, and replenishment, storage, and distribution procedures. All these changes must be considered in the control parameters implemented in the automatic warehouse system. At this stage, human errors made by service engineers may occur. The effects of these errors will affect the efficiency and correctness of operations in the warehouse system.

5 Conclusion

The intensification of the processes of automation and digitization of warehouse operations in recent years is due to the development of technology (development of the Industry 4.0 concept), demographic changes (demographic decline and difficulties in recruiting new employees), and the high employee turnover rate that we currently observe in many sectors. It is also a response to problems related to human errors, which may be caused by the inexperience of new employees and the routine of experienced people.

The results presented by us confirm that in the case of simple, repetitive warehouse operations, such as order picking, implementing cyber-physical systems reduces the number of human errors, such as picking incorrect products or in the wrong quantity. It is worth noting, however, that mistakes made by order pickers are individual - they concern the incorrect preparation of a single order. Meanwhile, in systems based on Industry 4.0, human error may relate to the work performed by an engineer who supports the complex cyber-physical system. In such a case, the nature of the estimated risk will change because the error's consequences concern the whole system's stoppage and not a single order-picking station.

The results presented in the article show the identification of adverse events carried out for the traditional system as part of the ex-post analysis and for the cyber-physical system following the ex-ante analysis. Thanks to this, the results of the analyzes carried out provide answers regarding current and future threats. Such a procedure allows for eliminating the causes of adverse events and developing emergency procedures for potential adverse events. Emergency procedures are critical in rapid response in the case of complex cyber-physical systems. It should be remembered that even minor errors or device interference can stop the entire functioning system. For this reason, preventive measures limiting the likelihood of their occurrence are significant for the service processes performed.

The results presented in the article were limited to the first phase of the risk assessment, namely identifying possible adverse events. The scope of this identification has also been limited to technical devices and human factors as the two primary sources of danger. Further research conducted by the authors will relate to risk analysis and estimation of risk indicators for each identified adverse event using qualitative and quantitative methods. It is also planned to extend the assessment scope to include events that are also caused by other reasons (e.g., incorrect procedures, environmental impact, type of stored material). Further research steps will allow for a comprehensive risk analysis related to the operation of the automatic storage system, and thus it will be possible to determine the resistance of the tested system to occurring disturbances.

References

1. Bordeleau, F.-E., Mosconi, E., de Santa-Eulalia, L.A.: Business intelligence and analytics value creation in Industry 4.0: a multiple case study in manufacturing medium enterprises. Product. Plan. Control. **31**, 173–185 (2020)
2. Stock, T., Obenaus, M., Kunz, S., Kohl, H.: Industry 4.0 as enabler for a sustainable development: a qualitative assessment of its ecological and social potential. Process Safety Environ. Protect. **118**, 254–267 (2018)
3. Falck, A.-C., Örtengren, R., Rosenqvist, M., Söderberg, R.: Proactive assessment of basic complexity in manual assembly: development of a tool to predict and control operator-induced quality errors. Int. J. Prod. Res. **55**, 4248–4260 (2017)
4. Grosse, E.H., Glock, C.H., Neumann, W.P.: Human factors in order picking: a content analysis of the literature. Int. J. Prod. Res. **55**, 1260–1276 (2017)
5. Wilson, J.R., Farrington-Darby, T., Cox, G., Bye, R., Hockey, G.R.J.: The railway as a socio-technical system: Human factors at the heart of successful rail engineering. Proc. Inst. Mech. Eng. Part F: J. Rail Rapid Transit. **221**, 101–115 (2007)
6. Hawkins, F.H.: Human Factors in Flight, 2nd edn. Routledge, London (2017)
7. Woods, D., Dekker, S., Cook, R., Johannesen, L., Sarter, N.: Behind Human Error, 2nd edn. CRC Press, London (2010)
8. Reason, J.: Human error: models and management. BMJ **320**, 768–770 (2000)
9. Read, G.J.M., Shorrock, S., Walker, G.H., Salmon, P.M.: State of science: evolving perspectives on 'human error.' Ergonomics **64**, 1091–1114 (2021)
10. . International Ergonomics Association (IEA). Council. The Discipline of Ergonomics. International Ergonomics Association (2014)
11. Karwowski, W.: A review of human factors challenges of complex adaptive systems. Hum. Fact. J. Hum. Fact. Ergon. Soc. **54**, 983–995 (2012)
12. Kolus, A., Wells, R., Neumann, P.: Production quality and human factors engineering: a systematic review and theoretical framework. Appl. Ergon. **73**, 55–89 (2018)
13. Neumann, W.P., Winkelhaus, S., Grosse, E.H., Glock, C.H.: Industry 4.0 and the human factor – a systems framework and analysis methodology for successful development. Int. J. Product. Econ. **233**, 107992 (2021)
14. Alsufyani, N., Gill, A.Q.: Digitalisation performance assessment: a systematic review. Technol. Soc. **68**, 101894 (2022)
15. Abdallah, Y.O., Shehab, E., Al-Ashaab, A.: Developing a digital transformation process in the manufacturing sector: egyptian case study. IseB **20**, 613–630 (2022)
16. Dhanpat, N., Buthelezi, Z.P., Joe, M.R., Maphela, T.V., Shongwe, N.: Industry 4.0: the role of human resource professionals. SA J. Hum. Resource Manage **18** (2020)
17. Giang, N.T.H., Hai, P.T.T., Tu, N.T.T., Tan, P.X.: Exploring the readiness for digital transformation in a higher education institution towards industrial revolution 4.0. Int. J. Eng. Pedagogy (iJEP) **11**, 4 (2021)
18. Matthyssens, P.: Reconceptualizing value innovation for Industry 4.0 and the Industrial Internet of Things. J. Bus. Indust. Mark. **34**, 1203–1209 (2019)
19. Sony, M., Naik, S.: Industry 4.0 integration with socio-technical systems theory: a systematic review and proposed theoretical model. Technol Soc. **61**, 101248 (2020)
20. Richnák, P.: Current trend of Industry 4.0 in logistics and transformation of logistics processes using digital technologies: an empirical study in the slovak republic. Logistics **6**, 79 (2022)
21. Lennon Olsen, T., Tomlin, B.: Industry 4.0: opportunities and challenges for operations management. Manufac. Serv. Oper. Manage. **22**(1), 113–122 (2019)
22. Ghobakhloo, M.: Industry 4.0, digitization, and opportunities for sustainability. J Clean Prod. **252**, 119869 (2020)

23. Tubis, A.A., Grzybowska, K.: In search of Industry 4.0 and Logistics 4.0 in small-medium enterprises—a state of the art review. Energies (Basel) **15**, 8595 (2022)
24. Glistau, E., Coello Machado, N.I.: Industry 4.0, Logistics 4.0 and materials - chances and solutions. Mater. Sci. Forum **919**, 307–314 (2018)
25. Amr, M., Ezzat, M., Kassem, S.: Logistics 4.0: definition and historical background. In: 2019 Novel Intelligent and Leading Emerging Sciences Conference (NILES), pp. 46–49. IEEE (2019)
26. Bigliardi, B., Casella, G., Bottani, E.: Industry 4.0 in the logistics field: a bibliometric analysis. IET Collabor. Intell. Manufac. **3**, 4–12 (2021)
27. Facchini, F., Oleśków-Szłapka, J., Ranieri, L., Urbinati, A.: A maturity model for Logistics 4.0: an empirical analysis and a roadmap for future research. Sustainability **12**, 86 (2019)
28. Winkelhaus, S., Grosse, E.H.: Logistics 4.0: a systematic review towards a new logistics system. Int. J. Product. Res. **58**, 18–43 (2020)
29. Naseem, M.H., Yang, J.: Role of Industry 4.0 in supply chains sustainability: a systematic literature review. Sustainability **13**, 9544 (2021)
30. Tubis, A.A., Poturaj, H.: Risk related to AGV systems—open-access literature review. Energies (Basel) **15**, 8910 (2022)
31. Simoni, M.D., Kutanoglu, E., Claudel, C.G.: Optimization and analysis of a robot-assisted last mile delivery system. Transp. Res. E Logist. Transp. Rev. **142**, 102049 (2020)
32. Tubis, A.A., Ryczyński, J., Żurek, A.: Risk assessment for the use of drones in warehouse operations in the first phase of introducing the service to the market. Sensors **21**, 6713 (2021)
33. Dördüncü, H.: Logistics, supply chains and smart factories. In: İyigün, İ., and Görçün, Ö.F. (eds.) Logistics 4.0 and Future of Supply Chains. AFSGFTA, pp. 137–152 (2022). Springer, Singapore. https://doi.org/10.1007/978-981-16-5644-6_9
34. Barreto, L., Amaral, A., Pereira, T.: Industry 4.0 implications in logistics: an overview. Procedia Manuf. **13**, 1245–1252 (2017)
35. Zoubek, M., Simon, M.: Evaluation of the level and readiness of internal logistics for Industry 4.0 in industrial companies. Appl. Sci. **11**, 6130 (2021)
36. Khan, M., Khan, M., Ali, A., Khan, M.I., Ullah, I., Iqbal, M.: Digitalization for fast, fair, and safe humanitarian logistics. Logistics **6**, 31 (2022)
37. Winkelhaus, S., Grosse, E.H., Morana, S.: Towards a conceptualisation of Order Picking 4.0. Comput. Ind. Eng. **159**, 107511 (2021)
38. Roodbergen, K.J., Vis, I.F.A.: A survey of literature on automated storage and retrieval systems. Eur. J. Oper. Res. **194**, 343–362 (2009)

Generating Complex Dynamic Forms in Low-Code Development Platforms

Robert Waszkowski(✉) and Maciej Dybowski

Faculty of Cybernetics, Military University of Technology, 2 Kaliskiego Street, 01-908 Warszawa, Poland
robert.waszkowski@wat.edu.pl

Abstract. This article highlights the potentialof creating a library for generating dynamic forms using the JSON Schema within the framework of the Low-Code platform. Low-Code platforms are designed to enable users to build workflows that reflect the business processes of their organizations. Forms play a key role in these workflows, as they are used to process data and execute business processes. As these processes grow more complex, users require more robust solutions to manage them, leading to the greater complexity of forms and the need to dynamically change the structure, scope of data, and layout. The library has a structure that divides forms into logical sections and individual form fields, making it easier for users to manage complex forms. The library was thoroughly tested and analysed in terms of efficiency in generating forms of various sizes. The article presents a comprehensive solution for generating dynamic forms on the Low-Code platform. The use of JSON Schema in combination with a well-designed library facilitates the management of complex forms and more effective implementation of business processes for users. The results of tests and analyses show that the library is a viable solution for generating dynamic forms, providing users with the tools they need to achieve their goals.

Keywords: Low-Code Platform · Dynamic forms · JSON-schema · Aurea BPM

1 Introduction

Nowadays, the Internet is an indispensable part of our lives, and its popularity is constantly growing. As technology develops, there is a growing number of tools that allow us to utilise its resources. One of the most important tools are the forms that constitute an underlying basis for the operation of many online platforms and applications. Currently, a properly designed form on a website or in a mobile application is a deciding factor in whether we will be able to acquire a customer, purchase an item on an auction site or transfer money to another bank account. In modern times, people are bombarded with a vast amount of information while our attention span keeps getting to shrink. The better the form is designed, the greater the chances that it will be filled out and accepted.

A new class of software development environments has emerged in recent years under the name "low-code", of which it is said that it not only offers an opportunity

A. Burduk et al. (Eds.): ISPEM 2023, LNNS 790, pp. 420–433, 2023.
https://doi.org/10.1007/978-3-031-45021-1_32

to significantly improve productivity in software development, but also provides new ways of promoting IT business alignment and empowering the user. These platforms are currently known as low-code platforms (LCP), low-code application platforms (LCAP) and low code development platforms (LCDP) [1]. On Low-Code platforms, the user builds a workflow that often reflects business processes that take place in their company or organisation. One of the development stages are forms. They are necessary for data processing and implementation of business processes. As a process grows more complex, the user requires solutions to enable them to handle that process, which entails greater complexity of forms and dynamic changes to structure, scope of data or layout [2, 3].

In order to meet the expectations of the users, it is necessary to deliver solutions that do not require programming skills to create forms. The proposed solution is software that constructs screen forms based on metadata described in the JSON Schema (JavaScript Object Notation) format [4]. The JSON Schema format is used in many different situations, such as validation of data sent from online forms or verification whether configuration files are correct. It can also be used to describe API interfaces, which enables better integration of different systems and services. This solution may be based on the "runtime" module developed in JavaScript. The JSON Schema format ensures that the generated forms are validated and compliant with expected standards. In order to verify the effectiveness of the proposed solution, the tester was configured, and scenarios were prepared. This will make it possible to compare the speed and efficiency of this solution against other commercial solutions. By providing a simple and efficient way to create forms, this solution has the potential to improve productivity and efficiency in different industries [5]. In addition, this solution can be easily integrated into existing software systems, making it a valuable addition to each organisation.

The purpose of the paper is to provide the low-code platform developers with the efficient library for the complex form generation. The structure of the article is as follows: the first part of this paper provides a brief overview of the current solution for form building automation. The second part is devoted to presenting the proposed structure of the form description based on JSON schema. Then the data synchronization mechanism is introduced to ensure the performance of the generation of the large-scale forms. At the end of the paper the method for testing the solution is proposed, the tests are performed, and the results are presented and discussed.

2 Overview of Current Automatic Form Building Solutions

Based on the "Magic Quadrant for Multiexperience Development Platforms" report by Gartner, published on 26 July 2021, leading products of leading businesses specialising in low-code software development were presented and are shown below in the form of a qualified diagram, divided into four groups [6] (Fig. 1).

- Leaders execute well against their current vision for changing market rules but do not yet execute well.
- Visionaries understand where the market is going or have a vision for changing market rules but do not yet execute well.
- Niche Players focus successfully on a small segment or are unfocused and do not out-innovate or outperform others.

Fig. 1. Magic Quadrant, comparison of companies according to Gartner, Inc.

- Challengers execute well today or may dominate a large segment but do not demonstrate an understanding of market direction [6].

Mendix

This platform is the leading low-code solution, recognised by analytical reports prepared, inter alia, by Gartner and large corporations, e.g. SAP and IBM. It is a platform designed to build applications quickly. The platform disregards and automates various application development layers. For example, the data structure is built using the Unified Modelling Language (UML). The business logic site uses BPMN (Business Process Modeling Notation), whereas the user interface is built from widgets according to the "what you see is what you get" (WYSIWIG) model. The platform enables business and IT to cooperate and build applications that add business [7, 8].

It supports 6 main functionalities:

- Collaboration,
- Data structure and domain model,
- Business logic,
- User interface and experience,
- User security and authentication,
- Implementation [9].

Outsystems

A low-code visual software development environment consisting of two parts: a server and a desktop application for software developers. The desktop application is used for designing mobile and online applications the simple drag-and-drop method. The platform allows to easily model databases, workflows and rules, and then deploy them to the production server or multiple digital distribution platforms with a single click [10].

3 Structure Based on JSON Schema

The JSON Schema format is used in many different situations, such as validation of data sent from online forms or verification whether configuration files are correct. It can also be used to describe API interfaces, which enables better integration of different systems and services. It is consistent with the standard published by IETF (Internet Engineering Task Force) [11], which is responsible for the development and promotion of standards concerning online technologies, and it is more and more frequently used as a tool for describing the structure and contents of data stored in the JSON format. Its application ensures the quality of data and its compliance with certain requirements, which is particularly important in the case of large projects where large amounts of data is processed [12–14]. Generally speaking, JSON Schema is a powerful tool that can help to ensure the quality and consistency of JSON documents while also ensuring a more user-friendly way of working with this type of data [15].

3.1 Validation Scheme (JSON Schema)

JSON Schema is a useful tool for ensuring that JSON documents comply with certain principles and standards. JSON Schema allows to define a set of restrictions and requirements that must be met by a JSON document to be considered valid. These rules are defined using key words presented in the schema specifications, which helps to ensure that a JSON document follows specific structure and contents (Fig. 2). Using JSON Schema makes it easier to work with and manipulate JSON documents. For example, if a JSON document does not meet the requirements specified in the schema, it may trigger an error that helps to capture problems at an early stage of the software development process [16].

3.2 Form Schema (UI Schema)

One of the most useful features of JSON Schema is its ability to generate an interactive user interface. This user interface allows to browse, edit and manipulate JSON documents, as it ensures a more user-friendly way to interact with data (Fig. 3). This may be particularly useful for software developers who may not be so familiar with JSON or who may have difficulties working in a traditional text format. UI Schema is one of the input parameters of the form. It specifies which fields will be generated by the form, their configuration, layout and dependencies between them. It makes it possible to precisely specify the layout of the form and what information it should contain.

```
{
  "title": "Formularz rejestracji",
  "description": "Schemat walidacyjny dla rejestracji",
  "type": "object",
  "properties": {
    "login": {"type":"string"},
      "password": {"type":"string"},
      "email": {"type":"string"},
  },
  "required": ["login", "password"]
}
```

Fig. 2. Sample document in the JSON Schema format

```
{
 type: "object",
 properties: {
  login: { "x-display": "text-field", label: "Login" },
  email: { "x-display": "text-field", label: "E-mail" },
  password: {
   "x-display": "text-field",
   label: "Password",
   hint: {
    text: "Minimum 8 characters.",
    persistent: true
   }
  },
  marketingOffers: {
   "x-display": "checkbox",
   label: "Tick the selected options",
   "x-source": {
    source: [
     {id: "terms", label: "I accept the terms and conditions"},
     {id: "commercialOffer", label: "Agreed to receive offers"},
     {id: "phoneContact", label: "Agreed to telephone contact"}
    ],
    itemText: "label"
   }
  }
 },
 required: ["login", "email", "password"]
}
```

Fig. 3. Form layout definition (UI Schema)

Based on the provided JSON document, the library may generate the form shown below (Fig. 4). The form will have a structure consistent with the hierarchy and types of data defined in the JSON schema. The form may contain input fields for each property as well as any necessary validation rules and formatting requirements.

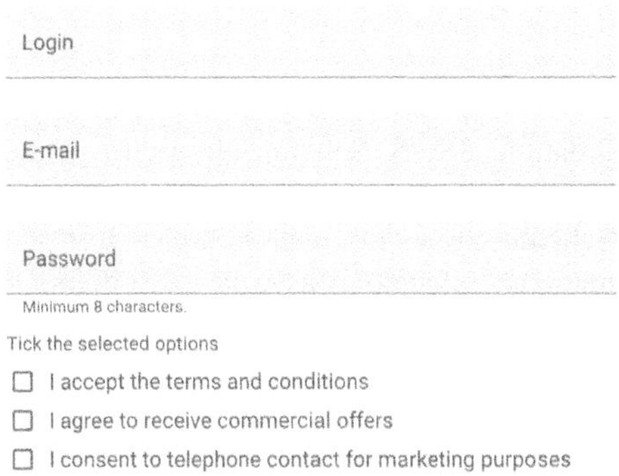

Fig. 4. Sample form generated using metadata

4 Ensuring Efficiency Through the Data Synchronisation Mechanism

Generating large forms in web-based applications is a problem as it may overload an application. Current solutions, both in native configuration and after downloading additional libraries, often fail to handle large and complex forms well. Form generation is prolonged and the user has difficulty with filling out the form at all. The process of creating and defining forms, including determining their components and views as well as implementing validation and submission procedures, is a labour-intensive and time-consuming undertaking. In order to solve this problem, it is worth to consider using tools that automate the process of creating and handling forms, such as libraries for dynamic form generation or form management solutions.

The proposed solution allows to dynamically generate form fields defined in the form of a JSON object. The definition of the form consists of two elements: a validation schema compliant with the JSON Schema standard that contains only those elements that are necessary to verify that transmitted data is correct and UI Schema, which defines the layout and behaviour of controls, their dependence and division into sections. This solution allows to quickly define fields, validation and layout of the form. Moreover, this approach ensures flexibility in the formulation of forms, as their structure and contents can be easily adapted using a JSON file. This is particularly useful when forms must be frequently changed or when it is necessary to create new forms [17].

The separation of data from the layout and behaviour of the form is a design principle according to which information included on the form are separated from the way in which this data is presented and how the form behaves when interacting with the user.

In practice, this means that the data is stored in a separate place and the user interface and business logic are separated from the data. This division makes it easier to manage and develop the form, as changes to one of these elements do not affect the others.

Advantages

- Automatic data validation: JSON Schema allows to define validation rules for individual form fields.
- Flexibility: JSON Schema allows to create complex forms with different types of fields (e.g. text, numerical, selectable) as well as to define dependencies between individual fields.
- With good application architecture, there is no need to recompile the application in order to change the layout of the form.
- Separation between data presentation method and the model
- Time needed to build a form
- Descriptive language of the definitions: JSON Schema is built using simple, easily understandable keys and values, which allows to quickly understand the structure of the form.

Disadvantages

- Designer or graphic building tool must have many validation rules for the schema alone
- Form must be divided into logic sections
- Two schemas are needed: validation and UI
- Data must be synchronised between sections, especially in cases where datum from section B depends on datum from section A

5 Tests and Measurements

In order to conduct tests on the dynamic form generation approach based on JSON metadata, a test environment was created using the Storybook tool [18]. This provided a controlled and isolated environment for performance tests that allows to accurately measure and analyse form rendering times. The efficiency of the developed solution was assessed in terms of whether it is able to generate forms with a varied number and type of fields. For this purpose, the forms were divided into three categories depending on their size:

- small (with 10 fields),
- medium (with 50 fields),
- large (with 100 or more fields).

In order to ensure a thorough comparison of results, form generation was tested against the reference standards set by Vue.js and Vuetify frameworks [17]. Furthermore, the tests also take into account the forms generated by solution developed by Mendix, thus ensuring a comprehensive assessment of various solutions. Mean results and diagrams were obtained after conducting 100 form generation samples and will be presented in subsequent chapters. Two methods were used to measure performance. The first consisted in using "Performance API" embedded in the web browser, which allowed to track and measure events over the life cycle of the component [19]. The second method used the "Profiler" tool embedded in the browser, which enabled an in-depth analysis of the rendered website.

Method I: Performance API

Using this method requires access to the source code of the solution. It consists in the performance of measurements using the available interface for software developers and testers, available in the JavaScript language (Fig. 5). This method is based on the life cycle of the component in the popular Vue.js framework. This means that it is possible to examine how individual application elements operate in different phases of its life, such as creation, mounting, update and deletion. Measurements begin when the component is created and continue until the component is mounted. The analysis shows that this

```
import { defineComponent } from "vue"
import debounce from "lodash/debounce"
export const measuringMixin = defineComponent({
  data() {
    return {
      debounced: { reload: () => null }
    }
  },
  methods: {
    forceReload() {
      window.location.reload()
    }
  },
  created() {
    performance.mark("startRender")
    this.debounced.reload = debounce(this.forceReload, 1000)
  },
  mounted() {
    performance.mark("endRender")
    performance.measure("rendering time", "startRender",
"endRender")
    const measure = performance.getEntriesByType("measure")
      .find((entry) => entry.name === "rendering time")
    const renderingTime = measure ? measure.duration : 0
    console.info(renderingTime)
    this.debounced.reload()
  }
})
```

Fig. 5. Form generation time measurement component

type of measurement can be considered accurate because the entire process of compiling the contents of the component and its embedding in the object HTML document model takes place between the two stages of its life cycle.

Method II: Browser Profiler

A separate testing methodology was used to examine the time necessary to generate forms using the Mendix solution. The "performance" tab in the browser console was used for that purpose, which allows to measure different critical factors affecting the performance and speed of the application, including script rendering and loading time. Performance analysis tool in the browser console enables an in-depth assessment of application performance.

5.1 Category I – Small Forms (10 Fields)

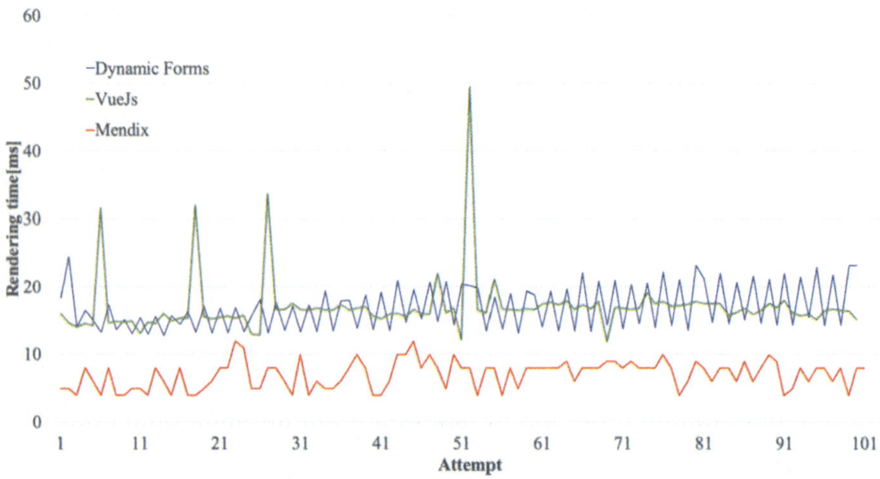

Fig. 6. Graph showing form rendering time depending on the solution for category I

Table 1. Mean generation time for category I.

Solution	Mean generation time [ms]
Dynamic Forms	16.91
VueJs	17.04
Mendix	7.03

The solution presented for small size forms performs similarly well compared to the native solution. The fastest solution is the solution developed by Mendix, which on average is 142% faster than the native solution (Fig. 6, Table 1).

Table 2. Mean generation time for category II.

Solution	Mean generation time [ms]
Dynamic Forms	16.11
VueJs	51.12
Mendix	11.40

5.2 Category II – Medium Forms (50 Fields)

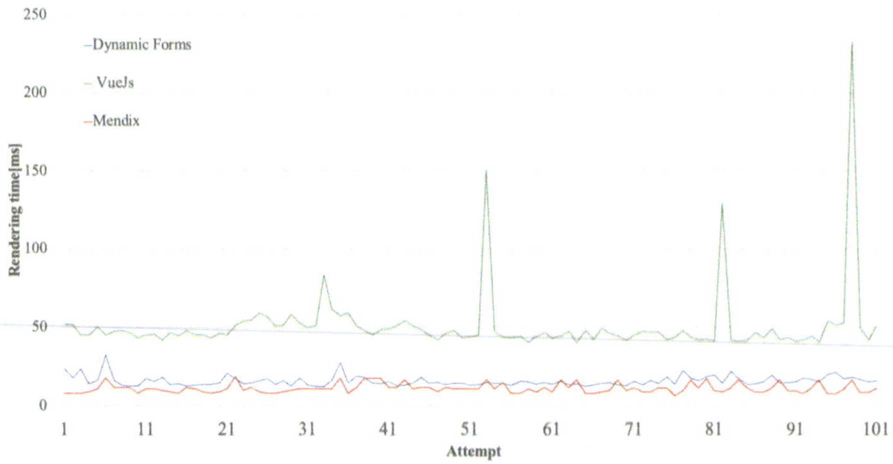

Fig. 7. Graph showing form rendering time depending on the solution for category II

Mean form generation time for the native solution increased by more than 300% compared to category I. For the proposed solution, there is a slight change in generation time but the solution still remains in the middle of the pack. In category II, the fastest solution continues to be the solution developed by Mendix, which generates a 50-field form in the mean time of 11.40 ms (Fig. 7, Table 3).

5.3 Category III – Large Forms (100 Fields)

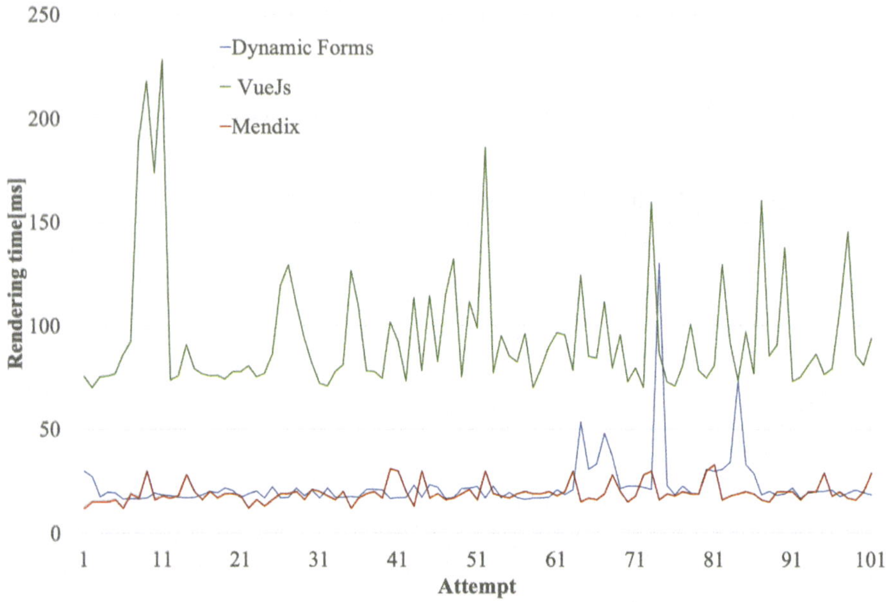

Fig. 8. Graph showing form rendering time depending on the solution for category III

Table 3. Mean generation time for category III.

Solution	Mean generation time [ms]
Dynamic Forms	22.80
VueJs	95.25
Mendix	19.35

Category III was given special attention during the tests due to the particular problem of generating forms containing a significant number of fields. The native solution requires on average as much as 95 ms to perform that task. As the number of generated fields increases, the result achieved by the proposed solution comes close to the solution developed by Mendix, which was 18% faster in this test (Fig. 8, Table 3).

5.4 Test of Generation Time in the Field Number Function (from 1 to 100)

The dependency between the increase in form generation time and the number of fields was analysed. The results indicate that the native solution showed linear increase in time as the number of fields grew, whereas dynamic forms showed significantly better performance, with a nearly linear characteristic (Fig. 9). This suggests that dynamic forms can be a more efficient solution for managing forms with a large number of fields, potentially saving a significant amount of time and resources.

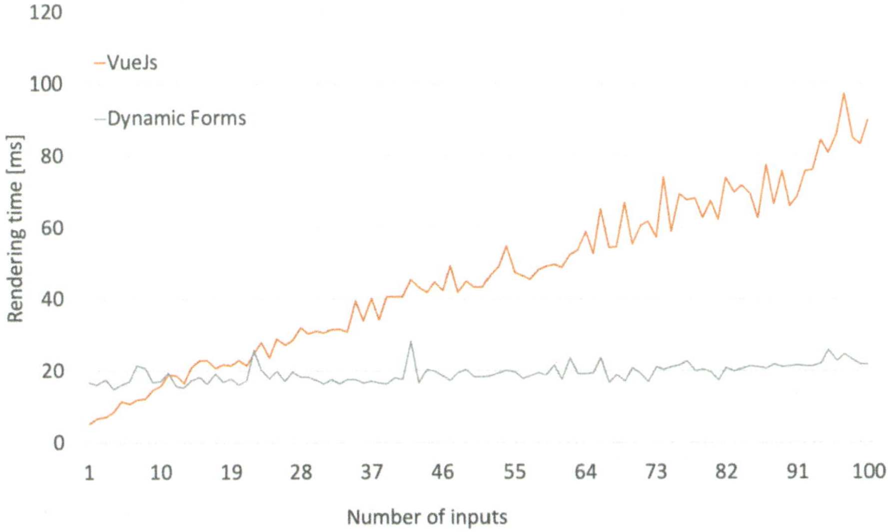

Fig. 9. Test of generation time in the field number function (from 1 to 100)

6 Conclusion

The article presents an innovative approach to dynamic form generation based on JSON metadata, which may be particularly beneficial in the context of low-code platforms. The proposed solution consists in using a set of rules and requirements defined in JSON metadata to dynamically generate forms, thereby reducing the number of manual coding required to create forms. One of the key benefits of this approach is that it can significantly improve the time necessary to generate forms compared to traditional, manual approaches. Tests carried out on the proposed solution have shown that it can generate forms faster and more efficiently than native solution. This may be particularly beneficial for low-code platforms, where the fast generation of forms is crucial for maintaining productivity and minimising development time. Using JSON metadata to generate forms dynamically, software developers can save time and effort while creating more flexible and configurable forms that can adapt to changing needs and requirements [20, 21]. The proposed solution has the potential for development in the following directions:

- Implementation of additional form fields, e.g. file type field
- Development of a graphic form modeler, based on the WYSIWIG model
- Integration with content management systems (CMS) to enable easy creation and management of forms from the CMS level.
- Expansion of libraries for other popular frameworks, e.g. Angular or React

- Addition of support for translations of labels and validation information into many different languages.
- Creation of data mapping from the database to UI Schema in order to facilitate the creation of forms based on database tables

Acknowledgements. This work was financed by Military University of Technology under research project UGB 810/2023.

References

1. Bock, A.C., Frank, U.: Low-code platform. Bus. Inf. Syst. Eng. **63**(6), 733–740 (2021). https://doi.org/10.1007/s12599-021-00726-8
2. Waszkowski, R., Bocewicz, G.: Visibility matrix: efficient user interface modelling for low-code development platforms. Sustainability **14**(13), 8103 (2022). https://doi.org/10.3390/SU14138103
3. Waszkowski, R.: Low-code platform for automating business processes in manufacturing. In: IFAC-PapersOnLine, Elsevier B.V., pp. 376–381 (2019). https://doi.org/10.1016/j.ifacol.2019.10.060
4. Bray, T.: The JavaScript Object Notation (JSON) Data Interchange Format, https://www.rfc-editor.org/rfc/rfc7159 (2014)
5. Waszkowski, R., Nowicki, T.W., Worwa, K.A.: Corporate efficiency improvement with business process automation. In: MATEC Web of Conferences (2018)
6. Analyst, B., Batchu, A., Leow, A., Scheibmeir, J.: Magic Quadrant for Multiexperience Development Platforms View All Magic Quadrants and Critical Capabilities Strategic Planning Assumption(s)
7. Jasiulewicz-Kaczmarek, M., Waszkowski, R., Piechowski, M., Wyczółkowski, R.: Implementing BPMN in maintenance process modeling. In: Information Systems Architecture and Technology: Proceedings of 38th International Conference on Information Systems Architecture and Technology–ISAT 2017: Part II, pp. 300–309 (2018)
8. Waszkowski, R., Kowalski, A.: Comparative analysis of business process management frameworks. In: Proceedings of the 30th International Business Information Management Association Conference, IBIMA 2017 - Vision 2020: Sustainable Economic development, Innovation Management, and Global Growth (2017)
9. Mew, L., Field, D.: A Case Study on Using the Mendix Low Code Platform to support a Project Management Course Low-code Hackathon View project Information Systems Undergraduate Program Development View project A Case Study on Using the Mendix Low Code Platform to support a Project Management Course (2018), [Online]. Available: http://iscap.info
10. Golovin, D.: OutSystems as a Rapid Application Development Platform for Mobile and Web Applications (2017)
11. "Internet Engineering Task Force," https://www.ietf.org/
12. Kiedrowicz, M., Nowicki, T., Waszkowski, R., Wesołowski, Z., Worwa, K.: Optimization of the document placement in the RFID cabinet. MATEC Web of Conferences (2016). https://doi.org/10.1051/matecconf/20167602001
13. Kiedrowicz, M., Nowicki, T., Waszkowski, R.: Business process data flow between automated and human tasks. In: 3rd International Conference on Social Science (ICSS 2016), pp. 9–11 (2016)

14. Waszkowski, R., Nowicki, T.: Efficiency investigation and optimization of contract management business processes in a workwear rental and laundry service company. Procedia Manuf. **44**, 551–558 (2020)
15. Understanding JSON Schema Release 2020–12 Michael Droettboom, et al Space Telescope Science Institute (2023)
16. Jackson, W.: The JSON schema: JSON structure validation. In: JSON Quick Syntax Reference (2016)
17. "VueJs," https://vuejs.org/
18. "Storybook," https://storybook.js.org/
19. "Performance API," https://developer.mozilla.org/en-US/docs/Web/API/Performance
20. Jasiulewicz-Kaczmarek, M., Antosz, K., Zhang, C., Waszkowski, R.: Assessing the barriers to industry 4.0 implementation from a maintenance management perspective-pilot study results. IFAC-PapersOnLine **55**(2), 223–228 (2022)
21. Antosz, K., Jasiulewicz-Kaczmarek, M., Paśko, Ł, Zhang, C., Wang, S.: Application of machine learning and rough set theory in lean maintenance decision support system development. Eksploatacja i Niezawodność **23**(4), 695–708 (2021)

Optimizing the Form Data Presentation in Low-Code Platforms with the Use of Artificial Intelligence

Robert Waszkowski(✉) and Marcin Sirant

Faculty of Cybernetics, Military University of Technology, 2 Kaliskiego Street, 01-908 Warszawa, Poland
robert.waszkowski@wat.edu.pl

Abstract. Low-code platforms contribute to the fast creation and implementation of applications. However, it is time-consuming to create forms and data presentation methods on the said platforms. Extensive forms may be difficult to understand. However, they can be improved by combining many attributes and presenting them in an accessible way. Current methods do not fully ensure the exploitation of such opportunities. Automating this process will significantly contribute to the time necessary to establish the information scope of forms and the way in which they are presented. In addition, using artificial intelligence methods, the application will expand its scope of operation and accelerate the development of the proposed product. The ability to reduce the information scope of the forms and combine data fields into sections means that the solution perfectly matches the current needs of web browser users on mobile phones and tablets. The product may be an innovative solution and change the way in which existing documents are presented on the aforementioned platforms.

Keywords: Low-code development platform · forms · LCDP · information scope · Artificial Intelligence · optimisation · summary · Aurea BPM

1 Introduction

Low-code platforms are a set of tools for programmers and non-programmers alike. It allows to quickly generate and deliver business applications with minimal coding effort and requires the minimum possible effort for installation and configuration of environments, training and deployment. With a rapidly growing number of businesses, the use of low-code solutions can be a significant step forward in creating the necessary business applications. So far, this is the fastest and most likely also the cheapest software development method. Businesses must change at a rate that allows them to keep up with the competition, sellers and impatient consumers [1, 11, 12].

Data is of key importance on such platforms. In this article, particular attention will be paid to data included in the forms, which are often prepared in order to pass business processes existing in a given business. Users define control and data flows in the created system through intuitive business logic specification mechanisms. Graphic workflows

A. Burduk et al. (Eds.): ISPEM 2023, LNNS 790, pp. 434–444, 2023.
https://doi.org/10.1007/978-3-031-45021-1_33

and text business rules are examples of business logic specifications [2, 13, 14, 15]. However, extensive forms are a frequent problem in such applications, as unnecessary amount of information is displayed, making documents difficult to read and unclear. In many cases, a given user does not require that much information and it could be hidden or not generated by the system, resulting in major improvement to system performance with extensive forms. Additionally, given that a growing number of people uses web browsers on mobile phones and tablets, reducing the information scope is an appropriate way to display only relevant information on such small working areas [16, 17, 18].

This article presents a solution to the problem of defining the information scope of the forms using an original product, which also uses artificial intelligence algorithms in order to improve efficiency and expand the scope of operation of the software. The solution is innovative and unprecedented in the materials available to me.

The formulation of the problem will therefore be as follows: *For a given set of variables whose values are to be displayed to the user on the form, minimize the area of the form needed to display full information by replacing the form consisting of labelled fields with summaries in the form of sentences in natural language.*

2 Analysis of Existing Methods for Summarising Information Based on Artificial Intelligence Algorithms

The processing of natural language is a subdiscipline of linguistic, IT and artificial intelligence dealing with interactions between computers and human language, in particular the programming of computers to process and analyse large quantities of data in natural language [3]. Human language is full of ambiguities, which make it extremely difficult to write software that accurately defines the intended meaning of textual or voice data. For the purposes of this article, we will focus only on text summaries. Summaries can be divided into two categories: extractive summary and abstract summary.

2.1 Extractive Summary

The extractive method of summarising a text in most cases consists of two stages:

- Sentence scoring [4]: the first step in many techniques is sentence scoring on the basis of their importance to the entire summary. Depending on the degree of advancement, certain methods utilise only the contents of sentences, while other methods also use any kind of metadata, such as position of a word in the sentence, length of word. Many approaches also utilise intermediate representation. For example, a table of important words or graphic representation of similarities may be created for a summary. In such case, sentences are assessed in terms of their ability to represent a key topic in a given document and then used to create the final summary.
- Sentence selection: Based on scoring, sentences are selected to represent the company. During this process, it is important not only to consider the outcome of the document, but also its redundance in relation to other selected sentences. Reducing sentence overlap is a key mechanism for controlling the length of the summary.

In many cases, the sentence scoring and selection processes are independent from each other, whereas in other cases (such as sentence similarity methods) scoring and selection processes are closely integrated. In cases where scoring and selection are independent, it is possible to reuse a specific sentence selection technique in several scoring methods [5].

2.2 Abstract Summary

As opposed to extractive summarisation, abstract summarisation is intended to create summary which do not necessarily need to use fragments of a document in a literal way. Extractive summaries are often non-liquid from the human perspective, as humans often completely rewrite document fragments to achieve greater clarity and it is a major problem for that type of summaries.

The design of completely fluid summaries that mimic human performance is quite difficult. Even if recent progress in deep learning has shown encouraging results. The high difficulty level of this setting also explains why the majority of modern systems is extractive.

There are two kinds of methods for abstract summarisation. The first kind of methods can be considered only partially "abstract", as it first creates extractive summaries from input data and then modifies portions of these summaries in order to improve presentation and output. However, these methods have achieved success only to a limited extent. In reality, study [6] showed that attempts to introduce abstract methods to extractive summaries can result in worse summary scoring. Despite the mixed results achieved by such methods, they have found practical applications.

The second kind of abstract summarisation methods usually constructs summaries directly using deep learning methods. These techniques have exhibited a much higher likelihood of generating high quality summaries.

People summarising documents shorten longer sentences in terms of linguistics and writing style. Unnecessary words are removed as well. Automated sentence compression methods are motivated by the exact imitation of human behaviour and efficiency. Sentence compression methods are also used to generate headlines [7, 8] where the length of a compressed summary is short enough to be considered a headline, but the problem is directly independent of text summarisation tasks.

3 Solution

The solution architecture indicates elements that are closely interlinked and demonstrates which technologies and tools are used to maintain and build them [9].

It is comprised of the following components (Fig. 1):

- Low-code platform
- Web browser application
- Application service
- Language model
- Database

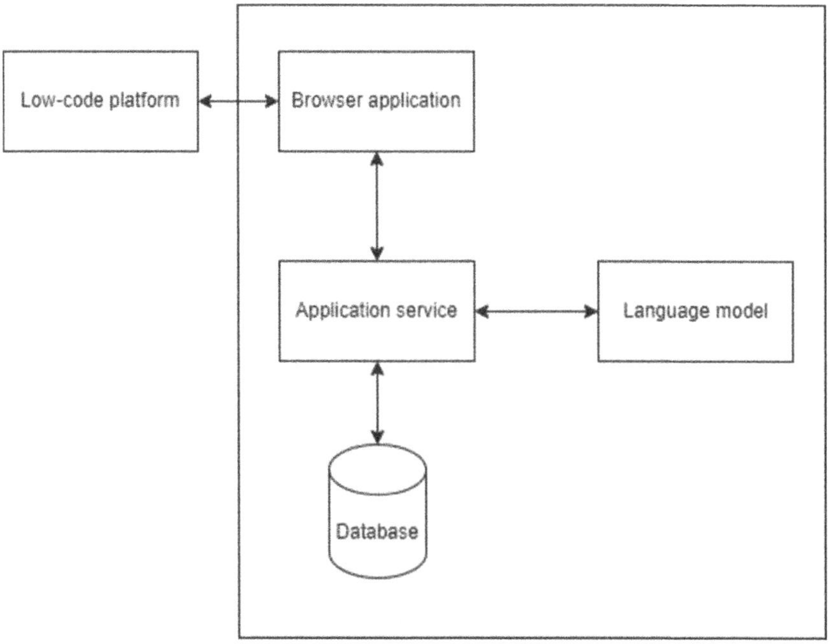

Fig. 1. Diagram presenting the system architecture

Due to the use of the language model [10], the solution is able to search for information scope templates and save the structure thereof [21]. It is also possible to manually add information scope structures. In such case, this is additionally supported by the language model in order to specify the template in greater detail.

Lets apply the solution to shorten the parental leave request form (Fig. 2):

The summary of the section presented above looks as follows:

As shown on the illustrations presented above (Fig. 2, Fig. 3), smaller forms are more concise and easier to browse through, which may increase the chances that users will be able to find the information they need faster. In addition, smaller forms occupy less space on the screen, which is particularly important in the case of mobile devices where the screen area is limited. However, lets also remember that the form should not be reduced at the expense of its functionality. Key information should continue to be included in the summary and users should have access to all necessary information. Furthermore, it is important that the form be transparent and intuitive.

4 Test

This chapter contains an analysis of the developed solution in terms of its performance and effectiveness. In order to carry out my experiment, I have prepared 50 forms of different sizes and with different information scopes that will be tested in order to determine the performance and effectiveness of the solution under analysis.

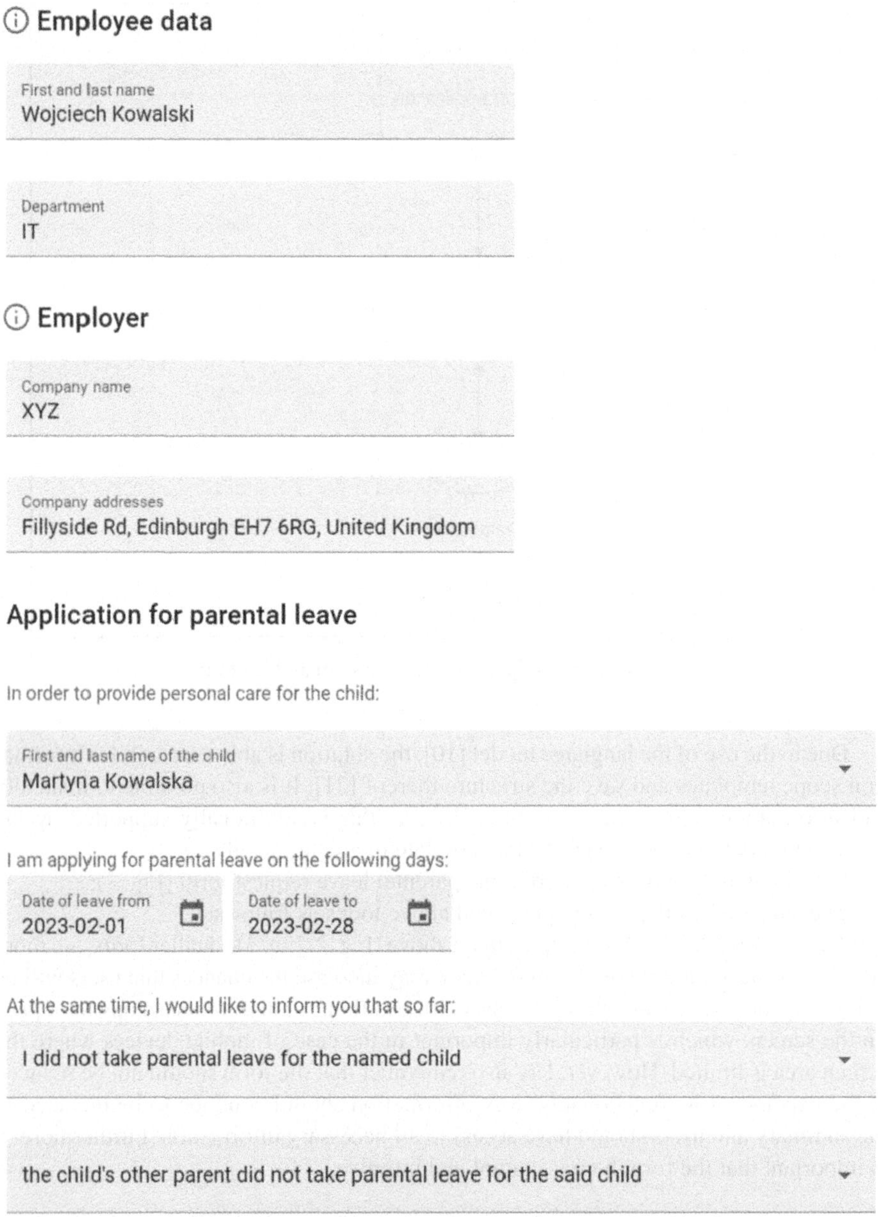
Employee data

First and last name
Wojciech Kowalski

Department
IT

Employer

Company name
XYZ

Company addresses
Fillyside Rd, Edinburgh EH7 6RG, United Kingdom

Application for parental leave

In order to provide personal care for the child:

First and last name of the child
Martyna Kowalska

I am applying for parental leave on the following days:

Date of leave from
2023-02-01

Date of leave to
2023-02-28

At the same time, I would like to inform you that so far:

I did not take parental leave for the named child

the child's other parent did not take parental leave for the said child

Fig. 2. Illustration presenting the parental leave form

4.1 Description of the Solution Performance Test

The forms were divided into groups according the scope of information included in them. Subsequently, sections that could potentially be shortened in each of the forms were counted by means of a manual check. After the developed solution was applied,

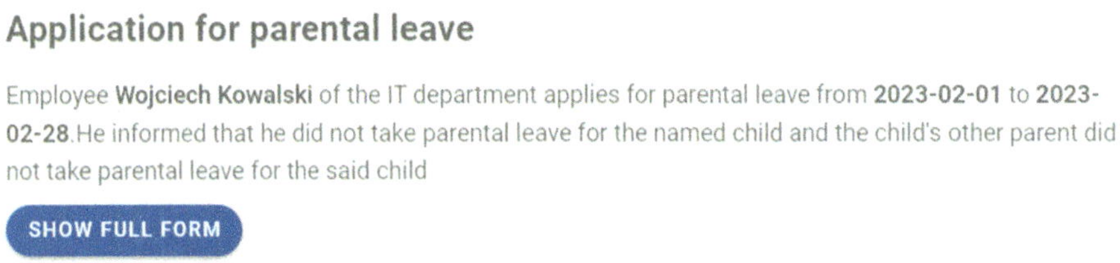

Fig. 3. Illustration presenting the form after its summarisation in a readable format

the accuracy of the functionality consisting in searching for and summarising the aforementioned potential sections has been verified. Finally, we generate the final result by dividing the number of sections to be summarised by the number of resulting summaries.

The following categories were identified:

- forms with addresses
- forms with personal data
- forms with company data
- leave request forms
- forms with date ranges

In total, there was 113 information scope to be included in the test in all 50 forms (Table 1).

Table 1. Number of forms with relevant sections

	Quantity
Forms with addresses	30
Forms with personal data	30
Forms with company data	30
Leave request forms	3
Forms with date ranges	20
Number of all sections	113

After the test was conducted, the following results were obtained (Fig. 4):

The algorithm performance test (Table 2) indicates that the algorithm identifies and summarises the specifies forms at a level of 96%. This is a high result indicating that the innovative solution is able to summarise the majority of forms that contain the relevant information scope. In order to thoroughly analyse the accuracy of the algorithm, we should process more production data from low-code platforms and conduct the aforementioned experiment on that data, as this would enable us to determine the accuracy of the algorithm on forms that are actually being used, thus providing a more objective result.

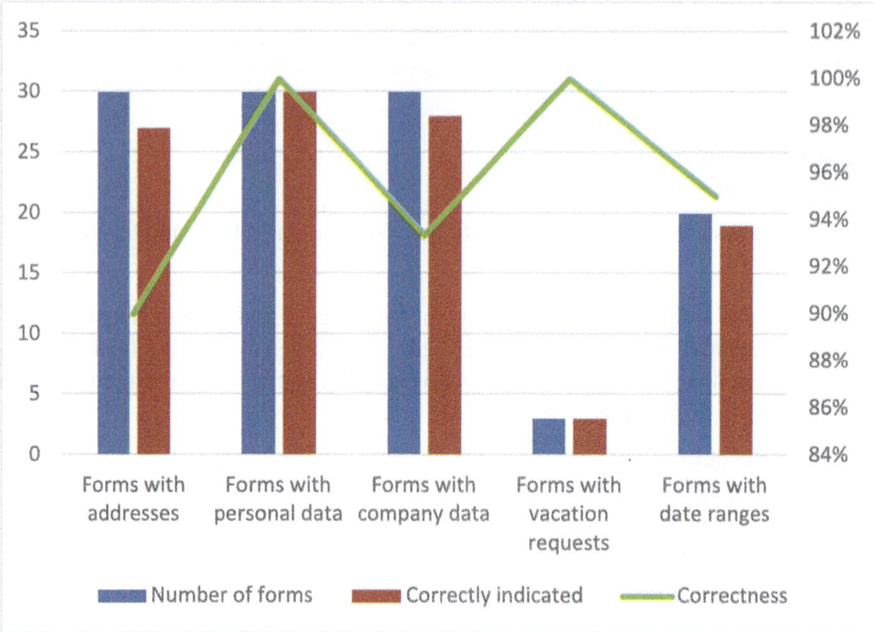

Fig. 4. Chart of algorithm performance test results

Table 2. Algorithm performance test results

	Number of sections to be summarised	Correctly indicated	Accuracy
Forms with addresses	30	27	90%
Forms with personal data	30	30	100%
Forms with company data	30	28	93%
Leave request forms	3	3	100%
Forms with date ranges	20	19	95%

4.2 Description of the Solution Efficiency Test

In order to test the efficiency of the solution [19, 20], forms were divided into 3 groups according to their size. The field area was calculated for each form. Next, all forms were subjected to the summarisation process performed by the solution under analysis, one by one. After summarisation, field areas were calculated once again and compared to initial results. The results are shown in the tables and on the diagrams presented below.

The following groups were identified:

- small forms (1 - 10 fields)
- medium forms (11 - 20 fields)

- large forms (more than 20 fields)

After the test was conducted, the following diagrams were generated (Fig. 5, Fig. 6, Fig. 7):

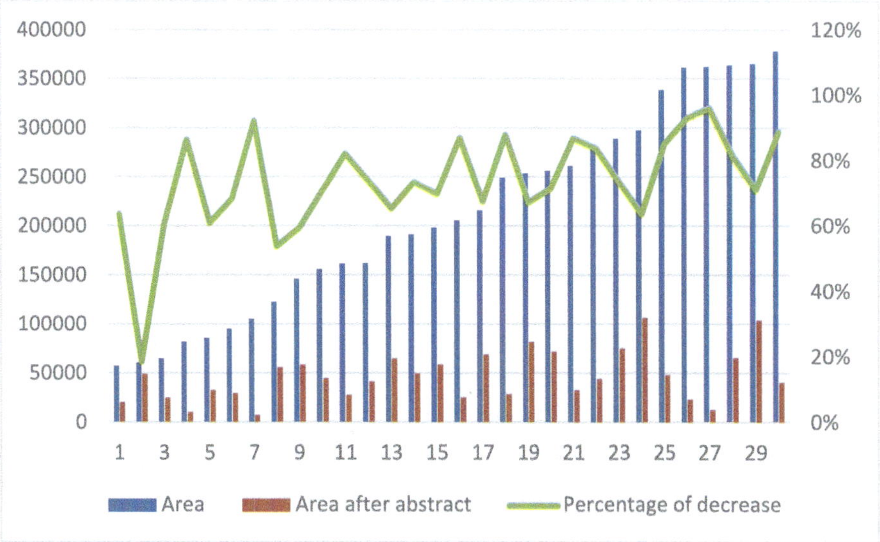

Fig. 5. Chart of the solution efficiency test for small forms

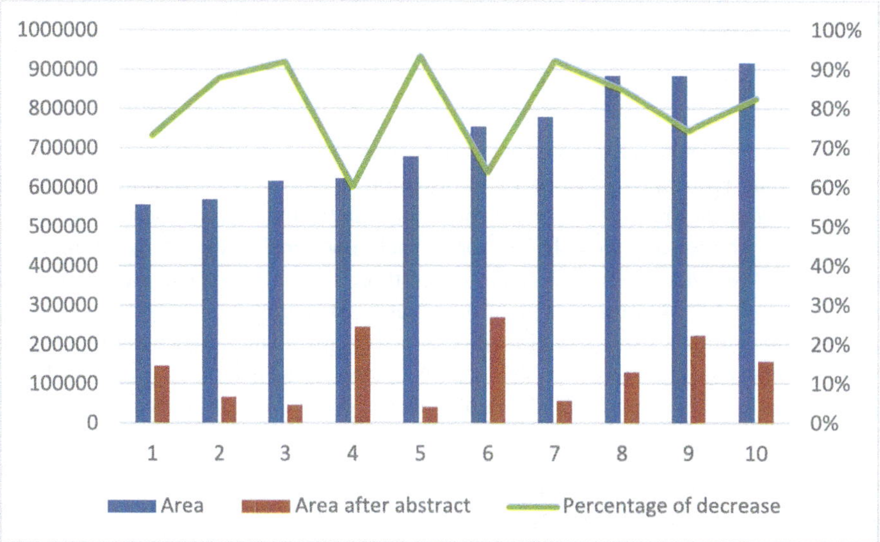

Fig. 6. Chart of the solution efficiency test for medium forms

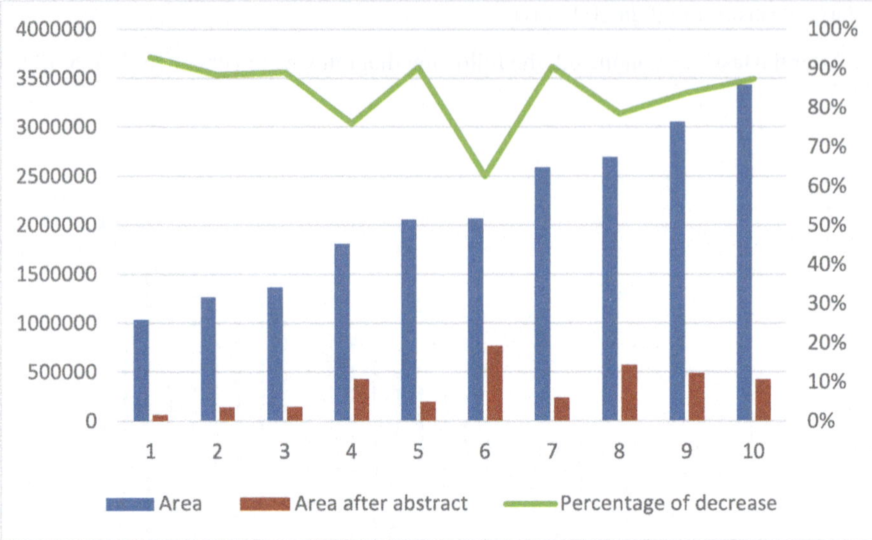

Fig. 7. Chart of the solution efficiency test for large forms

On average, small forms were reduced in size by 74%, medium forms – by 81%, whereas large forms – 84%. As expected, small forms showed a smaller percentage of reduction in the field area than medium and large forms. This is because a lot portion of data included on small forms contains one type of information, i.e. personal data.

5 Conclusion

The solution will be effective on platforms that depend on mobile device and tablet users. Form summarisation would be an innovative step in many IT systems oriented towards the flow of documents. The solution would significantly shorten readable document versions, thus reducing the scope of information to only crucial information in the relevant document. Should the need to check the full document arise, the user would be able to click a button and view the full form. Such functionality would, therefore, grant full control over what the user wants to check thoroughly. Test results are satisfactory and indicate that the solution has great potential for its application in real systems. By applying the solution, users would receive only key information which could improve working time and, therefore, the flow of documents in a given business would also generate some time savings [22, 23]. Further development opportunities may be related to:

- Software migrating already existing forms
- Improving artificial intelligence
- Adding the language of tags, allowing to define non-standard tags
- Translation of summaries into multiple languages
- Making it possible to refer to business facilities in the summary

Acknowledgements. This work was financed by Military University of Technology under research project UGB 810/2023.

References

1. Waszkowski, R.: Low-code platform for automating business processes in manufacturing. IFAC-PapersOnLine **52**(10), 376–381 (2019)
2. Chinosi, M.: BPMN: An introduction to the standard (2012)
3. Natural Language Processing and Information Systems (2018)
4. Shin, J., Lee, Y., Jung, K.: Effective Sentence Scoring Method Using BERT for Speech Recognition (2019)
5. Aggarwal, C.C.: Machine Learning for Text (2018)
6. Zajic, D., Dorr, B., Lin, J., Schwartz, R.: Multi-candidate reduction: sentence compression as a tool for document summarization tasks (2007)
7. Dorr, B., Zajic, D., Schwartz, R.: Hedge trimmer. a parse-and-trim approach to headline generation (2003)
8. Mittal, M.J., Witbrock, V.O.: Ultra-summarization: a statistical approach to (1999)
9. University Software Engineering Institute (Carnegie Mellon. Software Architecture) (2017)
10. GPT-NEO Documentation. https://www.eleuther.ai
11. Waszkowski, R., Bocewicz, G.: Visibility matrix: efficient user interface modelling for low-code development platforms. Sustainability **14**(13), 8103 (2022)
12. Waszkowski, R.: Multidimensional modeling and analysis of business processes. In: InIOP Conference Series: Materials Science and Engineering 2018 Aug 1, Vol. 400, No. 6, p. 062031 (2018)
13. Jasiulewicz-Kaczmarek, M., Waszkowski, R., Piechowski, M., Wyczółkowski, R.: Implementing BPMN in maintenance process modeling. In: Information Systems Architecture and Technology: Proceedings of 38th International Conference on Information Systems Architecture and Technology–ISAT 2017: Part II 2018, pp. 300–309. Springer International Publishing (2018)
14. Waszkowski, R., Kowalski, A.: Comparative analysis of business process management frameworks. In: Proceedings of the 30th International Business Information Management Association Conference, IBIMA 2017 - Vision 2020: Sustainable Economic development, Innovation Management, and Global Growth, Volume 2017-January, pp. 3542–3553, ISBN: 978–098604199–0 (2017)
15. Waszkowski, R., Nowicki, T.: Efficiency investigation and optimization of contract management business processes in a workwear rental and laundry service company. Procedia Manuf. **1**(44), 551–558 (2020)
16. Kiedrowicz, M., Nowicki, T., Waszkowski, R.: Business process data flow between automated and human tasks. In: 3rd International Conference on Social Science (ICSS 2016) December 2016 Dec 9, pp. 9–11 (2016)
17. Kiedrowicz, M., Nowicki, T., Waszkowski, R., Wesołowski, Z., Worwa, K.: Optimization of the document placement in the RFID cabinet. In: MATEC Web of Conferences 2016, Vol. 76, p. 02001. EDP Sciences (2016)
18. Tarapata, Z., Nowicki, T., Antkiewicz, R., Dudzinski, J., Janik, K.: Data-driven machine learning system for optimization of processes supporting the distribution of goods and services–a case study. Procedia Manuf. **1**(44), 60–67 (2020)
19. Nowicki, T., Pytlak, R., Waszkowski, R., Bertrandt, J.: The method for finding optimal sanitary inspectors schedules. In: Annals of Nutrition and Metabolism 2013 Jan 1, Vol. 63, p. 1027. ALLSCHWILERSTRASSE 10, CH-4009 BASEL, SWITZERLAND: KARGER (2013)

20. Waszkowski, R., Nowicki, T.W., Worwa, K.A.: Corporate efficiency improvement with business process automation. In: MATEC Web of Conferences 2018, Vol. 210 (2018)
21. Musiał, K., Kotowska, J., Górnicka, D., Burduk, A.: Tabu search and greedy algorithm adaptation to logistic task. In: Computer Information Systems and Industrial Management: 16th IFIP TC8 International Conference, CISIM 2017, Bialystok, Poland, June 16–18, 2017, Proceedings 16, pp. 39–49. Springer International Publishing (2017)
22. Jasiulewicz-Kaczmarek, M., Antosz, K., Zhang, C., Waszkowski, R.: Assessing the barriers to industry 4.0 implementation from a maintenance management perspective-pilot study results. IFAC-PapersOnLine **55**(2), 223–228 (2022)
23. Antosz, K., Jasiulewicz-Kaczmarek, M., Paśko, Ł, Zhang, C., Wang, S.: Application of machine learning and rough set theory in lean maintenance decision support system development. Eksploatacja i Niezawodność **23**(4), 695–708 (2021)

Algorithm for the Assignment of Courses and Instructors in Class Planning

Tadeusz Nowicki(✉) and Patryk Wicherek

Faculty of Cybernetics, Military University of Technology, 2 Kaliskiego St., 01-908 Warszawa, Poland

robert.waszkowski@wat.edu.pl

Abstract. The article focuses on the issue of assigning instructors to courses in class planning. Many universities and schools face the problem of assigning teachers to courses. These decisions largely contribute to ensuring an appropriate class schedule and the efficiency of work performed by teachers and the quality of teaching. As the number of teachers and courses increases, universities are facing a growing problem of how to assign teachers appropriately to instructors, so as to ensure the highest teaching efficiency possible. This article presents a mathematical model that expands on the classic staff assignment model by adding additional restrictions and introducing a new target function. The presented target function seeks to minimise variance in the result of the assignment of each instructor to requisitions, as it is the ratio of assigned hours to the overall salary of the instructor. A genetic algorithm was implemented on the basis of the mathematical model.

Keywords: genetic algorithm · assignment of instructors to courses · class planning

1 Introduction

The term "course" should be understood as taking into consideration that each course is divided into types of course assessments to which instructors are assigned. Courses may have the following types of course assessments:

- lecture,
- practical class,
- laboratory,
- project,
- seminar.

Therefore, instructors are assigned to specific types of course assessments made available by universities through requisitions. Not everyone instructor may conduct a course assessment for a given course, as they may not be adequately prepared to conduct a course assessment or they may not have a scientific degree. On the other hand, the instructor may facilitate the assignment process by indicating courses and types of course assessment in which they are interested and which they can conduct.

A. Burduk et al. (Eds.): ISPEM 2023, LNNS 790, pp. 445–460, 2023.
https://doi.org/10.1007/978-3-031-45021-1_34

When reviewing literature, no considerations were found that would correspond to the problem of assigning instructors to courses, however the generalised assignment problem is a subject of consideration [1]. This problem may be defined as finding the most cost-efficient assignment of tasks to machines in such a way that each task is assigned to exactly one machine and maximum utilisation of each machine is not exceeded [2]. This partially refers to the problem being considered in this article, as teachers may be assigned to multiple courses, therefore restrictions should be expanded.

The aim of the paper is to develop a genetic algorithm that solves the issue of assigning instructors to courses.

The first part of the article presents a description of the problem, its substance and related challenges. The problem under consideration is a specific case of a generalised assignment problem. Next, it presents a mathematical model formula that takes into consideration various criteria and limitations, such as instructor preferences, their salaries and the time consumption of the course. Then, on the basis of the mathematical model, it proposes a genetic algorithm that solves the issue of assigning instructors to courses. The next chapter of the article presents the results of tests performed using the proposed genetic algorithm. Using a data generator, the algorithm was verified under various problem scenarios. The test results compared the proposed genetic algorithm against the solution obtained using OpenSolver.

2 Assignment Problem

A specific example of the assignment problem can be found in literature, i.e. the staff assignment problem [3], which may be presented as a bipartite graph. Separate parts of the graph between which there are no connections form a set of nodes that represent employees and a set of nodes that represent workstations. There are arcs with assigned weights between employees and workstations. The weights of connections between employees and positions determine the productivity of the employee in performing tasks on a given position. In the classic problem, the number of workstations is equal to the number of assigned employees and each employee is connected to each workstation. Each employee is assigned to one workstation (performs a single task). Each workstation (task) is assigned exactly one employee. The solution being sought is a set of assignments (arcs connecting employees with workstations) where work efficiency, represented as the sum of weights assigned to selected connections, is the highest.

The generalised assignment problem is considered NP-hard, as the standard assignment problem is NP-hard [10]. In the generalised assignment problem, additional restrictions and costs are added, which makes it more difficult to solve.

This problem may be formulated as follows:

$$min \sum_{i=1}^{m} \sum_{j=1}^{m} c_{ij} x_{ij} \quad (1)$$

$$\sum_{j=1}^{n} a_{ij} x_{ij} \le b_i, i = 1, \ldots, m \quad (2)$$

$$\sum_{i=1}^{m} x_{ij} = 1, j = 1, \ldots, n \quad (3)$$

$$x_{ij} \in \{0, 1\}, i = 1, \ldots, m, j = 1, \ldots, n \quad (4)$$

where:

c_{ij} – is the cost of assigning i machines to j a task.

a_{ij} – is the authorisation of i machines to j the task.

b_i – is the maximum utilisation of i machines.

In the analysed literature [4–7], there are many different versions of a solution for the problem of assigning staff members to workstations. These solutions helped to establish restrictions, problems that may be faced and how to solve the problem of assigning instructors to courses.

Genetic algorithms are a type of optimisation algorithm, which means that they are used to find an optimal solution(s) to a given calculation problem that maximises or minimises a specific function [8]. Genetic algorithms are one of the fields of science that is based on the imitation of biological reproduction and natural selection processes in order to find the best solutions.

The results obtained by the genetic algorithm were compared to the spreadsheet created for research purposes together with OpenSolver [11], which solves the Mixed Integer Nonlinear Programming (MINLP) problem [9].

The algorithm proposed in the article is a base for implementing the schedule planning software based on the low-code application development platform Aurea BPM [14–16]. The tasks performed by different members of the planning process are supported by the business process management software [17–20]. All optimization tasks were implemented and evaluated [21, 22, 23, 24, 25].

3 Problem Formulation

When planning classes, the curriculum of the programme determines which courses are to be organised in the relevant year. These courses are divided into types of the course assessments for which requisitions are issued.

Let's assume that:

$\mathbb{Z} = \{1, 2, 3, \ldots, z, \ldots, Z\}$ - set of integers representing requisitions for courses

Requisition for a specific course is allocated a number of hours which are transferred onto the instructor during assignment.

$$\ell = (l_1, l_2, \ldots, l_z \ldots, l_Z), z \in \mathbb{Z}$$

Instructors may be assigned to each of the requisitions:

$$\mathbb{P} = \{1, 2, 3, \ldots p, \ldots, P\} - \textit{set of numbers representing instructors}$$

Each instructor has their own salary that limits the maximum number of hours during the academic year.

$$q = (q_1, q_2, \ldots, q_p, \ldots, q_P), p \in \mathbb{P}$$

Class instructors may specify their preferences for individual requisitions. Lets assume that this is described by the matrix:

$$\mathbb{U} = [u_{zp}]_{ZxP}$$

where:

$$u_{zp} = \begin{cases} 1 - where\ the\ p - instructor\ must\ be\ asssinged\ to\ requisition\ z \\ 2 - where\ the\ p - instructor\ must\ be\ asssinged\ to\ requisition\ z \\ 3 - where\ the\ p - instructor\ must\ be\ asssinged\ to\ requisition\ z \\ 0 - in\ any\ other\ case \end{cases}$$

$$p \in \mathbb{P}, z \in \mathbb{Z}$$

In order to satisfy the requirement related to the fact that instructors who specified in their preferences that they must teach a specific course must be assigned to that course, the following matrix is established:

$$\mathcal{B} = [b_{zp}]_{ZxP}$$

where:

$$b_{zp} = \begin{cases} 1 - where\ there\ is\ a\ \text{'must'}\ choice\ preference\ for\ any\ requisition \\ or\ where\ there\ is\ no\ \text{'must'}\ choice\ preference\ but\ there\ is\ a\ \text{'want/can'}\ preference \\ 0 - in\ any\ other\ case \end{cases}$$

In addition, the expected instructor salary utilisation value is specified:

$$\tau \in (0, 2 >$$

Having the above information, it is possible to determine the decision variable that refers to the assignment of an instructor to a course:

$$\mathbb{X} = [X_{zp}]_{ZxP}$$

where:

$$X_{zp} = \begin{cases} 1 - where\ p - instructor\ will\ be\ assigned\ to\ requisition\ z \\ 0 - in\ any\ other\ case \end{cases}$$

$$p \in \mathbb{P}, z \in \mathbb{Z}$$

The problem of assigning instructors to courses comes down to minimising variance between assigned hours and instructor salary.

$$\sigma^2 \to min \tag{5}$$

The ratio of assigned hours originating from the requisition to instructor salary is calculated for each instructor:

$$\chi_p = \frac{\sum_{i=0}^{z} x_{pi} * \ell_i}{g_p} \tag{6}$$

In order to calculate variances, it is necessary to calculate the average number of hours assigned to instructors:

$$\mu = \frac{\sum \chi}{p} \tag{7}$$

Therefore, assignment variance can be determined as follows:

$$\sigma^2 = \frac{\sum (\chi_p - \mu)^2}{p} \tag{8}$$

The problem of minimising variance is solved with the following restrictions:

- Each instructor should be assigned to at least one requisition:

$$\sum\nolimits_{i=1}^{z} \mathcal{B}_{zp} * X_{ij} \geq 1 \forall j \in < 1, p > \tag{9}$$

- One instructor should be assigned to each requisition:

$$\sum\nolimits_{j=1}^{p} \mathcal{B}_{zp} * X_{ij} = 1 \forall i \in < 1, z > \tag{10}$$

- Instructor cannot exceed the specified maximum number of hours during the year:

$$\frac{\sum_{i=0}^{z} \mathcal{B}_{zp} * X_{pi} * \ell_i}{\mathcal{g}_p} \leq \tau \tag{11}$$

In the example under consideration, the university issued 3 requisitions that require specific time consumption. The university employs 3 teachers who have an annual salary specified in their contracts Teachers received the list of requisitions and selected their preferences. When the university received the list of teacher preferences, it commenced assigning them to teach specific courses.

As regards requisitions, the following data is provided:

- Numbers representing course requisitions:

$$\mathbb{Z} = \{1, 2, 3\}$$

- Number of hours:

$$\ell = (14, 12, 16)$$

As regards instructors, the following data is provided:

- Numbers representing instructors:

$$\mathbb{P} = \{1, 2, 3\}$$

- Salaries for instructors:

$$q = (20,20,20)$$

- Instructors specified the following preferences:

$$\mathbb{U} = \begin{bmatrix} 1\,0\,2 \\ 3\,1\,2 \\ 0\,3\,2 \end{bmatrix}$$

which means that:

- requisition z_1 must be taught p_1 or is wanted by p_3
- requisition z_2 may be taught p_1, must be taught p_2 or is wanted by p_3
- requisition z_3 may be taught p_2 or is wanted by p_2

We obtain the following matrix from the instructor preference matrix:

$$\mathcal{B} = \begin{bmatrix} 1\,0\,0 \\ 0\,1\,0 \\ 0\,1\,1 \end{bmatrix}$$

which means that:

- requisition z_1 may be assigned to p_1,
- requisition z_2 may be assigned to p_2,
- requisition z_3 may be asssigned to p_2 or p_3

Decision Variables:

- Designated instructor-to-course assignment matrix:

$$\mathbb{X} = \begin{bmatrix} x_{11}\ x_{12}\ x_{13} \\ x_{21}\ x_{22}\ x_{23} \\ x_{31}\ x_{32}\ x_{33} \end{bmatrix}$$

Restrictions:

- Each instructor should be assigned to at least one requisition:

$$\sum\nolimits_{j=0}^{p} X_{ij} \geq 1 \forall i \in <1, z> \tag{12}$$

After calculating the sum for each instructor:

$$b_{11} * x_{11} + b_{21} * x_{21} + b_{31} * x_{31} \geq 1,\ for\ p = 1 \tag{13}$$

$$b_{12} * x_{12} + b_{22} * x_{22} + b_{32} * x_{32} \geq 1,\ for\ p = 2 \tag{14}$$

$$b_{13} * x_{13} + b_{23} * x_{23} + b_{33} * x_{33} \geq 1,\ for\ p = 3 \tag{15}$$

After substituting the value $\boldsymbol{b}$:

$$b_{11} * x_{11} \geq 1,\ for\ p = 1 \tag{16}$$

$$b_{22} * x_{22} \geq 1,\ for\ p = 2 \tag{17}$$

$$b_{23} * x_{23} + b_{33} * x_{33} \geq 1,\ for\ p = 3 \tag{18}$$

- One instructor should be assigned to each requisition:

$$\sum_{j=1}^{p} \mathcal{B}_{zp} * X_{ij} = 1 \forall i \in < 1, z > \tag{19}$$

After calculating the sum for each requisition:

$$b_{11} * x_{11} + b_{12} * x_{12} + b_{13} * x_{13} = 1, forz = 1 \tag{20}$$

$$b_{21} * x_{21} + b_{22} * x_{22} + b_{23} * x_{23} = 1, forz = 2 \tag{21}$$

$$b_{31} * x_{31} + b_{32} * x_{32} + b_{33} * x_{33} = 1, forz = 3 \tag{22}$$

After substituting the value ***b***:

$$b_{11} * x_{11} = 1, forz = 1 \tag{23}$$

$$b_{22} * x_{22} = 1, forz = 2 \tag{24}$$

$$b_{32} * x_{32} + b_{33} * x_{33} = 1, forz = 3 \tag{25}$$

- Instructor cannot exceed the specified maximum number of hours during the year:

$$\frac{\sum_{i=0}^{z} \mathcal{B}_{zp} * X_{pi} * \ell_i}{g_p} \leq \tau \tag{26}$$

After calculating the sum for each instructor:

$$\frac{(b_{11} * x_{11} * l_1) + (b_{21} * x_{21} * l_2) + (b_{31} * x_{31} * l_3)}{g_1} \leq \tau \; for \; p = 1 \tag{27}$$

$$\frac{(b_{12} * x_{12} * l_1) + (b_{22} * x_{22} * l_2) + (b_{32} * x_{32} * l_3)}{g_2} \leq \tau \; for \; p = 2 \tag{28}$$

$$\frac{(b_{13} * x_{13} * l_1) + (b_{23} * x_{23} * l_2) + (b_{33} * x_{33} * l_3)}{g_3} \leq \tau \; for \; p = 3 \tag{29}$$

After substituting the value ***b***:

$$\frac{(b_{11} * x_{11} * l_1)}{g_1} \leq \tau \; for \; p = 1$$

$$\frac{(b_{22} * x_{22} * l_2) + (b_{32} * x_{32} * l_3)}{g_2} \leq \tau \; for \; p = 2$$

$$\frac{(b_{33} * x_{33} * l_3)}{g_3} \leq \tau \; for \; p = 3$$

The optimal solution was calculated using OpenSolver and the result was presented in Fig. 1.

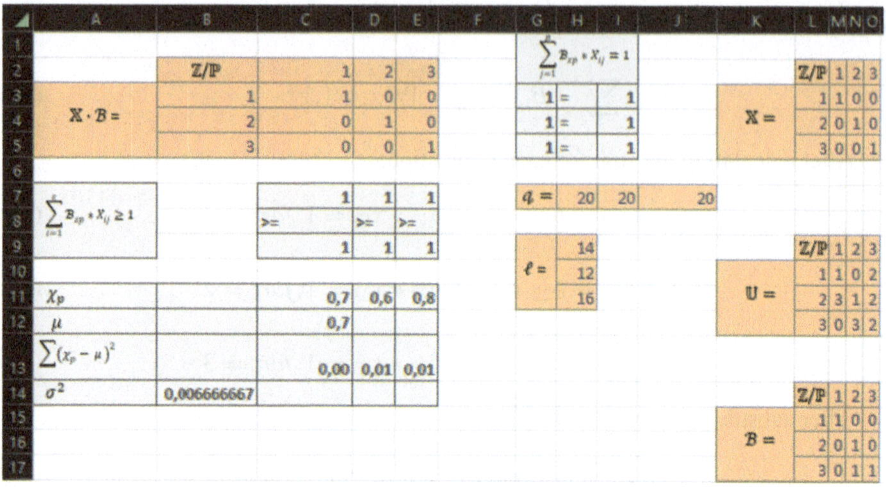

Fig. 1. Spreadsheet for the problem of assigning instructors to the course

4 Genetic Algorithm for the Problem of Assigning Instructors to the Course

The problem of assigning instructors to courses comes down to minimising variance between assigned hours and instructor salary. A required characteristic of the algorithm is that it must generate an optimal solution, while also always assigning all instructors in a way that best utilises their salary for the relevant year.

As the first step, the implemented algorithm generates an initial population whose size is specified by the user. The population is then subjected to quality assessment that takes into consideration the minimum variance of the assignment and the maximum number of assigned instructors. After the quality assessment, there is a verification of the stop condition that determines whether the best individual from the population meets the criteria specified by the user, which include:

- Variance value
- Maximum population size

If the population does not meet the stop criterion, the algorithm selects the best individuals. Selection consists in selecting a specific number (it is determined by the user) of the best solutions to be subjected to a crossover and mutation. Crossover in the algorithm is performed in two ways: one-point and two-point. When individuals in a population are subjected to a crossover and mutation, the population is re-assessed in terms of quality. When the quality assessment meets the stop criteria, the algorithm terminates the execution and displays board containing requisitions with instructors assigned to them. The algorithm was implemented in accordance with the diagram presented in Fig. 2.

In order to test the effectiveness of the implemented algorithm, a spreadsheet was created in the Microsoft Excel and the OpenSolver suite. The Bonmin engine, which solves the MINLP (Mixed Integer NonLinear Programming) generalised problem, is

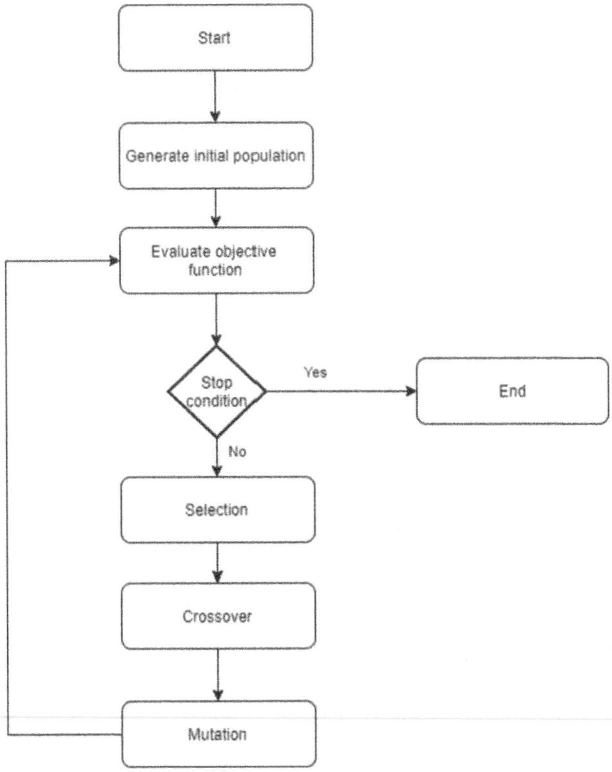

Fig. 2. Block diagram of the implemented genetic algorithm

used to find the most optimal solution. In addition, the solution is sought with the support of the NEOS Server, which is a free website for solving numerical optimisation problems. Problems are solved on distributed, high-efficiency machines, which results in finding in a relatively quick solution for the presented problem.

5 Results

In order to simulate different situations, a generator of data for the algorithm has been implemented. The said generator accepts parameters related to the number of instructors and the number of requisitions, and additionally allows to specify the probability of a preference occurring, i.e. must, want, may. The generator returns the list of instructors along with their salaries, the list of requisitions and the number of hours, and the list of preferences specified by instructors regarding requisitions. The genetic algorithm for each experiment has been launched with the various parameters shown in Table 1, and the algorithm will be launched 10 times.

Table 1. Algorithm parameters used during experiments

Configuration	Population size	Mutation Rate
Genetic 1	100	0.1
Genetic 2	100	0.2
Genetic 3	500	0.2

5.1 Experiment for a Small Population Size

The first experiment was carried out on a relatively small population size. Small population size refers to the number of requisitions and instructors, i.e. 62 instructors and 137 requisitions. The aim was to verify the behaviour of the algorithm and its effectiveness.

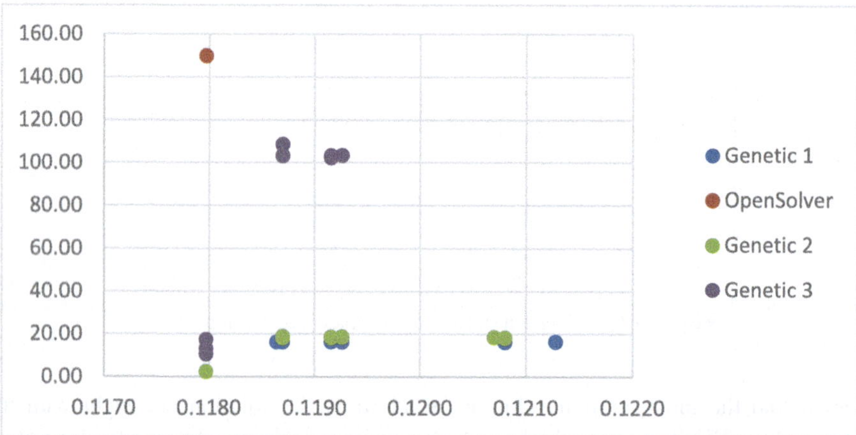

Fig. 3. Chart showing dependence between time and variance for a small population size

The results have been presented in Fig. 3. The chart shows that in two cases the genetic algorithm achieved the optimal solution in a significantly shorter time than using OpenSolver. The obtained results indicate that a higher mutation rate and higher initial population size affected the achievement of the optimal solution. However, there are iterations where the duration of the genetic algorithm is significantly longer. Detailed data on the obtained results is presented in Table 2. In conclusion, over a total of 30 runs the algorithm achieved the optimal solution three times. Including the time spent waiting for the solution and differences in the values of obtained variances, the algorithm is a more effective solution than OpenSolver.

Table 2. Results obtained for a small population size

OpenSolver		Genetic 1		Genetic 2		Genetic 3	
Variance	Time [s]	Variance	Time [s]	Variance	Time [s]	Variance	Time [s]
0.1180	150	0.1186	16.44	0.1180	2.61	0.1180	17.48
		0.1187	16.62	0.1187	19.03	0.1180	10.76
		0.1187	16.40	0.1187	18.51	0.1180	13.25
		0.1192	16.59	0.1187	18.62	0.1187	108.77
		0.1192	16.48	0.1192	18.62	0.1187	103.29
		0.1192	16.49	0.1192	18.74	0.1187	103.59
		0.1193	16.79	0.1192	18.58	0.1192	103.22
		0.1193	16.42	0.1193	18.89	0.1192	103.56
		0.1208	16.36	0.1207	18.62	0.1192	102.40
		0.1213	16.49	0.1208	18.46	0.1193	103.64

5.2 Experiment for a Large Population Size

The second experiment was carried out on a relatively large population size. Large population size refers to the number of requisitions and instructors, i.e. 62 instructors and 500 requisitions.

The results have been presented in Fig. 4. The chart shows that the genetic algorithm once again found the optimal solution in a significantly shorter period of time than using OpenSolver. For this population size, a higher mutation rate was of greater relevance than a higher population size, which affected the achievement of the optimal solution.

Detailed data on the obtained results is presented in Table 3. For this population size, the genetic algorithm also achieved the optimal solution three times. The remaining results slightly diverge from the optimal solution, therefore it can be stated that the algorithm is effective for a large population size.

5.3 Experiment for a Case of High Instructor Workload

The test verifies how the algorithm behaves in the case of a high instructor workload, understood as a relatively large number of requisitions compared to the number of instructors.

The results have been presented in Fig. 5. The chart shows that the genetic algorithm once again found the optimal solution in a significantly shorter period of time than using OpenSolver. In this case, a change of parameters did not have a significant impact on the obtained results, longer time is a result of higher population size.

Detailed data on the obtained results is presented in Table 4. The results indicate that in this case the algorithm is much more effective, as the optimal solution was achieved 17 times.

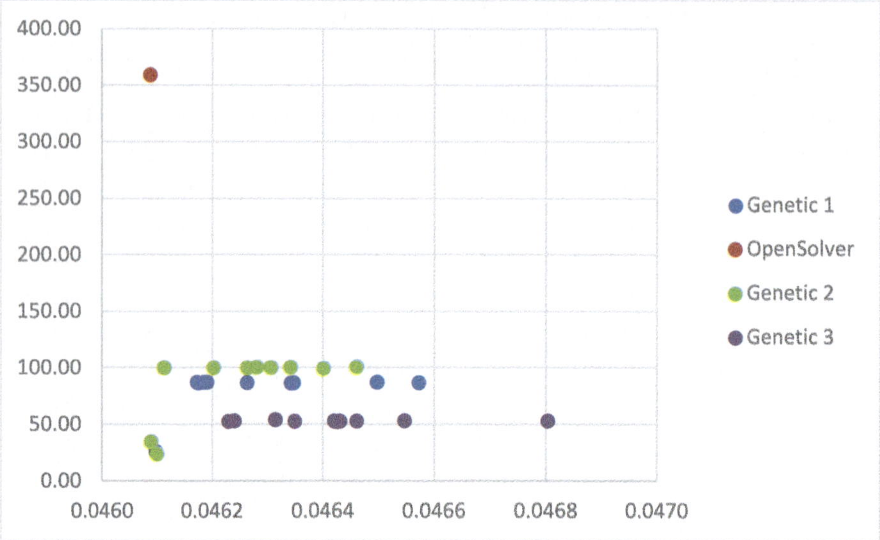

Fig. 4. Chart showing dependence between time and variance for a large population size

Table 3. Results obtained for a large size population size

OpenSolver		Genetic 1		Genetic 2		Genetic 3	
Variance	Time [s]	Variance	Time [s]	Variance	Time [s]	Variance	Time [s]
0.0461	360	0.0461	27.09	0.0461	35.75	0.0462	53.47
		0.0462	87.89	0.0461	24.72	0.0462	53.86
		0.0462	87.48	0.0461	100.89	0.0463	55.01
		0.0462	88.01	0.0462	101.03	0.0463	53.51
		0.0462	87.97	0.0463	100.96	0.0464	53.70
		0.0463	87.67	0.0463	101.39	0.0464	53.18
		0.0463	87.35	0.0463	101.14	0.0464	53.56
		0.0463	87.80	0.0463	101.55	0.0465	53.69
		0.0465	87.89	0.0464	100.35	0.0465	53.77
		0.0466	87.41	0.0465	101.87	0.0468	53.74

5.4 Experiment for a Case of Low Instructor Workload

The test verifies how the algorithm behaves in the case of a low instructor workload, understood as a relatively low number of requisitions compared to the number of instructors. The results have been presented on the chart below.

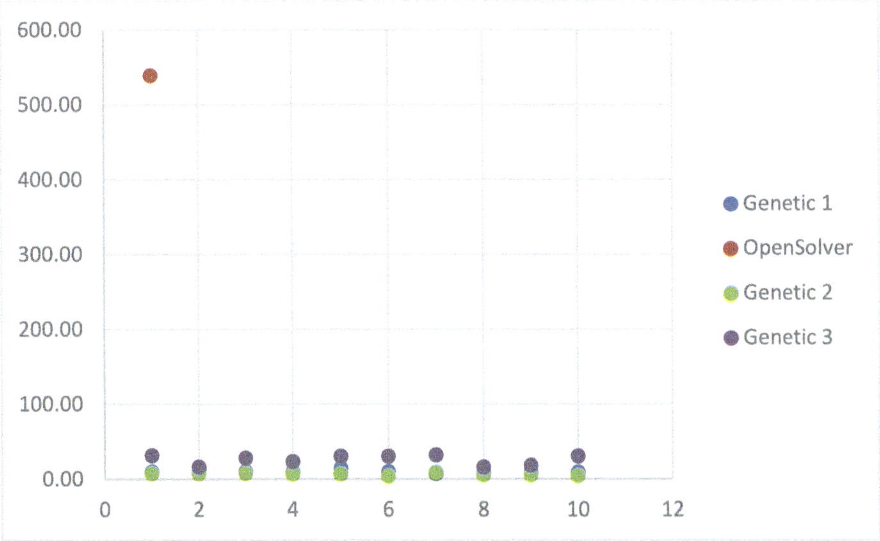

Fig. 5. Chart showing dependence between time and variance for a case of high workload

Table 4. Results obtained for a case of high workload

OpenSolver		Genetic 1		Genetic 2		Genetic 3	
Variance	Time [s]	Variance	Time [s]	Variance	Time [s]	Variance	Time [s]
3.0445	540	3.0445	11.05	3.0445	9.45	3.0445	32.68
		3.0445	13.89	3.0445	9.56	3.0445	17.72
		3.0445	11.54	3.0445	9.86	3.0445	29.40
		3.0445	10.79	3.0445	8.95	3.0445	24.79
		3.0445	16.11	3.0445	9.23	3.0446	31.78
		3.0445	11.53	3.0445	6.15	3.0446	32.07
		3.0446	9.04	3.0445	10.55	3.0446	33.53
		3.0446	10.20	3.0446	7.59	3.0446	17.34
		3.0446	10.99	3.0446	7.64	3.0446	20.22
		3.0446	11.21	3.0446	7.01	3.0446	32.36

The results have been presented in Fig. 6. The chart shows that the genetic algorithm was able to find the optimal solution only in one run, however taking the time into consideration it still significantly exceeds OpenSolver. In this case, a change of parameters did not have a significant impact on the obtained results, longer time is a result of higher population size.

Detailed data on the obtained results is presented in Table 5. In this case, the algorithm was not effective as it only found the optimal solution only once over 30 runs.

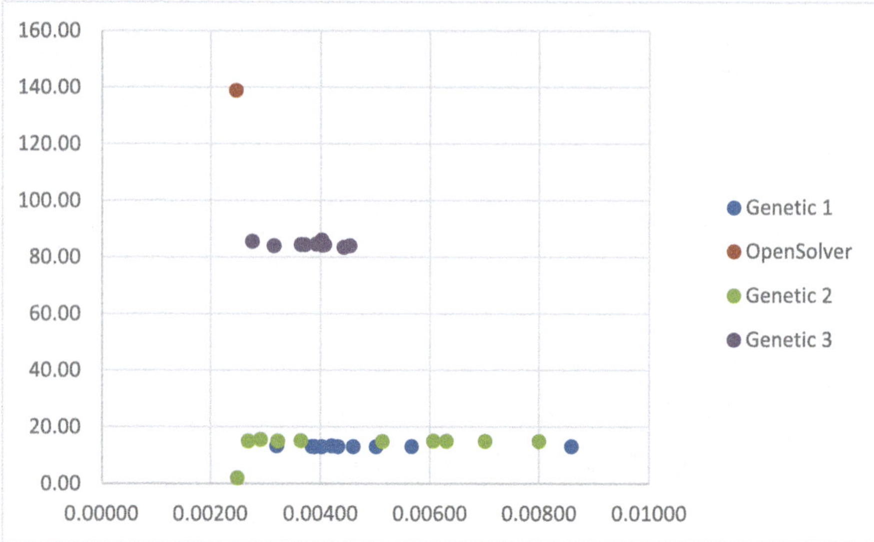

Fig. 6. Chart showing dependence between time and variance for a case of low workload

Table 5. Results obtained for a case of low workload

OpenSolver		Genetic 1		Genetic 2		Genetic 3	
Variance	Time [s]	Variance	Time [s]	Variance	Time [s]	Variance	Time [s]
0.0025	600	0.0032	13.39	0.0025	2.15	0.0028	85.58
		0.0038	13.17	0.0027	15.08	0.0032	84.06
		0.0039	13.09	0.0029	15.63	0.0036	84.52
		0.0040	13.10	0.0032	15.12	0.0037	84.42
		0.0042	13.40	0.0036	15.29	0.0039	84.67
		0.0043	13.13	0.0051	14.98	0.0040	86.08
		0.0046	13.12	0.0061	15.15	0.0040	84.28
		0.0050	13.12	0.0063	15.11	0.0041	84.42
		0.0057	13.14	0.0070	15.05	0.0044	83.48
		0.0086	13.14	0.0080	15.05	0.0045	84.06

6 Conclusion

The article examines the problem of planning classes, more specifically assigning courses and instructors to these courses. It presents a mathematical model that allows to identify the optimal solution for that problem. It also presents a genetic algorithm. The conducted tests showed that the algorithm is able to identify the optimal solution significantly faster than using OpenSolver.

In the future, this algorithm may be supported by artificial intelligence, which will help instructors to select appropriate courses to reach on the basis of their previous choices. Furthermore, artificial intelligence may be able to more effectively assign instructors to courses, thus allowing to plan classes even more effectively.

References

1. Cattrysse, D., Van Wassenhove, L.N.: A survey of algorithms for the generalized assignment problem. Eur. J. Oper. Res. **60**(3) (1992)
2. Nauss, R.M.: Solving the generalized assignment problem: an optimizing and heuristic approach. Informs J. Comput. **15**(3) (2003)
3. Kuhn, H.W.: The Hungarian method for the assignment problem. Naval Res. Logistics Quarterly 2 (1955)
4. Yagiura, M., Ibaraki T., Glover, F.: An ejection chain approach for the generalized assignment problem. Informs J. Comput. (2004)
5. Landowska, A.: Solution of the optimal allotment problem using classical Genetic algorithm (1993)
6. Pentico, D.: Assignment problems: a golden anniversary survey. Eur. J. Oper. Res. **176**, 774–796 (2007)
7. Klimek, M., Łebkowski, P.: Personnel assignment problem to workstations with Corruption risk (2012)
8. Carr, J.: An introduction to genetic algorithms. Senior Project **1**(40) (2014)
9. Tolle, J.W.: Encyclopedia of Physical Science and Technology (2003)
10. Yagiura, M., Ibarki, T.: The Generalized Assignment Problem and Its Generalizations
11. OpenSolver, https://opensolver.org/. Accessed 16 Feb 2023
12. Waszkowski, R., Bocewicz, G.: Visibility matrix: efficient user interface modelling for low-code development platforms. Sustainability **14**(13) (2022)
13. Waszkowski, R.: Multidimensional modeling and analysis of business processes. In: IOP Conference Series: Materials Science and Engineering, IOP Publishing 400(6) (2018)
14. Jasiulewicz-Kaczmarek, M., Waszkowski, R., Piechowski, M., Wyczółkowski, R.: Implementing BPMN in maintenance process modeling. In: Information Systems Architecture and Technology: Proceedings of 38th International Conference on Information Systems Architecture and Technology–ISAT 2017, Springer International Publishing, pp. 300–309 (2018)
15. Waszkowski, R., Kowalski, A.: Comparative analysis of business process management frameworks. In: Proceedings of the 30th International Business Information Management Association Conference, IBIMA 2017 - Vision 2020: Sustainable Economic development, Innovation Management, and Global Growth, pp. 3542–3553 (2017)
16. Waszkowski, R., Nowicki, T.: Efficiency investigation and optimization of contract management business processes in a workwear rental and laundry service company. Procedia Manufacturing (2020)
17. Kiedrowicz, M., Nowicki, T., Waszkowski, R.: Business process data flow between automated and human tasks. In: 3rd International Conference on Social Science (ICSS 2016), pp. 9–11 (2016)
18. Kiedrowicz, M., Nowicki, T., Waszkowski, R., Wesołowski, Z., Worwa, K.: Optimization of the Document Placement in the RFID Cabinet. In: MATEC Web of Conferences 2016, EDP Sciences 76, 02001 (2016)
19. Tarapata, Z., Nowicki, T., Antkiewicz, R., Dudzinski, J., Janik, K.: Data-driven machine learning system for optimization of processes supporting the distribution of goods and services–a case study. Procedia Manuf. (2020)

20. Nowicki, T., Pytlak, R., Waszkowski, R., Bertrandt, J.: The method for finding optimal sanitary inspectors schedules. Annals Nutrition Metabolism **63**, 1027–1027 (2013)
21. Waszkowski, R., Nowicki, T.W., Worwa, K.A.: Corporate efficiency improvement with business process automation. In: MATEC web of conferences 2018 210, (2018)
22. Sroka, B., Rosłon, J., Podolski, M., Bożejko, W., Burduk, A., Wodecki, M.: Profit optimization for multi-mode repetitive construction project with cash flows using metaheuristics. Archives Civil Mech. Eng. **21**(2), 67 (2021)
23. Kotowska, J., Markowski, M., Burduk, A.: Optimization of the supply of components for mass production with the use of the ant colony algorithm. In: Intelligent Systems in Production Engineering and Maintenance–ISPEM 2017: Proceedings of the First International Conference on Intelligent Systems in Production Engineering and Maintenance ISPEM 2017, Springer International Publishing, pp. 347–357 (2018)

Application of Clustering and SOM Analysis for Identification of Conveyor Belt Damage Based on Data from the Diagbelt + Magnetic System

Aleksandra Rzeszowska[1(✉)], Leszek Jurdziak[1], Ryszard Błażej[1], and Agata Kirjanów-Błażej[2]

[1] Faculty of Geoengineering, Mining and Geology, Wroclaw University of Science and Technology, 15 Na Grobli Street, 50-421 Wrocław, Poland
aleksandra.rzeszowska@pwr.edu.pl

[2] Faculty of Information and Communication Technology, Wroclaw University of Science and Technology, 11/17 Janiszewskiego Street, 50-372 Wrocław, Poland

Abstract. The article presents the application of two unsupervised learning methods, clustering and SOM (self-organizing maps) analysis, for grouping failures in conveyor belts based on data from the DiagBelt + magnetic system. Both methods were used to automatically group the obtained damage signals in order to enable their further identification. Data grouping was performed based on real signals obtained from scanning conveyor belts with the DiagBelt+ diagnostic system. The belt loops on eight conveyors in one of coal mines in Poland, with a total length of 18 472 m, were examined. In total, 96 belt sections from various manufacturers, including belts after refurbishment and at different ages, were studied. The conducted research indicates the effectiveness of using SOM analysis and clustering for grouping damages in conveyor belts based on data from the magnetic system.

Keywords: Conveyor belt · Clustering · Damage identification · Magnetic method

1 Introduction

Belt conveyor diagnostics is an extremely important element in maintaining the entire material transport system in good technical condition [1–5]. In recent years, more and more attention has been paid to it, resulting in the development of diagnostic methods [5–7]. Automatic non-destructive methods are becoming increasingly popular in belt diagnostics. Compared to traditional methods, such as visual inspection, automatic non-destructive methods allow for faster and more accurate assessment of the technical condition, which enables faster decision-making regarding the need for repair or replacement of conveyor elements [8]. The most commonly used non-destructive methods in belt conveyor diagnostics include vibration analysis [9–12], thermography [13, 14], magnetic method [4, 15–18], and ultrasonic method [19]. These methods allow for

A. Burduk et al. (Eds.): ISPEM 2023, LNNS 790, pp. 461–475, 2023.
https://doi.org/10.1007/978-3-031-45021-1_35

a precise analysis of various conveyor elements, such as drives [20, 21], rollers [22–27], bearings [28–30], belts [31–34], including the identification of damage to the steel core of conveyor belts, such as corrosion, cracking, cutting, or missing cords. As a result, the use of non-destructive methods in belt conveyor diagnostics allows for an increase in not only the efficiency but also the safety of belt conveyors. It is important that the diagnostic methods used are effective and precise, enabling early detection of belt damage and avoiding more serious breakdowns.

The diagnostics of conveyor belts is extremely important because it allows for early detection of damage, which in turn enables the avoidance of more serious breakdowns [35]. However, just stating that the belt is damaged is not enough to take appropriate repair measures. Proper analysis of the obtained signals is equally important, as it allows for the identification of the type of damage and the determination of the causes of its occurrence. Measurement of the technical condition of the steel core of the belt using the DiagBelt+ magnetic system can provide information about irregularities in the belt [8] (cuts, deformations, or missing links), however, without proper analysis of this data, it would be difficult to determine what type of damage is occurring, and therefore, what repair steps need to be taken. Hence, proper analysis of obtained signals and recognition of the type of damage are of crucial importance. This allows for appropriate planning of repairs, using appropriate materials and tools.

2 Non-Destructive Testing (NDT) Methods for Assessing the Technical Condition of the Belt Core

Automatic systems for monitoring the technical condition of conveyor belts are increasingly used in industry [1–3], because they allow for the detection of damages and the prediction of the time of trouble-free operation of the conveyor. This enables the prevention of failures and unplanned downtime of production lines, which in turn translates into lower maintenance costs and greater efficiency [4, 5]. Non-destructive testing (NDT) methods are an indispensable tool in diagnosing the technical condition of conveyor belts. Among the most commonly used are visual, ultrasonic, magnetic, and X-ray methods. The advantage of NDT testing is that it allows for the assessment of the technical condition of the tested elements without compromising their integrity and without the risk of damage.

Visual method involves direct observation of the conveyor belt surface and is one of the simplest and cheapest NDT methods. It involves a visual inspection to detect visible defects such as mechanical damage, cracks, and material loss. The advantage of this method is that it is relatively simple and does not require specialized equipment, but its disadvantages are that it can be difficult to apply in the case of invisible defects, and its results may be subjective and dependent on the operator's experience.

The ultrasonic method uses high-frequency sound waves that are sent into the belt. Reflected ultrasonic waves are analyzed to detect failures such as cracks, internal material loss, etc. The advantage of this method is that it allows detection of internal failures and is relatively fast and precise. The disadvantage is that it can be costly and requires specialized knowledge and equipment, as well as often requiring lengthy calibration of the measuring tool.

The magnetic method uses a magnetic field to detect failures in the belt. The disadvantage of this method is that it will not detect defects unrelated to the magnetic field. However, the magnetic method is relatively fast and cheap, and it allows detection of internal defects.

The X-ray method uses X-ray radiation to penetrate through the belt material. The radiation is recorded on a special film or detector and then analyzed to detect damage. The advantage of this method is that it allows detection of even the smallest failures. However, the X-ray method can be costly, requires specialized knowledge and equipment, and can be harmful to health if not used properly.

2.1 DiagBelt + Diagnostic System

By enabling fast testing and offering high accuracy, the magnetic method is one of the popular techniques used in non-destructive testing (NDT) to assess the technical condition of various materials, including steel cord conveyor belt cores. The magnetic method uses the magnetic properties of materials, particularly their ability to generate a magnetic field. In the case of testing conveyor belt cores using magnetic methods, this property is utilized since the core of the belt is made of a material with magnetic properties.

During the testing of the belt core, typically one of two magnetic techniques is used. The first one is the flow method, which involves passing an electric current through the belt core. As a result of the current flow, a magnetic field is generated, which can be recorded using magnetic sensors. These readings allow for the identification of any defects in the belt core, such as cracks, voids, or changes in magnetic properties. The second technique is the magnetization method, which involves magnetizing the belt core from the outside and observing its magnetic properties. During the test, the strength of the magnetic field generated by the magnetized belt core is measured. In the presence of defects, such as cracks or changes in magnetic properties, the magnetic field undergoes changes, allowing for their detection.

Both of these methods enable non-invasive and precise assessment of the technical condition of the belt core. In the case of the DiagBelt + system, the second technique was used. The system utilizes a measuring head consisting of 112 coils placed at equal distances of 25 mm and two permanent magnet heads located above and below the tested belt at a suitable distance (about 30 mm) from its covers [39, 40].

Coils read the change in magnetic field that occurs at locations where discontinuities are present in the steel cords, allowing for an assessment of the conveyor belt's condition. Measurement signals from the coils are amplified and collected in a data recording module, which then transmits them to a computer. The system allows for coil sensitivity and measurement frequency to be adjusted, making it possible to customize measurements to individual requirements. Measurement results are saved in a CSV file and can be used to analyze the conveyor belt's condition and plan maintenance.

In mines, steel cord conveyor belts are widely employed for efficient material transportation. Below, we present a cross-sectional diagram of such a belt, illustrating the key components and the construction utilized in these applications (Fig. 1).

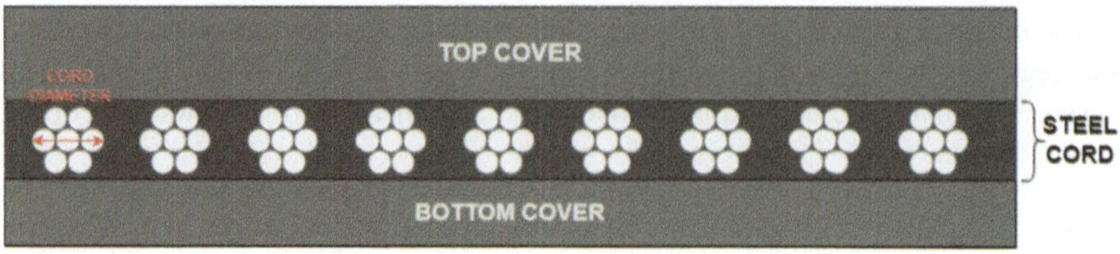

Fig. 1. Cross-sectional diagram of a steel cord conveyor belt commonly used in mines.

Figure 2 illustrates an example of the obtained measurement data visualized as a 2D image. This image is the result of scanning a reference belt with simulated failures in the Belt Transport Laboratory at Wrocław University of Science and Technology.

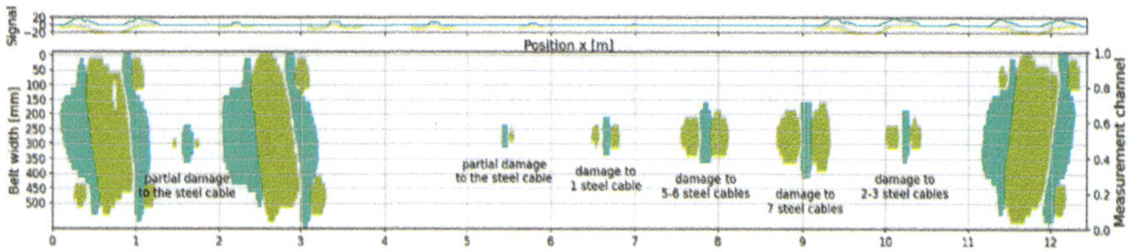

Fig. 2. Example of 2D visualization of measurement results from the DiagBelt+ system for a belt with modeled failures.

In conveyor belts operating in industrial conditions, the accumulation of failures is often much greater. An example of 21-m section of conveyor belt from one of the tested belts, along with numbered failures, is shown in Fig. 3.

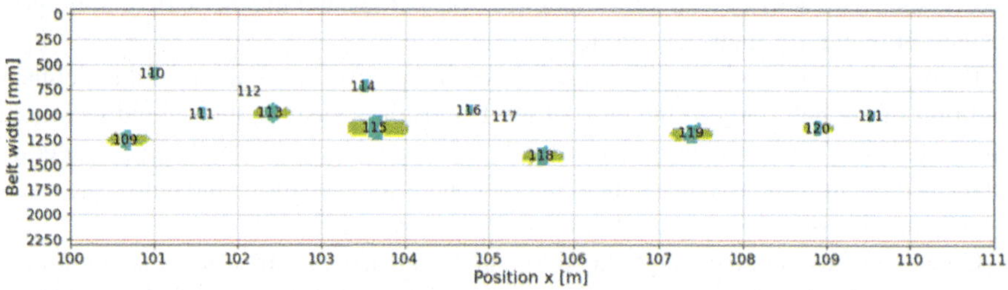

Fig. 3. Example of 2D visualization of measurement results from the DiagBelt + system for a conveyor belt operating in a mine.

3 Preparing Measurement Data for Further Processing

Preparing measurement data for further processing is essential because unprocessed or unprepared data can lead to incorrect conclusions and decisions based on them. Preparing measurement data involves various stages, including data collection, selection, processing, normalization, and analysis. In the case of failure signal analysis, preliminary data analysis allows defining key parameters describing failures, which enables reducing the size of the input vector during clustering and SOM network learning. This leads to a more detailed and unambiguous analysis.

Another advantage of preliminary data analysis is the possibility of faster and more efficient analysis of large data sets, which enables the use of more advanced data processing techniques such as machine learning, deep neural networks, and artificial neural networks. Furthermore, preliminary data analysis allows identifying potential errors and irregularities in the data, which enables correcting them before further processing.

The results obtained after preliminary data analysis enabled defining parameters that were used for further analysis. Instead of processing the entire input matrix (which includes many millions of values when inputting the entire recorded signal, or different data sizes for each failure), only a few key parameters describing failures were used during clustering and SOM network learning. This allowed maintaining a constant and small input size during the learning process and minimizing the computational time.

These parameters, namely the length, width, and surface area of the failure signal, were selected based on the previous analysis of the locations of failure signals. Moreover, binary information (0/1) about the presence of auxiliary clouds around the failure signal was added to the input vector. The histograms of the distribution of these values are presented in Fig. 4.

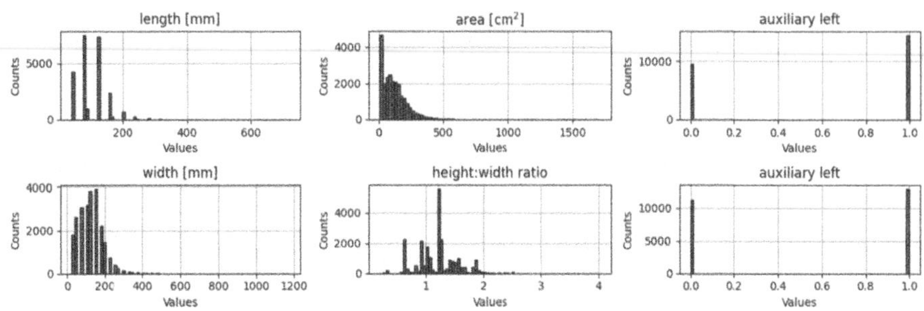

Fig. 4. Histograms of the determined damage parameters.

Normalization of input data is necessary in the case of measurement data analysis. Thanks to this process, it is possible to proceed to further stages of data processing in a more efficient and reliable way. Without normalization, it is not possible to compare the values of individual parameters, because their ranges may differ significantly, which in turn affects the analysis results.

In the case of measurement data, normalization allows obtaining values on a comparable and standardized scale, which is necessary in the process of data processing. Additionally, normalization enables avoiding situations where certain parameters are treated as less important than others, only because they have a different range of values. It also helps to avoid the so-called "dominance effect", in which one feature has a greater impact on the results than others.

In the context of specific measurement data described above, normalization is particularly important as it allows for maintaining the correct relationship between the parameters of damage signals (which differ from each other in both range and unit), and thus for a more accurate determination of the degree of damage advancement. Normalization also allows for reducing the influence of extreme values, which can significantly impact the results of analysis without normalization. There are several ways to normalize input data in data analysis, and the choice of the appropriate approach depends on the characteristics of the data and the purpose of the analysis. Among the most popular methods, the following are distinguished:

1. Min-max normalization - involves scaling the values in such a way that the minimum and maximum values correspond to 0 and 1, respectively. Intermediate values are calculated based on the formula:

$$\frac{x - min_x}{max_x - min_x}$$

where: x is the data vector, min_x is the minimum value from the set x, and max_x is the maximum value from the set x.

This method is simple to implement and works well when the data is evenly distributed throughout the range, but it can lead to issues in case of outliers, as the minimum and maximum values are calculated based on the entire data set. Moreover, in case the data set contains multiple groups with significantly different values, min-max normalization can lead to interpretation issues of the results.

2. Z-score (standardization) - this method involves scaling the values in such a way that they have a mean of 0 and a standard deviation of 1. Intermediate values are calculated based on the formula:

$$\frac{x - mean_x}{std_x}$$

where: x is the data vector, $mean_x$ is the mean value from the set x, and std_x is the standard deviation value from the set x.

In this method, outliers can have a significant impact on the mean and standard deviation, which can distort the normalization results. Additionally, this method works best for normally distributed data. For data that does not follow a normal distribution, Z-score normalization may not be appropriate.

In the analysis of the presented parameters, Z-score normalization was used. After normalizing the data, the defined parameters were used for clustering and training a self-organizing map (SOM) neural network to obtain a more detailed and accurate analysis of system failures. As a result, the obtained results become more comparable and enable the identification of patterns in the data and grouping them into similar clusters. Finally, the distribution of parameters used for further processing (clustering) has a distribution as shown in Fig. 5.

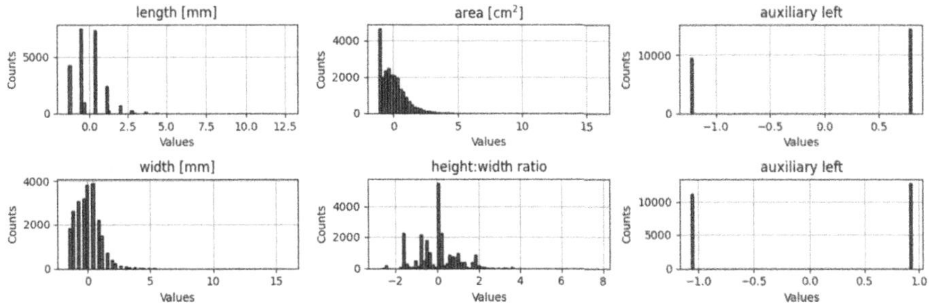

Fig. 5. Histograms of the calculated damage parameters after Z-score normalization.

4 Methods of Data Clustering

Automatic methods of data clustering are a popular approach in the field of data analysis, which involves dividing a dataset into groups or clusters based on similarities between data points. These methods allow for the clustering process to be performed without user interaction with the system. There are various clustering techniques, such as k-means, hierarchical clustering, Fuzzy C-means, and many others [41]. Each of these methods has its advantages and disadvantages, depending on the input data and the analysis goals. The advantages of using automatic methods for data clustering include the ability to process large datasets quickly and identify hidden patterns or structures in the data [42, 43].

Among automatic methods of data clustering, two main categories can be distinguished: supervised and unsupervised methods. Supervised methods require labeled training data, which is used to train the clustering algorithm. In the case of unsupervised methods, the data is not labeled, and the algorithm looks for natural structures and patterns in the data to group them. Supervised methods tend to achieve better results in specific clustering tasks, but they require a larger amount of labeled data and complex machine learning algorithms. On the other hand, unsupervised methods are more versatile because they do not require prior knowledge about the data, but they may struggle with discovering subtle differences between groups.

In the data analysis presented in the article, two methods of automatic data clustering were used: k-means clustering and Self-Organizing Maps (SOM). The k-means method is one of the most popular unsupervised clustering techniques. It involves dividing a dataset into k clusters in such a way that the elements within clusters are similar to each other and different from elements outside the clusters. The k-means algorithm is an iterative algorithm, in which centroids (i.e., the centers of the clusters) are updated at each step [44]. In the first step, the centroids are randomly initialized. Then, the data points are assigned to one of the clusters (by calculating the distance to each centroid and selecting the closest one). In the next step, after the data is assigned to clusters, the centroids' positions are updated - the average of all points belonging to the cluster is calculated, and the process of calculating the distance of each point to the new centroids is repeated. This process continues until a stopping criterion, such as stabilization of the assignment of elements to clusters or reaching the maximum number of iterations, is met.

Self-Organizing Maps (SOM) are another popular unsupervised clustering method that uses the structure of a neural network to group data. SOMs consist of an input layer, an output layer, and a hidden layer that forms a map of neurons. Each neuron on the map is assigned to one cluster, and its position on the map corresponds to the assigned weight vector. The learning process of SOMs involves iteratively modifying the weights of neurons so that similar elements are assigned to neighboring neurons on the map, and dissimilar elements are assigned to neurons that are further apart. Ultimately, after the learning process is completed, the clusters are represented by groups of neurons on the map. The difference from classical neural networks is that in SOMs, during each weight update, only the weights of the winning neuron and its neighboring neurons on the map are updated (though to a lesser extent). SOMs are resistant to noise and outliers in the input data. They are also capable of visualizing high-dimensional data in a lower-dimensional space, making them useful for data visualization and exploratory data analysis.

Both clustering and Self-Organizing Maps (SOM) allow for automatic grouping of data without the need for user supervision. They are useful in cases where the number of classes or their features are not known in advance, or when we aim for quick and efficient clustering of large datasets. The main difference between clustering and SOM lies in the way data is grouped. In clustering, data is divided into groups based on similarity between elements, and these groups are not organized on any spatial map. On the other hand, SOM creates a spatial representation of data groups on a two-dimensional map, which can facilitate analysis and interpretation of results. Clustering is more universal and can be applied to different types of data, while SOMs are particularly useful for multi-dimensional data where obtaining a visual representation is challenging.

5 Obtained Results

Preliminary analysis of the obtained results from measurements conducted on 8 conveyors with a total length of 18 472 m revealed that individual segments of the conveyor belt are diverse among themselves, as they come from different manufacturers and have been in operation for varying lengths of time. Some of the installed belts are new, while others have undergone one or two refurbishments, during which repairs, cutting, and splicing of steel links were performed in the core of the belt. Therefore, the data obtained were isolated for the previously refurbished segments and areas where belt joints were installed (not subject to analysis for damage recognition), resulting in 83 segments of the belt with a total length of 16 503 m (there were a total of 13 refurbished segments with a combined length of 1 969 m). In the next stage, based on the results of scanning with the DiagBelt + diagnostic system, signals corresponding to core damage were identified, and 23 945 clusters of points corresponding to core damage were located.

The KMeans method was implemented using Python programming language and the sklearn library for the available data, with variations of clustering into 2, 4, 6, or 8 clusters. The size of each cluster is presented in Table 1.

One of the most important challenges in clustering is determining the optimal number of clusters, or groups, into which objects should be divided. On one hand, choosing too few clusters may result in the creation of groups that are too large and too heterogeneous,

Table 1. Cluster Sizes in Clustering.

Number of clusters	Cluster 1	Cluster 2	Cluster 3	Cluster 4	Cluster 5	Cluster 6	Cluster 7	Cluster 8
2	10 866	13 079	-	-	-	-	-	-
4	10 125	5 147	5 679	2 994	-	-	-	-
6	8 257	4 852	642	3 695	4 686	1 813	-	-
8	1 790	5 855	3 678	500	2 728	5 390	3 046	960

meaning they contain objects with too diverse properties. In such cases, the analysis may be less precise, and detecting subtle relationships in the data may be impossible. On the other hand, choosing too many clusters may result in the creation of groups that are too small and too homogeneous, meaning they contain objects with very similar properties.

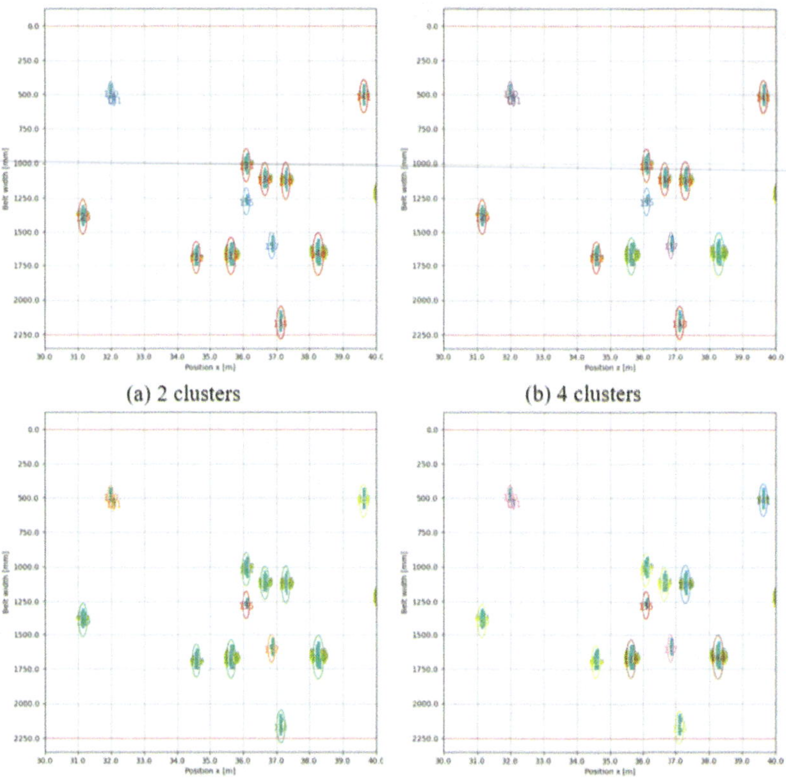

(a) 2 clusters (b) 4 clusters

(c) 6 clusters (d) 8 clusters

Fig. 6. Assignment of failures resulting from clustering – Example

In such cases, the analysis may become overly detailed and complex, and detecting meaningful patterns may become more difficult.

In the case of image analysis, similar to other fields, choosing the optimal number of clusters is crucial for obtaining the best results. An experienced operator, while analyzing the obtained signals, is able to detect more than just two groups, hence dividing the data into only two clusters in clustering may be misleading. At the same time, choosing too many clusters may result in the creation of noise, and the analysis process may become overly complicated. Figure 6 shows the visualization of the obtained results of assigning individual failures to groups on selected 10-m segments.

In the next part of the study, the same data was subjected to grouping using a self-organizing neural network. The implementation of the network was done using the "minisom" package. The network was initialized with a two-dimensional grid of neurons, with sizes of 2 × 2, 3 × 3, 4 × 4, or 5 × 5. As a result, classes with their corresponding heatmaps, as shown in Fig. 7, were obtained.

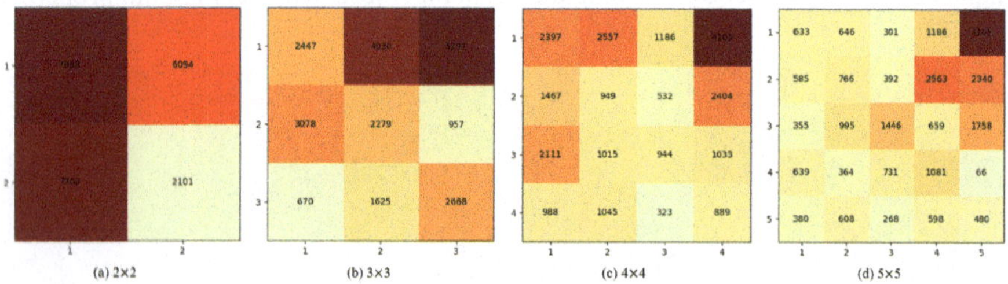

Fig. 7. Class frequencies obtained from the operation of the SOM (Self-Organizing Map).

On Fig. 8, the visualization of the obtained results of assigning individual failures to groups on selected 10-m sections is presented.

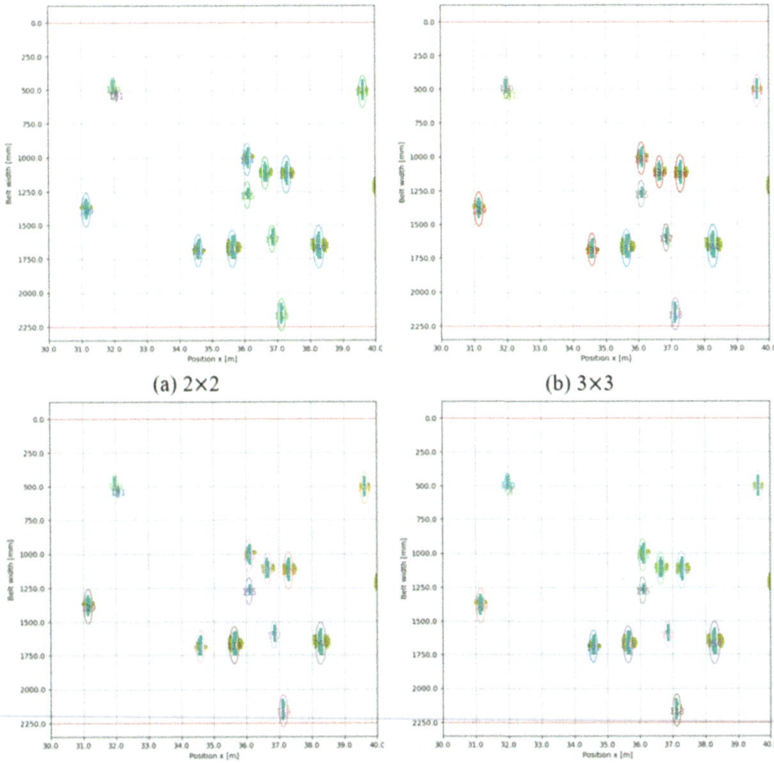

(a) 2×2 (b) 3×3

(c) 4×4 (d) 5×5

Fig. 8. Assignment of failures resulting from the operation of SOM network - Example

6 Summary

The recognition of types of failures is a crucial element of equipment and machinery failure analysis. Knowledge about the type of identified failures during scanning allows for the application of appropriate preventive repair actions, which in turn shortens down-time and reduces repair costs. In case of incorrect identification of the type of failures, there is a risk of implementing inappropriate repair actions, which can lead to further failures and increased costs.

Recognition of the type of failures can be carried out in various ways, including visual analysis of images or photographs, analysis of mechanical vibrations, or analysis of electrical signals. In the case of image analysis, various image processing methods such as image segmentation, clustering, or neural networks are used, allowing for effective identification of failures and proper recognition of their different types.

Incorrect identification of the type of failures can result in errors in decision-making regarding repair actions. Therefore, the proper application of analysis and classification methods for failures, as described in the article, is of crucial importance. At the same time, it is important to continuously develop and refine these methods, which will allow for even more accurate and precise identification of equipment and machinery failures.

In the case of core damage concentration, the local strength of the belt may be significantly reduced, which directly threatens the safety of transport continuity. That's why it is important not only to control the location of damage, but also to determine its type. Damage associated with corrosion, even with partial loss of cross-section of cords concentrated in one section, does not pose as much threat as multiple cords being cut or missing in that area. The cutting of multiple cords is a damage that requires immediate intervention, while developing corrosion requires further observation. Over time, it may progress to the absence of cords in the corrosion area, but until then, it does not pose an immediate threat of rupture. Fragments of belt with extensive corrosion are unlikely to be suitable for refurbishment due to the reduced adhesion of rubber to corroded cords. Therefore, identification of the areas of its occurrence is also important in making decisions about the further fate of the belt. Recognizing the type of damage enables faster and more effective repairs, which translates into reduced downtime and associated costs. Moreover, a thorough analysis of the type of damage allows for a better understanding of the belt wear process and identification of factors affecting its durability, and an increase in belt working time affects transportation costs. Dismantling the belts at a moment that allows for effective refurbishment also significantly reduces these costs. Identification of types of damage and the process of their development is knowledge that can be used to introduce changes in production processes to reduce or delay the occurrence of such failures in the future. It is also important to pay attention to the possibility of different types of damage occurring simultaneously or in close proximity to each other, resulting in overlapping measurement signals. Preliminary data analysis and accurate classification of damage allows for quick detection of such situations, appropriate classification of failures, and implementation of appropriate repair actions.

Accurate diagnosis of conveyor belt damage, including identification of their types, is therefore a necessary direction for the development of conveyor belt diagnostics. Only in this way is it possible to carry out effective repairs and minimize the negative consequences associated with downtime in the production process. Monitoring the technical condition of conveyor belts is extremely important from the perspective of ensuring their reliable operation. The methods of classification and grouping used allow for quick detection of failures, which enables their prompt repair or replacement, and thus minimizes downtime in mining operations. Furthermore, the analysis of the technical condition of conveyor belts allows for the prediction of time to the next failure, which enables planning of maintenance work and minimization of costs associated with unplanned downtime [38, 45, 46]. All of this translates into increased productivity and profitability of mining operations, where transportation costs have always been a significant component.

The article presents methods for classifying failures using clustering and neural networks. The application of these methods allows for fast and accurate detection of failures, enabling prompt corrective actions to be taken. Preliminary analysis of measurement data allowed for the determination of parameters describing the damage, which enabled the use of clustering to group similar failures. The application of the K-means algorithm allowed for the extraction of several clusters, enabling a more precise determination of the degree of damage and indicating the appropriate repair method. The use of a self-organizing map (SOM) neural network allowed for accurate categorization of failures based on the previously determined parameters. This made it possible to identify the most frequently occurring damages, allowing for faster response to problems in the transportation system.

The conducted research confirmed that proper preparation of input data and the use of effective classification methods are crucial for detecting failure in transportation systems. However, it should be noted that both too few and too many clusters can lead to classification errors, so operator experience and proper adjustment of classification algorithm parameters to specific cases are important. It is also worth noting that the methods used in the article for classifying failures employed unsupervised learning, thus not requiring labeled input data. The clustering was conducted only based on a few parameters describing the damage, rather than the entire image. This way, satisfactory results were achieved without the need for manual labeling of classes for each damage.

In the case of the decision to send belts for refurbishment, milling of the belt cover is usually performed, which exposes the steel links. This allows for a precise determination of the type and degree of damage. Based on this information, data labeling can be performed and used for further machine learning algorithms. Data labeling allows for an even more precise determination of the type of damage and for the alignment of the classification model to the specificity of the data.

References

1. Webb, C., Sikorska, J., Khan, R., Hodkiewicz, M.: Developing and evaluating predictive conveyor belt wear models. Data-Centric Engineering (2020)
2. Fedorko, G.: Implementation of Industry 4.0 in the belt conveyor transport in MATEC Web of Conferences (2019)
3. Boslovyak, P., Lagerev, A.: Optimization of the conveyor transport cost. IFAC-PapersOnLine **52**(25), 397–402 (2019)
4. Błażej, R., Jurdziak, L., Kirjanów-Błażej, A., Kozłowski, T.: Identification of damage development in the core of steel cord belts with the diagnostic system. Sci. Rep. **14**(11), 1–13 (2021)
5. Kirjanów-Błażej, A., Błażej, R., Jurdziak, L., Kozłowski, T., Rzeszowska, A.: Innovative diagnostic device for thickness measurement of conveyor belts in horizontal transport. Sci. Rep. **12**(1) (2022)
6. Bzinkowski, D., Ryba, T., Siemiątkowski, Z., Rucki, M.: Real-time monitoring of the rubber belt tension in an industrial conveyor. Mech. Eng. **3**(1), 1–10 (2022)
7. Shiri, H., Wodecki, J., Ziętek, B., Zimroz, R.: Inspection robotic UGV platform and the procedure for an acoustic signal-based fault detection in belt conveyor idler. Energies **14**(22), 1–17 (2021)
8. Błażej, R.: Ocena stanu technicznego taśm przenośnikowych z linkami stalowymi, Wrocław: Wydział Geoinżynierii, Górnictwa i Geologii Politechniki Wrocławskiej (2018)
9. Karabacak, Y.E., Özmen, N.G., Gümüşel, L.: Intelligent worm gearbox fault diagnosis under various working conditions using vibration, sound and thermal features. Appl. Acoustics, 186 (2022)
10. Bortnowski, P., Król, R., Ozdoba, M.: Roller damage detection method based on the measurement of transverse vibrations of the conveyor belt. Eksploatacja i Niezawodność – Maintenance and Reliability, **24**(3), 510–521 (2022)

11. Martínez-Parrales, R., del Carmen Téllez-Anguiano, A.: Vibration-based fault detection system with IoT capabilities for a conveyor machine. Acta Polytechnica Hungarica **19**(9), 7–24 (2022)
12. Wodecki, J., Shiri, H., Siami, M., Zimroz, R.: Acoustic-based diagnostics of belt conveyor idlers in real- life mining conditions by mobile inspection robot in ISMA2022. Belgium, Leuven (2022)
13. Dąbek, P., Szrek, J., Zimroz, R., Wodecki, J.: An Automatic procedure for overheated idler detection in belt conveyors using fusion of infrared and RGB images acquired during UGV robot inspection. Energies **15**(2), 1–20 (2022)
14. Uhryński, A., Bembenek, M.: The thermographic analysis of the agglomeration process in the roller press of pillow-shaped briquettes. Materials, **15**(8) (2022)
15. Kozłowski, T., Wodecki, J., Zimroz, R., Błażej, R., Hardygóra, M.: A diagnostics of conveyor belt splices. Appl. Sci. **10**(18), 1–15 (2020)
16. Mazurek, P., Kwaśniewski, J., Roskosz, M., Siwoń-Olszewski, R.: The use of a magnetic flux leakage in the assessment of the technical state of a steel wire rope subjected to bending. J. Konbin **48**(1), 493–513 (2018)
17. Błażej, R., Jurdziak, L., Kozłowski, T., Kirjanów-Błażej, A.: The use of magnetic sensors in monitoring the condition of the core in steel cord conveyor belts – tests of the measuring probe and the design of the DiagBelt system. Measurement, 123 (2018)
18. Kozłowski, T., Błażej, R., Jurdziak, L., Kirjanów-Błażej, A.: Magnetic methods in monitoring changes of the technical condition of splices in steel cord conveyor belts. Eng. Failure Anal. **104** (2019)
19. Ericeira, D., Rocha, F., Bianchi, A., Pessin, G.: Early Failure Detection of Belt Conveyor Idlers by Means of Ultrasonic Sensing in International Joint Conference on Neural Networks (IJCNN), Glasgow, Scotland (2020)
20. Szurgacz, D., et al.: Thermal imaging study to determine the operational condition of a conveyor belt drive system structure. Energies, **14**(11) (2021)
21. Kulinowski, P., Kasza, P., Zarzycki, J.: Identification of the operating parameters of the friction drum drive. Eksploatacja i Niezawodnosc – Maintenance and Reliability, **23**(1), 94–102 (2021)
22. Skoczylas, A., Stefaniak, P., Anufriiev, S., Jachnik, B.: Belt conveyors rollers diagnostics based on acoustic signal collected using autonomous legged inspection robot. Appl. Sci. **11**(5) (2021)
23. Liu, X., Pei, D., Lodewijks, G., Zhao, Z., Mei, J.: Acoustic signal based fault detection on belt conveyor idlers using machine learning. Adv. Powder Technol. **31**(7), 2689–2698 (2020)
24. Król, R., Gładysiewicz, L., Kisielewski, W., Kaszuba, D.: New quality standards of testing idlers for highly effective belt conveyors in World Multidisciplinary Earth Sciences Symposium, Prague (2017)
25. Król, R.: Studies of the durability of belt conveyor idlers with working loads taken into account. IOP Conference Series: Earth and Environmental Science (2017)
26. Dąbek, P., Wodecki, J., Zimroz, P., Krot, P.: Visual method of detection of the frequency of rotation of rollers in diagnostics of belt conveyors using drones in Vibroacoustics and diagnostics machinery, devices and structures. In: Proceedings of 2nd Scientific Conference VibDiag 2021, Poznań, Poland (2021)
27. Gładysiewicz, L., Król, R., Kisielewski, W.: Measurements of loads on belt conveyor idlers operated in real conditions. Measurement **134**, 336–344 (2019)
28. Chen, W., Li, J., Wang, Q., Han, K.: Fault feature extraction and diagnosis of rolling bearings based on wavelet thresholding denoising with CEEMDAN energy entropy and PSO-LSSVM. Measurement, 172 (2021)
29. Surajkumar, K., Sudhagar, E.: Fault diagnostics of roller bearings using dimension theory. ASME J. Nondestructive Eval. **4**(1), 1–10 (2021)

30. Wodecki, J., Zdunek, R., Wyłomańska, A., Zimroz, R.: Local fault detection of rolling element bearing components by spectrogram clustering with semi-binary NMF. Diagnostyka **18**(1), 3–8 (2017)
31. Kirjanów-Błażej, A.: Model rozwoju uszkodzeń rdzenia taśm przenośnikowych z linkami stalowymi, Wrocław University of Science and Technology: doctoral dissertation (2018)
32. Kirjanów-Błażej, A., Rzeszowska, A.: Conveyor belt damage detection with the use of a two-layer neural network. Appl. Sci. **11**(12) (2021)
33. Andrejiova, M., Grincova, A., Marasova, D.: Measurement and simulation of impact wear damage to industrial conveyor belts. Wear, 368 (2016)
34. Błażej, R., Jurdziak, L., Kirjanów-Błażej, A., Kozłowski, T.: A device for measuring conveyor belt thickness and for evaluating the changes in belt transverse and longitudinal profile. Diagnostyka, 97–102 (2017)
35. Jurdziak, L., Bajda, M., Błażej, R.: Estimation of purchase and replacement costs of conveyor belts and their splices in an underground mine based on their durability. IOP Conference Series Earth and Environmental Science (2019)
36. Martyna, R., Martyna, M.: Polish technology for testing wire ropes of the largest rope devices in the world. Nondestructive Testing Diagnosics **4**, 11–16 (2017)
37. Jurdziak, L., Błażej, R., Bajda, M.: Conveyor Belt 4.0. In: ISPEM 2018: Intelligent Systems in Production Engineering and Maintenance (2019)
38. Kirjanów-Błażej, A., Jurdziak, L., Burduk, R., Błażej, R.: Forecast of the remaining lifetime of steel cord conveyor belts based on regression methods in damage analysis identified by subsequent DiagBelt scans. Eng. Failure Anal. 100 (2019)
39. DiagBelt+, https://diagbeltplus.pwr.edu.pl/. Accessed 25 Aug 2022
40. Olchówka, D., Rzeszowska, A., Jurdziak, L., Błażej, R.: Statistical analysis and neural network in detecting steel cord failures in conveyor belts. Energies, **14**(11) (2021)
41. Singh, V.K., Tiwari, N., Garg, S.: Document clustering using k-means, heuristic k-means and fuzzy C-Means. In: International Conference on Computational Intelligence and Communication Networks, IEEE (2011)
42. Rokach, L., Maimon, O.: Clustering Methods. Data Mining and Knowledge Discovery Handbook, pp. 321–352, Springer-Verlag (2006)
43. Milligan, G.W., Hirtle, S.C.: Clustering and Classification Methods. John Wiley & Sons, Inc., Handbook of Psychology (2003)
44. MacKay, D.: Chapter 20. An Example Inference Task: Clustering. Information Theory, Inference and Learning Algorithms. Cambridge University Press, pp. 284–292 (2003)
45. Liu, X., He, D., Lodewijks, G., Pang, Y.: Integrated decision making for predictive maintenance of belt conveyor systems. Reliability Eng. 347–351 (2019)
46. Błażej, R., Jurdziak, L., Kirjanów-Błażej, A., Bajda, M., Olchówka, D., Rzeszowska, A.: Profitability of conveyor belt refurbishment and diagnostics in the light of the circular economy and the full and effective use of resources. Energies **15**(20), 1–15 (2022)

Performance Marketing in the Management of Supply Chains Sustainable Development

Shulhina Liudmyla(✉)

National Technical University of Ukraine "Igor Sikorsky Kyiv Polytechnic Institute", Beresteisky Avenue (Former Peremohy), 37, Kyiv, Kyiv 03056, Ukraine
shulm@ukr.net

Abstract. An analysis of scientific publications was carried out to outline the essence and factors of the sustainable development concept emergence. A brief overview of definitions, types, kinds and models of supply chains is presented, the impact of digitization and globalization of the economy on their development is indicated, as well as the emergence of new opportunities for their further development due to these trends. The logic of the emergence and formation of the newest concept of performance marketing, its essence and composition of Key Performance Indicators are outlined. Limitations in the application of performance marketing for evaluating the effectiveness of sustainable development of supply chains are identified, while the relevance and feasibility of finding ways to implement this concept to stop the practice of ignoring environmental and social problems of today is substantiated. A sketch of the program for ensuring the implementation of the concept of sustainable development has been developed, which should have a multi-level structure and contain a set of mandatory restrictions taking into account the industry specifics of the enterprises and supply chains that serve them. It is proposed to choose the concept of marketing cooperation or co-creating value together with customers as the ideological and methodological basis of the Program. An approach to the development of a methodology for evaluating the effectiveness of the implementation of the provisions of the Program is also proposed, which should include three groups of indicators: used in the concept of performance marketing together with indicators of environmental and social impact.

Keywords: Sustainable Development · Supply Chains · Performance Marketing

1 Introduction

The basis of the emergence of almost all inventions is the aggravation of a certain problem and the need to solve it. This statement is true not only in relation to engineering and technical developments, but also in relation to the emergence of new philosophical ideas or currents, the development of general theoretical scientific principles or methods of human activity, which offered an innovative way to overcome certain problems at each stage of social evolution.

A. Burduk et al. (Eds.): ISPEM 2023, LNNS 790, pp. 476–489, 2023.
https://doi.org/10.1007/978-3-031-45021-1_36

It is also natural that the level of the problem determines always the level of relevant tasks and methods for its solution. So, the global problem of the rapid deterioration of the human habitat, caused primarily by costly management and irrational social organization, gave impetus to the emergence and development of the concept of sustainable development, called the most promising ideology of the 21st century.

The international, even global level of the problem requires the appropriate unification of efforts and the development of methods of the same – that is, mega – level. After all, a global problem cannot be solved by the efforts of a single country, so the concept of sustainable development applies to all economic entities from all countries of the planet Earth. At the same time, each individual enterprise at the level of its functioning on the market - that is, at the micro level - inevitably encounters negative manifestations of this problem, and therefore needs to level it with the help of adequate inventions, which include, in particular, the development of appropriate business formats, methods, approaches and concepts for its implementation.

Among the wide range of innovative solutions that we can consider as a kind of invention, as mentioned above, and with the help of which enterprises are trying to withstand the challenges of the globalized economy, we believe that the following two deserve special attention:

1. the formation of various kinds of associations (alliances, networks, supply chains, etc.) on the basis of deepening specialization in all types of marketing activities, including ensuring the production of goods and bringing them to consumers [1, 2, etc.];
2. development and implementation of marketing concepts that would be most relevant to a particular situation, specialization and type of market (e.g., holistic, affiliate and other types of marketing) [3, 4, etc.].

The tools that are necessary to implement these solutions have emerged and are rapidly evolving along with another global trend – the digitalization of all spheres of human life, including entrepreneurship - which is significantly changing the operating environment for businesses of all sizes, types and specializations. After all, information technology allows the development and implementation of new concepts at an unprecedented level and also systems of relevant indicators, innovative strategies and algorithms for their implementation. All of the above is combined in one of the most promising concepts of enterprise performance management called performance marketing.

The practical application of the principles of performance marketing by enterprises of various industries has already shown a significant increase in the efficiency of their activities on the market [5–7, etc.]. Based on this fact, the author of the presented article considered it expedient to investigate the essence, features and conditions of successful implementation of the newest concept of performance marketing into the supply chain sustainability management system.

It should be noted that each of the categories mentioned in the title of this article (sustainable development, supply chains, and performance marketing), as well as the phenomena and processes they denote, have been sufficiently researched and described in the scientific and popular literature. However, the author did not find any studies of these phenomena in the proposed context and combination.

The logic of the above considerations formed the basis for organizing and conducting a study aimed at filling this gap, and also became the essentials for choosing methods of collecting, systematizing and analyzing information, shaping the content and structure of this article.

The purpose of the study is to test the hypothesis that performance marketing can be used to implement the concept of sustainable development in supply chains.

The article is aimed at presenting the results of the study and, on their basis, forming a sketch of a program for ensuring the implementation of the concept of sustainable development of supply chains using extended system of performance marketing key indicators.

Given that the study was conducted in 2022–2023 as a search study, at this (initial) stage, information was collected using two methods: desk research and expert interviews. The desk analysis was used to review scientific and popular science publications that examined the essence of the concepts of "sustainable development", "supply chains" and "performance marketing", as well as the features and evolution of the phenomena they denote. Expert interviews were used as a source of insider information from supply chains on the application of the principles of the sustainable development concept and performance marketing tools in their work. This theoretical analysis, adjusted for the opinions of market experts, is presented in the first part of the article (Subsects. 2.1–2.4). To develop the program sketch (Subsect. 2.5) general scientific methods were used: analysis and synthesis, induction and deduction, comparative analysis, analogy.

2 Investigation Results

2.1 Sustainable Development

Since 1980, when this category was first mentioned in the International Union for Conservation of Nature («World Conservation Strategy» document), the concept of "sustainable development" has become a central idea of a number of major UN conferences and national strategies [8–10, etc.]. Given the significant number of studies and publications that describe the essence of the concept [11, 12, etc.]; the history of its development [13, 14, etc.]; the main pillars [15, 16, etc.]; as well as the arguments for and against [17, 18, etc.] in this article, we will limit ourselves to a brief summary of its current perception.

The analysis of widely available detailed literature reviews on sustainable development proves that today the relevance of the concept has not only diminished, but has even increased [19, 20, etc.]. According to one of the most common definitions "sustainable development is development that meets the needs of the present, without compromising the ability of future generations to meet their own needs" [21]. Despite the fact that attempts continue to be made to formulate a more specific definition of the concept essence, the author of this article believes that the above definition is the most meaningful. It implies that priority attention should be focused on the development of management methods that do not harm humanity and its environment, which are called the three fundamental pillars of sustainable development - economic, environmental and social spheres in all studies on this topic. The very existence of each of these spheres in modern realities, not to mention their effective functioning, is impossible without the activities of supply chains, which is the next category we have taken for analysis.

2.2 Supply Chains

The dictionary of the world-famous Oxford University offers the following definition: "A supply chain is the series of processes involved in the production and supply of goods, from when they are first made, grown, etc. until they are bought or used" [22]. A more extended and detailed, yet identical in content, interpretation can be found in popular science publications: "A supply chain is the network of all the individuals, organizations, resources, activities and technology involved in the creation and sale of a product. A supply chain encompasses everything from the delivery of source materials from the supplier to the manufacturer through to its eventual delivery to the end user. The supply chain segment involved with getting the finished product from the manufacturer to the consumer is known as the distribution channel". Accordingly, supply chain management (SCM) is the oversight of materials, information and finances as they move in a process from supplier to manufacturer, to wholesaler, to retailer and then to the consumer [23].

There is an alternate approach to SCM presented in a literature, explaining why the traditional materials planning environment, typically embodied by an Enterprise Resource Planning (ERP) system, is an ineffective support system for a company that wants to adopt Lean practices [24, 25]. Very close in meaning to the traditional definition, but taking into account the processes of globalization, is proposed in [26]: "A supply chain is a global network that ensures the delivery of products and/or services domestically and internationally. A supply chain ensures product delivery from raw materials to the final customers using a definite structure involving an efficient network and productive information channel. The primary supply chain of any product involves raw materials, producer, distributor, retailer, and the customer."

In the age of digitalization, more and more explanatories investigation aim to discuss current status and challenges, especially in aspect of security and trust problems, of digital supply chain management system with applying some advanced information technologies, such as Internet of Things, cloud computing and blockchain, for improving various system performance and properties, i.e. transparency, visibility, accountability, traceability and reliability [27, 28].

A new approach has also introduced that was designed to understand the importance of supply chain considerations for a suitable technology assessment. It is aimed at proposing a quantitative model for the evaluation of different structure of supply chains based on different production technologies. By defining a set of Key Performance Indicators and applying a multi-criteria decision method, a final score is computed, giving important information about both environmental and economic aspects [29, etc.].

As commodity and market relations develop, the diversity of supply chains is also growing, with new types and models emerging. In particular, market experts discuss the classification of supply chain models based on the criteria of "stage of the product life cycle" and "market demand features" [30]:

- continuous flow model – for companies working with homogeneous products for which there is a high constant demand. The objectives of the model are to optimize production time, strictly control inventory, and regularly replenish raw materials to prevent production interruptions;

- fast chain model – for companies that sell innovative products for which demand is growing rapidly. The model's objectives are to bring products to market quickly, make the most of fast-changing fashion, and shorten the path from producer to consumer;
- flexible model – for companies that produce seasonal or holiday goods. The objectives of the model are to anticipate surges and downturns in demand for their products, to establish processes for rapid preparation for the start of production and effective shutdown, and to accurately predict their needs for raw materials, inventory, and labour.

In the context of the growing relevance of sustainability issues, special attention should be paid to studies that propose to classify supply chains into first-, second-, and n-th order chains depending on the type of resources they use. A first-order supply chain means that a significant part of it is based on the use of non-renewable and non-recycled resources, while subsequent n-th order chains (e.g., second-order) are based on a larger share of renewable and/or recycled resources. According to the authors of this classification, it is not enough today to simply match supply and demand between points of consumption and origin in first-order supply chains, but a clear extension to second- and n-th order supply chains is needed in the future [31].

Based on the review results [32], the researchers call for further studies in the following directions that are important but received little or no attention: design for circularity, procurement and circle supply chain management (CSCM), biodegradable packaging, circular supply chain collaboration and coordination, drivers and barriers of CSCM, circular consumption, product liabilities and producer's responsibility, and technologies and CSCM.

However, the development of supply chains has reached a fundamentally new level with the emergence of new digital technologies that have enabled more efficient supply chain management through the introduction of technologies such as blockchain, the Internet of Things, machine learning, etc. These technologies actually increase and expand the possibilities for optimizing planning, sourcing and procurement strategies [33, etc.].

Among the most obvious consequences of the rapid development of digital technologies in the context of our study are the following: easier access to information, the emergence of opportunities to operate with large amounts of data and, as a result, the acceleration of various processes (including any business processes). In such circumstances, the Management by Objectives approach, which was dominant until recently, is rapidly giving way to the promising concept of Performance Management, with Performance Marketing as its central part.

2.3 Performance Marketing

The concept of Performance Marketing is quite new. It emerged and began to spread actively in the markets of developed countries in the early 2000s, and in Ukraine – about ten years later [34]. The history of the development of this concept is traditional for marketing theory, all the provisions of which originated from the systematization of best practices:

1. at the first stage, under the pressure of competition and the need to reduce unreasonably high operating costs, entrepreneurs began to actively study and implement the capabilities of Internet marketing tools, including digital analytics;
2. over time, the use of innovative methods became more and more complex, terminology developed, and indicator systems were formed;
3. having gained experience, market experts began to publish analytical articles highlighting the methods and advantages of analysing big data sets, as well as the possibility of adjusting management decisions based on them in real time.

An important point is that today these publications are mostly applied and common in business publications. The concept of performance marketing is still extremely underrepresented in scientific works.

Given the novelty of the analysed concept, it is worth explaining the meaning of its name. After all, the term "performance" has a large number of meanings in its language of origin – English, but in nowadays marketing theory and practice, the expression "performance marketing" is used exclusively to refer to effective online marketing.

In ISO 9000:2015, IDT, there are two terms to explain the content of the "performance" category: effectiveness and efficiency [35]:

- effectiveness is the degree of the planned results achievement, the ability of the company to be result-oriented;
- efficiency is the ratio between the results achieved and the resources spent, the company's ability to realize its goals and plans with a given quality level, expressed by certain requirements – time, costs, degree of goal achievement.

The following definitions of the studying concept essence, proposed by practicing marketers, best correspond to the above terminological explanations:

- performance marketing is a concept of Internet marketing aimed at achieving specific, financially measurable business results in the shortest possible time [36];
- performance marketing is a type of digital marketing campaign in which budgeting and decision-making are driven primarily by the measurable results of that campaign. Performance marketing is based on an active, iterative feedback loop: run campaigns, see what works, double down, repeat;
- performance marketing is a digital marketing strategy that's driven by results. It's ideal for companies that are looking to reach their audience at scale, because payment is based on how users interact with the content;
- performance marketing refers to a form of digital marketing in which brands only pay marketing service providers after their business objectives have been met or when specific actions have been taken, such as a click, sale, or lead. In other words, it is performance-based marketing [37].

From all the above definitions, it follows that performance marketing has certain advantages over other concepts, primarily the following:

- the ability to access the necessary data in order to analyze it, evaluate business indicators (CPA, ROI, etc.) and influence them in real time;
- measuring the effectiveness of the channels used and the ability to select them (abandon ineffective ones and increase the use of effective ones);

- synchronization of all effective tools (SEO, SMM, email marketing, display advertising, etc.) to achieve maximum results;
- focus on the result, as performance marketing tools encourage the consumer to make a specific decision, for example, to make a purchase;
- payment for the result – customers of performance marketing services pay exclusively for the achievement of planned indicators (KPIs).

KPIs, or Key Performance Indicators, are numerical performance indicators that help measure the degree to which goals are achieved or a process is optimal, namely, efficiency and effectiveness. The most commonly used cost indicators are: Customer Acquisition Cost (CAC) and Cost per Action (CPA). The latter includes the following: Cost per Click (CPC), Cost per Millennium (CPM) or Cost Per Thousand (CPT), Cost per Install (CPI), Cost per Lead (CPL), Cost per Order (CPO), Cost per Sale (CPS). Return on Investment (ROI) and Return On Marketing Investment (ROMI), as well as site performance metrics: Average Session Duration (ASD), Bounce Rate (BR) and Conversion Rate (CR) play an important role.

2.4 Limitations in the Application of Performance Marketing for the Sustainable Development of Supply Chains

The presented list of indicators used by performance marketing is far from complete, but it is sufficient to demonstrate the limitations of this concept, of which the following two are the most important in the context of our research:

- performance marketing is closely related to online activities and very little to offline activities;
- measurable efficiency relates exclusively to the economic and financial performance of a particular enterprise, and has no connection with either social or environmental effects.

Of course, it can be assumed that the efficient operation of each supply chain participant (including producers, vendors, warehouses, transportation companies, distribution centres, and retailers) should have a positive impact on the efficiency of the entire system. However, this assumption is likely to be true only in terms of the economic outcome, but does not provide a basis for conclusions about social or environmental impacts. In other words, the key question is how to use the significant benefits of performance marketing, which have been proven in practice, to implement the concept of sustainable development in a period when the complex of growing challenges increasingly resembles a closed circle: 1) the problems of all aspects of sustainable development (economic, social and environmental) have intensified → 2) at the same time, funds are needed to solve any social and environmental problems → 3) the development of the economic component becomes a priority again → 4) and this often leads to ignoring the social and ecological components → 5) all aspects of sustainable development are further complicated.

Thus, it should be recognized that social and environmental responsibility has not yet become the first vital necessity of business and the most important criterion for its success. Such a conclusion, without exaggeration, characterizes the work of a significant part of entrepreneurs, not only in Ukraine, but also in the world. We will give only a few common examples from practice: during the transportation of goods, there are known cases of exceeding the permissible weight of trucks, which damages the roads; mining is often accompanied by the destruction of the entire ecosystem around mining enterprises; industrial breeding of animals leads to soil, water and air pollution, etc. We emphasize that the last two examples illustrate situations that, as a rule, turn the regions of localization of the mentioned types of enterprises and their supply chains into places practically unsuitable for human life.

The given short list of examples can illustrate the extreme relevance and complexity of the raised problem, as well as to draw attention to some logical conclusions that are obvious:

- similar examples of business ignoring social and environmental needs can be found in the activities of almost all branches of the economy, i.e. the scale of the problem is the entire economy of Ukraine (as well as of many other countries);
- in the struggle for survival, enterprises developed supply chains necessary for their activity, which in many cases separated into independent associations of various organizational and legal forms and business formats;
- supply chains exert an increasingly greater influence on the functioning of the economy, including international one, since their activity is quite profitable, and therefore "effective" (the term "effective" is enclosed in quotation marks to remind the reader once again about the simultaneous ignoring of environmental and social problems).

2.5 Sketch of the Program for Ensuring the Implementation of the Sustainable Development Concept

Based on the above, the author of this article strongly believes that the "polluter pays" formula is no longer sufficient. From the moment of its emergence to today, the level of threats has grown significantly, therefore, in order to maintain the balance between the three "pillars" of sustainable development, declarations are no longer needed, but a detailed multi-level Action Program with clearly defined responsibilities (disciplinary, material, administrative, financial and criminal). It is the legal restrictions that should prevent further deterioration of the situation, which requires not only the appropriate improvement of the legal field, but also the creation of mechanisms for strict control over the implementation of clearly defined norms.

The working title can be "Program for ensuring the implementation of the sustainable development concept". The status of the Program should not be advisory, but mandatory, as further deterioration of the situation threatens the very existence of humanity. Program levels are interstate, state, regional and local. At each level, the Program should have sectoral and inter-sectoral sections. The brief review of scientific and popular publications on supply chains presented in this article showed that the characteristics of the industry, product, market, etc., are significant factors in the size and type of the chain. Therefore, it is in the branch sections that there should be warnings about the types of risks that may be provoked by the activities of the branch subjects, as well as the necessary measures aimed at preventing the occurrence of these risks (see Fig. 1).

At each level, the program has corresponding tasks, and therefore, performers with their functions (see Fig. 2 and Fig. 3).

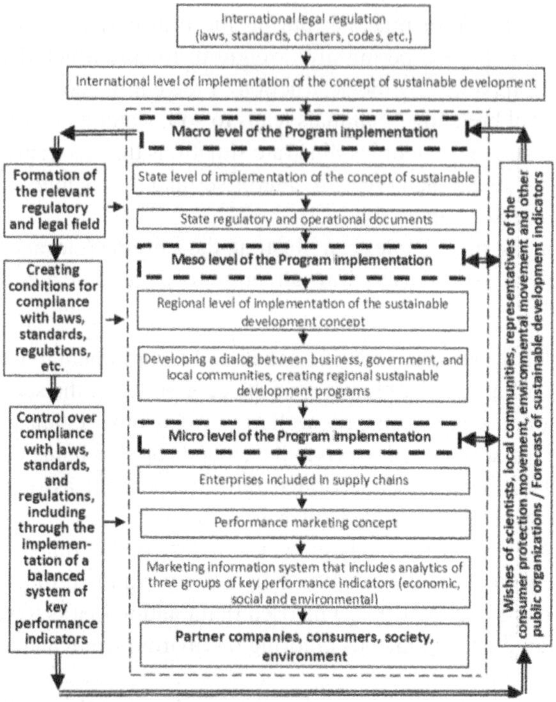

Fig. 1. A systematic approach to the recommended Program design. *Author's development*

The ideological and methodological basis of the sustainable development program should be the concept of marketing cooperation [38] or – very close to it in essence – the concept of co-creating value together with customers [39]. Both the concepts are aimed at creating "lasting", "eternal", "permanent" value for all participants in the process of economic and other types of human life – that is, value that has a long-term and versatile positive effect.

The methodology for evaluating the implementation effectiveness of the Program provisions should be based on the indicator systems used in the concept of performance marketing. However, simultaneously with the above indicators of economic efficiency, indicators of environmental and social impact should be measured. Given the diversity of business entities, for each industry, type and species of enterprise, its own balanced system of indicators should be developed, which would prevent the deepening of the mentioned problems at the first stage, and ensure their gradual and irreversible solution at the next stages (see Fig. 3).

Before concluding, it is worth emphasizing three critical points.

First. Outlining the general appearance of the proposed Program, the author in no way tried to question its extreme complex nature. The proposals presented here currently have the form of a possible sketch to the problem solution, each element of which is an important direction of a separate prospective study.

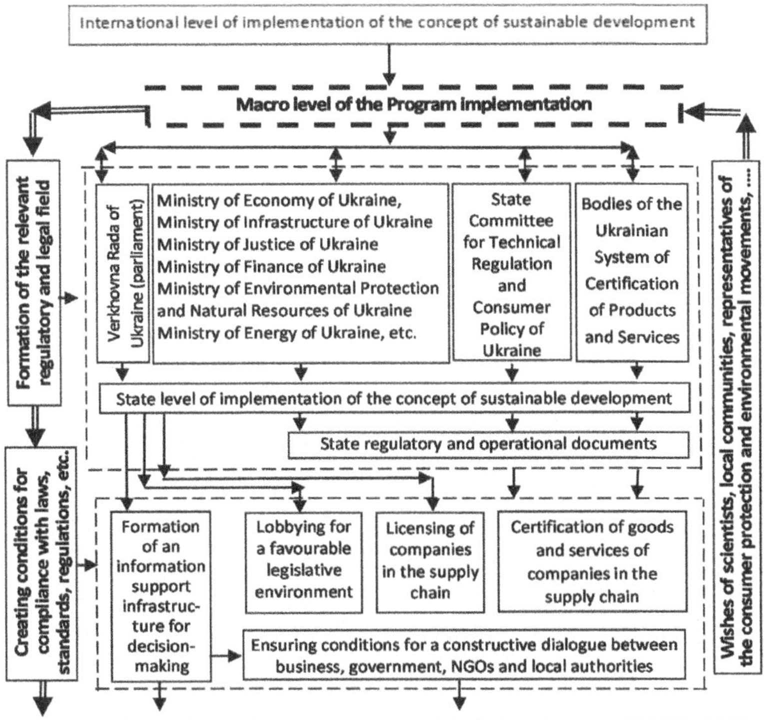

Fig. 2. The macrolevel of the recommended Program design. *Author's development*

Second. Pointing to the need to create a system of stricter restrictions and control over compliance with the provisions of the sustainable development concept, the author warns against the creation of additional administrative or bureaucratic barriers. On the contrary, restrictions should have the force of law and the framework of industry standards, and the control system should be as transparent and automated as possible.

Third. Representatives of local communities, social movements, science and business, who would be experts in this particular field, should be involved in the development of the Program, especially in the part of defining limitations and benchmarks for a certain field. Representatives of specialized universities, research institutes, and talented student youth, who would create initiative groups interested in preserving the environment and harmonious development of society, should become the driving force of such activities.

Separately, it is worth emphasizing the importance of such two elements of the Program as enlightenment and control. The author is convinced that a well-argued explanation of the level of danger that has already loomed over humanity, but can still be controlled, as well as strict restrictions that would be able to stop the growth of this danger, are key factors in successfully solving the problem.

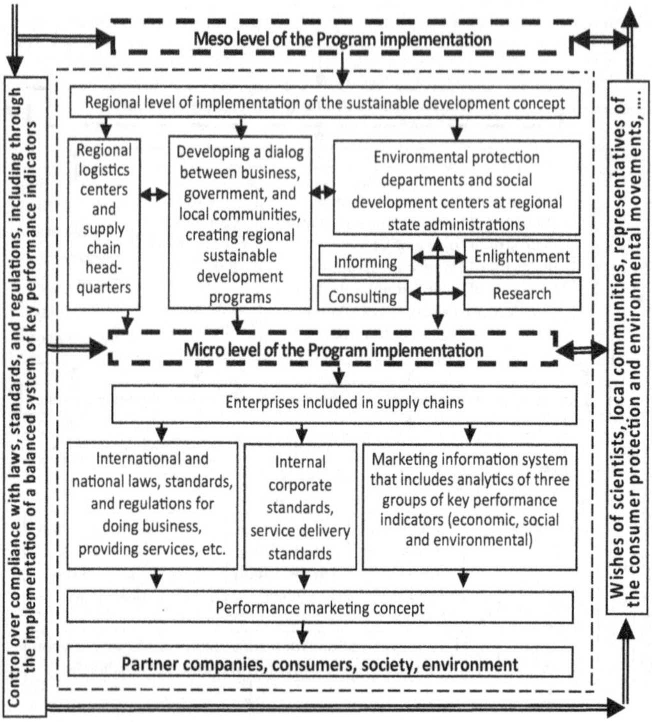

Fig. 3. The meso- and microlevels of the recommended Program design. *Author's development*

3 Conclusions

The global level of the problem of the rapid deterioration of the human habitat requires an appropriate combination of efforts for effective management and the development of adequate measures that would be able to stop negative trends at the first stage, and in the future to ensure a real transition to sustainable development.

A review of scientific publications showed that one of the most effective is the concept of performance marketing, which, thanks to a set of advantages (measurability, relevance, focus on results and payment for results), allows enterprises to achieve specific and financially measurable business results in the shortest possible time. However, while creating conditions for enterprises to achieve economic efficiency, this concept ignores two more "pillars" of sustainable development - the responsibility of business for the state of the social and environmental spheres.

At the same time, the process of globalization of the economy includes the rapid development of international companies, the service of which requires powerful supply chains. This has stimulated the emergence of a large number of types, sizes and models of supply chains, the characteristics of which continue to diversify, and their impact on the environment (largely negative) is increasing.

The author suggested moving to a more demanding position regarding the introduction and implementation of the concept of sustainable development by developing a program for its support, which would have the following levels: interstate, state, regional and local. The main features that, in the opinion of the author, should ensure the effectiveness of the Program are also proposed:

- status – mandatory;
- ideological and methodological basis – the concept of marketing cooperation or the concept of co-creating value together with customers;
- structure – sections by industry, containing warnings about possible types of risks for ecology and society, as well as necessary measures aimed at preventing the occurrence of these risks;
- evaluation method – balanced systems of indicators used in performance marketing with the addition of environmental and social impact indicators, developed for each industry considering their specifics.

In the opinion of the author, there are at least three promising directions for further research: 1) analysis of the legislative field with the aim of bringing it into line with the real needs of ensuring sustainable development, which in many cases requires greater restrictions and clear mechanisms for monitoring implementation; 2) development of balanced indicator systems for supply chains taking into account the specifics of each industry and all pillars of sustainable development, i.e. including environmental and social components, not just financial and economic ones; 3) creation and popularization of educational programs for all segments of the population, which at every level of perception would explain the vital necessity of changing the attitude to the living environment, and therefore the change of understanding of the very concepts of "efficiency" and "sustainability".

References

1. Emmett, S., Crocker, B.: The Relationship-Driven Supply Chain: Creating a Culture of Collaboration Throughout the Chain, CRC Press: 208. https://books.google.pl/books?id=8oAfwQEACAAJ&dq=scientific%20arti-cles%20on%20supply%20chains%20creating&hl=uk&lr&source=gbs_book_other_ver-sions (2018)
2. Helo, P., Hao, Y.: Blockchains in operations and supply chains: a model and reference implementation. Comput. Ind. Eng. **136**, 242–251 (2019). https://doi.org/10.1016/j.cie.2019.07.023
3. Graesch, J.P., Hensel-Börner, S., Henseler, J.: Information technology and marketing: an important partnership for decades. Ind. Manag. Data Syst. **121**(1), 123–157 (2021). https://doi.org/10.1108/IMDS-08-2020-0510
4. Gusti Ketut Adi Winata, I., Made Wulan Sari Sanjaya, N., Gusti Made Oka Astana, I.: Entrepreneurship orientation and holistic marketing mix in creating competitive advantages bumdes. J. Bus. Hospitality Tourism, **06**(02), 237–246 (2020), https://pdfs.semanticscholar.org/e236/09ef5b8cc7e4cf9ca51a1de6c706a45cf431.pdf
5. Desai, V.: Digital marketing: a review. Int. J. Trend Sci. Res. Dev. (IJTSRD), 196–200 (2019), https://www.ijtsrd.com/pa-pers/ijtsrd23100.pdf
6. Saura, J.R.: Using data sciences in digital marketing: framework, methods, and performance metrics. J. Innov. Knowl. **6**(2), 92–102 (2021), https://www.sciencedi-rect.com/science/article/pii/S2444569X20300329

7. Järvinen, J.: The use of digital analytics for measuring and optimizing digital marketing performance. Jyväskylä studies in business and economics, 170 (2016), http://urn.fi/URN:ISBN:978-951-39-6777-2
8. United Nations Conference on Environment and Development, Agenda 21: Table of Contents. Earth Summit, 1992. Habitat.igc.org. Archived from the original on 30 July 2014, https://web.archive.org/web/20140730190755/http://habitat.igc.org/agenda21/, (2014)
9. Summary of the United Nations Conference on Sustainable Development. Earth Negotiations Bulletin. A Reporting Service for Environment and Development Negotiations (UNCSD). International Institute for Sustainable Development (IISD). 27 (51), http://www.iisd.ca/uncsd/rio20/enb/, (2012)
10. The 24th Conference of the Parties to the United Nations Framework Convention on Climate Change, https://unfccc.int/process/conferences/pastconferences/marrakech-climate-change-conference-november-2016/events-and-programme/side-events-and-exhibits, (2016)
11. Giovannoni, E., Fabietti, G.: What is sustainability? a review of concepts and its applicability: Department of Business and Law, University of Siena, Siena, Italy Integrated Reporting (2014). https://doi.org/10.1007/978-3-319-02168-3_2
12. Griggs, D.J., Nilsson, M., Stevance, A., McCollum, D.: A guide to SDG interactions: from science to implementation. Paris, France: International Council for Science (ICSU), http://www.icsu.org/cms/2017/05/SDGs-Guide-to-Interactions.pdf, (2017)
13. Mensach, J.: Sustainable development: meaning, history, principles, pillars, and implications for human action. Literature Rev. Cogent Soc. Sci. **5**(1) (2019). https://doi.org/10.1080/23311886.2019.1653531
14. Springett, D., Redclift, M.: Sustainable development: history and evolution of the concept. In: Routledge international handbook of sustainable development (2015), https://www.taylorfran-cis.com/chapters/edit/https://doi.org/10.4324/9780203785300-2/sustainable-development-delyse-springett-michael-redclift
15. Murphy, K.: The social pillar of sustainable development: a literature review and framework for policy analysis. Sustain. Sci. Pract. Policy, 15–29 (2012). https://doi.org/10.1080/15487733.2012.11908081
16. Purvis, B., Mao, Y., Robinson, D.: Three pillars of sustainability: in search of conceptual origins. Sustain. Sci. (2018). https://doi.org/10.1007/s11625-018-0627-5
17. Department of Economic and Social Affairs. Sustainable Development Challenges, World Economic and Social Survey 2013, United Nations publication, https://sustainabledevelopment.un.org/content/documents/2843WESS2013.pdf, (2013)
18. Science and education for sustainable development: Monograph Publishing House of University of Technology, Katowice, http://www.wydawnictwo.wst.pl/up-loads/files/6ea021fc2983bafe60afd42fd5d707a2.pdf, (2022)
19. Ogryzek, M.: The sustainable development paradigm, geomatics and environmental engineering, 17(1) (2023). https://doi.org/10.7494/geom.2023.17.1.5
20. Ruggerio, C.A.: Sustainability and sustainable development: a review of principles and definitions. Sci. Total Environ. 786 (2021). https://doi.org/10.1016/j.sci-totenv.2021.147481
21. Sustainable Development Commission Official Site, https://www.sd-commis-sion.org.uk/pages/what-is-sustainable-development.html. Accessed 14 Apr 2023
22. Oxford Learner's Dictionaries. Supply Chain. https://www.oxfordlearnersdictionar-ies.com/definition/english/supply-chain. Accessed 12 Apr 2023
23. Lutkevich, B.: Supply Chain Definition, TechTarget Network, Tech Accelerator, Webinars. https://www.techtarget.com/whatis/definition/supply-chain, (2023)
24. Kerber, B., Dreckshage, B.: Lean supply chain management essentials: a framework for materials managers. CRC Press, https://books.google.pl/books?id=LRTSBQAAQBAJ&printsec=frontcover&hl=uk&source=gbs_ge_summary_r&cad=0#v=onepage&q&f=false, (2011)

25. Cooper, R., Slagmulder, R.: Supply Chain Development for the Lean Enterprise: Interorganizational Cost Management, https://books.google.pl/books?hl=uk&lr=&id=xNNBDwAAQBAJ&oi=fnd&pg=PT13&dq=related:0lJkGsUpXbkJ:scholar.google.com/&ots=ZTamchUnea&sig=68gTCHSn46mw7ReiW-QY6z2QBp8&redir_esc=y#v=onepage&q&f=false (2017)
26. Korenblit, V.N.: Traditional vs. Modern (Digital) Supply Chain Management, MA, Supply Chain as a Service, https://www.linkedin.com/pulse/traditional-vs-modern-digital-sup-ply-chain-management/, (2021)
27. LeMay, S., Helms, M.M., Kimball, B., McMahon, D.: Supply chain management: the elusive concept and definition. Int. J. Logistics Manag. **28**(4), 1425–1453 (2017). https://doi.org/10.1108/IJLM-10-2016-0232
28. Zhang, H., Nakamura, T., Sakurai, K.: Security and trust issues on digital supply chain. In: IEEE Intl Conf on Dependable, Autonomic and Secure Computing, Intl Conf on Pervasive Intelligence and Computing, Intl Conf on Cloud and Big Data Computing, Intl Conf on Cyber Science and Technology Congress (DASC/PiCom/CBDCom/CyberSciTech), Fukuoka, Japan, pp. 338–343 (2019), https://doi.org/10.1109/DASC/PiCom/CBDCom/CyberSciTech.2019.00069
29. Rinaldi, M., Caterino, M., Fera, M., Manco, P., Macchiaroli, R.: Technology selection in green supply chains-the effects of additive and traditional manufacturing. J. Cleaner Prod. 282 (2021). https://doi.org/10.1016/j.jclepro.2020.124554
30. Hayes, A.: The Supply Chain: From Raw Materials to Order Fulfillment, Investopedia, Business, Supply Chain (2023), https://www.investopedia.com/terms/s/supplychain.asp
31. Svensson, G.: Aspects of sustainable supply chain management (SSCM): conceptual framework and empirical example. Supply Chain Manag. **12**(4), 262–266 (2007). https://doi.org/10.1108/13598540710759781, https://www.emerald.com/insight/content/doi/https://doi.org/10.1108/13598540710759781/full/html
32. Farooque, M., Zhang, A., Thürer, M., Qu, T., Huisingh, D.: Circular supply chain management: a definition and structured literature review. J. Clean. Prod. **228**, 882–900 (2019). https://doi.org/10.1016/j.jclepro.2019.04.303
33. Bigliardi, B., Filippelli, S., Petroni, A., Tagliente, L.: The digitalization of supply chain: a review. Procedia Comput. Sci. **200**, 1806–1815 (2022). https://doi.org/10.1016/j.procs.2022.01.381
34. Shylko, T.: What is performance marketing and where did the usual one go? https://web-promo.ua/ua/blog/what-is-performance-marketing/ (2020)
35. Quality management systems: basic provisions and glossary of terms, Kyiv, SE "UkrNDNC". https://khoda.gov.ua/image/catalog/files/%209000.pdf, (2016)
36. Wezom. Blog, Internet marketing, https://wezom.com.ua/ua/blog/performance-marketing. Accessed 01 July 2022
37. What is Performance Marketing: How it Works, Channels, and Benefits. Digital Marketing. https://www.simplilearn.com/what-is-performance-marketing-article. Accessed 17 Apr 2023
38. Szulgina, L.: Marketing współpracy jako najnowsza koncepcja marketingu turystycznych agencji Ukrainy, Zarządzanie publiczne, **2**(26), 149–162 (2014), www.ejournals.eu/Zarzadzanie-Publiczne/
39. Prahalad, C.K., Ramaswamy, V.: Co-creating unique value with customers. Strategy Leadership **32**(3), 4–9 (2004). https://doi.org/10.1108/10878570410699249

The Use of Multi-criteria Decision-Making Methods in the Outsourcing of Logistics Services

Xenie Lukoszová[1], Elżbieta Jadwiga Szymańska[2(✉)], and Ewa Kulińska[3]

[1] VŠB – Technical University of Ostrava, Sokolská třída 2416/33, 70200 Ostrava, Czech Republic

[2] Department of Logistics, Institute of Economics and Finance, Warsaw University of Life Sciences - SGGW, 02-787 Warsaw, Poland
elzbieta_szymanska@sggw.edu.pl

[3] Faculty of Production Engineering and Logistics, Department of Logistics, Opole University of Technology, ul. Sosnkowskiego 31, 45-272 Opole, Poland

Abstract. The aim of the article is an assessment of the suitability of outsourcing selected logistics processes based on the criteria preferred by the customer. This can be achieved by applying appropriate decision support methods. In this case, it is an analysis of a multi-criteria decision-making process of the basic logistics activities, which are transportation and storage. By comparing the selected criteria, determining their weight and then arranging them from the most significant, i.e. the most preferred to the least significant, a specific variant of the outsourced activity or service is selected. At the same time, it depends on the value of the given criterion so that the logistics company ensures optimal fulfilment with regard to the perception of the customer - the decision maker. Using the resulting information, logistics companies can not only increase the level of quality of services provided to their customers, but also make appropriate innovations in the operating model of corporate outsourcing.

Keywords: multi-criteria decision-making · logistics processes · outsourcing · transportation · storage

1 Introduction

Multi-criteria decision-making (MCDM) deals with the analysis of decision-making situations. It is a discipline of operational research, within which decision variants are assessed according to several conflicting criteria. Multi-criteria decision problems are described by a set of variants, evaluation criteria and links between criteria and variants, which allow defining evaluation functions.

In almost every situation where a person is faced with any decision, this decision-making can be characterized as multi-criteria. The goal is to choose one variant that will be the final decision, i.e. a kind of compromise between individual decision options (criteria). Furthermore, it is no longer important which of the variants is in the second, third or next place in the order. In another case, a situation arises where an arrangement of

A. Burduk et al. (Eds.): ISPEM 2023, LNNS 790, pp. 490–505, 2023.
https://doi.org/10.1007/978-3-031-45021-1_37

variants is more desirable than the selection of a single final variant. It is directly required that the variants be ordered from best to worst. It should be added here that in multi-criteria decision-making, the terms best or worst depend on the definition according to the preferences of the decision-maker. In decision-making situations, several decision-making criteria need to be considered. A variant evaluated as the best according to one criterion may not be so evaluated according to another criterion. In this case, it is necessary to resolve the conflict between mutually opposing criteria, when the goal is to choose one of them. It is an analysis of a multi-criteria decision-making task.

2 Multi-criteria Decision-Making in the Literature

The MCDM method is a technique that combines the results of alternatives in multiple, conflicting, qualitative, and quantitative criteria and results in a solution that requires consensus [1–3]. The method uses knowledge from many fields, including behavioral decision theory, computer technology, economics, information systems, and mathematics. Since the 1960s, many MCDM techniques and approaches have been developed, proposed, and successfully implemented in many areas of application [1, 4, 5]. The purpose of MCDM is not to suggest the best decision but to help decision-makers choose alternative solutions that meet their requirements and preferences [1, 6–8].

Several MCDM methods are available, such as hierarchical analytical process (AHP), analytical network process (ANP), TOPSIS, data envelope analysis (DEA), and fuzzy decision-making [1, 9]. All methods are equally capable of making decisions under uncertainty, and each has its advantages [1]. The solutions generated by these methods compromise several goals, which leads to the failure to obtain the optimal point due to the nature of the problem [8]. The condition for using multi-criteria analysis is a more significant number of quantifiable criteria considered when making decisions [10, 11]. The method consists of four successive stages [12]:

- identifying alternatives and criteria,
- evaluating (quantifying) the requirements,
- assigning weights (normalization),
- calculating the rating.

The result of the multi-criteria analysis may be the selection of the optimal variant from the set of evaluated variants and the ordering of individual variants in descending (ascending) order in accordance with the given preferences [13].

"Multi-criteria decision-making methods are often used to help solve different categories of decision-making problems that may arise in different fields. In recent years, multi-criteria decision-making methods have been used to help decision-makers organize and synthesize information in a way that makes them feel confident in making decisions while balancing multiple criteria that tend to conflict with each other [14]. MCDM is among the fastest-growing problem areas in many disciplines [1, 15]. Over the last decade, many researchers have used these methods to make decisions in traffic engineering and urban logistics [16, 17]. Multi-criteria decision-making methods have been used in sustainable energy planning [18], supplier evaluation and selection [19], financial decision-making [20], natural resource management [21], and in construction [22].

3 Defining the Problem

The solved problem is the assessment of the suitability of outsourcing selected logistics processes based on criteria preferred by the customer (or decision maker). The focus is on the two core activities provided in the outsourcing of logistics services, which are transportation and warehousing, including all associated ancillary activities such as packaging, handling, picking, palletizing, customs clearance, labelling and other value-adding activities [23]. Assessment of the suitability of outsourcing selected logistics processes from the point of view of the customer of a logistics company can be achieved by applying appropriate methods to support decision-making [24]. The goal is therefore to evaluate the selected criteria according to their influence on decision makers when choosing to outsource a given logistics service. Determining their significance for the customer is an incentive for increasing the quality level of the outsourcing provider's services and introducing new trends in business, including the innovation of the operational model for the implementation of orders for outsourced activities.

4 Methods

Basic methods for decision support include multi-criteria decision analysis, which will be applied when deciding on the preferences of used services that are the subject of outsourcing. Part of the decision-making process is determining the importance of individual variants by assigning weights. Therefore, most methods of multi-criteria evaluation of variants require first determining the weights of individual evaluation criteria. These are a numerical expression of their importance, and the more important the criterion, the higher its weight. As a rule, these weights are standardized so that their sum is equal to one. They can be expressed in the form of a weight vector [25]:

$$v = (v_1, v_2, \ldots, v_n), \sum v_i = 1, v_i > 0 \tag{1}$$

whereas the condition that the sum of the components of the weight vector should be equal to one is not necessary, as some methods do not require it to be fulfilled [26].

As part of the multi-criteria decision analysis, the method of paired comparison and Saaty's matrix will be used [27].

4.1 Pairwise Comparison

Criteria weight estimation methods are generally simple procedures that, based on subjective information, quantify the importance assigned to individual criteria in order to enable the decision-maker to rank them according to importance. If the weights are determined by the method of pairwise comparison, also called Fuller's triangle, it is a process where the preference relations of pairs of criteria are determined. The decision-maker is presented with a triangular diagram in which pairs of criteria are marked in such a way that each pair occurs exactly once. The criteria in the columns are listed in the same order as the criteria in the rows, and the decision maker gradually subjectively determines which of the two criteria he prefers [28].

4.2 Saaty's Method

Saaty's method determines how many times one criterion is more significant than another. This method can potentially yield the most accurate expression of criteria preferences. On the other hand, it imposes the highest computational requirements. Saaty's method introduces additional information of a cardinal nature into the process, therefore it belongs rather to methods with cardinal information about the criteria.

Saaty's method differs from the pairwise comparison method in that instead of entering a matrix of preferences, a matrix of preference intensities is entered [29].

A scale is established for evaluating criteria in pairwise comparisons [28]:

1 – criteria i and j are equally important,

3 – criterion i is weakly preferred over criterion j,

5 – criterion i is strongly preferred over criterion j,

7 – criterion i is very strongly preferred over criterion j,

9 – criterion i is absolutely preferred over criterion j.

If this scale is not sufficient, it is still possible to use intermediate levels 2, 4, 6 and 8 [26]. The criteria are arranged by the evaluator in a table where identical criteria are listed in the same order in rows (i-th criterion) and columns (j-th criterion), ideally, they can be sorted by degree of significance. The expert compares each pair of criteria and writes the magnitude of the preference of the i-th criterion with respect to the j-th criterion according to the above-mentioned scale into the matrix. On the diagonal of the matrix, the rating of the criteria in the scale has the value 1.

It is therefore necessary to create tables to evaluate the criteria, partial evaluation of the variants and calculate the final evaluation of the variants. The first step is to create a table of dimensions that are given by the specifications of the given calculation. Ones are written on the main diagonal of the table. In the remaining cells of the table, the solver writes values by assigning weight to preferences with whole numbers from 1 to 9. If the solver prefers an element in a column to an element in a row, it writes the inverted value of the weight in the cell. Therefore, if the j-th criterion is more significant than the i-th criterion, then the entry in the matrix has the form [26]:

$$s_{ij} = 1/s_{ij} \tag{2}$$

A problem can arise if the expert is not consistent enough in his assessments. Consistency can be simply expressed as a state where the evaluator prefers criterion K1 to criterion K2, and criterion K2 to criterion K3, and then it logically follows that he also prefers K1 to criterion K3. We can express this relationship according to Grasseová et al. [30]:

$$K1 > K2 \wedge K2 > K3 \leftrightarrow K1 > K3 \tag{3}$$

5 Problem Solution

As the subject of the analysis is not the characteristics or evaluation of the logistics company's customers, it will not be further detailed the focus of the business and business activities of the decision-makers are specified. Decision-maker No. 1 is a representative of the first group of customers of the logistics company, because it is the most important manufacturer and distributor of pharmaceuticals, operating not only in the Czech Republic, but also throughout the world. Its mission is to provide quality innovative medicines to all people to improve their health and quality of life. The most important manufacturer and supplier of sustainable and innovative paints and coatings in accordance with the environment for the protection of future generations was chosen as decision-maker No. 2. The company's core product range includes decorative paints, powder paints, automotive, industrial, marine, yacht and specialty protective coatings. It is currently active in eight countries around the world and is, among other things, classified by the Top Employers Institute as a European top employer.

5.1 Application of the Paired Comparison Method

Table 1 records decision maker 1's preferred criteria for transportation services, including determining their scale and determining the order of importance.

Table 1. Pairwise comparison - shipping - decision maker 1

	K1	K2	K3	K4	K5	K6	K7	*p*	*p + 1*	*scale*		*rank*
K1								4	5	0.179	18%	2
K2	K1							0	1	0.036	4%	5
K3	K1	K3						1	2	0.071	7%	4
K4	K1	K4	K4					3	4	0.143	14%	3
K5	K5	K5	K5	K4				5	6	0.214	21%	1
K6	K1	K6	K6	K6	K5			3	4	0.143	14%	3
K7	K7	K7	K7	K7	K5	K7		5	6	0.214	21%	1
							Σ		28	1.000	100%	

From the established order of the individual criteria, it can be read that decision maker 1 most prefers criteria K5 and K7, which are the reliability of transport and the

Table 2. Pairwise comparison results - transportation

Evaluation	Criterion	Criterion definition
1	K5 a K7	time assurance and transport insurance
2	K1	offered services
3	K4 a K6	flexibility and cost
4	K3	knowledge and experience in the field
5	K2	good reputation and references

form of insurance. For a better overview, the results of the evaluated criteria were listed in the following Table 2.

Table 3 records the preferred criteria of decision-maker #2 for transport services, including determining their scale and determining the order of importance.

Table 3. Pairwise comparison - shipping - decision maker 2

	K1	K2	K3	K4	K5	K6	K7	*p*	*p + 1*	*scale*		*rank*
K1								5	6	0.214	21%	2
K2	K1							0	1	0.036	4%	7
K3	K1	K3						1	2	0.071	7%	6
K4	K1	K4	K4					4	5	0.179	18%	3
K5	K5	K5	K5	K5				6	7	0.250	25%	**1**
K6	K1	K6	K6	K4	K5			2	3	0.107	11%	5
K7	K1	K7	K7	K4	K5	K7		3	4	0.143	14%	4
							Σ		28	1.000	100%	

Table 4. Transportation - decision maker 2

Evaluation	Criterion	Criterion definition
1	K5	time certainty/reliability
2	K1	offered services
3	K4	flexibility
4	K7	transport insurance
5	K6	costs
6	K3	knowledge and experience in the field
7	K2	good reputation and references

For decision maker 2, the most important criterion is K5, which is time certainty and reliability. The overall results are shown in Table 4.

Table 5 records the preferred criteria of decision maker #1 for storage services including determining their scale and determining the order of importance.

The order of the individual criteria shows that decision-maker 1 most prefers criterion K3, followed closely by K5, as both criteria are closely related. The overall results are shown in Table 6.

Table 7 records the preferred criteria of decision-maker #2 for storage services, including determining their scale and determining the order of importance.

The most important criterion for decision maker 2 is K5, which are special storage conditions.

Table 5. Pairwise comparison - storage - decision maker 1

	K1	K2	K3	K4	K5	K6	K7	*p*	*p + 1*	*scale*		*Rank*
K1								1	2	0.071	7%	6
K2	K2							4	5	0.179	18%	3
K3	K3	K3						6	7	0.250	25%	**1**
K4	K4	K2	K3					3	4	0.143	14%	4
K5	K5	K5	K3	K5				5	6	0.214	21%	2
K6	K6	K2	K3	K4	K5			2	3	0.107	11%	5
K7	K1	K2	K3	K4	K5	K6		0	1	0.036	4%	7
							Σ		28	1.000	100%	

Table 6. Storage - decision maker 1

Evaluation	Criterion	Criterion definition
1	K3	certificates and certificates
2	K5	special conditions for storage
3	K2	price per area (PM, m3)
4	K4	object security
5	K6	time certainty/reliability
6	K1	regional availability
7	K7	due date of invoices

Table 7. Pairwise comparison - storage - decision maker 2

	K1	K2	K3	K4	K5	K6	K7	*p*	*p + 1*	*scale*		*rank*
K1								2	3	0.107	11%	5
K2	K2							5	6	0.214	21%	2
K3	K1	K2						1	2	0.071	7%	6
K4	K4	K2	K4					3	4	0.143	14%	4
K5	K5	K5	K5	K5				6	7	0.250	25%	**1**
K6	K1	K2	K3	K4	K5			0	1	0.036	4%	7
K7	K7	K2	K7	K7	K5	K7		4	5	0.179	18%	3
							Σ		28	1.000	100%	

Table 8 shows the results of a pairwise comparison of the evaluated criteria for the storage services of the second decision maker.

Table 8. Storage - decision maker 2

Evaluation	Criterion	Criterion definition
1	K5	special conditions for storage
2	K2	price per area (PM, m3)
3	K7	due date of invoices
4	K4	object security
5	K1	regional availability
6	K3	certificates and certificates
7	K6	time certainty/reliability

5.2 Application of Saaty's Method

The preferences of the first decision maker in the field of transport are shown in the following Table 9.

Table 9. Transport - decision maker 1

	K1	K2	K3	K4	K5	K6	K7	*geom. diameter*	*scale*		*rank*
K1	**1**	7	5	5	1/7	3	1/7	1.403258	0.159	16%	4
K2	1/7	**1**	1/3	1/3	1/4	1/4	1/6	0.28825	0.033	3%	7
K3	1/5	3	**1**	5	1/3	6	5	1.625614	0.185	18%	3
K4	1/5	3	1/5	**1**	5	1/5	1/7	0.559405	0.064	6%	6
K5	7	4	3	1/5	**1**	7	7	2.60916	0.296	30%	**1**
K6	1/3	4	1/6	5	1/7	**1**	1/5	0.61088	0.069	7%	5
K7	7	6	1/5	7	1/7	5	**1**	1.705661	0.194	19%	2
							Σ	8.802228	1.000	100%	

Using Saaty's method, more accurate values of the weights of individual criteria were determined and thus the order of their importance was also specified. K5 still remains in first place - time certainty/reliability. Table 10 gives the results of the comparisons.

Table 11 shows the preferences of the second decision-maker for the selected criteria and their comparison.

For decision maker 2, even when using Saaty's method, the most important criterion is K5, which is time certainty and reliability. The order of the criteria based on the evaluation of the second decision-maker is clearly presented in Table 12.

Table 13 shows a comparison of the analysed criteria of the storage processes of the first decision maker.

Table 10. Results of Saaty's method - transport - decision maker 1

Evaluation	Criterion	Criterion definition
1	K5	time certainty/reliability
2	K7	transport insurance
3	K3	knowledge and experience in the field
4	K1	offered services
5	K6	costs
6	K4	flexibility
7	K2	good reputation and references

Table 11. Transport - decision maker 2

	K1	K2	K3	K4	K5	K6	K7	*geom. diameter*	*scale*		*rank*
K1	**1**	7	5	5	1/7	3	7	2.446776	0.237	24%	2
K2	1/7	**1**	1/3	1/3	1/4	1/4	1/6	0.28825	0.028	3%	7
K3	1/5	3	**1**	1/5	1/3	1/6	1/5	0.388398	0.038	4%	6
K4	1/5	3	5	**1**	1/5	5	7	1.544858	0.150	15%	3
K5	7	4	3	5	**1**	7	7	4.132439	0.400	40%	**1**
K6	1/3	4	6	1/5	1/7	**1**	1/5	0.643545	0.062	6%	5
K7	1/7	6	5	1/7	1/7	5	**1**	0.888557	0.086	9%	4
							$\sum$	10.33282	1.000	100%	

Table 12. Results of Saaty's method - transport - decision maker 2

Evaluation	Criterion	Criterion definition
1	K5	time certainty/reliability
2	K1	offered services
3	K4	flexibility
4	K7	transport insurance
5	K6	costs
6	K3	knowledge and experience in the field
7	K2	good reputation and references

For storage services, it can be read from the established order of the individual criteria that decision maker 1 most prefers K3, this is about meeting the necessary certifications

Table 13. Storage - decision maker 1

	K1	K2	K3	K4	K5	K6	K7	*geom. diameter*	*scale*		*rank*
K1	1	1/7	1/3	1/4	1/7	1/6	5	0.39180	0.036	4%	6
K2	7	1	1/4	5	1/5	7	9	1.95780	0.179	18%	3
K3	3	4	1	7	3	4	8	3.61475	0.331	33%	**1**
K4	4	1/5	1/7	1	1/7	4	7	0.89420	0.082	8%	4
K5	7	5	1/3	7	1	6	7	3.19921	0.293	29%	2
K6	6	1/7	1/4	1/4	1/6	1	7	0.67295	0.062	6%	5
K7	1/5	1/9	1/8	1/7	1/7	1/7	1	0.187339	0.017	2%	7
							$\sum$	10.91805	1.000	100%	

and certificates. The order of the individual criteria based on the evaluation of the first decision-maker is clearly shown in the following Table 14.

Table 14. Results of the Saaty method - storage - decision maker 1

Evaluation	Criterion	Criterion definition
1	K3	certificates and certificates
2	K5	special conditions for storage
3	K2	price per area (PM, m3)
4	K4	object security
5	K6	time certainty/reliability
6	K1	regional availability
7	K7	due date of invoices

A comparison of the criteria of the second decision maker for storage is shown in Table 15.

For decision maker 2, when comparing the significance of the criteria of storage logistics activities, the most important criterion is K5.

Table 16 shows the results of the analysis of the storage service criteria of the second decision maker.

Table 15. Storage - decision maker 2

	K1	K2	K3	K4	K5	K6	K7	*geom. diameter*	*scale*		*rank*
K1	1	1/7	3	1/5	1/7	3	1/7	0.47237	0.042	4%	5
K2	7	1	5	9	1/4	5	5	2.95528	0.265	27%	2
K3	1/3	1/5	1	1/3	1/4	5	1/5	0.476232	0.043	4%	5
K4	5	1/9	3	1	1/8	3	1/6	0.723895	0.065	7%	4
K5	7	4	4	8	1	7	7	4.604843	0.414	1%	**1**
K6	1/3	1/5	1/5	1/3	1/7	1	1/5	0.277583	0.025	2%	6
K7	7	1/5	5	6	1/7	5	1	1.625614	0.146	15%	3
							$\sum$	11.13582	1.000	100%	

Table 16. Results of the Saaty method - storage - decision maker 2

Evaluation	Criterion	Criterion definition
1	K5	special conditions for storage
2	K2	price per area (PM, m3)
3	K7	due date of invoices
4	K4	object security
5	K1 a K3	regional availability, certificates and certificates
6	K6	time certainty/reliability

6 Discussion of Results

From the results of the pairwise comparison analysis and the Saaty matrix, it was found which of the investigated criteria are most significant for both decision makers for selected outsourced logistics processes such as transport and storage. By arranging the evaluated criteria in Table 17 and Table 18, their position according to the preference of decision-makers can be determined. Table 17 summarizes the evaluated criteria of the pairwise comparison method and the Saaty matrix.

By applying the paired comparison method and Saaty's method, it was proven that the most preferred criterion of both decision-makers for the outsourcing of transport services is K5, i.e. reliability. What matters most to both clients is that their shipment is delivered on time and in good order to the destination location and thus fulfils the basic logistical goal, which is to effectively overcome space and time while satisfying the customer's requirements. The second place is occupied by K1, which are services offered by an outsourcing company. It is followed by K7, K4, K6, K3, and the last position was occupied by criterion K2, from which it follows that the references of other customers have almost no influence on the selected decision-makers (Table 18).

Table 17. Summary of results - transportation

SUMMARY transportation	K1	K2	K3	K4	K5	K6	K7
decision maker 1 – pairwise comparison	2	5	4	3	1	3	1
decision maker 2 – pairwise comparison	2	7	6	3	1	5	4
decision maker 1 – Saaty's method	4	7	3	6	1	5	2
decision maker 2 – Saaty's method	2	7	6	3	1	5	4
Sum	10	26	19	15	4	18	11
final ranking	**2**	**7**	**6**	**4**	**1**	**5**	**3**

Table 18. Evaluation of the results of the synthesis of transport criteria

Evaluation	Criterion	Criterion definition
1	K5	time certainty/reliability
2	K1	offered services
3	K7	transport insurance
4	K4	flexibility
5	K6	costs
6	K3	knowledge and experience in the field
7	K2	good reputation and references

The following Table 19 summarizes the evaluated criteria of both methods, i.e. pairwise comparison and Saaty's matrix for the storage process.

Table 19. Summary of analysis results - storage

SUMMARY storage	K1	K2	K3	K4	K5	K6	K7
decision maker 1 – pairwise comparison	6	3	1	4	2	5	7
decision maker 2 – pairwise comparison	5	2	6	4	1	7	3
decision maker 1 – Saaty's method	6	3	1	4	2	5	7
decision maker 2 – Saaty's method	5	2	5	4	1	6	3
Sum	22	10	13	16	6	23	20
final ranking	**6**	**2**	**3**	**4**	**1**	**7**	**5**

For the outsourcing of warehousing and other logistics services, such as handling, picking, labelling, palletizing, packaging, etc., with the help of pairwise comparison analysis and Saaty's method, it was found that K5 is a very important criterion for both decision makers. It is about the creation and observance of special conditions for

Table 20. Evaluation of the results of the synthesis of storage criteria

Evaluation	Criterion	Criterion definition
1	K5	special conditions for storage
2	K2	price per area (PM, m3)
3	K3	certificates and certificates
4	K4	object security
5	K7	due date of invoices
6	K1	regional availability
7	K6	time certainty/reliability

the storage of assortments such as dangerous substances and medicines, which must meet strict principles for the storage and distribution of medicinal products. Second place belongs to criterion K2, i.e. price per area. Criterion K3, which is certificates and certifications, came third. The following Table 20 clearly shows the storage criteria according to their order.

7 Conclusion

The paired comparison method is a basic, ordinal method of multi-criteria decision-making. The use of Saaty's matrix represents a more sophisticated procedure of the method for calculating the weights of individual criteria. This is the cardinality and most widely used method of estimating the weights of selected criteria, where the decision maker compares combinations of all possible pairs of selected criteria. With the use of computer technology, the mentioned methods can be used in a similar way for a wider range of logistics services and a higher number of decision-makers.

Ensuring the required criterion in transport, which was ranked highest and was evaluated as the most preferred by the selected decision-makers, i.e. transport reliability, can be perceived as the level of quality of the individual elements of the transport process that participate in its implementation. It is a summary of temporally and substantively related actions to ensure the relocation of goods according to the expectations and requirements of the customer. The technical reliability of the vehicles used in transportation appears to be key. Vehicle breakdowns must be monitored and analysed, their severity, frequency, costs incurred to eliminate breakdowns and determine their cause. Then there is a synthesis of the entire process and subsequent proposals for measures that lead to the continuous improvement of vehicle maintenance processes with the aim of reducing related costs. Another essential measure of transport reliability is regular driver training, compliance with applicable legislation and other measures to achieve success in meeting selected criteria.

In order to ensure the highest chosen criterion in storage, which is the provision and observance of special conditions for the storage and distribution of medicines or products with the ADR label, regular training of employees, during which they are informed about the dangerous properties of the substances in question, about the conditions of handling and storage, and about disposal in in case of their damage and leakage. Further training will teach them how to provide first aid and fire protection. Furthermore, flammable substances cannot be exposed to the sun and high temperatures, as there is a risk of fire. Therefore, all regulations and safety instructions must be followed. It goes without saying that safety principles must be observed near flammable materials - in particular, no smoking and no open flames. To ensure safe handling and storage, it is necessary to have a guideline drawn up and safety regulations, operating and storage regulations visibly posted in all storage areas. At the same time, it is necessary to observe all the conditions for the proper storage and storage of medicines. The use of a drug that has been stored under inappropriate conditions does not guarantee quality, effectiveness or safety and may cause adverse effects of varying extent. Thanks to outsourcing, selected decision-makers do not have to worry about storage spaces, warehouse employees, or work safety in them, and save costs associated with the operation of the logistics department. Today, outsourced process providers no longer view the costs of using these services as a burden. On the contrary, they have calculated that the benefit of these spent funds is the opportunity to devote themselves fully to their own business activities while conquering new markets, consolidating their position on the existing market and gaining a competitive advantage. When using transport services, there is a synergistic effect, when mass transport is cheaper and more environmentally friendly, as emissions are reduced.

Outsourcing of business processes is nowadays a significant helper for companies in securing a competitive advantage of the company by allowing them to focus on their core business. On the other hand, it is necessary to foresee the risks associated with the outsourcing of the given activities. This is, for example, the risk of leakage or misuse of information, the risk of termination of cooperation when there is no direct control and the contracting authority and the intermediary of outsourced activities may have different ideas about the level of outsourced services and thus the credibility of the business partner is impaired [31]. However, these considerations are not the subject of this article, but they could be a useful stimulus for further processing.

References

1. Khan, S.A., Chaabane, A., Dweiri, F.T.: Multi-Criteria Decision-Making Methods Application in Supply Chain Management: A Systematic Literature Review. Multi-Criteria Methods and Techniques Applied to Supply Chain Management: Valerio A. P. Salomon, IntechOpen. Homepage (2018). https://www.intechopen.com/books/multi-criteria-methods-and-techniques-applied-to-supply-chain-management/multi-criteria-decision-making-methods-application-in-supply-chain-management-a-systematic-literatur. DOI:https://doi.org/10.5772/intechopen.74067. Accessed 21 Feb 2023
2. Seydel, J.: Data envelopment analysis for decision support. Ind. Manag. Data Syst. **106**(1), 81–95 (2006)
3. Kolios, A., Mytilinou, V., Lozano-Minguez, E., Salonitis, K.A.: Comparative study of multiplecriteria decision-making methods understochastic inputs. Energies **9**(7) (2016)
4. Dadda, A, Ouhbi, I.A.: Decision support system for renewable energy plant projects. 20 multi-criteria methods and techniques applied to supply chain management. In: Fifth International Conference on Next Generation Networks and Services (NGNS), pp. 356–362 (2014)

5. Mardani, A., Jusoh, A., Nor, K.M.D., Khalifah, Z., Zakwan, N., Valipour, A.: Multiple criteria decision-making techniques and their applications – a review of the literature from 2000 to 2014. Econ. Res. Istraživanja **28**(1), 516–571 (2015)
6. Brito, T.B., Silva, RCdS, Pereira, R.C., Medina, A.C.: Discrete event Simulation combined with multicriteria decision analysis applied to steel plant logistics system planning. In: Proceedings of the 2010 Winter Simulation Conference, pp. 2126–2137 (2010)
7. Belton, V., Stewart, T.: Multiple Criteria Decision Analysis: An Integrated Approach. Kluwer Academic Publishers, The Netherlands (2002)
8. Dooley, A., Smeaton, D., Sheath, G., Ledgard, S.: Application of multiple criteria decision analysis in the New Zealand agricultural industry. J. Multi-Criteria Decis. Anal. **16**, 39–53 (2009)
9. Stopka, O., Stopkova, M., Kampf, R.: Application of the operational research method to determine the optimum transport collection cycle of municipal waste in a predesignated urban area. Sustainability **11**(8), 2275 (2019)
10. Zacek, V.: Management Decisions: Theory, Examples, Solutions. Czech Technical University in Prague, Prague (2015)
11. Meszaros, F., Markovits-Somogyi, R., Bokor, Z:. Modelling and multicriteria optimization of road trafic flows considering social and economic aspects. Logi – Sci. J. Trans. Logist. **3**(1), 70–82 (2012)
12. In the right direction. Multi-criteria analysis. Homepage. http://spravnym.smerem.cz/Tema/Multikriteri%C3%A1ln%C3%AD%20anal%C3%BDza. Accessed 28 Feb 2023
13. Fiala, P., Jablonsky, J., Manas, M.: Multi-criteria Decision Making. University of Economics, Prague, Prague (1994)
14. Dožić, S.: Multi-criteria decision making methods: application in the aviation industry. J. Air Transp. Manag. **79**(8), 1–22 (2019)
15. Triantaphyllou, E.: Multi-Criteria Decision Making Methods: A Comparative Study. Springer, Boston (2013)
16. Lizbetinova, L., Kampf, R., Lizbetin, J.: Requirements of a transport system user. Commun. Sci. Lett. Univ. Zilina **14**(4), 106–108 (2012)
17. Fedorko, G., Rosova, A., Molnar, V.: The application of computer simulation in solving trafic problems. the urban trafic management in Slovakia. Theor. Empir. Res. Urban Manag. **9**, 5–17 (2014)
18. Pohekar, S., Ramachandran, M.: Application of multi-criteria decision making to sustainable energy planning-a review. Renew. Sustain. Energy Rev. **8**(4), 365–381 (2004)
19. Ho, W., Xu, X., Dey, P.K.: Multi-criteria decision making approaches for supplier evaluation and selection: a literature review. Eur. J. Oper. Res. **202**(1), 16–24 (2010)
20. Zopounidis, C. Doumpos M.: Multi-criteria decision aid in financial decision making: methodologies and literature review. J. Multi-Criteria Decision Anal. **11**(4–5), 167–186 (2002)
21. Mendoza, G., Martins, H.: Multi-criteria decision analysis in natural resource management: a critical review of methods and new modelling paradigms. For. Ecol. Manage. **230**(1), 1–22 (2006)
22. Jato-Espino, D.E., Castillo-Lopez, E., Rodriguez-Hernandez, J., Canteras-Jordana, J.C.: A review of application of multi-criteria decision making methods in construction. Autom. Constr. **45**, 151–162 (2014)
23. Lukoszova, X., Stopka, O.: Logistics Centers in the Global Market. Ekopress, Jesenice (2019)
24. Schwalbach, L.: Outsourcing der Einkaufsprozesse: Kostenreduktion und strategische Konzentration durch Einbringug von Dienstleistern im Einkauf. Books on Demand, Norderstedt (2020)
25. Fotr, J., Svecova, L.: Managerial decision making. Procedures, methods and tools. Ekopress, Prague (2010)

26. Jablonský, J.: Operations Research: Quantitative Models for Economic Decision Making, 3rd edn. Professional Publishing, Prague (2007)
27. Saats, T.: Decision making with the analytic hierarchy process. Int. J. Serv. Sci. **1**(1), 83–98 (2008)
28. Brožová, H., Houška, M., Šubrt, T.: Models for Multicriteria Decision Making. Czech University of Agriculture, Prague (2014)
29. Talašova, J.: Fuzzy Methods of Multicriteria Evaluation and Decision Making. Palacký University, Olomouc (2003)
30. Grasseová, M., Dubec, R., Řehák D.: Effective decision-making: analysis, decision-making, implementation and evaluation. 1st ed. Edika, Brno (2013)
31. Field, V.: Outsourcing: Principles and Practices. Larsen and Keller Education, New York (2018)

Strategic Activities in the Area of Intellectual Property Management in the Enterprise

Aldona Małgorzata Dereń and Jan Skonieczny(✉)

Faculty of Management, Wroclaw University of Science and Technology, ul. Wybrzeże Wyspiańskiego 27, 51-200 Wrocław, Poland
{aldona.deren,jan.skonieczny}@pwr.edu.pl

Abstract. Intangible assets are becoming increasingly important for enterprises. However, the question remains how they can make full use of them, especially intellectual property resources. Based on the knowledge and experience of cooperation with small and medium-sized enterprises in Poland, the authors believe that the condition for the effective use of intellectual property resources is to take appropriate strategic actions. The aim of the article is to identify strategic activities in the area of intellectual property resource management in a company. We recommend these activities in the proposed proprietary business model based on intellectual property resources. Such thinking and acting can help small and medium-sized enterprises not only in maintaining a competitive advantage but also in creating new products, processes, and organizational values that will contribute to their internal and market growth.

Keywords: intellectual property · enterprise · business model · strategic activities

1 Introduction

According to the assumptions of the resource school, the development possibilities of an enterprise are conditioned by its internal potential, i.e. a unique set of tangible and intangible resources. According to this concept, the profitability of a company depends on the type, size, and nature of the resources held, as well as on the imperfections of the market on which these resources were obtained. An appropriate method of shaping and controlling resources enables obtaining above-average economic benefits. The acquired market advantage of the company is also strengthened by the difficulty of competitors in obtaining analogous or similar resources.

In the opinion of J. Barney, for resources to be a source of competitive advantage, they should be valuable, i.e. strategic resources that take advantage of opportunities or minimize threats; rare, difficult to access by current and future competitors; difficult to imitate and copy by competitors; difficult to substitute, which results from their unique character [1]. Such features are possessed by intangible resources, e.g. brand, customer relations, reputation of the organization, knowledge of employees, intellectual property, etc. Nowadays, there is an increase in the importance of intangible resources in economic

A. Burduk et al. (Eds.): ISPEM 2023, LNNS 790, pp. 506–514, 2023.
https://doi.org/10.1007/978-3-031-45021-1_38

activity. It is recognized that in a knowledge-based economy, the source of success is not the allocation of scarce resources alone - achieving success is related to the creation of new knowledge, which comes from ideas and ideas, and these, as intangible goods, are an important component of creating value material organization. The rank of intangible goods is evidenced by the fact that they have been called the "fourth factor of production" and the basis of organizational power [2]. Enterprises not only base their operations on these resources but also offer intangible products that, thanks to the development of information and communication technology, can reach any place without major restrictions.

Intangible resources are closely related to the individual in the sense that they are "generated" by humans and develop in human minds, and their specific features give them the status of a factor in the enterprise management process that has a significant impact on the value and market position of the organization.

The purpose of this article is to present strategic activities in the area of intellectual property management as an intangible resource at the disposal of the company. Trademarks, inventions, and copyrights functioning in the enterprise - also those created by it - create a large reservoir of resources. Thanks to them, the company can build its position in the market. Effective management of these resources should lead to an appropriate level of protection and, at the same time, to their economic use.

2 The Concept of Intellectual Property

Intellectual property means various products of the human mind, such as inventions, literary and artistic works, as well as symbols, names, graphics, and patterns used in the broadly understood economy. It covers both products and markings applicable for the purposes of economic activity (industrial property) and existing works the subject of authorship, i.e. works of a scientific, literary, and artistic nature.

Intellectual property is closely related to the processes of creating, developing, and using the acquired knowledge, experience, and competencies and is the result of human creativity, entrepreneurial behavior, ideas, and inventions. Intellectual property is distinguished by the fact that its economic value can be determined and it is subject to legal protection [3]. This protection already at the level of creative results is an important strategic skill in the development of enterprises because thanks to it organizations can be partners in market competition and even leaders. The importance of this skill is growing in the modern economy, where intellectual resources require not only financial outlays, but also activities related to their protection.

In the discussion on intellectual property, economists, sociologists, lawyers, and specialists in organizational behavior emphasize social the nature of intellectual property. Such a term clearly emphasizes the resource context of intellectual property. Defining the relationship of individuals and communities as a special good - a product of metal work - established with the intention of using it, its usefulness, and specific features are emphasized.

The first of these is inappropriateness. An idea or a new solution being the result of human creative activity acquires value from the moment of its manifestation. Thus, the creator does not appropriate the products of his own creativity, unlike the owner of the land who, for example, prohibits access to it. The creator has the right to control the use of his externalized creative effect.

The second feature that allows for emphasizing the different properties of ideas, ideas, or solutions is their uncountability. In practice, we are dealing with a "world of ideas and ideas", and this uncountability shows the limitless creative possibilities of people and organizations.

The third feature shaping the specificity of intellectual resources is indivisibility. It is difficult to imagine, for example, the transferability of only part of the idea or only the right to use it. This is the basis of the economic problem of assessing the profitability of investments in innovations [4]. This natural consistency influences decision-making in any situation of economic choices. Intellectual property is characterized by the inability to wear out during use and perfect liquidity, which means that it can be transformed into any material resource [5].

Intellectual property is also characterized by the ability to be commercialized. It is a feature of bringing intellectual property into a form that allows it to be sold on the market and compete. This ability is dynamic, variable, and dependent on many factors. It is focused on the absorption of intellectual knowledge and creative skills in the manufacturing process, in order to successfully market them in the form of a product/products. Commercialization is also defined as "shaping added value for ideas, research results, technology, and a new product" [6].

Intellectual property is also strategic. In the most general terms, the strategic nature of the resource is to ensure the survival and development of the enterprise, which should be reflected in a change in the strategic position in the sector or industry. A resource-based strategy should therefore take into account all the organization's intangible resources conducive to achieving strategic goals. Today, intellectual property and the organization's ability to multiply and protect them play a key role in the set of intangible resources that determine the success of an enterprise.

The specific features of intellectual property also include the ability for protection (protection) [7]. Within this property, three aspects of securing the intellectual property as a resource can be distinguished. The first aspect is the state-defined legal framework for the market processes of exchange of goods and services for the area of intellectual property, which establish the legal order known as intellectual property law. It is a set of legislative forms of intellectual property with their specific jurisdictional forms shaped by case law and legal developments.

The essential elements of these legal regulations are patent law, right to industrial designs, right to utility models, trademark law, and copyright law. These regulations, based on the constitutional basis, form a specific whole, providing a fundamental guarantee of private property, which is the general and superior basis for the protection of private forms of property - the concept of intellectual property. In addition to this aspect, there are also legal regulations concerning competition, e.g. the Act on Combating Unfair Competition or Antimonopoly Law. The legal framework and control mechanisms they create are also aimed at the negative effects of the forms of using the intellectual property on the competition.

The second aspect of intellectual property protection concerns security at the organizational level. The choice of institutional forms is at stake here, such as patents, industrial design protection rights, utility model protection rights, integrated circuit topography registration rights, etc. They allow the holder to keep exclusive use of a specific solution, primarily when it is disclosed and disseminated. In addition to these forms, protection in the form of business secrets is also possible. It allows you to keep confidential information not disclosed to the public, which can be assigned a certain technical, economic, or economic usefulness (know-how).

The third aspect of securing intellectual property concerns behaviors and safeguards at the level of an individual and group functioning in an organization. There are different possible paths of behavior depending on internal factors (person) and factors from the closer and further environment. The indicated features of intellectual property as a resource imply the need to take strategic actions that determine the effective process of intellectual property management in the enterprise.

3 In Search of a Business Model in the Field of Intellectual Property in the Enterprise

The process of using intangible resources mentioned above requires thinking in terms of strategy and answering four key questions: where are these resources located in the company? how can they be used effectively? what is the value of intellectual resources for the company in the area of production, marketing, strategic planning, employee training, etc.? How to protect your intellectual property resources?

The answer to these questions is the basis for developing a business model based on intellectual property resources.

The concept of a business model in management sciences causes many controversies and different views. There are many definitions of business models, which are characterized by various parameters and interpretations. There is an ongoing scientific discourse on the essence of understanding a business model, both in terms of its reference to a given paradigm, as well as the place of its embedding in the criterion of the environment, as well as cause and effect relationships and correlations between other ontological entities. The business model sets new decision-making centers and creates new impulses in shaping value, appropriating it, and capturing it. The business model, being a model and a point of reference, ensure the materialization of strategic ideas through a constructive confrontation of various factors, often ambidextrous, becoming a platform for strategic accessories that often have a resource dimension [8].

A business model is a "method of operating on the market, unique for a given enterprise, which ensures that it maintains a long-term competitive advantage by providing buyers with added value, understood as meeting or even exceeding expectations as to the broadly understood quality of the product (good/service)" [9].

In the opinion of M. Duczkowska-Piasecka and others, a business model is used to create value for customers, engages company resources, is the subject of strategic choices and activities, it must be controlled according to indicators for strategic management, it must be modifiable, giving knowledge about migration values and knowledge of the direction in which it takes place must be managed, and for this, you need appropriate human resources structures [10].

In the literature on the subject, you can find many classifications of business models presented by various researchers. These classifications are located in three directions of thinking that we can observe in building business models.

In the first direction, researchers use the business model as an abstract concept to describe the creation, sale, and delivery of value to customers in a very general way [11–13].

The second direction is characterized by an emphasis on identifying the structure of the business model. The authors of this approach present a list of necessary elements of a business model [14].

On the other hand, the third direction of business model research focuses on a specific situation and analyzes business models of real companies, e.g. Intel, Microsoft, and Apple [15].

J. Niemczyk and R. Trzaska define business models in relation to Industry 4.0, pointing to:

- data-driven models;
- models based on an intelligent product and service (product-service systems);
- models based on seemingly unscaled activities;
- models based on the strategies of modern market competition.

The Table 1 shows the proprietary classification of business models. The basis for its development was R. M. Grant's considerations on the phases and characteristics of organization management in historical terms [16].

Table 1. Evolution of business models due to phases and characteristics of organization management

Management phase	Features of organization management	Business model based on
Financial budgeting	- capital budgeting based on discounted cash flows - financial control	controlling
Development planning	- medium-term economic forecasts - formal planning - diversification and striving for synergy in creating development departments	planning
Organization positioning	- sector analysis - market segmentation - experience curve - PIMS analysis - portfolio planning	competition

(continued)

Table 1. (*continued*)

Management phase	Features of organization management	Business model based on
Competitive advantage	- analysis of resources and skills - maximizing shareholder value - restructuring and re-engineering - strategic alliances	cooperation
New Economy	- strategic innovations - innovative businesses - breakthrough technologies	innovation
Third millennium	- CSR and business ethics - ecology - creativity - competition for standards - "winner takes all" - orientation to global and local markets - orientation towards information and communication technologies	intellectual property resources

Source: own study based on [16].

4 Typology of Strategic Activities in a Business Model Based on Intellectual Property Resources

Intellectual property resources are a business tool, and like any other should only be used when necessary. Strategy building includes the identification of competitive advantages of the company, differences between it and its products and competition, and how revenue is generated. In Fig. 1. an original business model based on intellectual property resources was presented. The dimensions of this model are:

- competitive goal;
- competitive advantage;
- M.E. generic strategies Porter - cost leadership, differentiation, focus on costs and focus on differentiation [17];
- internal and external development of intellectual property resources.

Such an approach to the dimensions of the discussed model requires undertaking certain strategic actions.

These activities are the protection of intellectual property; lending of intellectual property; providing intellectual property; acquisition of intellectual property. Intellectual property protection is related to the organization's goals through the use of patents, trademarks, trademarks, copyrights, trade secrets, and other intellectual property.

It is one of the mechanisms of turning innovation into income, using the transfer of research results, technologies, and products. They are an element of intellectual property management in an organization, which aims to: plan, acquire, and commercialize

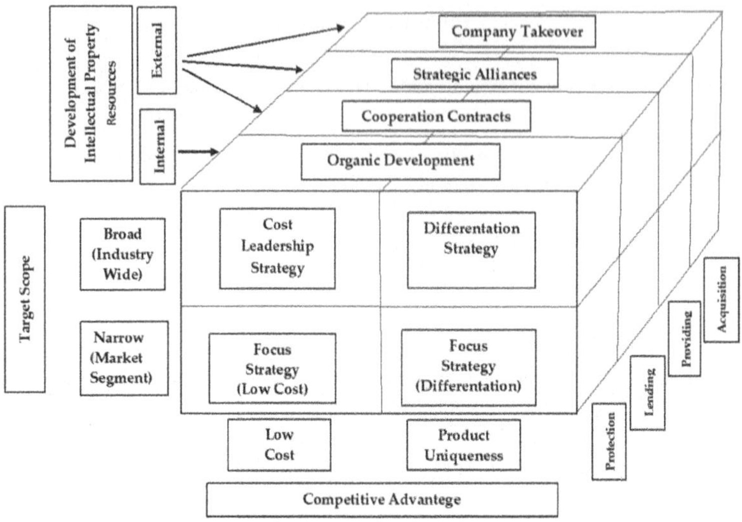

Fig. 1. Business model based on intellectual property resources.

intellectual property in its activities, in order to increase the competitiveness of the organization, generate income, and conquer markets. The possibility of combining research and development with practice is possible when transferring technology and know-how as well as own achievements.

The lending of intellectual property is a strategic activity that may take the form of, for example, rental, lease, or transfer of rights, thus enabling the exploitation of a specific intellectual asset for the time specified in the contract.

Sharing intellectual property is a strategic activity involving many different forms and tools, including licensing, leasing, crowdsourcing, crowdfunding, and franchising. The most popular form is a license agreement under which the creator (owner) authorizes the licensee to use the intellectual property in a specific way. Unlike agreements for the transfer of rights to intellectual property, licensing agreements do not result in the transfer of creative property rights, but only in authorizing the use of intellectual property in a specific field of exploitation.

The described strategic action is often used in the processes of creating a strategic alliance, which enables the cooperation of partners based on the intellectual resources held by one of the participants of the created alliance.

Another strategic action proposed in the model is the acquisition of intellectual property. The leading form of this action is a sales contract, which on the one hand allows for the acquisition of intellectual resources necessary for the organization, and on the other hand, is an instrument for selling unnecessary (due to the adopted goal, competitive advantage and further development of the organization) intellectual property. This strategic action is carried out as part of a friendly or hostile takeover. In practice, it means the acquiring organization supplementing and enriching its intellectual property with new resources, and sometimes it is a form of optimization and getting rid of ineffective or surplus resources.

5 Conclusion

In The dynamics of the ongoing competitive processes force enterprises to look for new, unique development strategies. In these strategies, intangible resources are of key importance, which, although they do not have a physical form, are treated as a unique asset, contributing to the success of the company. Among these resources, a special place is occupied by intellectual property, which is becoming more and more strategic. The growing importance of these "mind resources" raises questions about how best to use them to achieve the goals of the enterprise that creates and controls them. Therefore, it is necessary to develop strategies that use and maximize the value of the company's intellectual resources. This will allow us to determine the mutual relationship between intellectual property resources and the company's strategy.

In the opinion of the authors, the typology of strategic activities presented in the article within the recommended model based on intellectual property resources is universal. The synergy of these activities and intellectual property will arise when the use of intellectual property gives the company an advantage over the competition. Intellectual property assets require analysis and integration, constituting the same capital of the enterprise as other assets. The business model recommended in the article, consistent with the overall development strategy of the company, takes into account both the dynamics of changes within the company and external relations in a rapidly changing economy, as it allows for the selection of activities that will turn into economic value. What's more, this model leads to the creation of a sustainable competitive advantage for the company in a given market and takes into account the decision variables relating to the strategic, operational, and financial aspects.

References

1. Barney, J.: Firm resources and sustained competitive advantage. J. Manag. 2, 112 (2000). https://doi.org/10.1016/S0742-3322(00)17018-4
2. Sikorski, C.Z.: Zachowania ludzi w organizacji", Wydawnictwo naukowe PWN, Warszawa, p. 274 (2001)
3. Blair, R.D., Cotter, T.F.: Intellectual Property Economic and Legal Dimensions of Rights and Remedies, pp. 1–2. Cambridge University Press, Cambridge (2005)
4. Wiśniewska, S.: Skuteczność niekomercyjnych instytucji otoczenia biznesu we wspieraniu innowacji marketingowych małych i średnich przedsiębiorstw, Wyd. Uniwersytetu Ekonomicznego, Kraków, p. 10 (2013)
5. Koźmiński, A.K., Jemielniak, D.: Zarządzanie od podstaw, Wydawnictwa Akademickie i Profesjonalne, Warszawa, p. 110 (2008)
6. Rajan, Y.S.: Empowering Indians. With Economic, Business and Technology Strengths for the Twenty First Century. pp. 130–139. Har Anand Publications, New Delhi (2002)
7. Dereń, A.M., Skonieczny, J.: Cykl rozwoju przedsiębiorczości intelektualnej w organizacji – zarys problemu. In: Balcerak, A., Malara, Z. (eds.) Innowacyjność w gospodarce, organizacji i technice, pp. 62–71. Warszawa, Centrum Prawa bankowego i Informacji (2017)
8. Jabłoński, A.: Twórczy model biznesu w koncepcji gospodarki sieciowej, Kolegium zarządzania i Finansów, Zeszyt naukowy 162, Oficyna Wydawnicza SGH Warszawa, pp. 184–185 (2018)

9. Brdulak, H.: Nowoczesne modele biznesu w logistyce, w: Strategie i logistyka w sektorze usług. Logistyka w nietypowych zastosowaniach, red. J. Witkowski, A. Baraniecka, Prace Naukowe UE we Wrocławiu 234, Wrocław 2011, s. 31; por.: B. Skowron-Grabowska, Business Models in Transport Services, Przegląd Organizacji, no 1, p. 35–39 (2014)
10. Duczkowska-Piasecka, M., Poniatowska-Jaksch, M., Duczkowska-Małysz, K., (eds.), Model biznesu: Nowe myślenie strategiczne, Wydawnictwo Difin, Warszawa, pp. 153–154 (2013)
11. Chesbro, G.: Open business model. IP-Manag. Gen. **352** (2008)
12. Markides, C.: The new business model: strategy of painless innovation. the new business model: strategy of painless innovation, Alpina Publisher, Moscow, 298 (2010)
13. Osterwalder, A., Pine, I.: Building of Business Models. Alpina Publisher Series, Skolkovo, Moscow (2013)
14. Slywotzky, A.J.: Migration values. What will happen to your business the day after tomorrow, Mann, Ivanov and Ferber, Moscow (2006)
15. Yoffie, D.B., Cusumano, M.A.: Zasady strategii. Bill Gates, Andy Grove i Steve Jobs. Pięć ponadczasowych strategii, Dom Wydawniczy Rebis, Poznań (2016)
16. Grant, R.M.: Współczesna analiza strategii, Oficyna Wydawnicza Wolters Kluwer, Warszawa, p. 32 (2011)
17. Porter, M.E.: Przewaga konkurencyjna. Osiąganie i uzyskiwanie lepszych wyników, Wydawnictwo Helion, Gliwice, pp. 38–39 (2006)

A Systematic Literature Review of Management and Technological Approaches Integration with Industry 4.0

Danielle Barreto Blanco[1], Leonilde Varela[1], José Vicente[2], and José Machado[3](✉)

[1] Algoritmi Research Centre, University of Minho, 4804-533 Guimarães, Portugal
[2] InsideLimits, Canidelo, Vila Nova de Gaia, Portugal
[3] University of Minho, MEtRICs Research Centre, 4804-533 Guimarães, Portugal
jmachado@dem.uminho.pt

Abstract. Industry 4.0 (I4.0) presents a new era of production and process management, and in order for companies to take advantage of the benefits offered by this technological advancement, understanding how it can influence or be influenced by management approaches is a point of study. For this reason, this investigation aimed to analyze production and quality management approaches in order to identify synergies as well as disagreements with current requirements imposed by Industry 4.0. Therefore, this article presents a systematic literature review with article selection based on the PRISMA methodology, resulting in the analysis of 45 studies. The main conclusions indicate that agile management presents the highest synergy with Industry 4.0, followed by collaborative, real-time, and lean management, respectively. The requirement of virtualization presented synergy classified as very high for all approaches, while ecodesign was the only one classified as low synergy with management approaches. In conclusion, this study highlights the importance of a joint strategy for integrated management and technological approaches to take advantage of the benefits of Industry 4.0. This study is a base to be taken into account, and being used in industrial practice, in the context of iFixturing Project.

Keywords: Production and Quality Management Aproaches · Integrated Management Systems · Industry 4.0 · Lean Management · Agile Management · Collaborative Management · Real-time Management

1 Introduction

In order to achieve budget compliance, with business growth and consolidation, opportunities related to improving operational efficiency were identified; hence Industry 4.0 (I4.0) developed as the new era of production, a result of technological and digital advancement. Due to its widespread adoption in different contexts and industrial sectors, studies related to the fourth industrial revolution are increasingly in the spotlight (Fettermann et al. 2018).

A. Burduk et al. (Eds.): ISPEM 2023, LNNS 790, pp. 515–532, 2023.
https://doi.org/10.1007/978-3-031-45021-1_39

As noted by (Tortorella et al. 2019), the interest of professionals, academics, and even governments regarding its 4.0 technologies has grown as reports of benefits from integrating this technological advancement with production management approaches have emerged. In addition to creating new economic opportunities, (Wadhwa & Salkever 2017) and (Tapscott & Tapscott 2016) have also discussed the role of I4.0 in transforming society. In this scenario, companies are concerned with applying new integrated approaches to production and quality management that allow them, together with the usual requirements inherent to production processes and flows, to ensure the satisfaction of quality requirements (Christopher 2011; Costantino et al. 2012; Putnik et al. 2019).

The traditional approaches to production management have been primarily based on the Lean Management philosophy, which aims to promote a cycle of consistent and lasting transformations based on the elimination of waste and the maximization of efficiency (Liker 2004; Papadopoulou & Özbayrak 2005; Womack et al. 1992; Womack & Jones 1996).

More recently, management approaches have emerged that are strongly focused on meeting the current requirements imposed by Industry 4.0, such as Agile Management, a flexible and customer-oriented approach that quickly adapts to change, and Collaborative Management, which emphasizes collaboration, teamwork, and resource exchange to benefit all parties involved (Chukwu et al. 2013; Huang et al., 2008; Vieira et al. 2018).

In addition, Real-Time Management approaches based on the use of simulation, either alone or in combination with other more recent technologies, such as digital twin and augmented reality, are increasingly becoming a reality and a necessity (Boschert & Rosen 2016; Negri et al. 2017; Vachálek et al. 2017). Furthermore, other authors such as (Schwab 2016) and (Brynjolfsson & McAfee 2014) emphasize that understanding how Industry 4.0 can influence or be influenced by management approaches is relevant so that companies are prepared to take advantage of the benefits and opportunities offered, but also to deal with the challenges that accompany this change.

Given the objective of the research and the evolution of technologies embedded in today's industry, it is essential to understand the requirements of Industry 4.0, which are essential for the proper functioning of its technologies. In this context, (Hermann et al. 2015), as well as (Rodrigues et al. 2016) and (Mabkhot et al. 2018), indicate six requirements that stand out, derived from Industry 4.0 technologies: interoperability, virtualization, decentralization, real-time capability, service orientation, and modularity.

Considering that one way to promote the survival and sustainability of organizations in this next era of the fourth industrial revolution is through the efficient use of resources (Ferreirinha et al. 2019), (Habib & Chimsom 2019), in agreement with (Rio et al. 2010), add a seventh requirement that has gained prominence in the literature for the successful development of sustainable Industry 4.0 systems: ecodesign.

These requirements aim to make the factory intelligent and adaptable, with the ability to communicate and collaborate between systems, remote monitoring, autonomous decision-making, reliable and agile operations, integrated service provision, production flexibility, and environmental preservation. For these requirements, gathered in Fig. 1, to be met, they must be followed rigorously for the implementation of the fourth industrial revolution model, which demonstrates the need for deeper knowledge about each principle.

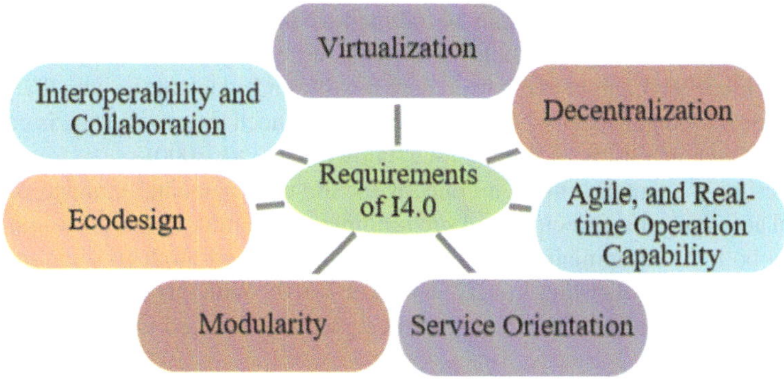

Fig. 1. Management requirements in the I4.0.

Interoperability and collaboration – is the principle that truly makes a smart factory is the ability of systems to communicate with each other through the Internet of Things (Sakurai & Zuchi 2018), as well as the permission to work together, easily accessing and using each other's functionalities, technologies, components, and human or robotic agents, aiming (Chen et al. 2008; Lu 2017; Van Overeem et al. 2007), aiming at a co-creation of something, based on co-learning (VER);

Virtualization – in order to track and monitor the industrial system remotely, the smart factory creates a virtual copy of all processes performed in the real world environment. Real-time sensor data, constantly synchronized and updated, is used not only to track and control physical processes but also as test banks (Lee 2019). This modernization is capable of simulating adverse situations and product prototypes, which allows for employee training and greater production flexibility, including possible customer inclusion in the project development (Habib & Chimsom 2019).

Decentralization – ensures the ability of cyber-physical systems to work independently; with autonomous modules making decisions about production needs in real-time (Brettel et al. 2014). (Habib & Chimsom 2019) state that with the delegation of possible production bottlenecks to higher (vertical integration) or similar (horizontal integration) modules, the system is capable of handling a more complex and personalized environment.

Agile, and Real-time based Operation Capability – virtually instantaneously, this principle allows for rapid intelligent decision making for operation regulation, based on real-time data acquisition and processing, along with the underlying need for a quick changeover in the manufacturing systems to enable changing from one production order to another one, in real time. By assisting in identifying faults, maintenance needs, and consequent operation safety, the real-time operation capability increases reliability and influences overall production effectiveness, minimizing waste and optimizing processes (Netto 2018; Sobreira 2018).

Service Orientation – as products reach competitive parity, manufacturing industries become service providers, with a focus on the integrated sale of products and services (Mabkhot et al. 2018). This paradigm shift allows for customer inclusion while increasing value creation and profitability in industries (Habib & Chimsom 2019).

Modularity – this principle can be defined as the ability of system components to be separated and combined quickly and easily, according to demand (Mabkhot et al. 2018). Modularity is characterized by a more flexible and adaptable system that, through increased development speed of new products and reduced time to market, is capable of delivering a greater variety of production (Baldwin & Clark 2000).

Ecodesign – by representing a concern for the environmental damage caused by industrial activities, this principle seeks to minimize or eliminate environmental impacts caused by both the equipment and tools used in the design of products and services and the products themselves (Lima Jr 2022; Rio et al. 2010). With this objective, production processes, products, and even services are designed and planned to achieve it (Jabbour et al. 2018).

Implementing a robust production process - despite dynamic environmental influences - is the task of production control (Schuh et al. 2011). Therefore, the relationships between the Fourth Industrial Revolution (I4.0) and the different production management approaches should be taken into account in the decision-making process by its executors to achieve the desired objectives. In this context, knowledge of different production and quality management approaches is essential: lean management, agile management, collaborative management, and real-time management.

For this purpose, a systematic literature review was conducted first. The methodology adopted in this work is described in the next chapter. Subsequently, the results, analyses, and gaps found are presented. The study concludes with the final considerations composed of the main conclusions, limitations, and research suggestions.

2 Method

The selected research method for the investigation was bibliographic research, a qualitative approach in which the data collection instrument is a systematic literature review. Due to its format, the method is capable of filling the gap within methodologies designed to analyze problems in a specific research area or literature (Adams et al. 2016; Odważny et al. 2018). The work was developed based on the proposal of Tranfield, Denyer and Smart, in their article published in 2003 in the British Journal of Management. The publication proposes a method for systematic reviews divided into three stages, as described in Fig. 2:

Stage 1, composed of three phases, concerns the planning of the review. Stage 2 is where the systematic review itself takes place and includes five phases. The third and final stage is about reporting and disseminating the collected information, which has two phases (Tranfield et al. 2003).

Phase 0: The first phase of the review is also the justification for this work. As exemplified in the introductory section, there is a growing interest in searches regarding Industry 4.0, therefore the study and exploration of approaches related to the theme still present potential added value in industrial management.

Phase 1: Based on the identification of the need for a review, the purpose of this research is to conduct a survey and study of different types of approaches associated with Integrated Production and Quality Management that fit the requirements of I4.0.

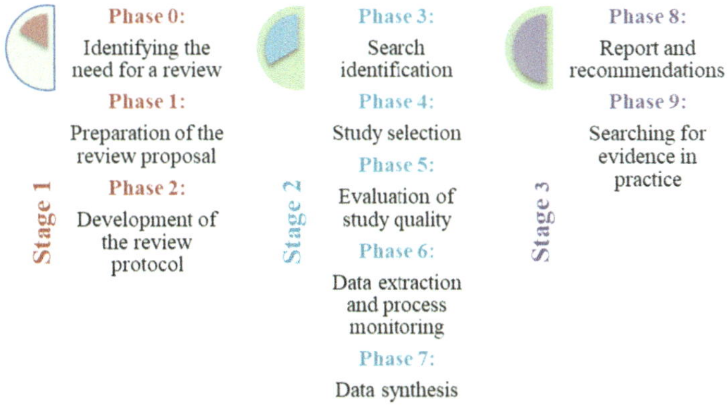

Fig. 2. Stages of the systematic review methodology.

Phase 2: Considering the established proposal, the use of the b-on electronic database was defined, due to its broad coverage in gathering data from the most respected content providers in academia.

The search in the international b-on database was conducted on January 23, 2022. Keywords from the five main blocks of this study were selected: Industry 4.0, Lean Management, Agile Management, Collaborative Management, and Real-Time Management. To ensure that the Industry 4.0 block is included and at least one of the other four blocks as well, the advanced search string for title and subject was developed: ("industry 4.0" OR "fourth industrial revolution") AND (("lean manufacturing" OR "lean thinking") OR ("agile manufacturing" OR "agile management") OR ("collaborative manufacturing" OR "collaborative management") OR ("real-time management" OR "real-time systems")).

Other options such as the French term industrie 4.0, the well-known acronym I4.0, the terms lean production, automation lean, and agile thinking, leagile, other nomenclatures that refer to collaborative management - flat structure, horizontal structure, horizontal management, open innovation - and the term real-time without a hyphen were tested and discarded due to null or low amounts of new results.

Phase 3: Initially, the Boolean search presented, only with the restriction of publication year up to 2021, 867 results. After excluding duplicate works identified with score divergences in the titles, 240 works were deleted by their DOI, leaving 627 documents.

Of these documents, 246 were academic journal articles, 306 conference materials, 44 books and e-books, in addition to 31 results from other sources such as theses/dissertations, reports, reviews, specialty publications, and non-academic magazine materials. Conference materials also recorded as articles were only considered in academic journals.

The quantity of articles by publishers is identified in Fig. 3. When filtering only academic journal articles, IEEE stands out as the main publisher, followed by Elsevier, MDPI, Taylor & Francis, Emerald, Springer, and Wiley. These are responsible for 190 articles, that is, almost 80% of the publications.

As visible in Fig. 4, researchers' interest in the various management approaches associated with Industry 4.0 has been steadily increasing, reinforcing the relevance of

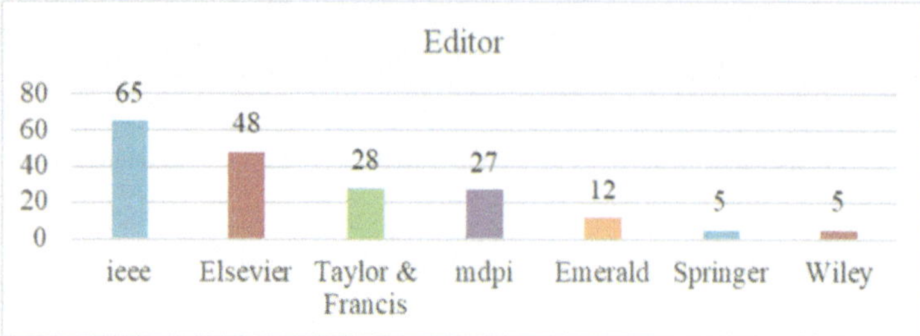

Fig. 3. Number of papers by editor.

the topic. The number of articles from the last two years reaches over 60% of the total publications.

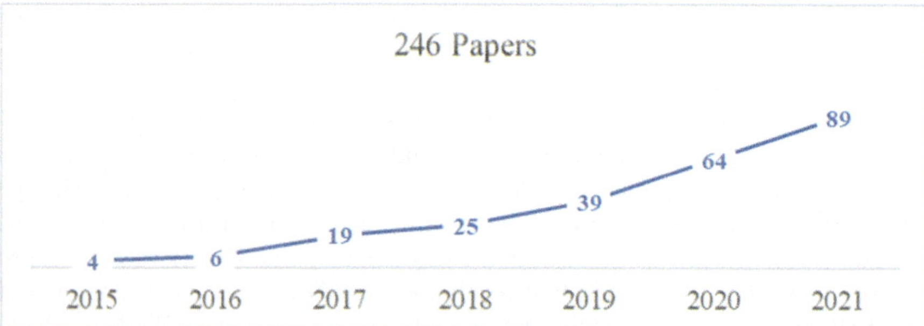

Fig. 4. Number of papers per year.

After reading the detailed record of the 246 documents, it was possible to identify 136 articles in academic journals that primarily mention the Lean block, 22 on the agile management approach, 14 on collaborative management, and 74 on real-time management. Thirty-nine articles mention more than one approach, and all blocks have at least one article that integrates with another block.

Phase 4: For data selection, articles written in languages other than English or Portuguese were disregarded due to the author's preference. Articles that were not available on the b-on platform or were not peer-reviewed were also excluded to maximize the quality of the selected works. Of the initial 246, 172 documents remained, with the lean management block still having the majority of works followed by the real-time block. Articles written in other languages that were not available and/or had not been peer-reviewed did not influence further eliminations.

Moving on with the study, the abstracts of the 172 articles were read on the b-on platform. In this phase, a pertinence analysis is carried out qualitatively to evaluate whether the articles found address the topic at hand, leading to the elimination of a considerable portion of articles with low alignment with the objective of this work, leaving 45 articles included.

The PRISMA flow diagram, shown in Fig. 5, was designed to help authors report more transparently why and how systematic reviews and meta-analyses were conducted, summarizing the phases of study identification and selection (Galvao et al. 2015; Moher et al. 2009; Page et al. 2021).

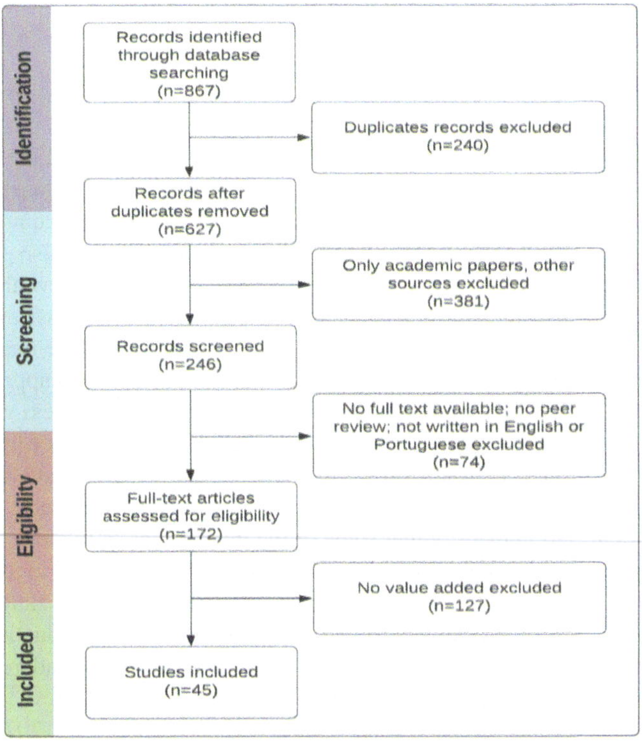

Fig. 5. Prisma diagram of the systematic review process.

Phase 5: All 45 articles selected after reading the abstracts followed an acceptable scientific rigor, possibly because they had already been filtered as peer-reviewed.

Phase 6: The articles were extracted for detailed reading.

Phase 7: Just like in the total publications, the number of articles from the last two years represents 60% of the total publications, with 27 out of the 45 selected articles. Regarding the main study block, 24 articles with added value included the Lean management approach, 7 included the Agile management approach, 7 included the Collaborative management approach, and also 7 included the Real-time management approach. Most of the included studies have a qualitative research approach, with 36% being literature reviews and 11% being case studies, totaling 47% of the selected articles. However, quantitative research studies were also present, representing 40% of the studies with 24% being experimental and 16% being survey studies.

Phase 8 of reporting and recommendations, as well as **Phase 9** of evidence searching in practice, are included in the next sections, which analyze the selected literature in detail and group the findings not only by each type of approach but also make the connection with the current requirements imposed by Industry 4.0.

3 Results and Discussions

This chapter presents the results of the systematic literature review, which involves Industry 4.0 and the four previously selected approaches to production and quality management: lean, agile, collaborative, and real-time. In order to understand the interactions, reports of the advantages and disadvantages of each approach, relationships between the approaches and Industry 4.0 integration, and synergies with the requirements were extracted from the articles. The gaps identified in the literature were also presented.

When searching for evidence in the systematic literature review that one approach supports requirements or that Industry 4.0 requirements can assist in integrated management, it is clear that all articles end up discussing performance improvements.

(Tortorella & Fettermann 2018) indicate that the lean management approach emphasizes simplicity and effectiveness, usually aligned with a shared business vision. In line with the requirements of decentralization and service orientation, according to (Ciano, Dallasega, et al. 2021), lean benefits still encompass flexibility, information sharing, and collaboration, to improve service quality, as well as customer satisfaction.

(Sanders et al. 2017) identify that some of the interactions of the lean management approach with the requirements of Industry 4.0 may not be as clear, such as the divergence of concepts between the modularity of Industry 4.0 and the standardization characteristic of lean. However, they emphasize that the principle of waste reduction should be applied before virtualization, in order to maintain consistent data standards for further processing, and also highlight how important standardization is to achieve interoperability. In agreement, the results of the study conducted by (Kamble et al. 2020), with data collected from 205 managers working in 115 Indian manufacturing companies, present Industry 4.0 as a facilitator of lean practices.

The importance of sustainable development is highlighted by (Lekan et al. 2020), who argue that the natural environment should be preserved in a way that does not compromise the chances of future generations meeting their needs. They also affirm that this development can be achieved through the integration of knowledge from economic and environmental variables. The relevance of the results obtained through the study conducted by (Varela et al. 2019) is also noteworthy, as they allow us to affirm that although there is a relationship between Industry 4.0 and Sustainability, the relationship between lean and the requirement for ecodesign is not confirmed.

On the other hand, (Kamble et al. 2020) and (Ghaithan et al. 2021) found that the impact of Industry 4.0 technologies on sustainability performance is amplified by the presence of the lean management approach as a strong mediating variable. Without the presence of lean manufacturing, the impact on sustainability performance would not be as noticeable or rapid. Similarly, (Tripathi et al. 2021) assert that the lean intelligent approach associated with sustainability techniques can offer a new opportunity for employees to improve their operational performance and environmental conditions.

According to (de Almeida et al. 2021), agile project management has emerged as a complement to traditional project management, with the need to update and improve the project management process and make it more flexible and interactive. In parallel, (Ding et al. 2021) argue that Industry 4.0 technologies serve as facilitators for agile management approaches to improve the company's current position in terms of its manufacturing system, control, resource optimization, data management, collaborative relationships, and product development.

(Gromova 2018) argues that the rapid growth of technological development takes into account the demands of the new industrial paradigm, with the agile approach and its methodology meeting the contemporary requirements of the new era as a promising model. Similarly, (de Almeida et al. 2021) also emphasize that the combination of agile management with Industry 4.0 can reduce time-to-market, production errors, and promote collaboration and dynamism to the product development team, meeting the requirements of decentralization, virtualization, and service orientation.

According to (Jovanović et al. 2015), Industry 4.0 is a true example of agile manufacturing, and its real-time operating capacity should significantly reduce the time for developing new products and provide a solution for highly demanding market demands. This is because, according to (Rauch et al. 2019), agile management favors the modular and scalable structure of the system, which is opposed to the well-known waterfall project management that values extensive planning and pre-production (Gromova 2018).

(Odważny et al. 2018) conclude that there is not enough evidence in the literature to confirm the relationship between sustainable development of the organization and the agile management approach. On the other hand, (Ding et al. 2021) make statements that both agile and lean management promote performance improvements in Industry 4.0 and can relate to the requirement for ecodesign.

(Xue et al. 2018) recognize that with the rapid development of technology, it becomes very convenient to integrate information from various manufacturers through collaborative management approaches. (Vianna et al. 2020) agree that Industry 4.0 technologies are assigning new possible roles to individuals to help organizations create value. Thus, the ecosystem envisioned by Industry 4.0 predicts the decentralization of organization management, where, according to (Bellavista et al. 2021), Industry 4.0 technologies should also be integrated horizontally: producers, suppliers, and customers should collaborate.

From this perspective, according to (Nakayama et al. 2020) manufacturing and innovation are now supported by value co-creation networks and collaborative service systems that go beyond simple data integration between companies, being able to enhance more ambitious and comprehensive manufacturing arrangements, especially when virtualized and supported by real-time capacity technologies. (Moghaddam & Nof 2018) also highlight that the collaborative integration of service components in cloud virtualized manufacturing can meet the demand of Industry 4.0 for modularity agility and service orientation.

With the emergence of Industry 4.0, technological advancements associated with the internet of things, big data analytics, sensor technology, and artificial intelligence have led to the design and implementation of new approaches to maximize data-driven solutions (Trovati et al. 2020). In this context, interoperability, i.e., the ability of a system to interact with another, in order to use different data platforms without hindering their synthesis, is a point of great attention for collaborative management approaches, especially when it comes to collaboration between companies (Bellavista et al. 2021; Mladineo et al. 2017; Moghaddam & Nof 2018).

(Fernandez-Carames & Fraga-Lamas 2018) recall that one of the principles of Industry 4.0 is to efficiently, quickly, and flexibly collect the maximum amount of real-time information from different parts of the value chain, in order to help increase quality. In this scenario, (Bousdekis et al. 2020) emphasize that predictive maintenance is also moving towards its next evolutionary step, achieved by exploring real-time information to generate significant data even before symptoms are manually detected.

(Alvarez et al. 2020) state that by merging physical and nectic space - which refers to the virtual or online environment in which there is interaction and sharing of information without geographical or time limitations - it ends up involving the virtualization of the production system and (Liu et al. 2021) add that the virtualized real-time system becomes capable of performing more efficient fault diagnosis and optimizing and better evaluating planning, so that simulation and process improvement are increasingly implemented.

As a consequence, (Alvarez et al. 2020) emphasize that the ideal behavior of real-time management approach depends heavily on its system architecture, and therefore, (Ciano, Pozzi et al. 2021) do not fail to highlight the great need for system interoperability. (Liu et al. 2021) also say that the real-time management approach should allow data, knowledge, and resource sharing to meet the decentralization requirement.

In order to make the interaction between management approaches and Industry 4.0 requirements more visible, a matrix of interdependence was developed between the two frameworks, represented by Fig. 6. Vertically, each requirement is correlated with each management approach horizontally. The percentage of requirements per block presented was given by the amount of citations, which was considered to measure the defined level of synergy identified in each article of each block.

Taking as an example the level of interaction or cooperation between the interoperability requirement and the lean management approach: out of a total of 24 lean articles, 9 cited that they would be valued with the interoperability requirement of I4.0, representing 37.50% of the articles. Of the 7 articles classified as the agile management approach, 6 cited their modularity, representing 85.71% of the articles. Similarly, all articles in the real-time management approach expressed their synergy with the requirements of interoperability, virtualization, and real-time capability, as all of them have cited these key terms, achieving a 100% mark.

The last column referring to the total presents the percentage equivalent to the acquired requirements within the possible ones, for example, out of the 45 articles of all approaches, 30 referenced or cited the decentralization requirement representing 66.67%. The last row referring to the total presents the percentage equivalent of the management approach for all requirements. For instance, the collaborative approach achieved a 65.77% mark of synergy with Industry 4.0, as out of the 49 possible synergies (citations) with requirements - that is, 7 articles multiplied by 7 requirements - 31 were identified. Likewise, the last number in the row and column of totals represents that out of a total possible of 315 synergies - 45 articles multiplied by 7 requirements - 176 were identified in the 45 selected articles, achieving a percentage of 55.87%.

Requirements / Approaches	Lean	Agile	Collab.	Real-time	Total
Interoperability	37,50%	57,14%	52,86%	100,00%	71,11%
Virtualization	75,00%	85,71%	85,71%	100,00%	82,22%
Decentralization	58,33%	71,43%	100,00%	57,14%	66,67%
Real-time Operation	62,50%	57,14%	52,86%	100,00%	64,44%
Service Orientation	33,33%	85,71%	100,00%	42,86%	53,33%
Modularity	41,67%	85,71%	57,14%	42,86%	51,11%
Ecodesign	20,83%	42,86%	14,29%	14,29%	22,22%
Total	47,02%	69,39%	65,77%	65,31%	55,87%

0% <= x% < 25% (low synergy)

25% <= x% < 50% (medium synergy)

50% <= x% < 75% (high synergy)

75% <= x% < 100% (very high synergy)

Fig. 6. Management interdependence matrix: requirements versus approaches.

Tortorella et al. (2021) emphasized that validating which practices and principles of integrating the approach with Industry 4.0 have the greatest synergy can reduce efforts and achieve expected benefits. For this reason, ecodesign being the least referenced Industry 4.0 requirement and the only one to show low synergy, regardless of the management approach, can be considered a serious issue indicating the need for greater care towards the requirement, as well as the need for further studies relating it.

On the other hand, according to the results of the interdependence matrix, virtualization is indicated as a main requirement, being classified as having very high synergy by all approaches and in an exclusive way. Although not in the best classification, all other requirements were able to achieve the milestone of high synergies with management.

Individually, interoperability with lean and collaborative management; real-time capability with collaborative management; service orientation with lean and real-time management; modularity with lean, collaborative and real-time management, and ecodesign with agile management were classified as medium synergy, also indicating points of attention.

Interoperability with agile management; decentralization with lean, agile and real-time management; real-time capability with lean and agile and collaborative management achieved the classification of high synergy. Furthermore, beyond the aforementioned virtualization with all management approaches; interoperability with real-time management, decentralization with collaborative management, real-time capability with real-time management, service orientation with agile and collaborative management, and modularity for agile and collaborative management achieved the best results, some of them even representing 100%, being classified as very high synergy, indicating where the greatest integration benefits can arise.

Overall, with a focus on approaches, the results showed agile and collaborative management with the highest percentage of interdependence with Industry 4.0, reaching 69.39%, and 65,77% of synergy with these requirements. Followed by the real-time based management approach, with 65.31%.

On the other hand, lean had the lowest overall percentage, retaining 47.02% of synergy. Although it achieved the mark of high synergy for decentralization and real-time capability and very high for virtualization, the results suggest that, among the selected approaches, lean management would be the least compatible with Industry 4.0 requirements.

Literature Gaps

The study revealed the relationship between production and quality management approaches and Industry 4.0, highlighting that there are still gaps and promising study opportunities involving these themes.

Most of the articles found in the literature for integrating the lean management approach with Industry 4.0 are conceptual or literature reviews, so, in agreement with (Ejsmont et al. 2020), empirical research and case studies confirming the synergy of the concepts are lacking.

The literature related to agile or collaborative management approaches in integration with I4.0 lacks both qualitative theoretical models and case studies or results from quantitative empirical research. In general, studies tend to only to make preliminary considerations about possibilities of interaction. Some articles presented benefits of agile and collaborative management among stakeholders; however, few reports correlated an approach with Industry 4.0 technologies.

The supposedly related literature on the real-time management approach presented many works that only mentioned that real-time data were extracted, but in a general context, most did not extend concepts about the type of management. Regarding the approach, greater investigation needs were identified on issues of trust, cyber security, and level of autonomy, according to (Moghaddam & Nof 2018).

Longitudinal studies are also recommended to perceive the real contributions of management approaches, as well as their results over longer periods. In addition, studies on the requirements of the fourth industrial revolution, mainly the few relationships with ecodesign, stand out as an opportunity for research development in this field of knowledge.

4 Conclusions

The present study focused on investigating production and quality management approaches in order to identify synergies and discrepancies with current requirements imposed by Industry 4.0. After analyzing different types of philosophies, methods, and technologies associated with integrated management, a systematic literature review was conducted to identify the interactions between Industry 4.0 and lean, agile, collaborative, and real-time approaches.

According to the main findings, ecodesign was the least referenced requirement and the only one to show low synergy, which is concerning and highlights the need for greater attention to this requirement. On the other hand, virtualization was indicated as the requirement that satisfied the most, as it was classified as having very high synergy by all approaches. The other requirements addressed achieved a total milestone of high synergy, although they did not reach the best classification.

Regarding the results focused on the approaches, agile and collaborative management were found to have the highest percentage of interdependence with Industry 4.0, achieving 69.39%, and 65,77% of synergy with the requirements, correspondingly. This is because these agile and collaborative management approaches provide added value for Industry 4.0, using its adaptability and quick response, along with the focus on the co-construction of solutions, especially in the development of new products and services, with a focus on value and quality for the customer.

Moreover, the collaborative management approach showed synergy, in general, with the main requirements, carrying many concepts from Industry 4.0, with a focus on decentralization and service orientation, which were able to achieve their maximum values. Its value co-creation, through information sharing, enables innovation that can better contribute to the further development of companies, while enabling to solve production problems, and further contributing for increasing the agility and stability of the production systems in meeting demand without major investments in additional resources, despite the difficulty in finding ideal partners.

Next, real-time management approach achieved 65.31% of synergy with the requirements, demonstrating that many of its practices are already part of the fourth industrial revolution, by allowing early detection of deviations and future diagnoses, which minimize losses and increases the reliability of products and services.

On the other hand, the lean management approach had the lowest overall percentage of synergy, retaining only 47.02%, suggesting that, among the selected approaches, lean would be the least compatible with the requirements of I4.0. This result is supported by some claims that authors mention about the lean management approach being outdated and no longer applicable (Buer et al. 2018; Ciano, Dallasega, et al. 2021; Ejsmont et al. 2020; Müller et al. 2017).

However, there are also some works that provide some evidence to enhance the lean approach continues reveling some relevance in the current digital age, mainly in the preparation of traditional companies to overcome some major issues, for instance, regarding lack of preparation for a transition to the Industry 4.0, and the global manufacturing and circular economy requirements satisfaction. Therefore, requiring the application of lean management to be able to further be prepared for the next level, to embrace the exigent conditions and requirements currently imposed by the fourth industrial revolution. Furthermore, the high synergy for decentralization and real-time capability, and very high synergy for virtualization, show that integration with Industry 4.0 promotes the beginning of the new management era.

Difficulties and Limitations

During the process of the systematic literature review, some inconveniences were identified, such as some bugs in the server of the b-on extraction platform, and difficulties in handling the extracted data files for analysis. Additionally, the selection and inclusion of articles are carried out through a qualitative procedure, which carries a certain level of subjectivity, so some studies may have been neglected.

Most of the combinations of some production and quality management approach with Industry 4.0 found are presented at very general levels; it is also noted that the sample of articles related to lean is significantly larger, so care must be taken to verify that the results obtained are not biased.

Despite the limitations presented, the results obtained can be considered robust since, as the first systematic review to involve concepts of the fourth industrial revolution with concepts of the four selected production management approaches, the present work provides a presentation of the current literature and theoretical viewpoints.

Future Work

Based on the contributions achieved in this study and the gaps in the literature identified, the need for more empirical studies of the interactions of I4.0 with production management approaches is expressed, as well as studies on the requirements of the fourth revolution applied to organizations, with a focus on ecodesign. Longitudinal studies are also recommended in an attempt to understand the contributions of management approaches, as well as their results in longer terms. Additionally, a comparative analysis between organizations or sectors that applied or did not apply integrated management practices is suggested, highlighting the possibility of taking into account the little-cited financial aspect; the real cost-benefit of investments in technologies.

Acknowledgement. This work was partially supported by national funds through the FCT - Fundação para a Ciência e Tecnologia through the R&D Units Project Scopes: UIDB/00319/2020 and EXPL/EME-SIS/1224/2021; and it is co-funded by the Euro-pean Regional Development Fund (ERDF) through the North Regional Operational Program (NORTE 2020) of the Portugal 2020 Program [Project No. 43922, with acronym "iFixturing"; Funding Reference NORTE-01-0247-FEDER-043922].

References

Adams, R., Jeanrenaud, S., Bessant, J., Denyer, D., Overy, P.: Sustainability-oriented innovation: a systematic review. Int. J. Manag. Rev. **18**(2), 180–205 (2016). https://doi.org/10.1111/ijmr.12068

Alvarez, I., Moutinho, L., Pedreiras, P., Bujosa, D., Proenza, J., Almeida, L.: Comparing admission control architectures for real-time ethernet. IEEE Access **8**, 105521–105534 (2020). https://doi.org/10.1109/ACCESS.2020.2999817

Baldwin, C., Clark, K. B.: Design Rules: The Power of Modularity, vol. 1. MIT Press (2000)

Bellavista, P., Esposito, C., Foschini, L., Giannelli, C., Mazzocca, N., Montanari, R.: Interoperable blockchains for highly-integrated supply chains in collaborative manufacturing. Sensors **21**(4955), 4955 (2021). https://doi.org/10.3390/s21154955

Boschert, S., Rosen, R.: Digital twin — the simulation aspect. Mechatron. Futures 59–74 (2016). https://doi.org/10.1007/978-3-319-32156-1

Bousdekis, A., Apostolou, D., Mentzas, G.: Predictive maintenance in the 4th industrial revolution: benefits, business opportunities, and managerial implications. IEEE Eng. Manag. Rev. **48**(1), 57–62 (2020). https://doi.org/10.1109/EMR.2019.2958037

Brettel, M., Friederichsen, N., Keller, M., Rosenberg, M.: How virtualization, decentralization and network building change the manufacturing landscape: an industry 4.0 perspective. Int. J. Mech. Aerosp. Ind. Mechatron. Manufact. Eng. **8**(1), 37–44 (2014)

Brynjolfsson, E., McAfee, A.: The Second Machine Age: Work, Progress, and Prosperity in a Time of Brilliant Technologies. W. W. Norton & Company (2014)

Buer, S.-V., Strandhagen, J.O., Chan, F.T.S.: The link between Industry 4.0 and lean manufacturing: mapping current research and establishing a research agenda. Int. J. Product. Res. **56**(8), 2924–2940 (2018). http://10.0.4.56/00207543.2018.1442945

Chen, D., Doumeingts, G., Vernadat, F.: Architectures for enterprise integration and interoperability: past, present and future. Comput. Ind. **59**, 647–659 (2008). https://doi.org/10.1016/j.compind.2007.12.016

Christopher, M.: Logistics e Supply chain Management (4th ed.). Pearson (2011)

Chukwu, U.C., Member, S., Mahajan, S.M., Member, S.: Real-time management of power systems with v2g facility for smart-grid applications. 1–9 (2013)

Ciano, M.P., Dallasega, P., Orzes, G., Rossi, T.: One-to-one relationships between industry 4.0 technologies and lean production techniques: a multiple case study. Int. J. Product. Res. **59**(5), 1386–1410 (2021). http://10.0.4.56/00207543.2020.1821119

Ciano, M.P., Pozzi, R., Rossi, T., Strozzi, F.: Digital twin-enabled smart industrial systems: a bibliometric review. Int. J. Comput. Integr. Manuf. **34**(7/8), 690–708 (2021)

Costantino, N., Dotoli, M., Falagario, M., Pia, M., Marcello, A.: A model for supply management of agile manufacturing supply chains $. Int. J. Product. Econ. **135**, 451–457 (2012). https://doi.org/10.1016/j.ijpe.2011.08.021

de Almeida, J.F., Amaral, D.C., Coelho, R.T.: Innovative framework to manage new product development (NPD) integrating additive manufacturing (AM) and agile management. Procedia CIRP **103**, 128–133 (2021). http://10.0.3.248/j.procir.2021.10.020

Ding, B., Ferras Hernandez, X., Agell Jane, N.: Combining lean and agile manufacturing competitive advantages through Industry 4.0 technologies: an integrative approach. Product. Planning Control (2021). https://doi.org/10.1080/09537287.2021.1934587

Ejsmont, K., Gladysz, B., Corti, D., Castaño, F., Mohammed, W.M., Lastra, J.L.M.: Towards lean industry 4.0 – current trends and future perspectives. Cogent Bus. Manag. **7**(1), 1781995 (2020). https://doi.org/10.1080/23311975.2020.1781995

Fernandez-Carames, T.M., Fraga-Lamas, P.: A review on human-centered IoT-connected smart labels for the industry 4.0. IEEE Access **6**, 25939–25957 (2018). https://doi.org/10.1109/ACCESS.2018.2833501

Ferreirinha, L., et al.: An industry 4.0 oriented tool for supporting dynamic selection of dispatching rules based on Kano model satisfaction scheduling. FME Trans. **47**, 757–764 (2019). https://doi.org/10.5937/fmet1904757F

Fettermann, D.C., Cavalcante, C.G.S., Almeida, T.D. de Tortorella, G.L.: How does industry 4.0 contribute to operations management? J. Ind. Product. Eng. **35**(4), 255–268 (2018)

Galvao, T., Andrade Pansani, T., Harrad, D.: Principais itens para relatar revisões sistemáticas e meta-análises: a recomendação PRISMA. Epidemiol Serv Saude **24** (2015). https://doi.org/10.5123/S1679-49742015000200017

Ghaithan, A., Khan, M., Mohammed, A., Hadidi, L.: Impact of industry 4.0 and lean manufacturing on the sustainability performance of plastic and petrochemical organizations in Saudi Arabia. Sustainability **13**(11252), 11252 (2021). https://doi.org/10.3390/su132011252

Gromova, E.: Agile management in the context of Russian industrial sector. In: MATEC Web of Conferences, vol. 178, p. 8007 (2018). https://doi.org/10.1051/matecconf/201817808007

Habib, M.K., Chimsom, C.: Industry 4.0: sustainability and design principles. In: 2019 20th International Conference on Research and Education in Mechatronics (REM), pp. 1–8 (2019). https://doi.org/10.1109/REM.2019.8744120

Hermann, M., Pentek, T., Otto, B.: Design principles for industrie 4.0 scenarios: a literature review. Technische Universität Dortmund, Dortmund **45**, 1–15 (2015)

Huang, G., Zhang, Y., Jiang, P.: RFID-based wireless manufacturing for real-time management of job shop WIP inventories. Int. J. Adv. Manufactur. Technol. **36**, 752–764 (2008). https://doi.org/10.1007/s00170-006-0897-4

Jabbour, C.J.C., Jugend, D., Jabbour, A.B.L. de S., Govindan, K., Kannan, D., Leal Filho, W.: There is no carnival without samba: revealing barriers hampering biodiversity-based R&D and eco-design in Brazil. J. Environ. Manag. **206**, 236–245 (2018). https://doi.org/10.1016/j.jenvman.2017.10.019

Jovanović, M., Latić, B., Mas, A., Mesquida, A.: The agile approach in industrial and software engineering project management. J. Appl. Eng. Sci. **13**(4), 213–216 (2015). https://doi.org/10.5937/jaes13-9577

Kamble, S., Gunasekaran, A., Dhone, N.C.: Industry 4.0 and lean manufacturing practices for sustainable organisational performance in Indian manufacturing companies. Int. J. Product. Res. **58**(5), 1319–1337 (2020). https://doi.org/10.1080/00207543.2019.1630772

Lee, H.: Real-time manufacturing modeling and simulation framework using augmented reality and stochastic network analysis. Virtual Reality **23**(1), 85–99 (2018). https://doi.org/10.1007/s10055-018-0343-6

Lekan, A., Clinton, A., Fayomi, O.S.I., James, O.: Lean thinking and industrial 4.0 approach to achieving construction 4.0 for industrialization and technological development. Buildings **10**(221), 221 (2020). https://doi.org/10.3390/buildings10120221

Liker, J.K.: Toyota Way: 14 Management Principles from the World's Greatest Manufacturer. McGraw-Hill Education (2004)

Lima, J.O.: Practices and benefits of innovation and sustainability in project management within SMEs context [University of Minho] (2022). https://hdl.handle.net/1822/77806

Liu, J., Yu, D., Hu, Y., Yu, H., He, W., Zhang, L.: CNC machine tool fault diagnosis integrated rescheduling approach supported by digital twin-driven interaction and cooperation framework. IEEE Access **9**, 118801–118814 (2021). https://doi.org/10.1109/ACCESS.2021.3106797

Lu, Y.: Industry 4.0: A survey on technologies, applications and open research issues. J. Ind. Inf. Integr. **6**, 1–10 (2017). https://doi.org/10.1016/j.jii.2017.04.005

Mabkhot, M.M., Al-Ahmari, A.M., Salah, B., Alkhalefah, H.: Requirements of the smart factory system: a survey and perspective. Machines **6**(2), 23 (2018). https://doi.org/10.3390/MACHINES6020023

Mladineo, M., Veza, I., Gjeldum, N.: Solving partner selection problem in cyber-physical production networks using the Humant algorithm. Int. J. Product. Res. **55**(9), 2506–2521 (2017). http://10.0.4.56/00207543.2016.1234084

Moghaddam, M., Nof, S.Y.: Collaborative service-component integration in cloud manufacturing. Int. J. Product. Res. **56**(1/2), 677–691 (2018). http://10.0.4.56/00207543.2017.1374574

Moher, D., Liberati, A., Tetzlaff, J., Altman, D.G., Group, P.: Preferred reporting items for systematic reviews and meta-analyses: the PRISMA statement. Ann. Intern. Med. **151**(4), 264–269 (2009)

Müller, R., Vette, M., Hörauf, L., Speicher, C., Burkhard, D.: Lean information and communication tool to connect shop and top floor in small and medium-sized enterprises. Procedia Manufactur. **11**, 1043–1052 (2017). http://10.0.3.248/j.promfg.2017.07.215

Nakayama, R.S., Spínola, M., de M., Silva, J.R.: Towards I4.0: a comprehensive analysis of evolution from I3.0. Comput. Ind. Eng. **144**, 106453 (2020). http://10.0.3.248/j.cie.2020.106453

Negri, E., Fumagalli, L., Macchi, M.: A review of the roles of digital twin in CPS-based production systems. Procedia Manufact. **11**, 939–948 (2017). https://doi.org/10.1016/j.promfg.2017.07.198

Netto, R.S.: Framework em Tempo Real para Sistema de Gerenciamento de Energia de uma Smart Microgrid Utilizando Sistema Multiagente (2018)

Odważny, F., Wojtkowiak, D., Cyplik, P., Adamczak, M.: Smart factory within sustainable development and green growth concepts. LogForum **14**(4) (2018). https://widgets.ebscohost.com/prod/customerspecific/ns000290/authentication/index.php?url=https%3A%2F%2Fsearch.ebscohost.com%2Flogin.aspx%3Fdirect%3Dtrue%26db%3Dedsdoj%26AN%3Dedsdoj.1ee6e5958724114bdc124113dd5127a%26amp%3Blang%3Dpt-pt%26site%3Deds-live%26

Page, M.J., et al.: Updating guidance for reporting systematic reviews: development of the PRISMA 2020 statement. J. Clin. Epidemiol. **134**, 103–112 (2021). https://doi.org/10.1016/j.jclinepi.2021.02.003

Papadopoulou, T.C., Özbayrak, M.: Leanness: experiences from the journey to date. J. Manufactur. Technol. Manag. **16**(7), 784–807 (2005)

Putnik, G., Ferreira, L., Lopes, N., Putnik, Z.: What is a cyber-physical system: definitions and models spectrum. FME Trans. **47**, 663–674 (2019). https://doi.org/10.5937/fmet1904663P

Rauch, E., Dallasega, P., Unterhofer, M.: Requirements and barriers for introducing smart manufacturing in small and medium-sized enterprises. IEEE Eng. Manage. Rev. **47**(3), 87–94 (2019). https://doi.org/10.1109/EMR.2019.2931564

Rio, M., Reyes, T., Roucoules, L.: A framework for ecodesign : an interface between LCA and design process. Ann. Fac. Eng. Hunedoara – Int. J. Eng**9**(1), 121–126 (2010)

Rodrigues, L.F., de Jesus, R.A., Schützer, K.: Industrie 4.0–a literature review. Revista de Ciência & Tecnologia **19**(38), 33–45 (2016)

Sakurai, R., Zuchi, J.D.: Revoluções industriais até a industria 4.0. Revista Interface Tecnológica **15**(2), 480–491 (2018). https://doi.org/10.31510/infa.v15i2.386

Sanders, A., K. Subramanian, K.R., Redlich, T., Wulfsberg, J.P.: Industry 4.0 and lean management – synergy or contradiction? In: Lödding, H., Riedel, R., Thoben, K.-D., von Cieminski, G., Kiritsis, D. (eds.) APMS 2017. IAICT, vol. 514, pp. 341–349. Springer, Cham (2017). https://doi.org/10.1007/978-3-319-66926-7_39

Schuh, G., Kampker, A., Stich, V., Kuhlmann, K.: Prozessmanagement. In: Strategie und Management produziernder Unternehmen, pp. 327–382. Springer (2011). https://doi.org/10.1007/978-3-642-14502-5_5

Schwab, K.: A Quarta Revolução Industrial (Levoir, Ed.) (2016)

Sobreira, J.V.: Desafios para a manutenção na perspetiva da indústria 4.0. Instituto Superior de Engenharia de Lisboa (2018)

Tapscott, D., Tapscott, A.: Blockchain revolution: how the technology behind bitcoin is changing money, business, and the world. Portfolio/Penguin (2016)

Tortorella, G.L., Fettermann, D.: Implementation of Industry 4.0 and lean production in Brazilian manufacturing companies. Int. J. Product. Res. **56**(8), 2975–2987 (2018). http://10.0.4.56/00207543.2017.1391420

Tortorella, G.L., Rossini, M., Costa, F., Portioli Staudacher, A., Sawhney, R.: A comparison on Industry 4.0 and Lean Production between manufacturers from emerging and developed economies. Total Qual. Manag. Bus. Excellence **32**(11–12), 1249–1270 (2019). https://doi.org/10.1080/14783363.2019.1696184

Tortorella, G.L., Saurin, T.A., Filho, M.G., Samson, D., Kumar, M.: Bundles of lean automation practices and principles and their impact on operational performance. Int. J. Product. Econ. **235**, N.PAG-N.PAG (2021). http://10.0.3.248/j.ijpe.2021.108106

Tranfield, D., Denyer, D., Smart, P.: Towards a methodology for developing evidence-informed management knowledge by means of systematic review. Br. J. Manag. **14**(3), 207–222 (2003). https://doi.org/10.1111/1467-8551.00375

Tripathi, V., et al.: An innovative agile model of smart lean–green approach for sustainability enhancement in industry 4.0. J. Open Innov. Technol. Mark. Complex. **7**(215), 215 (2021). https://doi.org/10.3390/joitmc7040215

Trovati, M., Zhang, H., Ray, J., Xu, X.: An entropy-based approach to real-time information extraction for industry 4.0. IEEE Trans. Ind. Inform. **16**(9), 6033–6041. https://doi.org/10.1109/TII.2019.2962029

Vachálek, J., Bartalský, L., Rovný, O., Sismisova, D., Morhac, M., Loksik, M.: The digital twin of an industrial production line within the industry 4. Concept , 258–262 (2017). https://doi.org/10.1109/PC.2017.7976223

Van Overeem, A., Witters, J., Peristeras, V.: An interoperability framework for Pan-European e-government services (PEGS). In: Proceedings of the Annual Hawaii International Conference on System Sciences, p. 7 (2007). https://doi.org/10.1109/HICSS.2007.68

Varela, L., Araujo, A., Ávila, P., Castro, H., Putnik, G.: Evaluation of the relation between lean manufacturing, industry 4.0, and sustainability. Sustainability **11**(5), 1439 (2019). https://doi.org/10.3390/su11051439

Vianna, F.R.P.M., Graeml, A.R., Peinado, J.: The role of crowdsourcing in industry 4.0: a systematic literature review. Int. J. Comput. Integr. Manufact. **33**(4), 411–427 (2020). http://10.0.4.56/0951192X.2020.1736714

Vieira, G.G., Varela, M.L.R., Putnik, G.D., Machado, J.: Intelligent platform for supervision and production activity control in real time. In: Hamrol, A., Ciszak, O., Legutko, S., Jurczyk, M. (eds.) Advances in Manufacturing. LNME, pp. 151–159. Springer, Cham (2018). https://doi.org/10.1007/978-3-319-68619-6_15

Wadhwa, V., Salkever, A.: The Driver in the Driverless Car: How Our Technology Choices Will Create the Future. BK Berrett-Koehler Publishers, Inc., a BK Business book (2017)

Womack, J.P., Jones, D.T.: Lean Thinking: Banish Waste and Create Wealth in Your Corporation (Elsevier, Ed.) (1996)

Womack, J.P., Jones, D.T., Roos, D.: The Machine that Changed the World: The Story of Lean Production (1992). http://search.ebscohost.com/login.aspx?direct=true&db=edsebk&AN=1965187&site=eds-live

Xue, X., Kou, Y., Wang, S., Liu, Z.: Computational experiment research on the equalization-oriented service strategy in collaborative manufacturing. IEEE Trans. Serv. Comput. Serv. Comput. **11**(2), 369–383 (2018). https://doi.org/10.1109/TSC.2016.2569082

Impact Analysis and Evaluation of the COVID-19 Pandemic on Decision-Making Processes in Enterprises Based on Just-In-Time Methodology

Anna Woźna(✉), Małgorzata Rusińska, and Mateusz Bryłkowski

Wroclaw University of Science and Technology, 50-371 Wrocław, Poland
anna.wozna@pwr.edu.pl

Abstract. This research analyzes the impact of the COVID-19 pandemic on companies implementing the Just In Time (JIT) methodology. It examines the wide-ranging effects on enterprises, adaptive measures taken, and long-term problem-solving strategies. The study employs a comprehensive methodology, including a detailed survey and diverse research group. Four key sectors—labor, supply chain, operations, and corporate investment—were identified as significantly affected by the pandemic. Subsequently, a comprehensive situational assessment was synthesized, and a set of recommendations was developed to aid organizations in future decision-making processes. The findings and recommendations outlined in this article provide valuable guidance for organizations navigating the uncertainties and challenges posed by the pandemic, ensuring their resilience and adaptability in the post-pandemic era.

Keywords: just-in-time · COVID-19 · logistic chains

1 Introduction

The reason for writing this article are significant deviations from the standard operation of enterprises operating on the basis of the "Just In Time" methodology (hereinafter referred to as JIT), which was caused by the COVID-19 pandemic and activities related to preventing the spread of infection. The purpose of the analysis was to define the impact of the pandemic on companies operating under the JIT philosophy and how their management staff make decisions related to daily operations, as well as how they adapt the organization's mode of operation to the current market situation as part of a delayed supply chain, rising prices, limited availability of materials and a highly competitive labour market. The result of the research work will be recommendations for enterprises indicating the required improvements in decision-making processes (including operational and logistic activities of companies) in order to improve their financial results, flexibility to changes, resistance to external problems, as well as consistency and timeliness of activities.

A. Burduk et al. (Eds.): ISPEM 2023, LNNS 790, pp. 533–547, 2023.
https://doi.org/10.1007/978-3-031-45021-1_40

It is important to accurately identify the factors why some companies solve emerging problems more effectively than others, how their decision-making processes have changed and how this has affected their supply chain, for example. It is to be suspected that better-performing companies have greater flexibility to change in the form of management based on live data, checking supply chain susceptibility to sudden risk situations and reducing supply liquidity. In addition, they immediately manage risk in all possible sectors of activity - increasing the number of key warehouse components, equipping with smarter software that anticipates potential problems with the liquidity of supplies, investing in the qualifications of employees as part of handling current production adapted to the latest practices and using the implemented forecasting software. The examples presented are just some of the proposed solutions, their exact effectiveness at this stage is and only hypothetical.

2 Literature Review

2.1 Just-In-Time Concept

"Just-In-Time" as a concept was originally applied in the years from 1950 to 1960 in Toyota's Japanese production plants. The creator of the method is considered to be Taiichi Ohno, and the concept was based on a production system, e.g. American companies Ford and Morris [1].

The "Just-In-Time" work methodology consists in delivering a website, service or product at the right time, in the right quantity, to the right place. It allows for effective organization and management of business activities (e.g. production) of the company. It is assumed that raw materials will be ordered based on the resulting orders, minimizing inventory and labor input. According to the creator's words, the company's activities were to resemble the human body, which in the event of a need (e.g. for food) satisfies the shortages. The advantage of this method is a significant reduction of costs to the required minimum and the elimination of waste - thus increasing the market competitiveness of the organization through conscious cost management and unit maximization in the form of a service/product. The most important aspect of JIT is to ensure a seamless and continuous flow of services or materials to the right places for business activities. If a single element of the structure does not arrive on time, the entire business process is delayed, exposing the company to high losses related to stopping the provision of services or the production of elements. It should be understood that the JIT method applies to all activities of the company, i.e. it is necessary to constantly work on optimization in cooperation with the supply system and operations. This includes appropriate actions:

1. Reducing the distance between successive operating points to a minimum.
2. Maximizing the effectiveness of operations using the best possible work tools.
3. Use of maximally standardized services/products and high repeatability of existing processes.
4. Care for the constant improvement of employees' qualifications in order to notice and implement possible improvements with each subsequent process.
5. If there is a supply chain, the closest possible location of the plants of entities cooperating with the organization should be considered.

6. Optimizing the flow of information between employees vertically and horizontally, and if possible, minimizing redundant information in all information channels.

It has been shown [2] that the vast majority of enterprises do not achieve satisfactory results from the implementation of the JIT system due to, among others, with the lack of decision-making independence of the organization's departments (in favor of decisions at the level of the management or the board of the organization), failure to use the rights to optimize processes by employees, insufficiently high level of elimination of waste and low level of employee discipline.

2.2 The Impact of COVID-19 on Enterprise Operations

COVID-19 is an infectious disease of the respiratory system resulting from infection with the SARS-CoV-2 virus. In Europe, the first cases were detected at the end of January 2020, while the WHO organization recognized the spread of the disease after 2 months (March 2020) as a pandemic threatening the health and life of all people [3, 4]. In the first months of the pandemic, citizens were advised to stay at home regardless of the type of work they do, excluding key workers (such as doctors, drivers or the army). Each country of the world acted individually, introducing a series of restrictions and recommendations regarding working from home or remote teaching in schools. In the long term, this led to the destabilization of timeliness and production/service capabilities of companies around the world. Therefore, the impact of COVID-19 on the operation of JIT enterprises should be recognized as crucial [2, 5, 6].

Research has shown that the pandemic has led many companies to change their strategic models and operations to accommodate the sudden changes in demand caused by working from home and new consumption trends. Therefore, as many as 93% of the surveyed organizations [7] intended to change their supply chains to be more flexible, resistant to market changes and agile. In addition, the introduction of remote work resulting from the recommendations of citizens' health protection required changes in the operating models of the organization and adaptation to the new reality [8, 9].

However, you should be aware of the global impact of the pandemic on any type of business. Comarch's research on the basis of 39 enterprises shows that about 90% of organizations are affected by the coronavirus pandemic [10]. Which may mean that most operating models and the way these enterprises make decisions currently differ significantly from those before 2020. In turn, over 56% experienced a decrease in the number of orders or customers as a result of the pandemic [10], so it is important to understand what factor in the negative impact. And what factors made the remaining approx. 44% of enterprises not lose or record profits as a result of the new situation. Furthermore, long supply chains have been proven to be less able to recover from economic crises like the one in 2008/9 [11].

As many as 89% of companies were forced to adapt to the new operating conditions [10]. The ways in which companies order materials, offer services or products have changed - because e.g. demand has changed in all sectors.

It is worth emphasizing that people are behind the smooth course of the entire operational process, i.e. employees responsible for responding to errors and early noticing, reporting and solving problems. For this reason, employees are an even more crucial resource than ever before – so it is necessary to invest in qualified employees in the labour market, as well as understand their needs and requirements. Unfortunately, due to the emergence of sudden demand on the labour market, the increasing number of offers makes the behaviour of the most important employees an increasing challenge for entrepreneurs. Unfortunately, a significant part of the organization is not interested in understanding the reasons behind the employee's departure or is unable to compete with the offers of other employers who can offer better employment conditions, higher remuneration, remote work or a reduced number of working hours in order to take care of the employee's mental health [12].

3 Characteristics of the Research Subject and Research Procedure

3.1 Characteristics of the Research Subject

Understanding the phenomenon of the COVID-19 pandemic and its wide-reaching impact on organizations requires careful analysis. Literature reports indicate that a significant part of enterprises cannot cope with the changes resulting from the pandemic [7, 13–17] therefore it is important to examine organizations on the market that are doing better than others in the existing conditions, and thus record an increase in economic activity and profits.

Therefore, in the context of the analysed topic, the issue is as follows:

"Analysis and assessment of the impact of the COVID-19 pandemic on the decision-making processes of enterprises based on the Just-In-Time methodology"

The main focus should be on the search for the cause (as a function of the objective), however, focusing only on JIT companies may lead to a limited perspective. On the other hand, examining a wider research group will allow for the search for unconventional or untypical solutions for the JIT methodology, but it is highly probable that they will still be effective, considering the improvement of the conditions and functioning of the organization after the coronavirus pandemic.

The following general problem has been defined:

- "What changes has COVID-19 caused in organizations since its onset?"
- In addition, the following question was defined as complementary and defining the final products:
- "What actions should be taken to prevent or improve surgery after the corona virus pandemic?"

Supply chain analysis allows you to determine whether there have been problems with deliveries, delays or changes in the prices of delivered products. This section focuses on determining the extent of the phenomenon, root causes and potential prevention. The study recognizes whether organizations already had operations monitoring systems or, as a result of the COVID-19 pandemic, implemented appropriate mechanisms for early response to potential problems with daily business activities. Changes related to the operations of enterprises were also analysed, e.g. to what extent their operations have changed, whether the number of orders/customers has decreased or increased. Organizational investments were also examined, it allows to examine the development strategy of the surveyed enterprises as a result of changes after the coronavirus pandemic.

3.2 Research Method

The adopted research method will present the impact of the pandemic on the decision-making processes of enterprises and their operation in an analytical and real way. It was based on the analysis of publicly available research in the form of articles developed by organizations (such as McKinsey Company), then defining the general problem and its dependencies using preliminary hypotheses, ending with their verification with appropriate analysis of data obtained from the completed survey. The analysis in question was carried out in the Dolnośląskie and Mazowieckie voivodeships in Poland for 2022. Therefore, the presented analysis is not a complete or sufficient analysis exhausting the issue to a full extent.

3.2.1 Preliminary Hypotheses

14 hypotheses were selected for the study, divided between four elements of the structure: employees, supply chains, operations and investments, which outline the research area. The adopted initial hypotheses were developed on the basis of publicly available articles summarizing research and recommendations for enterprises regarding functioning after the COVID-19 pandemic [7, 10, 12, 18]. The hypotheses are presented in the collective Table 1.

3.2.2 Survey Structure and Content

The survey contains 23 questions divided into five sections, of which the first four correspond to the elements of the structure, in accordance with the outlined area of research, additionally general questions were also included, completing the picture of the collected data. Each of the sections examines the scale of changes in phenomena related to the issue and indicates key changes in the sector under study. It is completely anonymous, but at the end of the survey, the age, gender and sector of the company's activity are examined. In addition, the survey was supplemented with the results of entrepreneurs surveyed in person. The study was conducted in March 2022 for a period of about 4 weeks. The questions were developed on the basis of the most frequently discussed issues in scientific articles, publicly available research and on the basis of my own and the environment's experience.

3.2.3 Correlation of Survey Questions and Initial Hypotheses

In order to increase the transparency of the procedure at the analysis stage, a summary was created showing which question from the survey will be an argument when assessing the truth or falsity of the initial hypothesis. This structure will make it easier to track and check the compliance of the analysis with the adopted preconditions.

As can be seen, some of the questions will be used more than once in justifications, even among elements of a different structure studied. This is due to the fact that there are correlations among the data or additional argumentation for the hypothesis consisting in observing visible deviations.

Table 1. A list of questions answering the relevant initial hypotheses

	The question	Hypothesis
EMPLOYEES	Has the COVID-19 pandemic affected working conditions in your organization?	Working conditions have improved significantly
	To what extent do you think working conditions have improved or deteriorated during and after the pandemic?	Working conditions have improved significantly
	Which parts of your work have improved?	Working conditions have improved significantly Employment conditions became the most important for employees
	Which parts of the job have deteriorated?	Working conditions have improved significantly Employment conditions became the most important for employees
	Has the number of employees in your work environment changed?	The number of employees has been reduced The probability of changing jobs is very high
	How likely are you to change your place of employment if you receive a competitive job offer?	The number of employees has been reduced The probability of changing jobs is very high
	What is most important to you in the post-pandemic work environment?	Employment conditions became the most important for employees
	Have you considered changing your place of employment since the beginning of the coronavirus pandemic?	The probability of changing jobs is very high
SUPPLY CHAIN	Has your organization experienced supply chain issues as a result of the pandemic?	COVID-19 has led to significant deviations across all sectors of the supply chain
	In what parts of the supply chain has the organization experienced change?	COVID-19 has led to significant deviations across all sectors of the supply chain
	To what extent (indicatively) did the resulting changes affect the financial results of the organization?	The financial results of enterprises deteriorated significantly Increased latency tolerance

(*continued*)

Table 1. (*continued*)

	Has the organization taken action to prevent future shortages or availability of products/services?	COVID-19 has led to significant deviations across all sectors of the supply chain Companies have implemented mechanisms or software to forecast future shortages and orders Increased latency tolerance
OPERATIONS	Has the way business activities and their organization changed as a result of the pandemic?	Only some of the operational activities have changed
	What elements of the daily organization of operational activities in your work have changed?	Only some of the operational activities have changed
	Has the organization introduced new tools/methodologies for controlling processes and operations or monitoring work among employees?	Only some of the operational activities have changed Process control mechanisms have been introduced
	Has the organization increased the flexibility of its business activities?	Process control mechanisms have been introduced Increased latency tolerance
	Has the number of orders or customers changed?	The number of orders/customers decreased significantly
INVESTMENTS	Has the coronavirus pandemic affected the level or amount of your organization's investments?	Investments in new tools for remote control and assessment of operations Most often, investments were made in increasing the competences of employees The number of investments has been limited
	Indicate which sectors your organization has invested the most in?	Most often, investments were made in increasing the competences of employees
	Has the organization invested in software/tools to remotely execute or control operations?	Investments in new tools for remote control and assessment of operations.
GENERAL	Please tick which age group you belong to.	
	Please select a gender.	
	Please indicate the sector in which your organization operates.	

3.2.4 Research Trial

Ultimately, the research sample contains 25 sets of answers from 13 different sectors, with 2 different types of activity. All surveyed enterprises operate as JIT organizations or have elements of this methodology. Which shows how universal its use is among the study group. More than 50% of the respondents account for only 4 sectors (finance, chemistry, plastics processing and automotive), therefore the results of the analysis are to a limited extent dependent on the specificity of the industries with the greatest impact on the answers to the survey. The nature of the surveyed enterprises should also be taken into account, which for the purposes of the survey distinguishes manufacturing companies (52%) and service companies (48%). Such a division allows for a more accurate understanding of which elements of the Just-In-Time methodology are used by the surveyed enterprises. For statistical purposes, the respondents' age and gender were also examined. There are no dominant gender groups among the types of activity in the received data, but there is a noticeable lack of representatives of the 24–40 age group, which leads to the limitation of the survey to the positions of junior specialists and lower-level positions.

4 Results

The analysis and verification of the survey results was carried out based on the adopted research method. Each hypothesis was analyzed in the context of the results of the survey questions correlated with it. The interpretation of the gathered answers and the summary of the results led to the determination of the correctness of the initial hypothesis.

Before presenting recommendations for Just-In-Time enterprises, it is required to present a uniform picture of the situation for each of the analyzed sectors. Summing up the research carried out as part of:

a) **Employee sector:**
   Despite the general improvement of working conditions in the opinion of employees (96% of employees report changes in the way they work and 59% report improvements in working conditions – Fig. 1), the level of rotation and willingness to change jobs is higher compared to the period before 2020 (72% of employees since of the coronavirus pandemic considered changing jobs, while 60% of respondents are able to accept a new job offer if it is competitive). Of course it is also influenced by more job offers and the growth rate of employees' salaries in a given company, employees noticed that the salary grows faster when changing jobs. The greatest improvement in working conditions is visible in the flexibility of employment (72% of responses) and working time (64% of responses) - which is directly related to the introduction of the remote working mode. For employees at work, the most important is the ability to perform tasks in remote working conditions (77% of responses), employment flexibility (65.5%), employment conditions (65.5%) and "work-life" balance (53.8%). In general most of the respondents declared improvement in their working conditions (Fig. 2).

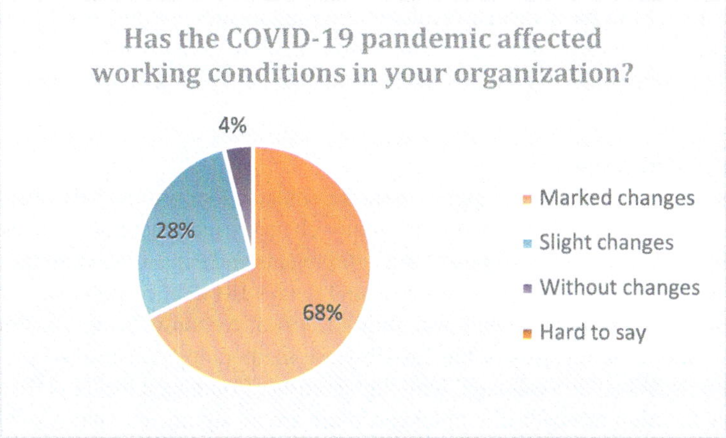

Fig. 1. Survey results based on question #1.

Unfortunately, employees have lower job satisfaction due to the deterioration of the atmosphere between employees (declared by 80% of respondents), resulting from the reduction of direct contact and integration opportunities, and the number of duties increased for 40% of respondents. For this reason, enterprises are exposed to increased costs associated with high turnover, loss of expertise and a decrease in quality due to the loss of valuable employees, which in the long run lengthens the process.

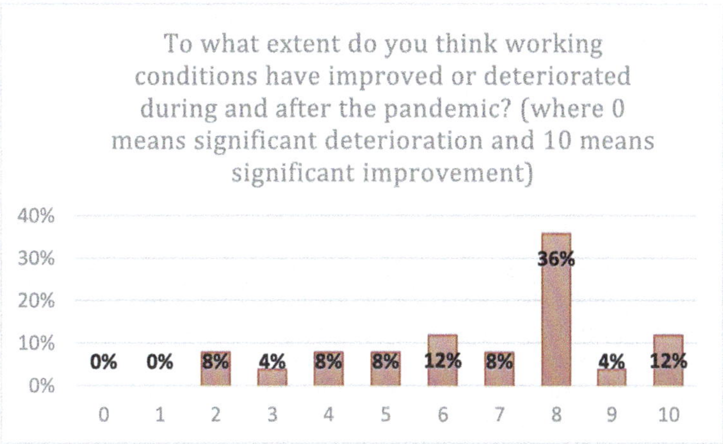

Fig. 2. Survey results based on question #2.

The labor market signals favoring the employee and his needs. Those entrepreneurs who do not adapt to the prevailing situation will experience problems with recruiting new ones. In the current situation, the level of willingness of employees to change jobs, if only a competitive offer appeared, is worrying. In addition, a correlation was found between changes in the financial results of the organization and the level of readiness of employees to change jobs - i.e. organizations that are doing worse as a result of the new situation and problems are less likely to change their jobs than those that are doing better.

b) **Supply chain sector:**
The issue of problems with suppliers concerns most organizations (77% of enterprises have experienced problems with the supply chain), which increases process unpredictability and expenditures on storing components and expanding warehouse space. The most problems concern timeliness and delays (81% of responses), component prices (65% of responses) and availability (54% of responses) (Fig. 3). Enterprises tend to adopt the features of the Just-In-Case methodology, instead of just-in-time operations. Which is associated with higher costs of running a business. In addition, despite the high turnover of employees, there are no significant changes in the supplier base, which means that there is a lack of search for new delivery sources, but it is expected to solve the problems with current ones.

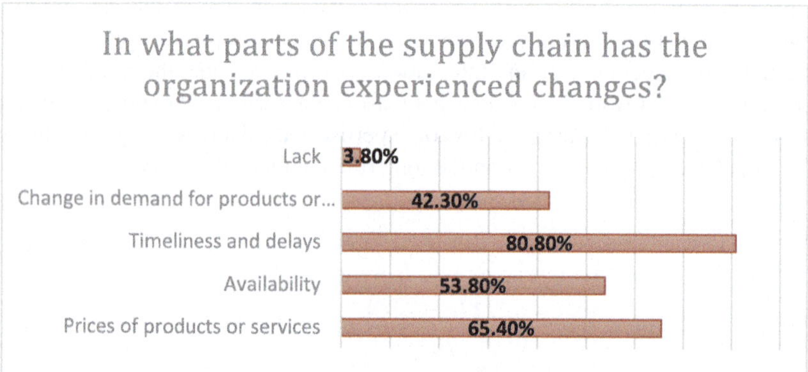

Fig. 3. Survey results based on question #10.

Despite the described problems, organizations do not experience negative or positive changes in financial results (70% of responses), which may mean that the prices of their products have increased to an average degree. After all, most organizations have changed the way they operate. The actions undertaken by companies mainly focus on preventing shortages and reducing the financial impact of delays, but do not focus on solving the problem. As the main methods of combating the problems, the creation of greater stock reserves, prioritizing more profitable products and adjusting the pace of production to the supply of components was indicated as the main methods.

c) **Sector of operation:**
As much as 85% of organizations have changed the way they operate. The main changes include remote work, digitization of processes, work with clients and contact with suppliers, increased control of warehouses and minimum stocks, and finally

the way employees are treated. Only 23% of companies have introduced online process monitoring tools, 43% of organizations have introduced new process control mechanisms, and 45% allow longer delays in the delivery of products or services. Around 60% of enterprises report no change in the size of their customer base or the number of orders. It is stated that changes in the way of operating are not directly related to the realized profits or losses. It has been confirmed that the organization's profits continue to increase with the growing number of customers.

d) **Sector of investment:**
First of all, the respondents declare no significant changes in the number of investments (approx. 73% of the respondents). The main investment motives are the digitization of all processes, tools for remote work (50% of organizations use new remote tools) (Fig. 4), greater control of warehouses and in a few cases "nearshoring" (searching for closer sources of components). The largest investment sector is operations (70% of responses), which, compared to other sectors, are an initiative undertaken at least 25% more often. 43% of organizations declare investments in employees and 35% of the answers concern investments in the supply chain.

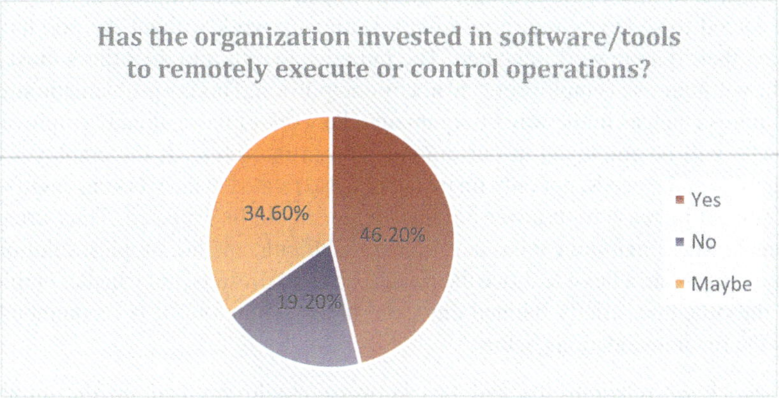

Fig. 4. Survey results based on question #20.

This can lead to a decrease in quality and control over business continuity due to unsustainable development. In the development activities undertaken, there are again declarations of reducing the side effects of problems, instead of looking for potential solutions. With the current level of demand maintained and no changes in supply, the situation of shortages and delays will remain as a long-term effect. There is a trend that those of the companies that invest more experience greater increases in financial results above the forecasted trend line. Based on the above, the next chapter will present recommendations and best solutions for JIT enterprises. A condensed situational picture will make it easier to combine data into a shortened database of comprehensive and strategic solutions, the application of which will be universal and can be used in most enterprises.

Referring to the defined problem, the created recommendations should answer the previously set question:

"What actions should be taken to prevent or improve sector of companies operations after the corona virus pandemic??"

Taking into account the synthesis from the previous chapter, a list of recommendations was created, which can be shortened to three leitmotifs.

For the first theme, "Understanding the needs and bonding of employees", it was noted that while the vast majority of employees believe that working conditions have improved, unfortunately, turnover and willingness to change jobs have continued to increase. Why is this happening? Because the basic needs of contact and integration with other co-workers are neglected, which employers did not consider as a problem during the pandemic. As it turns out, interpersonal contact is very important, although a large number of employees do not admit to wanting to return to the office, due to the fear of losing the acquired attributes, such as remote work or flexible working hours. Returning to the office at certain times to increase ties and contact between employees would entail the loss of the basic indicated advantages of working conditions after the outbreak of the coronavirus pandemic. Therefore, from the point of view of employers, it is important to find a golden mean, dictated by the needs of employees - they should not be forced to suddenly return to offices, stationary work without the possibility of adjusting their remote work schedule. The initiative to return to the offices must result from the willingness of employees, and not by compulsion. This is a problematic situation as it requires changes in the way of organizing the work of those already employed and those who will be employed in our organization in the future. If the employer is not willing to meet the needs, not only financial, but also related to the working environment - a significant increase in turnover inorganization should be expected. Thus, creating a sustainable work environment has never been so difficult, and the simplest solution tells employers to create a flexible office that focuses on work results, not whether employees are in the office on strictly defined days and hours. Therefore, it is recommended to follow the recommendations below:

- Create closer relationships and ties between employees and on the employee-employer line;
- Regular recognition of employees' needs through periodic interviews and understanding the reasons for resignation of the remaining ones through, for example, resignation interviews;
- The current environment is a market for meeting the needs of employees, so it is necessary to adapt to their requirements - it should be understood that employees no longer perceive the work environment as a relationship "I do work for which I receive a salary". Many other factors are indexed higher than the salary received.

Then, to address current supply chain issues and ever-changing operations, "Seek and experiment with unusual solutions." This means that we should find the real source of the problem and then fight it. Delays in deliveries do not mean that production should slow down - what can we do to prevent shortages on production lines? JIT companies rely on a small group of suppliers to such an extent that when there is the slightest delay in one of them, then the end customer is reliant on not receiving the product or service in the expected time. Unfortunately, there has been a shift from Just-In-Time to

Just-In-Case methodology, so the current work environment needs to be dealt with in a different way. If the current situation and the approach to accepting delays remain, the long-term problems of delays and lack of availability will not be solved - we will only minimize the financial impact of the current situation, not developing as organizations. Therefore, the following are suggested:

- Searching for the root problem and understanding what we as a company can do to fight it.
- Experimentation with new, previously unused solutions, in order to eliminate problems, i.e. expanding the base with emergency/additional suppliers, redesigning products so that it is possible to use several different materials from as many sources as possible (not limited to only one supply structure) or the universal construction of production lines, so that one product can be produced by 100% of employees or only 20% depending on market demand.

The conclusion is clear - let's look for long-term solutions, and not minimize the short-term effects of emerging problems.

However, pointing out that employers must work on the "sustainable development of industry 4.0", it is not obvious to develop all sectors of activity at a uniform pace. If activities - especially for JIT production companies - do not concern the development of employees and the supply chain, at the same pace as we focus on digitization and automation. This creates a new group of potential problems and production downtimes related to the lack of competence to solve them - which exposes us to increased production costs and line suspension. Therefore, it is recommended:

- To continue the digitization and automation of business activities to the highest possible extent in order to anticipate the demand and supply for the organization's products or services;
- To ensure the sustainable development of all sectors of the company, as extensive investment in only one of them leads to the creation of potential problems, the solution of which will be associated with higher costs in the future;
- To undertake more investments, however, monitor their progress on an ongoing basis, in order to increase the effects of work from the invested capital - i.e. ongoing monitoring and increasing CAPEX and OPEX investments.

The list of recommendations compiled in this chapter constitutes collective development directions for JIT organizations - it should be remembered that some of them have already introduced new solutions in these sectors and their application requires a thorough understanding of the work environment. It is important to prevent the duplication of strategic activities and to consider whether they will bring the expected results, because investment in the employee sector in highly automated organizations may not be as effective as increasing expenditure on automation and digitization. Therefore, the recommendations are a strategic picture of the main problems faced by organizations based on a research group of 25 companies.

5 Conclusion

The research has in-depth presented and analyzed the changes in the decision-making processes of Just-In-Time enterprises by answering the question:

"What changes has COVID-19 caused in organizations since its occurrence?"

And also investigating following theme:

"What actions should be taken to prevent or improve companies' sector of operation after the corona virus pandemic?"

The adopted investigation resulted in a full situational picture for a wide range of enterprises, limited to changes in enterprises in more than ten sectors. The approach was adopted due to the repeated pattern of problems in as many as 25 companies. The results of the analysis and recommendations created, constitute a complete list of actions required to achieve better results in the future and how to prevent the main problems that exist. The findings create a map of the steps required to verify intra-organizational problems and potential solutions at the strategic level.

The general conclusion for the analysis is that all of the enterprises, as a result of the changes resulting from the coronavirus pandemic, focused on minimizing losses in their operations and financial results. While forgetting about the sustainable development of the organization - which was reflected in the neglect of the employee sector and the supply chain. None of the companies showed a forward-looking approach in the design of smart or universal products, or in the context of new solutions/customer approach services. Which can only have prolonging effects in the form of further price increases and availability problems, if the same level of demand among consumers is maintained, creating a dangerous environment for JIT companies, highly focused on the smoothness of operations and orders.

The result of the research is an analytical list of recommendations, which maintains a general and sufficiently detailed level to be applicable to the widest possible range of enterprises. The presented analysis is a complete process in terms of the strategic level of operation and decision-making - it shows a uniform picture of how decision-making processes have changed and should change in the future. Entrepreneurs, based on this analysis, should consider their business activities in the context of their long-term, not guided by short-term profit or lack of losses, the effects of which, unfortunately, are noticeable. Finally, within the organization, a question should be asked:

"Has our company really become more resilient and flexible in the long term as a result of the adopted changes, or maybe only in a limited and short period of time?"

References

1. Pinto, J.L.Q., Matias, J.C.O., Pimentel, C., et al.: Just in time factory : implementation through lean manufacturing tools 119
2. Morris, J., Wilkinson, B.: Special issue of journal of management studies the transfer of Japanese management to alien institutional environments. J. Manag. Stud. **32**, 719–730 (1995). https://doi.org/10.1111/J.1467-6486.1995.TB00148.X
3. Hu, B., Guo, H., Zhou, P., Shi, Z.L.: Characteristics of SARS-CoV-2 and COVID-19. Nat. Rev. Microbiol. **19**, 141–154 (2021). https://doi.org/10.1038/s41579-020-00459-7

4. Khalaf, A.T., Wei, Y., Wan, J., et al.: How did the pandemic affect our perception of sustainability? enlightening the major positive impact on health and the environment. Sustain **15**, 892 (2023).https://doi.org/10.3390/SU15020892
5. Mondal, S., Baldwin, R.E., Evenett, S.J., (eds.), Covid-19 and trade policy: why turning inward won't work. J. Asian Econ. Integr. **3**, 267–270 (2021). https://doi.org/10.1177/26316846211029373/ASSET/26316846211029373.FP.PNG_V03
6. Just in time – Encyklopedia Zarządzania. https://mfiles.pl/pl/index.php/Just_in_time. Accessed 25 May 2023
7. How COVID-19 is reshaping supply chains|McKinsey. https://www.mckinsey.com/capabilities/operations/our-insights/how-covid-19-is-reshaping-supply-chains. Accessed 25 May 2023
8. Taghipour, A., Hoang, P., Cao, X.: Just in time/lean purchasing approach: an investigation for research and applications. J. Adv. Manag. Sci. 43–48 (2019). https://doi.org/10.18178/JOAMS.8.2.43-48
9. Medyński, D., Bonarski, P., Motyka, P., et al.: Digital standardization of lean manufacturing tools according to industry 4.0 concept. Appl. Sci. **13**, 6259 (2023).https://doi.org/10.3390/APP13106259
10. The Impact of COVID-19 on Small Business: Comarch Survey Results|COMARCH SA. https://www.comarch.com/trade-and-services/ict/news/the-impact-of-covid-19-on-small-business-comarch-survey-results/. Accessed 25 May 2023
11. Just in time (JIT) | Business|tutor2u. https://www.tutor2u.net/business/reference/just-in-time-jit. Accessed 25 May 2023
12. What You Need To Know About The Job Market In 2022. https://www.forbes.com/sites/nextavenue/2022/02/25/what-you-need-to-know-about-the-job-market-in-2022/?sh=185a62117bf7. Accessed 25 May 2023
13. Marcus JS COVID-19 and the shift to remote work
14. Brakman, S., Garretsen, H., van Witteloostuijn, A.: The turn from just-in-time to just-in-case globalization in and after times of COVID-19: an essay on the risk re-appraisal of borders and buffers. Soc. Sci. Humanit. Open **2**, 100034 (2020). https://doi.org/10.1016/J.SSAHO.2020.100034
15. Mondal, S., Baldwin, R.E., Evenett, S.J., (eds.), Covid-19 and trade policy: why turning inward won't work **3**, 267–270. https://doi.org/101177/26316846211029373
16. Bown, C.P.: COVID-19: Demand spikes, export restrictions, and quality concerns imperil poor country access to medical supplies (2020)
17. Yao, X.: COVID-19 pandemic and economic stimulus policies: evidence from 156 economies. In: Mobility and Inequality Trends, vol. 30, pp. 243–266 (2023). https://doi.org/10.1108/S1049-258520230000030011
18. Risk, resilience, and rebalancing in global value chains|McKinsey. https://www.mckinsey.com/capabilities/operations/our-insights/risk-resilience-and-rebalancing-in-global-value-chains. Accessed 25 May 2023

Digital Twins of the Manufacturing System Based on a Simulation Model with Bi-directional Process Data Exchange

Damian Krenczyk(✉)

Faculty of Mechanical Engineering, Silesian University of Technology, Gliwice, Poland
damian.krenczyk@polsl.pl

Abstract. Effective implementation of the Industry 4.0 concept requires searching for effective methods of creating digital twins - digital models connected and able to communicate in real-time, directly with mapped real objects. These methods must also take into account the need to obtain reliable and up-to-date data from the facilities of the generation system and its individual elements. The paper presents an architecture proposal for a solution of a cyber-physical system integrating the discrete events simulation model with the control system, An example of the practical implementation of the concept of a digital twin of a production system based on a dynamic digital simulation model is shown. Thanks to the possibility of mapping the current state of the production system object through a data access platform that collects, analyzes, processes and stores process data and the digital model connected to it using communication protocols, it is possible to compare and analyze the behavior of a system with nominal parameters. Mechanisms of automatic updating of the state of objects through the model-production system interfaces were used, using the built-in so-called emulation tools of industrial automation components, ensuring support for the OPC UA, OPC DA and Modbus communication protocols.

Keywords: digital twin · discrete-event simulation · cyber-physical system · simulation model

1 Introduction

The effective implementation of the Industry 4.0 concept requires taking action to enable the implementation of new methods of production and organization of manufacturing enterprises. Achieving a high level of flexibility and reconfigurability of production systems, as well as their autonomy, requires the development of the next generation of solutions that are a combination of advanced information and production technologies, based on cyber-physical systems, in which one of the basic elements of the cyber layer is the so-called digital twin of the production system [1–5].

In this context, the evolution of the approach to the concept of a smart factory has recently been observed. It is related to the transition from knowledge-based intelligent manufacturing to data-driven and knowledge-enabled smart manufacturing, where the

A. Burduk et al. (Eds.): ISPEM 2023, LNNS 790, pp. 548–559, 2023.
https://doi.org/10.1007/978-3-031-45021-1_41

term "smart" refers to the creation, acquisition and use of data that uses both advanced information and communication technologies and advanced data analytics. As a result, future production systems and their management methods are to be supported by real-time data exchange and analysis technologies along with simulation and optimization based on digital models [3, 4, 6].

Therefore, it becomes necessary to search for effective methods of creating digital twins - digital models connected and able to communicate in real-time, directly with mapped real objects. These methods must also take into account the need to obtain reliable and up-to-date data from the facilities of the manufacturing system and its individual components. This is particularly important and is a current challenge faced by researchers and designers of production and information systems. Research in this area is supported by the development of key technologies closely related to the Industry 4.0 concept, such as the Internet of Things (IoT), cloud computing, embedded systems, big data analytics (BDA) and information and communications technology (ICT) [2, 3, 7]. The concept of Industry 4.0 and the method of implementing the cited techniques is also the subject of many development works, implementations and reports relating to manufacturing companies. The report [8] on the results of surveys on experiences in implementing digitization and the concept of Industry 4.0 among manufacturing companies from 26 countries around the world provides a portrait of the so-called Digital Champions, i.e. companies that have introduced digital transformation to the highest degree. It presents, among others an example of digital transformation for manufacturers from the automotive industry. They use digital twins of products and manufacturing systems as a basis for data capture and real-time operational evaluation. However, the results of analyzes and decisions in feedback loops should be sent back to the appropriate physical systems via the network (Fig. 1). The results of analyzes and simulations will allow for quick adaptation to the current conditions and adjustment of the configuration, first in a virtual environment, and then reconfiguration and adaptation of production resources.

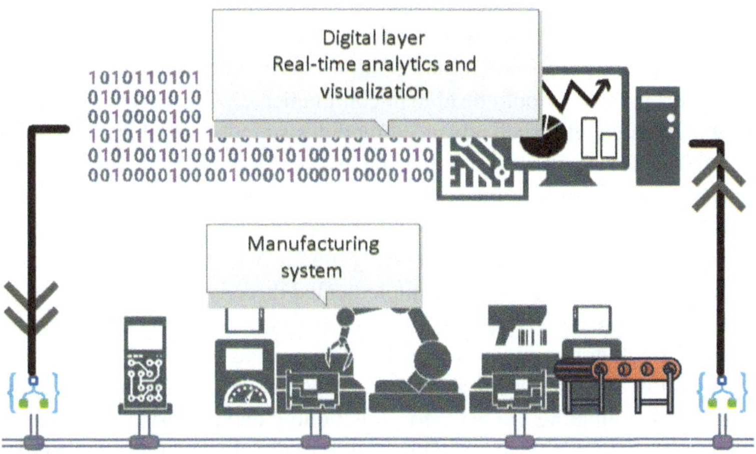

Fig. 1. Cyber-Physical System in the concept of Industry 4.0.

2 The Digital Twin as the Central Element of a Cyber-Physical System

The process of digital twins creation, which are the central element of a cyber-physical system, can be supported with the use of discrete-event simulation models [9–12]. Computer simulation, and in particular the advanced Discrete Event Simulation - DES software, has been developed for many years and, thanks to the ability to reflect almost every aspect of the process, successfully used to support decision-making at various levels and in various areas of production management, e.g. in the analysis of the functioning of production systems, identification of limitations, verification of new and improved concepts or their optimization. The development of modeling and simulation technologies and their application areas can be characterized by Rosen, Wichert, Lo et al. [12], four main phases of development:

- Individual application of simulation models (Since the 1960s) – simulation limited to very specific problems and areas;
- Simulation tools (Since the 1985s) - simulation as a standard tool for answering specific design and engineering questions;
- Simulation-based systems design (Since the 2000s) – simulation allows a systemic approach to multi-level and multi-disciplinary systems with an extended range of applications;
- The concept of a digital twin (Since the 2015s) - simulation as the core functionality of systems thanks to direct support throughout the entire life cycle.

With regard to the last of the above-mentioned phases of development, the next phase can now be distinguished, related to the use of computer simulation models as the central element of the cyber-physical system. In this context, postulated in [12], the need to connect the model directly with operational data must be complemented by two-way communication, allowing the transition from a digital model, through a digital shadow, to a digital twin. Data flow, in this case, should be possible in both directions between existing physical and digital objects, so that a change in the state of a physical/digital object directly leads to a change in the corresponding object [13, 14]. In addition, a simulation model as a digital twin, in order to be useful and included in the decision support process, should be able to communicate with analytical modules or be directly supplemented with such advanced modules supporting optimization processes, e.g. thanks to the use of artificial intelligence and machine learning methods [4, 5, 14].

In this paper, an example of the architecture and practical implementation of the digital twin of the production cell, which is a living lab for testing industry 4.0 technology, is shown. The described approach in creating digital twins is based on the use of discrete computer simulation models and implemented directly in the simulation software mechanisms for automatically updating the state of system objects (the so-called emulation of industrial automation components). The created model is a proof-of-concept for the application of the concept and methodology of creating dynamic digital models, mapping and simulating the behavior, functioning and performance of the physical counterpart, adapting to the changes taking place in the objects of the analyzed system.

Section 3 briefly discusses the proposed methodology and shows the schemes for data exchange and simulation model creation, which are an extension of the author's concept of simulation model generators in applications of creating cyber-physical systems. Section 4 is devoted to the presentation of the production cell being the subject of modeling, the simulation model in the FlexSim software, together with the created communication modules. Finally, Sect. 5 concludes the research with areas and directions for future research work.

3 Methodology for Creating a Digital Twin with Direct Access to Operational Data

The practical implementation of the digital twin of the production system was made using the developed proprietary method of semi-automatic generating and automatic updating of simulation models using data mapping and transformation methods [10, 14].

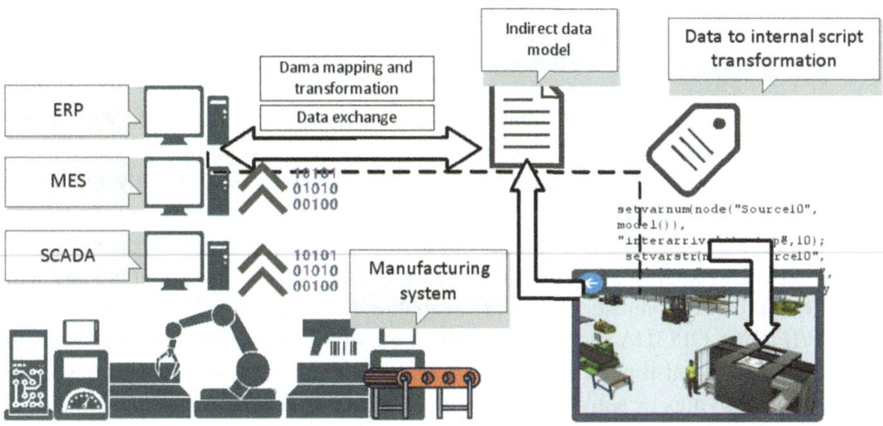

Fig. 2. The digital twin concept - the use of indirect data exchange [10].

The processes of automatic generation and updating of the simulation model usually require the preparation of data downloaded from the information systems. Therefore, their prior acquisition is required and transfer to one of the source systems (SCADA/MES/ERP), using a neutral data model. Integration of the simulation model through data exchange interfaces with information systems allows for direct transformation of data into code in the internal programming language of the simulation system or parameterization of objects directly in the model. A simplified diagram showing the idea of this method is shown in Fig. 2. Such generators, although creating virtual (digital) representations of real objects and systems, have so far been mainly tools used in the process of supporting decisions on newly designed investments or identifying limitations and searching for ways to improve already existing. Thus, they did not meet one of the requirements for simulation models used to create digital twins, which is a constantly updated representation of real objects through data exchange and real-time processing.

For this purpose, a model-control system architecture was proposed with the use of the so-called emulation of industrial automation components. The created model, or rather the individual objects that makeup it, are directly connected to external sensors and PLCs or communication clients/servers through special modules. The architecture of such a solution is shown in Fig. 3 [10].

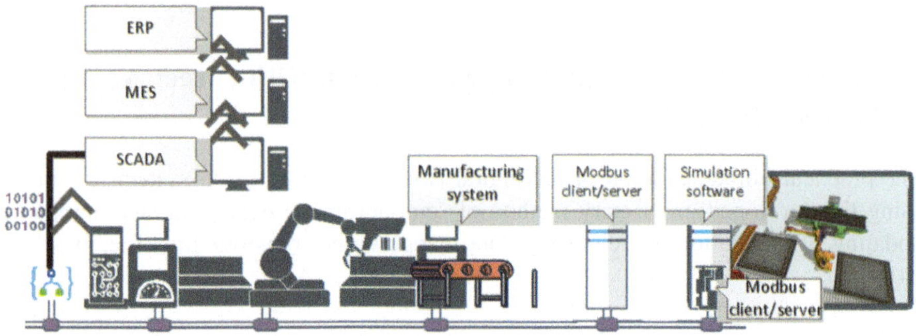

Fig. 3. Digital twin concept - direct data exchange with sensors and controllers.

Thanks to the direct implementation of interfaces for data exchange with industrial networks in simulation software (e.g. OPC, Modbus) and emulation modules for automation devices, data on the states of real system objects are sent directly and update the parameters of objects in the simulation model. This functionality is currently offered by the Emulation tool of the commercial simulation software FlexSim. It creates a connection between FlexSim and external PLCs or clients/servers that communicate with PLCs. The following connection types are available: Allen-Bradley, Modbus RTU, Modbus TCP/IPv4(6), OPC DA and OPC UA. Thanks to this, it is possible to synchronize the state of objects in real-time. This gives the opportunity to test and verify proposals for optimization activities or activities related to adapting to changing external conditions (e.g. changes in demand). In this case, it is possible to emulate real controllers or sensors and verify the possible effects, even before their implementation or connection to a real system. In addition, the digital twin is able to send data to the control system or partially take over the role of some elements of the control system, e.g. in order to verify the designated control procedures or adapt the production system to new conditions.

The presented concept, and the practical implementation of a digital twin based on it, shown in the following sections, is an extension and combination of the concept of automatic and semi-automatic model generators and data access methods in the process of creating and "operating" simulation models. Their overview and a broader description can be found in [10] or [15]. The presented solution also meets the current needs of ensuring a high level of flexibility and automatic reconfiguration of systems in smart factories, by ensuring the possibility of autonomous adaptation to changes in the production environment and its surroundings.

4 Duplicate Data Access Model in Creating a Digital Twin

The digital twin developed as part of the research is defined as a dynamic digital simulation model that maps and simulates the behavior, functioning and performance of a physical counterpart based on process data, also in real-time. Therefore, it must be integrated with both the platform for access to planning data or normative process parameters relating to all subsystems, as well as interfaces for access to data and signals of automation components, allowing for the mapping of the current state of facilities and the stage of advancement of the production process. The diagram of the proposed duplicate data access model is shown in Fig. 4.

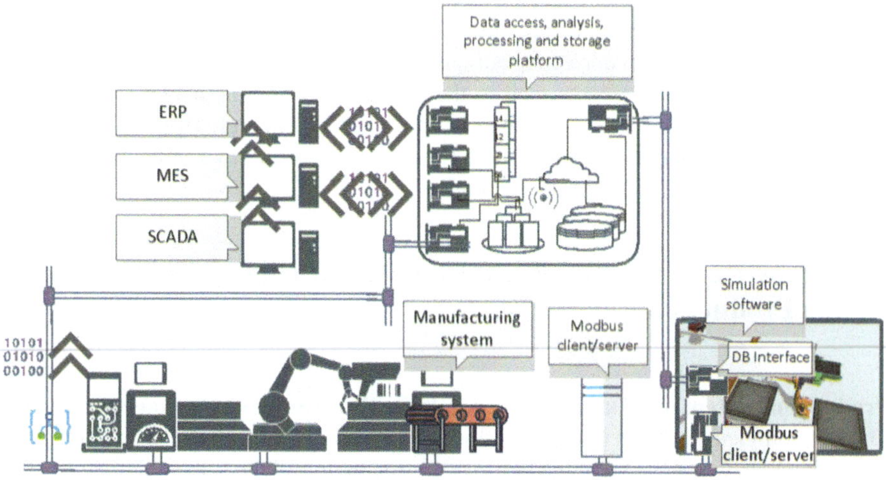

Fig. 4. The architecture of a bi-directional cyber-physical system.

The ability to create and implement a simulation model that meets the above assumptions and acts as a digital twin in a complete cyber-physical environment obviously requires access to other components of this environment. Production and logistics systems constituting the "physical" layer must meet the requirements for an appropriate level of automation, in which the technical layers, information systems and management systems are integrated. Individual actuators together with devices of modern industrial automation should be equipped with communication and identification technologies that allow for independent communication with each other, combining production, logistics, storage and distribution in a complete value chain (network of interconnections).

The flow of information should be possible both horizontally (horizontal integration - along the value chain) and vertically (vertical integration - from the operational level to the management level). In a word, it should meet the requirements of the Industry 4.0 concept by implementing IoT elements, wireless sensors (WSN), embedded systems and edge computing modules. The next component is a data access, analysis, processing and storage platform with which the simulation model could communicate. It allows access to historical data and current operational process data, which, thanks to integrated stream processing and data aggregation solutions, can supply the model with pre-processed data. For this to be possible, the platform should provide e.g. advanced ITC technologies, allowing to combine individual technical layers.

In a system that meets the above requirements, the simulation model (or models) thanks to duplicated access to data will be able to map both the current, nominal behavior of the production system and the real picture of changes in its states. It should be noted that the nominal "twin" should also be viewed as a dynamic model. This is due to the fact that, especially in multi-assortment production systems, planning decisions are also made in real-time, and "nominal" parameters or device settings often change during the operation of the production system. Comparison of indicators and results of simulations and analyses obtained for both models described above will allow identifying in real time both areas and values of observed deviations. On the other hand, the feedback sent to the management systems and analytical modules will allow for an immediate reaction and decision to implement corrective actions or reconfigure the system in order to adapt the way the processes are performed to the new conditions. The above activities and decisions, thanks to the use of machine learning techniques and artificial intelligence methods, can be carried out autonomously in the discussed environment, within the limits of specific system operating parameters.

5 A Practical Example of Implementing a Digital Twin in a Production Environment

Production System and Data Access, Analysis, Processing and Storage Platform
The technological platform that is the subject of modeling is a complete production cell - the Living Lab, which is part of the Industry 4.0 Technology Test Center (CTTP4.0) [16]. CTTP40 has been designed and manufactured by APA Group, an integrator and supplier of intelligent solutions for industrial automation and management systems, which is a strategic partner of the Center of Industry 4.0 of the Silesian University of Technology. It consists of components from dozens of leading suppliers of production line equipment, which have been connected using advanced communication technologies. The production station has been equipped with the latest Industry 4.0 technologies and, being a fully functional, automated and autonomous production cell, meets all the requirements regarding the implemented technologies presented in the previous section (Fig. 5).

The production cell under CTTP4.0 is integrated with the "NAZCA 4.0" platform provided by APA Group, which collects, consolidates and visualizes data from all machines and devices by means of communication protocols [17]. Includes cutting-edge and end-to-end applications for integrating production management systems in factories and enterprises. The platform integrates e.g. big-data technologies, cloud computing as well as advanced interfaces and data access servers through a whole range of communication protocols (including OPC, MODBUS, S7-PROTOCOL, KUKA KRL) used in industrial automation systems.

Simulation Model
In accordance with the concept and architecture of the designed digital twin described

Fig. 5. Industry 4.0 Technology Test Center – APA Group [15].

in Sect. 4, a simulation model was built to represent all subsystems and objects of the CTTP4.0 production cell. The model was made in the FlexSim software, which is an advanced 3D simulation package that provides a rich library of predefined objects with built-in handling of events and their behavior [18]. The model uses modified objects of the simulation environment so that they reflect the real geometry, and paths of movement, as well as drive elements of individual components of the station. In addition, in order to meet the requirements for mapping the current state of objects in real-time, it was necessary to replace the built-in predefined event handlers in the simulation software objects. It was also necessary to modify the procedures for controlling the flow of objects and the flow of material in the model. For this purpose, individual modules controlling the behavior of objects were prepared using the internal FlexScript programming language and the Process Flow tool. The created complete model of the CCTP4.0 production system is shown in Fig. 6.

The model has also been supplemented with mechanisms for automatically updating the state of objects. For this purpose, data exchange interfaces were designed using the built-in tool for emulation of industrial automation system components, providing support for OPC UA, OPC DA and Modbus communication protocols.

An example of mapping and connection with a selected sensor placed on a roller conveyor is shown in Fig. 7. The image of the location of the photocell on the CTTP4.0

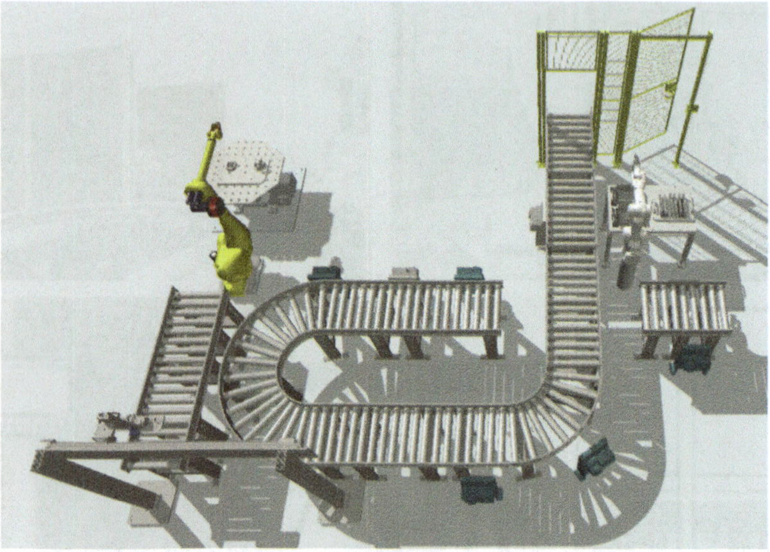

Fig. 6. Simulation model of CTTP4.0 production cell.

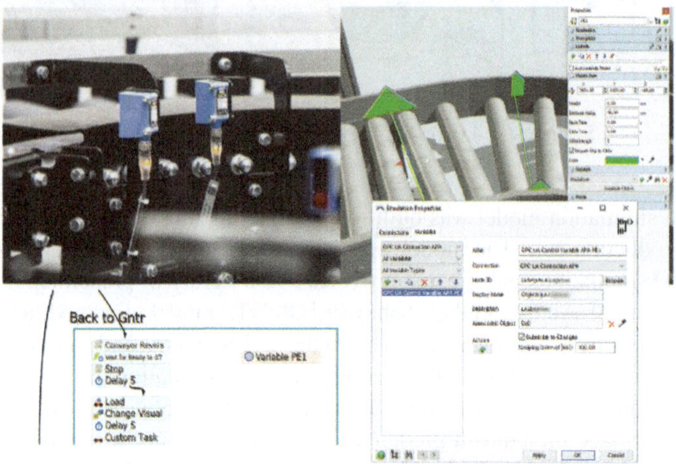

Fig. 7. Cyber-Physical System in the concept of Industry 4.0.

station is shown on the left, the middle part shows the model of the corresponding sensor in the digital twin, while the definition of the connection in the described emulation module above (in this case via the OPC UA protocol).

Thanks to the interface activation/deactivation option, the model can reflect the current real state of the object (active connection) or internally emulate the control logic at the level of the simulation system (inactive connection).

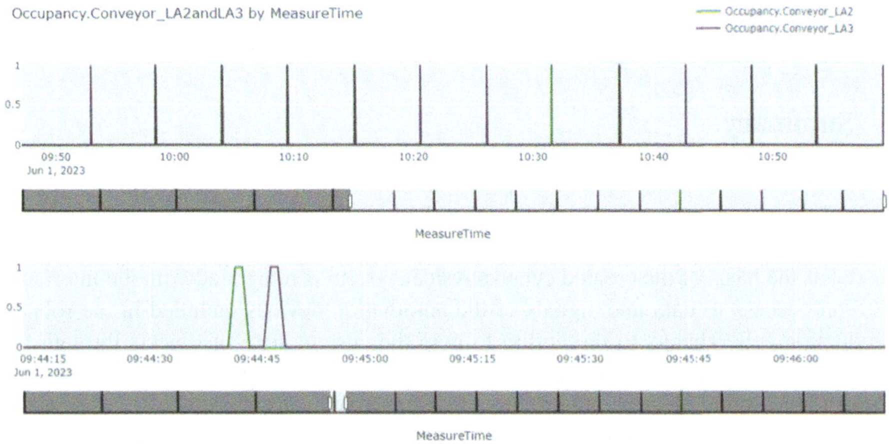

Fig. 8. States of sensors related to LA2 and LA3 roller conveyors.

Data Acquisition

The created simulation model together with the modules that allow updating the model to process data and the data analysis modules from the NAZCA platform, which make up the digital twin of the production system, make it possible to compare nominal and operational parameters in real-time. They are the basis for the created analytical modules, which, using machine learning algorithms, are to be able to identify deviations, constraints and areas of their occurrence in the production system, in accordance with the architecture of the proposed solution shown in Fig. 9.

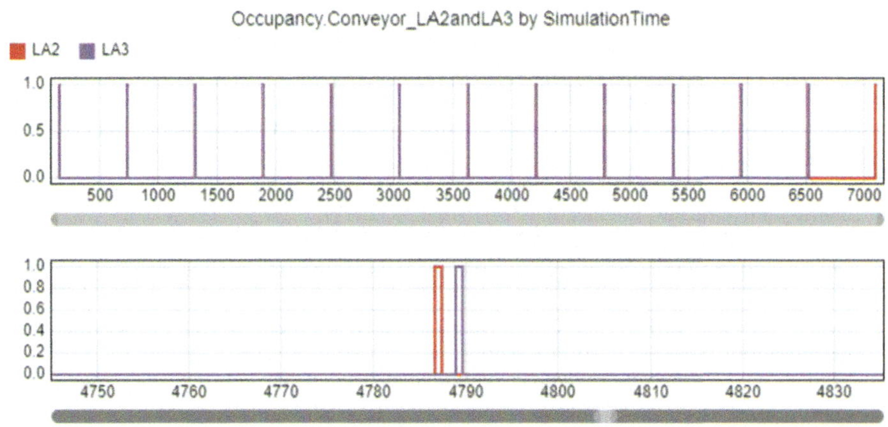

Fig. 9. Visualization of the state of the sensors obtained from the simulation model.

Figure 8 shows process data relating to the states of sensors related to roller conveyors in the system (see Figs. 5 and 6) obtained directly from the data acquisition module using data obtained from the NAZCA platform. Figure 9, in turn, shows a graph showing the states of the same conveyors obtained from a simulation model driven by the parameters of the production cell operation.

6 Summary

The paper presents an example of the practical implementation of the concept of a digital twin of the production cell, which is part of the Living Lab of Industry 4.0 Technology Test Center based on a dynamic digital simulation model. The discrete-event simulation model as the basis of the created cyber-physical system is integrated with the interfaces of direct access to data and signals of the automation devices included in the robotic production cell. Thanks to the ability to map the current state of objects through the NAZCA data access platform, which collects, analyzes, processes and stores process data and the digital model connected to it using communication protocols, it is possible to compare and analyze the behavior of the system with nominal parameters. Presented solutions are currently being tested in laboratory conditions similar to operational ones in CTTP4.0. At the current stage of the project, functional and verification tests of the simulation model and the acquired and generated data were carried out, which confirmed compliance with the real equivalent.

Based on the obtained results and conclusions, further work will be carried out related to the integration of all modules and testing the usefulness and effectiveness of selected machine learning algorithms in creating complete reasoning and analytical solutions. They are part of the designed architecture of the cyber-physical system supporting the functions of reconfiguration and self-adaptation of production systems by influencing the actuators directly from the digital twin.

References

1. Wang, S., Wan, J., Li, D., Zhang, C.: Implementing smart factory of industrie 4.0: an outlook. Int. J. Distrib. Sens. Netw. **12**(1), 3159805 (2016)
2. Chen, G., Wang, P., Feng, B., Li, Y., Liu, D.: The framework design of smart factory in discrete manufacturing industry based on cyber-physical system. Int. J. Comput. Integr. Manuf. **33**(1), 79–101 (2020)
3. Zhong, R.Y., Xu, X., Klotz, E., Newman, S.T.: Intelligent manufacturing in the context of industry 4.0: a review, Engineering **3**(5), 616–630 (2017)
4. Tao, F., Qi, Q., Wang, L., Nee, A.Y.C.: Digital twins and cyber–physical systems toward smart manufacturing and industry 4.0: correlation and comparison. Engineering **5**(4), 653–661 (2019)
5. Jwo, J.-S., Lee, C.-H., Lin, C.-S.: Data twin-driven cyber-physical factory for smart manufacturing. Sensors **22**, 2821 (2022)
6. Burduk, A., Łapczynska, D., Popiel, P.: Simulation modeling in production effectiveness improvement – case study. Manage. Product. Eng. Rev. **12**(2), 75–85 (2021)
7. Patalas-Maliszewska, J., Posdzich, M., Skrzypek, K.: Modelling information for the burnishing process in a cyber-physical production system. Int. J. Appl. Math. Comput. Sci. **32**(3), 345–354 (2022)
8. Geissbauer, R., Lübben, E., Schrauf, S., Pillsbury, S.: Global Digital Operations Study 2018 Digital Champions. How industry leaders build integrated operations ecosystems to deliver end-to-end customer solutions PwC. strategyand.pwc.com/gx/en/insights/industry4–0/global-digital-operations-study-digital-champions.pdf (2018). Accessed 05 Jan 2023

9. Flores-García, E., Kim, G.Y., Yang, J., Wiktorsson, M., Do, N.S.: Analyzing the characteristics of digital twin and discrete event simulation in cyber physical systems. IFIP Adv. Inf. Commun. Technol. **592**, 238–244 (2020)
10. Krenczyk, D.: Dynamic simulation models as digital twins of logistics systems driven by data from multiple sources. J. Phys. Conf. Ser. **2198**, 012059 (2022)
11. Ricondo, I., Porto, A., Ugarte, M.: A digital twin framework for the simulation and optimization of production systems. Procedia CIRP **104**, 762–767 (2021)
12. Rosen, R., von Wichert, G., Lo, G., Bettenhausen, K.D.: About the importance of autonomy and digital twins for the future of manufacturing. IFAC-PapersOnLine **48**(3), 567–572 (2015)
13. Kitzinger, W., Karner, M., Traar, G., Henjes, J., Sihn, W.: Digital twin in manufacturing: a categorical literature review and classification. IFAC-PapersOnLine **51**(11), 1016–1022 (2018)
14. Krenczyk, D., Paprocka, I.: Integration of discrete simulation, prediction, and optimization methods for a production line digital twin design. Materials. **16**(6), 2339 (2023)
15. Wenzel, S., Stolipin, J.,Rehof, J., Winkels ,J.: Trends in automatic composition of structures for simulation models in production and logistics. In: 2019 Winter Simulation Conference (WSC), National Harbor, MD, USA, pp. 2190–2200 (2019)
16. Industry 4.0 Technology Test Center (CTTP4.0) - APA Group. https://www.apagroup.pl/cttp40/. Accessed 01 May 2023
17. NAZCA 4.0 - end-to-end application for integrating production management systems in factories and enterprises. https://nazca40.pl/en/nazca-4-0-industry-iiot-platform-english/. Accessed 01 May 0223
18. Beaverstock, M., Greenwood, A., Nordgren, W.: Applied Simulation: Modeling and Analysis using FlexSim, Fifth Edition. FlexSim Software Products, Inc (2017)

Determination of the Scope and Parameters of Testing the Total Resistance of Military Equipment in Class N.11 to the Impact of Critical Environmental Conditions

Kamil Dworak[1], Cezary Grabowik[1](✉), Oleh Pihnastyi[2], and M. Anthony Xavior[3]

[1] Silesian University of Technology, Konarskiego 18a Street, 44100 Gliwice, Poland
cezary.grabowik@polsl.pl

[2] National Technical University, Kharkiv Polytechnic Institute, Kyrpychova 2 Street, Kharkov 61002, Ukraine

[3] Mechanical Engineering Faculty, VIT University, Vellore 632014, Tamil Nadu, India

Abstract. The article presents problems related to conducting research on the influence of various environmental factors on devices for special and military purposes. Acts regulating issues related to conducting this type of tests in the territory of the Republic of Poland, in the form of Defense Standards, were discussed. The work also contains a concise division of devices of this type into classes in accordance with applicable standards. As an example of interest in the work, a hydropneumatic suspension arm, currently at the stage of research and development related to determining the impact of environmental conditions on its operational properties, was selected. It was intended to show possible weaknesses of the current design solution of the hydropneumatic suspension arm. Considering the function performed by the suspension arm, a suspension element of a light armored tracked vehicle, this element was classified as class N.11 in accordance with the relevant standard. According to the applicable standard, class N.11 applies to ground vehicles. The paper presents the methodology and scope of tests that should be performed for vehicles of this class. In addition, a description and method of checking individual critical climatic conditions are presented. The knowledge obtained in this way allowed for the formulation of design and construction assumptions for the test stand, i.e., the climatic chamber. At the stage of defining the design and construction assumptions, the postulate to ensure the possibility of conducting a wide range of tests on the stand without the need to change its equipment was also considered.

Keywords: Environmental Conditions · Defense Standards · Hydropneumatic Suspension Arm · Climatic Chamber · Environmental Tests

1 Introduction

The military industry, due to the recent turmoil on the geopolitical scene, has developed significantly. Western countries, including Poland, want to protect their citizens against a potential military attack. The previous times of military stagnation in Europe, after

A. Burduk et al. (Eds.): ISPEM 2023, LNNS 790, pp. 560–574, 2023.
https://doi.org/10.1007/978-3-031-45021-1_42

the end of the so-called "Cold War", caused some nations to postpone defense spending to the background. Unfortunately, events in eastern Europe have led to increased unrest in the region. This is why as part of the old proverb: "If you want peace, prepare for war", states successively enlarge and modernize their military equipment. The Republic of Poland is a good example here. This country has been modernizing its troops for several years, but for the last year it has been possible to notice a significant acceleration in the pace of changes in the Polish Army. Investments in new technical solutions to replace the old post-Soviet equipment, such as the multi-task fully amphibious armored personnel carrier "Opal" or the Infantry Fighting Vehicle "Borsuk" will allow to replace the archaic designs of the BWP-1 and BWP-2 vehicles. New solutions generate tendencies to develop and improve quality issues. Thanks to various types of research on the impact of environmental factors on materials and entire components, it is possible to determine how a given element will withstand the test of time. During their operation in various terrain, environmental and climatic conditions, tracked vehicles are subjected to fast-changing dynamic loads from the interaction of the ground and vehicle components. These loads negatively affect the performance of the vehicle and significantly shorten the service life of vehicle components. For this reason, tracked vehicles are equipped with complex suspension systems that enable effective performance of tasks in hostile operating environment conditions, the battlefield. Striving to ensure adequate driving comfort and the possibility of precise and quick implementation of operational tasks is the main factor influencing the progress of development work on innovative solutions in the suspensions of modern tracked vehicles. So far, the tests of suspension elements of tracked vehicles have been carried out in a way that does not sufficiently reflect the realities of the modern theater of operations of this type of vehicle, i.e., at rest and in the absence of external forcing. The development of an innovative methodology for conducting research on the impact of critical climatic and environmental conditions, in simulated operation conditions, on the behavior of tracked vehicle suspension elements will allow for an in-depth understanding of the limits and possibilities of currently used construction solutions. Thus, it will be possible to determine the impact of individual climatic and environmental factors, with long-term exposure, on the operational properties and durability of suspension components. The knowledge and experience gained in this way will allow for the optimization and improvement of the suspension design in the context of eliminating the negative impact of climatic and environmental factors, contributing to the improvement of the quality and service life of the manufactured suspension elements.

2 Military Standards

A systematic set of legal provisions regulating aspects related to military equipment in the Republic of Poland are the Defense Standards. These standards define the requirements for the quality of equipment, materials and services used in defense of the country in products for military purposes. The main purpose of using these Defense Standards is to ensure that the tools and materials used are fully effective and safe for users. These standards cover all issues related to defense, including the subject of interest of the work, issues regarding general technical requirements related to inspection methods and research in the field of:

- reliability requirements,
- environmental requirements,
- construction requirements,
- general rules for testing prototypes and mass-produced devices.
- reliability testing methods,
- test methods for total resistance to environmental factors,
- methods of assessing compliance with construction requirements.

Table 1 contains a list of Defense Standards for military equipment, which define the requirements, and testing methods presented in the text above.

Table 1. The list of Defense Standards for military equipment, requirements

Standard No.	Standard scope
NO-06-A101:2021	Military equipment - General technical requirements, methods of inspection and testing - General provisions
NO-06-A102:2021	Military equipment - General technical requirements, methods of control and testing - Reliability requirements
NO-06-A103:2021	Military equipment - General technical requirements, methods of inspection and testing - Environmental requirements
NO-06-A104:2021	Military equipment - General technical requirements, methods of inspection and testing - Design requirements
NO-06-A105:2021	Military equipment - General technical requirements, methods of control and testing - General rules for testing prototypes and mass-produced devices
NO-06-A106:2021	Military equipment - General technical requirements, inspection, and testing methods - Reliability testing methods
NO-06-A107:2021	Military equipment - General technical requirements, control, and testing methods - Test methods for total resistance to environmental factors
NO-06-A108:2021	Military equipment - General technical requirements, methods of inspection and testing - Methods of assessing compliance with construction requirements

2.1 Division of Military Equipment Due to Climatic Conditions

The concept of military equipment is a general concept that needs to be clarified. And so, in the defense standards, there is a division of this type of equipment into two groups, where the factor determining belonging to a specific group are the climatic conditions to which the devices may be exposed in the process of operation. Accordingly, two groups are distinguished:

- O - general climate, in the case of devices operated in macroclimatic areas of the globe with a temperate, cold and tropical dry and humid climate,

- UZ - moderate (cold), in the case of devices operated in macro-climatic areas of the globe with a terrestrial cold temperate climate.

In the case of the design and construction process of military equipment intended for use only in a temperate climate area, the following input parameters should be adopted in the tactical and technical assumptions of the device:

- increased working environment temperature up to +50 °C, with the limit value of the limiting temperature parameter being +65 °C,
- reduced temperature of the working environment −30 °C, with a limit value of the limiting temperature parameter of −50 °C,
- working environment temperature variability range from −50 °C to +65 °C.

2.2 Division of Military Equipment Due to the Intended Use

Military devices, due to their intended use, are classified into 5 classes, i.e.:

- class N – ground equipment,
- class M – marine devices,
- class S – on-board aviation equipment,
- class R – on-board missile devices,
- class T – artillery ammunition devices.

In accordance with the NO-06-A103:2021 standard, each class is divided into subclasses. Considering the subject of interest of the work, i.e. the hydropneumatic suspension arm of a light tracked armored vehicle belonging to class N, Table 2 presents a detailed description of this class of (ground) devices.

3 Hydropneumatic Suspension Arm as an Example of Group N.11 Device

Tracked vehicles, in accordance with the tactical and technical assumptions adopted for them, are vehicles whose task is to ensure the possibility of effective movement in difficult terrain, such as swampy terrain, desert terrain, snow or off-road roads. One of the key components of a tracked vehicle from the point of view of the need to ensure its off-road capability is the vehicle suspension assembly/system. Suspension designs for tracked vehicles have evolved over the years, from simple suspension designs for tracked vehicles used during WWI, which used wheeled bogies with steering wheels, to advanced suspension designs that use hydro-pneumatic rockers, similarly to automotive vehicles. The main task performed by the suspension of a tracked vehicle is to ensure constant contact between the track and the ground, which allows to maintain stability and control the vehicle's trajectory. Currently, the most commonly used suspensions in tracked vehicles are pendulum, bar, semi-valve and independent suspensions. Each type of suspension has its advantages and disadvantages, which depend on the terrain and use of the vehicle. Currently, the suspensions of tracked vehicles are very advanced and often use many elements such as springs, shock absorbers and various types of pneumatic cushions to ensure the best performance and driving comfort [2].

Currently, the following design solutions are most often used in the construction of track suspensions:

Table 2. Division of N class into subclasses according to the NO-06-A103:2021 standard

Subclass of the N class	Device operating conditions
N.1	Devices intended for installation in above-ground stationary rooms and buildings
N.2	Devices intended for installation in wooden fortifications with small parts of protection against exposure, placed without suspension, and devices placed on cushioned bases (blocks) and platforms in all special fortification structures, as part of protection
N.3	Devices intended for installation in special fortification structures with an average degree of protection against exposure, without the use of cushioned bases (blocks) and platforms
N.4	Devices intended for installation in special fortification structures with an increased level of protection against exposure without the use of amortized bases (blocks) and platforms
N.5	Devices intended for installation in special underground fortification structures (launchers, command posts and others)
N.6	Devices intended for installation in temporary rooms and hiding places, transported by all types of transport, not working in motion
N.7	Devices intended for installation on ground objects (including rocket launchers) on wheeled chassis, without artillery and mortar armament
N.8	Devices intended for installation on ground objects (including rocket launchers) on tracked chassis with anti-ball armor, without artillery armament and mortars
N.9	Devices intended for installation in unarmored objects (including rocket launchers) on tracked chassis, without artillery and mortar armament
N.10	Devices intended for installation in tanks and devices built on their basis
N.11	Devices intended for installation in light tanks, infantry fighting vehicles, wheeled armored personnel carriers and objects built on their basis
N.12	Devices intended for installation on self-propelled objects with artillery (including anti-aircraft) armament and anti-ball mortars
N.13	Devices intended for installation in unarmored self-propelled and towed artillery (including anti-aircraft) and mortar systems
N.14	Devices intended for work in the open air, including portable (backpack), transported by all types of transport, operating in place and/or on the move

- swinging suspension, consists of a single point where the end of the rocker arm is attached, which moves in a circle with the center at the point of attachment. This suspension ensures good contact of the tracks with the ground. The disadvantages of this design solution include: inability to ensure adequate stabilization of the vehicle movement,

- link suspension, consists of rods connected to each other by means of rocker arms, which allows the crawler wheels to swing,
- semi-oscillating suspension is a combination of a pendulum and a link suspension. This is a good solution for vehicles with higher weight and better stability,
- independent suspension, each of the track wheels is suspended independently, which allows for better adaptation to uneven terrain. This suspension is used in vehicles with higher speed and lower weight,
- hydraulic suspension uses a hydraulic system to absorb vibrations and shocks while driving. This suspension is used in high-speed vehicles and provides high driving comfort.

A special type of independent suspension is the suspension based on hydropneumatic swinging arms. It is an advanced suspension system that combines hydraulic and pneumatic elements to ensure the best working conditions. The operation of hydropneumatic arms consists in the use of two systems: hydraulic and pneumatic. Hydropneumatic suspension is designed to provide the best driving comfort by adjusting the level of damping to road conditions. The principle of operation of the hydropneumatic suspension is to use hydraulic fluid and compressed air to maintain a constant level of the vehicle height, thus guaranteeing an even pressure of the tracks on the ground. While driving, the air pressure in the tank increases, which causes hydraulic fluid to flow from the cylinder to the tank, thus lowering the vehicle level. In turn, when braking or driving on uneven roads, the control valve is controlled by pressure sensors and causes the flow of hydraulic fluid, which results in raising the level of the vehicle. The use of hydropneumatic suspension in tracked vehicles is particularly useful in conditions where high quality traction and high stability are required, for example in the mining or construction industries. Hydropneumatic suspension also provides a much higher ride comfort than other types of suspension, which is why it is often used in passenger cars and buses.

The advantages of hydropneumatic suspension include:

- perfect adaptation to road conditions - hydropneumatic suspension automatically adjusts the level of damping to road conditions, resulting in better stability and control over the vehicle.
- high level of driving comfort - thanks to the use of compressed air and hydraulic fluid, the hydropneumatic suspension provides much better driving comfort than other types of suspension.
- less wheel wear - even distribution of track pressure on the ground results in less wheel wear.
- lower risk of damage to the vehicle - hydropneumatic suspension reduces vibrations and shocks, which results in a lower risk of damage to the vehicle (Fig. 1).

4 Environmental Factors

Due to their origin, environmental factors can be divided into:

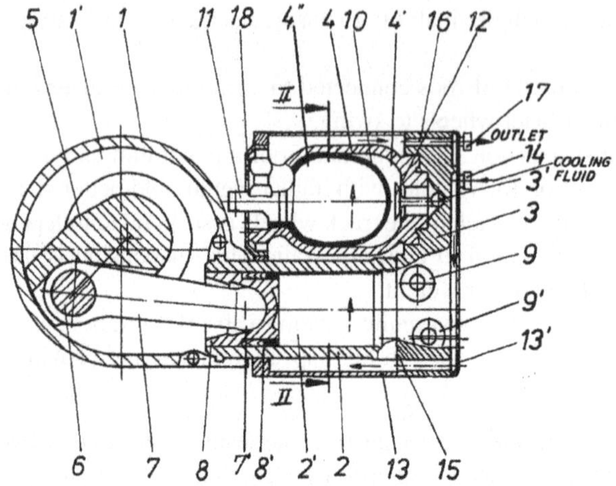

Fig. 1. Cross-section of the hydropneumatic suspension arm.

- mechanical,
- climatic,
- biological,
- other appropriate for operating conditions.

Mechanical environmental factors are a group of factors affecting the tested elements through the impact of physical forces, such as pressure, friction, vibrations, impacts, etc. The term impact is understood as sudden changes in speed that can cause dynamic effects on the tested elements. This group of environmental factors includes, among others, road accidents, falls, seismic shocks, explosions, etc.

The concept of friction is understood as the totality of physical phenomena accompanying the movement of two physical bodies relative to each other, in contact with each other - external friction or elements of the same body - internal friction and causing energy dissipation during movement. External friction occurs at the interface between two solids. Internal friction occurs in the flow of both fluids and the deformation of solids. From the point of view of operation processes, friction has a strong negative impact on the service life of the operated object, being a source of accelerated wear of machine elements, in the analyzed case of accelerated wear of vehicle suspension components and transmission systems, including caterpillar drive, tracks, pins, etc.

Vibrations are vibrating movements that can affect the functioning and life of structural elements as well as the nervous system and the human hearing system. Potential sources of vibration can be vehicle drive systems, earthquakes, and others.

The pressure consists in the impact of static forces on the structural elements of the vehicle, which may cause changes in their structure, shape and functioning. An example of this type of load can be the self-weight of the structure, pressure on support feet, pressure of liquid in tanks, and the like.

In the case of class N, i.e., terrestrial devices, the following mechanical environmental factors are tested:

- sinusoidal vibrations,
- acoustic vibrations,
- single mechanical impact,
- multiple mechanical impacts.

During the sinusoidal vibration resistance test, the tested device is subjected to the influence of sinusoidal vibrations for less than 2 min. During the test, the vibration frequency is changed smoothly in a given frequency range from the lowest value of the vibration frequency to the highest and vice versa. However, these changes are made at a speed not greater than 1 octave per minute. The range of vibration frequency for devices of class N.11 is from 5 Hz to 500 Hz, the displacement amplitude is 0.5 to 2.5 mm. In the Fig. 2 an example of the testing stand used in sinusoidal vibration resistance test is presented.

Fig. 2. The exemplary sinusoidal vibration testing stand.

The acoustic vibration resistance test is performed by placing the element in the acoustic chamber in such a way as to ensure its appropriate position in relation to the generated exposure. The required exposure values are the frequency of the acoustic wave from 50 Hz to 10000 Hz and the total sound pressure level from 100 dB to 170 dB.

During the test of resistance to single mechanical shocks, the tested element is subjected to single mechanical impact with: (i) peak acceleration 200 m/s^2, impulse duration from 5 ms to 15 ms, (ii) peak acceleration 750 m/s^2, impulse duration from 1 ms to 5 ms.

The test of resistance to multiple mechanical impacts is performed to check whether the device can resist the destructive effects of multiple mechanical impacts and whether after exposure to multiple mechanical impacts it retains its functional properties and is able to maintain the assumed operating parameters. For devices belonging to class N.11, the following test parameters are assumed: (i) impacts with a peak acceleration of 200 m/s^2 for an impulse duration of 5 ms to 15 ms, the test is performed 10 000 times, (ii) impacts with a peak acceleration of 1500 m/s^2 for pulse duration from 1ms to 5 ms, test performed 600 times.

The group of **climatic environmental factors** includes the following factors: ambient temperature, precipitation, air humidity, wind, and solar radiation.

The ambient temperature is the primary climatic factor that affects the test items. Extreme ambient temperatures may cause damage or lead to a loss of the device's ability to maintain proper operating parameters. The influence of temperature on devices depends on many factors, such as: latitude, altitude, location by the sea, etc. Precipitation, i.e., rain, hail, snow, affects the life of the tested elements and their ability to work properly. The impact of precipitation on equipment depends mainly on its intensity and frequency of precipitation.

Air humidity affects the life of the materials used in the manufacturing process of the device. It affects the ability of the systems to regulate the operating temperature. Extremely high air humidity can cause damage to the tested elements by intensifying iron oxidation processes.

In the operating environment of the device, the wind is responsible for causing air movements that enable the transport and lifting of various types of pollutants, e.g., dust, plant pollen, etc. which can negatively affect the life of the device, in particular bearing nodes, and moving surfaces of cooperating elements. Strong air blasts can lead to catastrophic hazards related to the loss of structural stability.

Exposure to solar radiation has an extremely negative effect on the durability of the test items. Extremely high exposure to solar radiation can cause damage to painted surfaces, surfaces of plastic elements and the surface of composite elements.

Class N ground equipment is tested for resistance to the following climatic environmental factors:

- Increased ambient temperature.
- Decreased ambient temperature.
- Changes in ambient temperature.
- Increased humidity.
- Precipitation (rain)
- Condensation atmospheric deposits (dew, frost, ice).
- Salt fog.

The study of the impact of increased ambient temperature is performed to check whether the device will retain its external appearance and operational efficiency after the exposure ceases. The elevated ambient temperature test consists in placing the test item in a climatic chamber in real or equivalent operating conditions. During the test, the temperature in the climatic chamber is gradually raised to the limit temperature value required by the test. After reaching the limit temperature, the tested element is kept in the chamber for a period of not less than 24 h. In the case of the elements of interest, i.e., hydropneumatic rocker arms of self-propelled tracked vehicles, equipment class N.11, the limit temperature of the test is 70 °C. In the Fig. 3 an example of the testing stand used in increased ambient temperature test is presented.

The study of the impact of decreased ambient temperature is performed to check whether a given device will retain its external appearance and operational efficiency after the exposure ceases. The test consists in placing the test item in a climatic chamber in real or equivalent operating conditions. During the test, the temperature in the climatic chamber is gradually lowered to the limit temperature value required by the test. After reaching the limit temperature, the tested element is kept in the chamber for a period of not less than 24 h. In the case of the elements of interest in the work, i.e.,

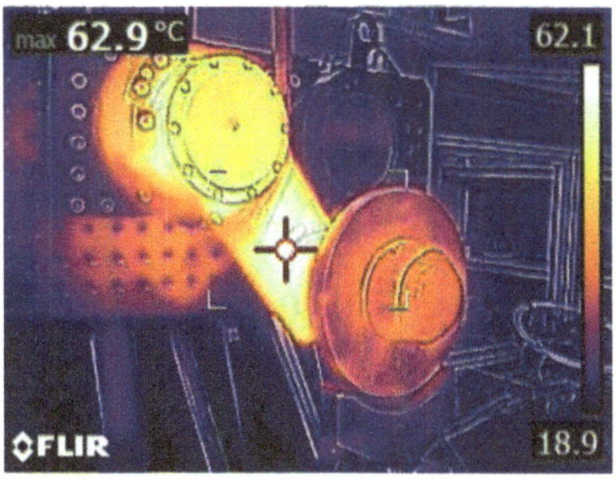

Fig. 3. The exemplary of the increased ambient temperature testing stand.

hydropneumatic rocker arms of self-propelled tracked vehicles, equipment class N.11, the limit temperature of the test is −60 °C. In the Fig. 4 an example of the testing stand used in decreased ambient temperature test is presented.

Fig. 4. The exemplary of the decreased ambient temperature testing stand.

When testing the influence of changes in ambient temperature on the behavior of the functional properties of the tested object, the temperature in the climatic chamber is subject to a rapid change between the values of limit temperatures from the upper, high limit temperature of 70 °C to the value corresponding to the lower limit temperature, i.e. −60°. The element storage time for temperatures corresponding to the temperature limit values is at least 2 h. The test is performed for the number of temperature change cycles equal to 2.

The study of the impact of increased humidity is performed to verify whether the tested object is fit for use and has not changed its external appearance during exposure. This test can be performed for 3 different test scenarios. In a considered case, the third test scenario is used, the conditions of which are specified in Fig. 5. The test is carried out for at least 5 test cycles.

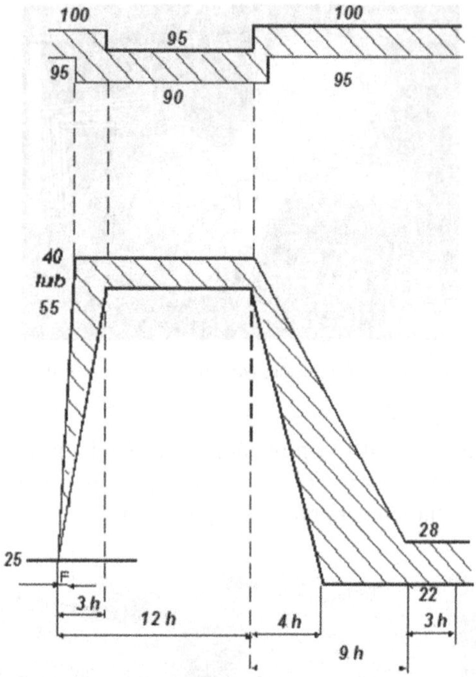

Fig. 5. Test cycle at decreased temperature and humidity

Testing the object's resistance to precipitation, rain, is aimed at assessing the impact of rainfall on the ability of the tested object to maintain its full functional properties. The test procedure in this case consists in spraying the tested object with a stream of water for at least 2 h. The water stream is directed successively from the side and from the top of the object at an angle of 40° to 45°, and the intensity of the water stream should be 5 ± 2 mm/min, respectively. The water temperature during the test should be kept between 5 and 10 °C. The duration of the test is 1.5 h, respectively.

The test of atmospheric condensation deposits (dew, frost, ice) consists in placing the tested item in the climatic chamber and then, after lowering and stabilizing the temperature to −20 °C, the device is removed from the chamber so that it is covered with frost. Then, the influence of dew, frost and ice on the behavior of the object is assessed. In the Fig. 6 frost deposited on the element tested during the test of resistance to condensation precipitation is shown.

The salt fog test consists in placing the tested element in a chamber with a temperature of 35 °C and then subjecting it to a salt solution with a concentration of 33 g of sodium chloride per 1 dm3 of distilled water. The intensity of the sprayed mist is from 2 to 3 g/m^3 for the first 15 min. The total exposure time is 48 h. This test makes it possible to assess the anti-corrosion properties of the structure, including the mechanical properties of paint coatings, etc.

Fig. 6. Frost deposited on the element tested during the test of resistance to condensation precipitation.

5 Design Assumptions of the Climatic Chamber for Testing Hydropneumatic Swinging Arms

The previous chapter presents a concise description of the tests that allow the assessment of the behavior of the structure in terms of the impact of mechanical and climatic environmental factors. The information contained in this chapter shows that it is practically impossible to develop a climatic chamber design that would allow to conduct all tests. The preparation of the concept of the construction of the climatic chamber involved the need to develop the design and construction assumptions of the chamber, while at this stage it was assumed that the construction of the chamber should be as universal as possible, and thus it should enable research on the impact of climatic factors on ground devices class N.11, in particular hydropneumatic swinging arms. Based on the literature search, the following design and construction assumptions for the climatic chamber were defined for the purposes of the research, the fulfillment of which is necessary:

- limit value of increased temperature: +70 °C,
- limit value of decreased temperature: −60 °C,
- smooth temperature control in the full range of limit temperatures,
- high resistance to corrosive factors,
- ability to generate air humidity of 98%,
- the ability to generate salt fog and spray rain with an intensity of 5 ± 2 mm/min,
- an environmentally safe system for desalination and purification of water used for testing.

Taking into account fact that currently used in the armaments industry swinging arms designs are characterized by a large variety of dimensions, it was necessary to carry out a literature study to determine the correct overall dimensions of the climatic chamber. It is assumed that the proposed design solution of the climatic chamber will allow testing of all swinging arms designs for special purposes included in the production program of individual industrial plants providing for military needs. The dimensions of the chamber must be large enough to allow testing under simulated operating conditions corresponding to field trials. Based on the analysis, it was found that the internal overall dimensions of the chamber should be as follows:

- climatic chamber height H: 1005 mm,
- climatic chamber width W: 750 mm,
- climatic chamber depth D: 1290 mm.

Fulfillment of these requirements will allow conducting tests in the field of resistance to environmental climatic factors, taking into account the suspension load profile resulting from the profile of the track on which the vehicle moves in off-road conditions, it is:

- increased ambient temperature,
- reduced ambient temperature,
- changes in ambient temperature,
- increased humidity,
- precipitation it is rain,
- condensation of atmospheric deposits, dew, frost, ice,
- salt fog.

5.1 The Concept of the Climatic Chamber Based on Design Assumptions

Based on design assumptions developed based on environmental tests described in defense standards, it was possible to create a concept of a universal test stand. In the case of the proposed design solution, it was decided to go a step further than required by the standards in question. It was assumed that the construction of the stand should ensure the possibility of conducting research on the impact of critical climatic factors during the simulation of normal suspension operation. The existing design solutions of climatic chambers, developed in accordance with the requirements of the relevant regulations and standards, assumed that the elements tested during the tests were fixed and tested only in a stationary position. Such a way of conducting tests did not allow to gain full knowledge about how a given design solution behaves in real working conditions and whether the durability criteria of the structure are met. The new approach proposed in the paper will allow for a more in-depth analysis of the test results, which in turn will allow for the introduction of the design modifications to the existing design solutions of the suspension arms, and will also contribute to the development of new, more advanced suspension systems for tracked vehicles. In the picture 7 the conceptual 3D model of the climatic chamber which meet formulated in the paper design assumptions is shown (Fig. 7).

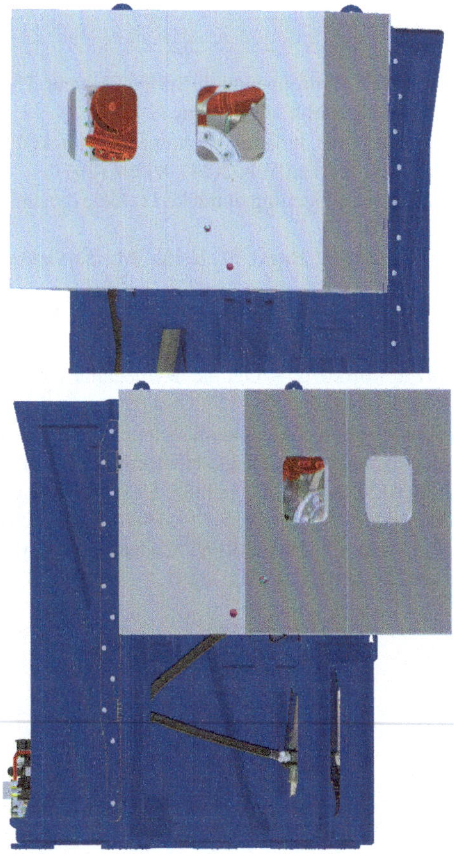

Fig. 7. Conceptual model of the test stand for testing the impact of critical climatic conditions on the suspension elements of tracked vehicles.

6 Conclusion

Defense standards are a very good source of knowledge on the type and methods of conducting climate research. Depending on the type of the tested element, its class affiliation can be specified and thus the specific quality requirements it faces. Based on the Defense Standards, on the example of hydropneumatic rocker arms as a class N.11 device, design assumptions can be set as precise guidelines that should be met in the process of designing and constructing machines that test the impact of individual conditions. The designated design assumptions were used to define the concept of the test stand in the form of a climatic chamber allowing for the maximum number of tests to be carried out. An additional advantage of the created stand is the possibility of conducting research on the impact of climatic conditions during the simulated operation of the tested element.

References

1. Hausenblas, H., Holzforster, H.: Hydropneumatic Spring Element, Thyssen Industrie Aktiengesellschaft. Essen, Federation Republic of Germany (1976)
2. Kciuk, S., Mura, G.: Modelling the dynamic phenomena of tracked vehicle suspension based on PT-91 chassis. High Speed Tracked Veh. **2**(28), 89–98 (2011)
3. Kciuk, S., Mężyk, A., Mura, G.: Modelling of tracked vehicle dynamics. J. KONES Powertrain Transp. **17**(1), 223–232 (2010)
4. Pawar, P., Bhujbal, P., Neknale, R., Bagal, S., Narhe, M.: Salt spray corrosion test chamber. JETIR **6**(2), 202–204 (2019)
5. Dobrzański, L.A., Brytan, Z., Grande, M., Rosso M.: J. Mater. Process. Technol. **192–193**(1), 443–448 (2007)
6. NO-06-A103:2021 Sprzęt wojskowy – Ogólne wymagania techniczne, metody kontroli i badań – Wymagania środowiskowe
7. NO-06-A107:2021 Military equipment - General technical requirements, control methods and tests - Methods of testing the total resistance to environmental factors (in Polish)
8. Darehshouri, S., Michelsen, N., Schüth, C., Schulz, S.: A low-cost environmental chamber to simulate warm climatic conditions. Vadose Zone J. **19**(1), 1–6 (2020)
9. Valdiserri, P., Rossi, E., Schio, D., Rondelli, V., Capacci, E.: A numerical analysis for the design of a climatic chamber. AIP Conf. Proc. **2191**, 1–7 (2019)

Implementation of Industry 4.0 Solutions in the Agricultural Field of Ukraine

Olga Tsesliv(✉) and Igor Sikorsky

Kyiv Polytechnic Institute, National Technical University of Ukraine, Beresteyskyi Avenue (Peremohy), 37, Kyiv 03056, Ukraine
cesliv.olga@lll.kpi.ua

Abstract. This article focuses on Industry 4.0 technologies implemented in the Ukrainian agriculture sector. The agricultural sector was one of the first to feel the effects of Russia's full-scale aggression in Ukraine. The Cabinet of Ministers of Ukraine approved an action plan to ensure food security under martial law. Using the example of three crops: wheat, buckwheat and peas, the paper proves the possibility of optimizing the area to implement the food security program and to ensure stable profits. To fulfill the state program, it is necessary to optimize all the lands of Ukraine, which is possible only with the use of innovative capabilities of Industry 4. For Ukraine, online services ensure civilized use of land, increase land value and market intelligence, and reduce the risk of corruption to zero. Trade with Europe is a successful direction for Ukrainian farmers. The scientific novelty of the paper is the application of the Lagrange multiplier method to solve the area optimization problem. The paper considers the case of price risk, when future market prices are not deterministic. Therefore, it is necessary to be guided by two criteria when making a decision: to maximize the expected total net income and to minimize the variance of the total net income. Food security programs are a multi-criteria problem. Further consideration should be given to the digitalization of agricultural enterprises, the use of communication innovations, ports, and many other factors that will contribute to the solution of the food security program.

Keywords: Industry 4.0 · food security · Lagrange multipliers

1 Introduction

This article analyzes and explores the potential of the National Food Security Program of Ukraine to leverage the Ukrainian agricultural sector to implement Industry 4.0.

The Russian invasion has dealt a real blow to Ukrainian agriculture. The total direct and indirect losses to agriculture exceed $40 billion, according to analysts at KSE Agrocenter and the Department of Agricultural Policy. Both crop and livestock production have suffered from the war. A large amount of farmland is still being exploited.

Although the grain export corridor is open, it can hardly be called reliable, given the unstable behavior of the Russian Federation, and it has not led to a significant increase in domestic prices for major crops.

A. Burduk et al. (Eds.): ISPEM 2023, LNNS 790, pp. 575–584, 2023.
https://doi.org/10.1007/978-3-031-45021-1_43

World recognition of the benefits offered by the concept of "Industry 4.0" makes it necessary to transition the national economy to "Industry 4.0" technologies. Satellite technologies, electronic weather stations, automatic irrigation and soil moisture control systems - as well as mobile applications and special software help Ukrainian farmers.

The basis of Industry 4.0 is composed of the following scientific and technical developments: large databases and their analysis; autonomous robots are the robots that are able to perform tasks independently without human intervention; modeling is currently used to a large extent only while designing the production and business processes, individual new production equipment or new products; horizontal and vertical system integration; the Industrial Internet of Things will connect all production components into a single network of real-time information exchange; cyber security involves taking measures related to the protection of data storage and processing facilities, data transmission networks; cloud technologies; additive (additional) production based on 3-D printing; extended (or virtual) reality will be used by employees for training, decision-making, etc. The following Table 1 summarizes information on innovative projects in the agricultural sector of Ukraine.

More and more companies offering farmers technological innovations at affordable prices began to appear on the Ukrainian market. Design bureaus in Kyiv, Zhytomyr, and Lviv are currently working on this, developing and assembling unmanned aerial vehicles for the army and farmers.

Thanks to cartographic solutions by Ukrainian developers, the following become possible:

- visualization of soil analysis dynamics;
- analysis of satellite images;
- creation of crop yield maps;
- determination of the most productive land lots;
- tracking the dynamics of culture development;
- control of the fields state;
- yield forecast, etc.

The Soil Lines startup demonstrated the technology of laser soil analysis created in Ukraine and based on micro lasers. Developed by a US company, the Climate Corporation 17, it analyzes weather and soil data and provides farmers with recommendations for increasing yields under the given conditions. Digitization ensures the minimization of the human error factor, obtaining real data and making management decisions based on them. A promising direction is the use of artificial intelligence. Software development is ongoing. Agricultural companies implement CRM (Customer Relationship Management) and BPM (Business Process Management) systems. The leading agricultural holding of Eastern Europe "Myronivskyi Hliboprodukt" cultivates one of the largest land banks in Ukraine - about 370,000 ha in 14 regions of Ukraine. Over 20 years of activity, the company has constantly improved its production, opened up to innovations and actively implemented the latest scientific and technical achievements and methods. We can consider the process of Ukrainian industry automation using the example of one of the largest agricultural holdings in the country - Ukrlandfarming. The company uses a technological system borrowed from foreign collegues, which allows them to easily

Table 1. Innovative projects in the agricultural sector of Ukraine.

Project name	Description
MHP	The company consistently implements innovative projects in various businesses. In 2017, MHP and Radar Tech technology cluster launched the country's only agriculture startup incubator - MHP accelerator. The largest innovative project is the biogas line using chicken droppings as the raw material
Kernel	Kernel is a leading investor in innovation. Back in 2016, the company decided to combine all the applied innovative products into a single #DigitalAgriBussiness project. The system integrates Kernel's internal and external IT services for global automation of production processes. Among the production innovations is the expansion of the company's port capacities (the port of Chornomorsk) and the formation of its own wagons fleet
Nibulon	The company has created a network of 27 modern cargo handling terminals and complexes for receiving, storing and shipping grain and oil crops. In terms of simultaneous grain storage capacity (2.1 million tons), this is currently the largest terminal network in Ukraine. In 2012 Nibulon opened a shipbuilding and ship-repairing plant - the only enterprise in Ukraine that builds full-fledged ships. In 2019, the company put into operation the longest specialized crane vessel of the "river-sea" class built during the independence of Ukraine, the 140-m Nibulon Max. Now the company's fleet includes 76 vessels. The total amount of investments in the implementation of the company's projects in this marketing year reached 85 million dollars
Astarta	The company has developed an agribusiness management portal that includes consolidated data on the company's land bank and helps various departments to communicate. The company also develops IT solutions which helps solving a number of production tasks
IMC	IMC initiated the Ukrainian Farm Management courses. The project was supported by 16 of the country's largest agricultural companies

receive data from all devices updating them on fuel consumption, process speed, engine load level and other factors with the help of the latest GPS trackers [3] (Fig. 1).

We will analyze the volume of export-import operations in the pre-war period. In recent years, imports into Ukraine have stably exceeded exports, and their difference sometimes reached 8% of GDP [3].

According to the results of 2022, the GDP drop in Ukraine is estimated to be 30.4% [1] due to the hostilities. This is the worst indicator in the entire history of the country's independence. Since some territories are temporarily occupied, it became necessary to optimize agricultural areas in order to support the state food security program.

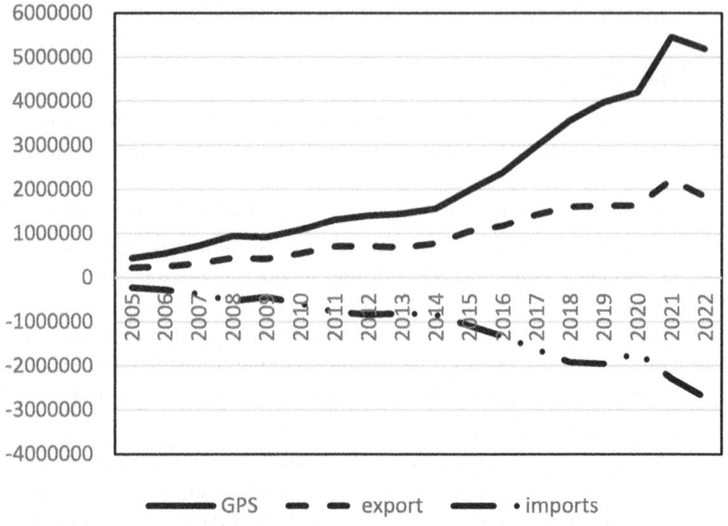

Fig. 1. Foreign trade balance of Ukraine from 2005 to 2021 (million hryvnias).

2 Presenting Main Material

Optimization of agricultural areas in order to support the state food security program. According to the data of the State Statistics Service [1] for 2021, the most profitable were agriculture, forest and fish industries, with the profitability comprising 18.4%.

The largest profit was obtained by agricultural enterprises that grew: sunflower - about 21.1 billion hryvnias, wheat - 19.0 billion hryvnias, corn for grains - 11.9 billion hryvnias, rapeseed - 6.6 billion hryvnias, as well as milk - 3. 8 billion hryvnias.

At the same time, according to the data of the State Statistics Service, sunflower oil of Ukrainian production on the domestic market rose in price by 64%, sugar - by 68%, eggs - by 34%, bread - by 17%, meat - by 10.5%. At the same time, according to the statistics for 2020, the income of Ukrainian families increased by only 2.6% [3]. Among the agricultural crops of Ukraine, wheat occupies more than half of the cultivated areas. In recent years, Ukraine has entered the top ten major producers and has become one of the world's leading wheat exporters.

Wheat imports to Africa, Southeast Asia and the Western Hemisphere were expected to increase in 2022/23. Wheat is one of the most important crops for food security globally, as well as a source of income for part of the Ukrainian population.

Export of flour in 2022 decreased to 79 thous. tons (25% less compared to 2021 y.). The main reason for the decrease in export volumes was the blockade of sea ports, which created the necessity to the search for new alternative sales channels. Since the beginning of 2022/23 year 80 thousand tons were exported, where 36.6 thous. tons - to EU countries, which is 437 times more than in the same period of the previous year.

Buckwheat is an important crop for Ukrainians. In 2021, 86.4 thous. tons of buckwheat were harvested from an area of 65.3 thous. hectares. The average yield is 1.32 t/ha. In 2020, Ukrainian farmers harvested 99,000 tons of buckwheat. In 2019, buckwheat production reached 85.1 thous. tons, compared to 137 thous. tons in 2018 y. The average yield of buckwheat in 2019 was 1.23 tons/ha, compared to 1.21 tons/ha in 2018 (in 2017 y.—0.97 t/ha, in 2016 y.—1.15 t/ha).

This year, farmers harvested an area of 69.3 thous. hectares (113.3 thous. hectares in 2018). In 2018, buckwheat production amounted to 137,000 tons compared to 180,400 tons in 2017.

Buckwheat prices in Ukraine have already dropped by 15%, however, some experts believe that the price for the popular crop will not decrease any more. In addition, as early as in 2023, a shortage of buckwheat in Ukraine is expected and the price for the product will increase.

The domestic market was always assumed to be 40 million people, but now it is actually 30 million people. The domestic buckwheat market is 120,000 tons. 3 kg of buckwheat are consumed per capita in a year, which means that only 90,000 tons are necessary for 30 million population.

It was noted that in peacetime the surplus of the buckwheat could be exported, but only some countries of the former USSR and China consume buckwheat. It is very far to take buckwheat to China, and Europe's consumption is rather small.

Therefore, producers will have to sell the entire harvest in Ukraine, in this case they will earn almost nothing or just get the amount of the cultivation costs.

The area of buckwheat cultivation in Ukraine has decreased 20 times over the past twenty years - from 570,000 ha in 2000 to 60,000 ha this year.

The Cabinet of Ministers of Ukraine by its order approved the action plan to ensure food security in the conditions of martial law.

First of all, it envisages the monitoring of the conditions of food security and agricultural infrastructure as a whole. Secondly, a state support will be provided to national food producers. Thirdly, the most socially vulnerable population will receive targeted food benefits. In addition, monitoring and appropriate regulation of prices for food will be under special control of the state.

Special attention is paid to the optimization of areas for certain crops.

This year, the area for buckwheat cultivation will increase both due to the need to implement the state food security program and due to the expected demand for the product.

In recent years, the demand for buckwheat comprised 3.5 kg per capita in a year. For comparison, the highest level of this crop consumption is in Latvia and comprises more than 7 kg per capita in a year.

Ukraine's need for buckwheat as a strategic product comprises 250,000 tons. Due to hostilities in the south-east, the area of peas cultivation has decreased from the usual 240,000–250,000 ha to 118,000 ha.

Drought in Canada has had a devastating effect on the pea crops, causing prices' increase for the product globally. Meanwhile, steady international demand and dwindling stocks could push up pea prices (see Fig. 2).

As the demand for pea protein has increased dramatically mainly due to the growing interest in plant protein alternatives as a substitute for meat. The protein market continues to grow every day, and that's what creates the demand.

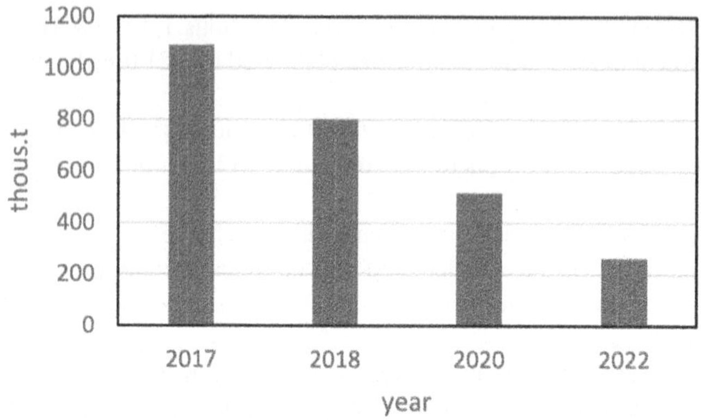

Fig. 2. Peas production 2017–2022 y.

In the current situation, it is much more profitable for Ukrainian farmers to sell more expensive goods or products with added value, therefore, the legumes production in Ukraine can still have new developments.

Theoretical issues of land management have been developed in the works of such foreign scientists: Lee, C [5], Sánchez-Fung J. [6], Rasool, H [7], Dorward, A. [8], Sitikantha, P., [9], Emmanuel, M. [10], Biber, P [11], Bhattacharya R, [12], Nwaolisa E. [13]. The works of Takha Kh. [14].

Let's consider the method of the land distribution optimizing in wartime. Let's consider the following agricultural crops: wheat, buckwheat, peas. Since the future prices on the market are not fixed, the decision must be taken based on the following two criteria: maximizing expected total net income and minimizing the variance of total net income. The Lagrange method is used to solve the problem. The methodology is developed for the case of probabilistic nature of the future market prices for products. The methodology makes it possible to take inflation into account.

Since future market prices are not fixed, the decision must be taken based on two criteria: maximizing expected total net income and minimizing the variance of total net income. The Lagrange method is used to solve the problem. The developed method is for cases when the price for agricultural products is of a probabilistic nature. We find the maximum income according to formula (1):

$$\overline{z}_{max} = \sum\nolimits_{i=1}^{N} (p_i - c_i)x_i \tag{1}$$

$\overline{z}_{max}$- maximum profit, p_i– the selling price of a product unit at a moment in time, c_i- expenses, x_i - volume of agricultural crops production; $\sigma(z)_{\min}$- minimum variance, a- product scopes, N- the number of agricultural crops. Then maximum variance makes (2):

Next, the worst values of the criterion indicators are calculated for a set of effective variants of the calendar plan:

$$\sigma(z)_{min} = \frac{a}{\sqrt{\sum_{i=1}^{N} \frac{1}{\sigma_i^2}}} \tag{2}$$

$$\overline{z}_{min} = \frac{a}{\sqrt{\sum_{t=1}^{T} \frac{1}{\sigma_t^2}}} \sum_{t=1}^{T} \frac{\overline{p}_t - c_t}{\sigma_t^2},$$

We calculate the range of variation for criterion indicators (2):

$$(\overline{z}_{min} \leq \overline{z}_0 \leq \overline{z}_{max}, \sigma(z)_{min} \leq \sigma_0 \leq \sigma(z)_{max}) \tag{3}$$

The optimal calendar plan for the sale of agricultural products stocks is determined. This plan $x^* = \left(x_1^*, \ldots, x_T^*\right)$ is computed by solving a convex programming problem (3):

$$s \to max,$$

$$\sum_{t=1}^{T} (\overline{p}_i - c_i)x_i \geq \overline{z}_0 + s(\overline{z}_{max} - \overline{z}_0)$$

$$\sum_{i=1}^{N} \sigma_i^2 x_i^2 \leq \sigma_0^2 - s\left(\sigma_0^2 - \sigma_i^2(z)_{min}\right), \tag{4}$$

$$\sum_{t=1}^{T} x_i = a,$$

$x_2 \leq 250;$

$$x_i \geq 0, i = \overline{1, N}$$

It shall be mentioned that the optimum value of s^* will indicate whether the product owner chose realistic admissible levels for criterial indices (at $s^* \geq 0$), or not (at $s^* < 0$).

With the help of the Lagrange multipliers method, the necessary conditions are established that allow the identification of optimal points in optimization tasks with constraints-equalities. At the same time, a problem with constraints turns into an equivalent problem of unconditional optimization with some unknown parameters, called Lagrange multipliers.

The Lagrange multiplier method is the most developed and most often used in economics.

The idea behind the Lagrange multiplier method is to replace the original problem with a simpler one. To do this, replace the objective function with another function with more variables, i.e. one that contains conditions represented as constraints. After such a transformation, the further solution of the problem is to find the extremum of the new function, which has no constraints on its variables. Every local minimum is a global minimum, the optimal set is convex, and if the objective function is strictly convex, the problem has at least one optimal point. Table 2 provides statistics on the harvest of crops as of December 1, 2021.

The general problem of conditional optimization with constraints-equalities is reduced to the problem of unconditional minimization using the Lagrange function, which is written down in the form of:

$$F(x1, x2, x3; \lambda_1, \lambda_2) = f(x1, x2, x3) + \sum_{i=1}^{K} \lambda_k h_k \tag{5}$$

Table 2. Harvesting of agricultural crops on the 1st of December, 2021.

Crops	Harvested area, thous. ha	Volume of production, thous. tons	Yield, centners from 1 hect. of harvested area/	Prices for crops UAH/t	Standard price variance, UAH
wheat	7, 055.5	327, 196.7	46.4	8,030.00	491.00
buckwheat	87.8	1, 133.5	13.0	42,000.00	1,339.00
peas	239.2	5, 819.3	24.3	13,000.00	520,00

An obligatory condition for using an expression (4) is that the number of constraints must be less than the number of variables. If this condition is not met, then there is no optimization problem, because the number of variables' connections exceeds their number. Thus, the task is reduced to finding the minimum of the function (6):

$$\frac{\partial F(x_j; \lambda_k)}{\partial x_j} = 0 \tag{6}$$

So, at this stage, we will first find the best values of the criterial indicators for expected total net income:

$$\bar{z}_{max} = 32719{,}67 * 8030 + 113{,}33 * 42000 + 581{,}9 * 13000 = 275064{,}740(\text{mln. UAH})$$

standard variance of total net income:

$$\sigma(z)_{min} = \mathbf{18346{,}9(\text{млн. грн})}$$

and then their worst values on the set of effective calendar plan options

$$\bar{z}_{min} = 114033{,}8021(\text{mln.UAH})$$

Let the acceptable levels of our problem be as follows:

$$\bar{z}_0 = 114033{,}8021(\text{mln. UAH}); \quad \sigma_0 = 18346{,}9(\text{mln. UAH})$$

We solve our problem using the method of Lagrange multipliers.

To calculate it in MathCAD, write down the maximum income using the formula:

$$F(x1, x2, x3, \lambda_1, \lambda_2) = 8030\text{x}1^2 + 42000\text{x}2^2 + 13000\text{x}3^2$$

where dispersion

$$\text{u}(\text{x}1, \text{x}2, \text{x}3) = 491\text{x}1 + 1339\text{x}2 + 520\text{x}3$$

System of partial differential equations:

$$\frac{dF(x1, x2, x3, \lambda_1, \lambda_2)}{dx1} \rightarrow 491 - 16060\lambda_1 x1 - \lambda_2$$

$$\frac{dF(x1, x2, x3, \lambda_1, \lambda_2)}{dx2} \rightarrow 1339 - 84000\lambda_1 x2 - \lambda_2$$

$$\frac{dF(x1, x2, x3, \lambda_1, \lambda_2)}{dx3} \rightarrow 520 - 26000\lambda_1 x3 - \lambda_2$$

$$\frac{dF(x1, x2, x3, \lambda_1, \lambda_2)}{d\lambda 1} \rightarrow 7436x1^2 - 1132x2^2 - 13000x3^2$$

$$\frac{dF(x1, x2, x3, \lambda_1, \lambda_2)}{d\lambda 2} \rightarrow 275064 - x1 - x2 - x3$$

The results of the calculations were as follows.

Optimal areas for: wheat - 7,000 thous. hectares; buckwheat - 250 thous. hectares, peas - 240 thous. hectares; expected total net profit: 2.820 billion UAH.

We also observe that there has been an improvement in the acceptable levels of criterial indicators originally chosen by the owner: the amount of expected total net income has increased by 6,813,507 thous. UAH.

Such results can be obtained by using the Industry 4.0 innovative technologies, which will bring undeniable progress in all areas. As a result, we will increase production and effectively manage resources; make quality decisions. An integrated production process formed on the basis of real data saves production time. Speaking about Ukraine, it should be noted that although there are some lags behind other countries in the process of the latest technologies introduction and creation, the national economy still has the potential for development. Not only large agricultural holdings, but also small agricultural companies in Ukraine are massively implementing robotic processes, quadcopters, motion sensors and sensor technologies. This indicates an increase in the level of advanced technologies availability for all agricultural companies.

The Scientific Novelty of this Study

The paper analyzes the wartime economy of Ukraine, its losses and problems. Following the order of the Cabinet of Ministers of Ukraine on the plan of measures to ensure food security under martial law, the optimization of areas for different crops is proposed. For this purpose, an economic and mathematical model was developed using the Lagrange multiplier method.

3 Conclusions

The paper analyzes the wartime economy of Ukraine, its losses and problems. The Russian invasion has dealt a real blow to Ukraine's agriculture. The total direct and indirect losses to agriculture exceed $40 billion, according to analysts at KSE Agrocenter and the Department of Agricultural Policy. Both crop and livestock production have suffered from the war. A large number of farmlands are still being exploited.

Following the order of the Cabinet of Ministers of Ukraine on the action plan to ensure food security under martial law, the optimization of areas for different crops was proposed. An economic and mathematical model was developed using the Lagrange multiplier method.

Innovative projects in the agricultural sector of Ukraine are indicated in the paper. The optimization of the area of the whole Ukraine should be combined with the use of the latest Industry 4.0 technologies in the agricultural sector.

Although the grain export corridor is open, it can hardly be called reliable, given the unstable behavior of the Russian Federation, and it has not led to a significant increase in domestic prices for major crops.

Food security programs are a multi-criteria problem. Further consideration should be given to the digitalization of agricultural enterprises, the use of communication innovations, ports, and many other factors that will contribute to solving the food security problem.

References

1. Ministry of Agrarian Policy and Food of Ukraine. Official website. https://minagro.gov.ua/en
2. FAOSTAT: Statistical database of food and agriculture organization of the United Nations (2022). http://faostat.fao.org/faostat/. Accessed 15 Mar 2022
3. Statistical Yearbook of Ukraine 2021. State Statistics Service of Ukraine. http://www.ukrstat.gov.ua. Accessed 2 Feb 2022
4. Socio-economic characteristics of households in Ukraine 2021. State Statistics Service of Ukraine. http://www.ukrstat.gov.ua. Accessed 2 Feb 2022
5. Lee, C., Schluter, G., O'Roark, B.: How much would increasing the minimum wage affect food prices? Agric. Inf. Bull. **3**, 747 (2000). https://doi.org/10.22004/ag.econ.33598
6. Sánchez-Fung, J.: Institutions for macroeconomic stability: a review of 'Monetary policy in low financial development countries'. Macroecon. Finance Emerg. Mark. Econ. **15**(1) (2022). https://doi.org/10.1080/17520843.2022.2096913
7. Rasool, H., Tarique, Md.: Testing the link between rural wages and food inflation in india a co-integration approach. IOSR 2 (IOSR-JBM) **20**(1), 87–98 (2018). https://doi.org/10.9790/487X-2001018798
8. Dorward, A.: The short- and medium-term impacts of rises in staple food prices. Food Secur. **4**, 633–645 (2012). https://doi.org/10.1007/s12571-012-0210-3
9. Sitikantha, P., Silu, M., Soumyajit, R.: Inflation expectations of households: do they influence wage-price dynamics in India? Macroecon. Finance Emerg. Mark. Econ. **13**(3), 244–263 (2020). https://doi.org/10.1080/17520843.2020.1720264
10. Emmanuel, M.: Optimisation of benefits from agriculture and wildlife land uses by wards in CAMPFIRE areas in Zimbabwe. AFBM J. **14**(2), 120–136 (2019). https://doi.org/10.22004/ag.econ.292469
11. Biber, P, Duckett, T.: Dynamic maps for long-term operation of mobile service robots. In: Robotics: Science and Systems I, June 8–11, pp. 17–24 (2005). https://doi.org/10.15607/RSS.2005.I.003
12. Bhattacharya, R., Sen Gupt, A.: Drivers and impact of food inflation in India. Macroecon. Finance Emerg. Mark. Econ. **11**(2), 146–168 (2018). https://doi.org/10.1080/17520843.2017.1351461
13. Nwaolisa, E.: Effects of exchange rate fluctuations on the balance of payment in the Nigerian economy. Approach IOSR J. Bus. Manag. (IOSR-JBM) **5**(12), 7576–7583 (2017). https://doi.org/10.18535/ijsrm/v5i12.01
14. Taha, H.: Operations Research: An Introduction Pearson, 10th edn., p. 848 (2016)

Optimization of Control Activities Using Poka Yoke Tool on the Example of a Selected Production Company

Paweł Jokiel[1], Magdalena Dąbrowska[1](✉), Phatchani Srikhumsuk[2], Jozef Huzar[3], Piotr Motyka[1], Daniel Medyński[1], and Krzysztof Kolbusz[1]

[1] Faculty of Technical and Economic Sciences, Witelon Collegium State University, Sejmowa 5A, 59-220 Legnica, Poland
magdalena.dabrowska@collegiumwitelona.pl

[2] Faculty of Science and Technology, Rajamangala University of Technology Krungthep, Bangkok 10120, Thailand

[3] Department of Industrial Engineering and Informatics, Faculty of Manufacturing Technologies, Technical University of Košice, Bayerova 1, 08001 Prešov, Slovakia

Abstract. The concept of waste is inseparable from the concept of Lean Management. The literature on the subject has many definitions of the term Lean Management. The essential feature of this concept is the pursuit of economic resource management, and as a result, to reduce costs, as well as improve quality and shorten the time of order completion by consistently eliminating waste. According to reference literature, the essential tools in Lean Production are 5S, Kaizen, Just in Time, Just in Sequence, Kanban (pull system), Single Minute Exchange of Die (SMED), Total Productive Maintenance (TPM), Value Stream Mapping (VSM), standardization, muda, Total Quality Management (TQM), takt time, heijunka, jidoka, andon (visual management), Poka-Yoke. The article aims to analyze the control activities performed by the employee at the measuring station and its optimization using Poka-Yoke tools to shorten the time and amount of performed activities.

Keywords: Lean management · Production process · Poka Yoke · Waste

1 Introduction

The concept of waste is inseparable from the concept of Lean Management. There are many definitions of Lean Management in the literature on the subject. Based on selected definitions, the Lean Management concept is perceived as a strategy, concerning others as a philosophy and still others as a method [2, 4–6, 12, 14, 17, 21, 32, 34, 47, 48]. The origins of the Lean concept date back to the 1950s, although the concept of Lean itself was first used in 1988 by J. Krafcik and was used to show the difference between the classic system of mass production (traditional management) and management based on the pillars of Toyota Production System - TPS (lean management) [15], whereas the concept of Lean Management was first officially named and defined in 1990 [45]. This

A. Burduk et al. (Eds.): ISPEM 2023, LNNS 790, pp. 585–603, 2023.
https://doi.org/10.1007/978-3-031-45021-1_44

concept was initiated in Japan in the automotive industry. Its founder was the Toyota company, which formed a new way of not only acting but also thinking [18, 19]. The Toyota Production System was based on: the complete elimination of waste, striving to achieve the lowest possible costs, ensuring the highest quality, and executing the process in the shortest possible time [28, 29]. The evolution of the Toyota production system was presented, among others, in the following: [7, 8] (Table 1).

Table 1. The development of Toyota Production System

Year	Task	Characteristics
1948	Implementation of the suction system in the engines department	• raising an order to go back to the prior functional position in order to accumulate a stock of parts crucial for production • seizing only the amount of goods, which is required once in the machining process and in the moment when it is mandated • the elimination of the interoperation stocks through the insightful commodities control
1949	Usage of the suction system in the treatment department	• the exclusion of the supplies appearance in the machining department • the necessity to operate several machines by one worker • employees conduct personal control, which increases the quality of manufactured products
1950	Advancing of the pull system to the marketing activities, departments synchronization, and implementation of the triple lights system	• adaptation of the production policy exclusively for vehicles ordered by customers from dealers • further reduction of the interoperation commodities • shortening the time of notifying the supervision services about the occurring issues
1953	Introduction of the kanban system (more about the kanban system see e.g. in: [13, 26, 41])	• exchange of the cards, which is a signal to renew production • to simplify the production, purchasing, and transport process, the company implemented parts and components standardization

(*continued*)

Table 1. *(continued)*

Year	Task	Characteristics
1955	Heijunka (more about the heijunka system see e.g. in: [10, 20]	• further reserves reduction through the controlling delivery of the parts to the production lines system • balancing production (both in terms of quantity and assortment) with a simultaneous better use of machines and devices
1957	Implementation of the triple light system in all production lines	
1961	Introduction by Toyota of the Kanban system to the factories of selected suppliers	
1962	Approach to the management in small sections	• extension of the existing kanban system to all branches and production corps • introduction of the system preventing errors, which assignment is to eliminate shortages and prevent overproduction • more effective usage of machines and devices, reduction of reserves, and reduction of producing lead times
1963	Boosting the employee productivity	• - employees operate an average of five machines
1965	Expansion of the Kanban system on the external suppliers	
1971	Reduction of the retooling of the stamping presses	• the reduction of the changeover time of stamping press to three minutes was obtained • the practices of transferring assembly workers between various operating positions were executed
1973	Integration of suppliers	

The purpose of the Lean Management concept, to show the difference between the classic mass production system and the management based on the Toyota Production System pillars, encounters its reference in many fields and areas of the company's activity.

According to J.K. Liker, the paramount feature of this concept is the pursuit of resource efficiency and, as a result, cost reduction, as well as improvement of quality and shortening of order fulfilment time by consistently eliminating waste as part of management based on a flat, flexible organizational structure [17].

Lean Management has many tasks, among which there is, e.g., levelling 3M (jap. muda, mura, muri). Muda means waste, mura means irregularity and changeability, and muri means overload [35].

T. Ohno, the creator of the Toyota Production System, TPS, originally distinguished seven primary sources of waste (Japanese muda) (Table 2). It should be emphasized that waste can be either necessary or unnecessary losses. Necessary losses include those activities that must be performed considering the implementation of the process but do not add any value to the product. They include, e.g., machine start-up, changeover, internal transport, workplace preparation, or document printing. These activities are often accepted but should be minimized. Necessary losses are a set of activities that should not occur at all. Examples include moving materials, excessive transport, looking for tools, waiting for a component, or making corrections.

Table 2. Primary sources of waste in the production process

Waste source	Causes	Impact on the company's functioning
Overproduction	• producing an assortment for which there are no orders • completing steps in the process faster than necessary • producing enough to use the machine's capacity	• overstaffing • increased storage costs • increased transport costs • surplus stocks • freezing of capital • accumulation of stocks at various stages of production
Inventory	• unbalanced production • a surplus of raw materials, work in progress, and finished products • - hiding problems, including those related to e.g., with defects, supplier delays, downtime	• - damage to products • aging of products • increasing transport costs • increasing storage costs • slowing down processes • uneconomic processes
Transport	• moving production in progress over long distances • placing materials, parts, or finished products in the warehouse and extracting them from it or moving them between different processes	• inefficient means of transport • the destruction of materials/products • deterioration of materials • loss of materials
Delays	• depletion of stocks • tracking of automatic devices • waiting for the next stage of processing • machine downtime • "narrow throats" • inactivity • lack of materials, information, delays in deliveries, failures of machines and devices • late or unprepared delivery of materials/parts, device failure, waiting for inspection, or waiting for information for the process • delays in the previous process	• reduction of production efficiency • deterioration of the flow • extending the delivery time

(*continued*)

Table 2. (*continued*)

Waste source	Causes	Impact on the company's functioning
Redundant movement	• searching for parts • searching for tools • relocation of tools • walking at the workplace • quasi-activity of employees	• extending the operation time • poor organization of workstations • improper ergonomics of workstations
Excessive processing	• taking unnecessary steps in the processing process • poor tool or product design • using too complicated procedures • use of inappropriate, inadequate, or outdated technologies • multiple parts cleaning • excessive documentation of the process • use of non-ergonomic tools • storing redundant copies of documents	• inefficient processing • unnecessary movements • defects in production • ensuring higher quality than necessary • taking unjustified, unnecessary steps in a given process
Deficiencies or defects	• production of defective parts • repairing defective parts • correcting defective parts	• additional control • additional employee involvement • increased costs of repairs, alterations, withdrawal of defective products, • the scrapping of products • the costs associated with defects increase with the time it takes to detect them • the product is not for sale

The article aims to analyze the control activities performed by the employee at the measuring station and its optimization using Poka-Yoke tools to shorten the time and amount of performed activities.

2 Poka-Yoke as an Optimization Tool in Lean Management

2.1 Poka Yoke - Definitions

According to reference literature, the most important tools in Lean Production are [22, 24, 27, 36, 43, 44, 45]: 5S, Kaizen, Just in Time, Just in Sequence, Kanban (system pull), Single Minute Exchange of Die (SMED), Total Productive Maintenance (TPM), Value Stream Mapping (VSM), standaryzacja, muda, Total Quality Management (TQM), takt time, heijunka, jidoka, andon (visual management), Poka-Yoke.

Practical elements of the implementation of Poka Yoke solutions and the importance of this issue about the practice of operating enterprises are introduced in many publications, including: [1, 3, 9, 16, 25, 31, 37, 40, 49].

In the literature, Poka-Yoke is defined as (Table 3):

Table 3. Chosen Poka-Yoke definitions

Author	Definition	Source
J. Grout	*"A poka-yoke is the use of process or design features to prevent errors or the negative impact of errors"*	[11]
P. Middleton	*"A poka-yoke is the systematic practice of eradicating errors, by locating their root cause."*	[23]
F. Pakdil et al.	*"Poka Yoke is a simple and economical device used at service and manufacturing process for mistakes prevention, which does not allow employees to fail."*	[30]
F. E. Plonka	*"A poka-yoke is a mechanism for detecting, eliminating, and correcting errors at their source, before they reach the customer"*	[33]
N. Shimbun	*"A poka-yoke is a technique for avoiding simple human error at work"*	[38]
S. Shingo	*„A poka-yoke is a mechanism for detecting errors and defects, which inspects 100% of the pieces, working independently on the operator's attention span"*	[39]
J.C. Ysou, W.J. Chen	*"Poka-Yoke uses devices on process equipment to provide 100% inspection and to prevent causes that results in defects"*	[42]

2.2 Poka Yoke - Methodology

Creator of the Poka-Yoke method – S. Shingo has classified Poka-Yoke tools, taking into account their functions and methods [39] (Table 4).

Table 4. Poka-Yoke functions and methods

Function	Method	Characteristic
Regulatory functions	Control methods	It consists of stopping the machine/process in case of a defect
	Warning methods	It consists in warning the operator that a defect has occurred through an audible alarm, a light alarm, etc.
Setting functions	Contact methods	It detects specific irregularities (changes in shape, color, weight, temperature, etc.)
	Fixe value methods	It detects errors by checking the number of moves and/or elements in operations where it is required to make the correct number of moves or use a certain number of elements
	Motion step methods	It detects irregularities when a given move is to be made at a specific time or in a particular sequence about subsequent operations

The stages of implementation of the Poka-Yoke tools are presented in the Fig. 1 [9]. SPT is a set all the documents, standards, instructions and guidance.

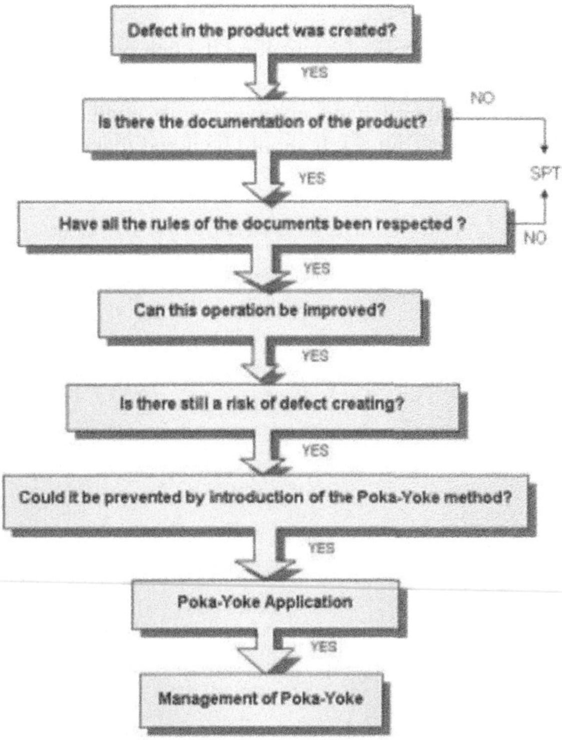

Fig. 1. The steps of the implementation of the Poka-Yoke tools

3 Description of the Research Problem

The surveyed company was originally part of a German plastics processing group, a supplier for the automotive, household appliances, and construction industries. The group's product range includes:

- in the automotive industry - superfinishing and sealing systems for acoustic and thermal insulation and absorption of engine linings, chassis, and additionally plastic structural elements and interior fittings for commercial vehicles;
- in the field of household appliances – products ensuring adequate acoustic insulation, mainly in dishwashers, refrigerators, and other household appliances;
- in the construction field - sound insulation, mainly in bridges and viaducts.

Currently, the company produces systems and products for bonding, sealing, damping, strengthening, and protection for the chemical, construction, and automotive industries. The audited company mainly cooperates with production plants located in Germany, Spain, the United States, and China and with partners located in South Korea and Japan.

The article presents the process of controlling the first piece of a manufactured item for an MB order. According to the customer's requirements, in the software used, the following values should be entered: batch number, length, width, thickness, weight, area, edges, and comments (if any) (Fig. 2). It should be noted that the language used in the software is Polish.

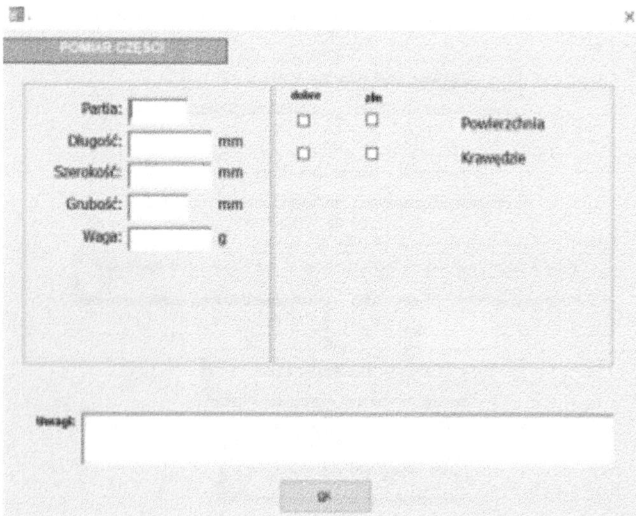

Fig. 2. Screenshot of the software used in the control process of the first piece of the manufactured element

Batch

The batch determines the order of the prepared recipe for the production of the bitumen mat and is assigned at the stage of creating the order. The employee inspecting the item sees the lot number on the information screen (Fig. 3).

Length, width

The temperature of the bitumen mat in the production process is 180 °C. It shrinks after cutting. For this reason, the element's length and width need to be controlled. The measurement is made on a control-measuring table using rulers mounted on the table. Length control requires precise positioning of the inspected item. Thus, the farthest edge determined the value. Measuring the width requires adding a ruler to extend the edge line of the part. Only then is the measurement made. The length of the elements required by the customer is 1200 ± 7 mm, and the width is 600 ± 5 mm. Figure 4 presents the method of measuring the length and width of the element.

Fig. 3. Screenshot of the software used in the control process of the first piece of the manufactured item - the batch number

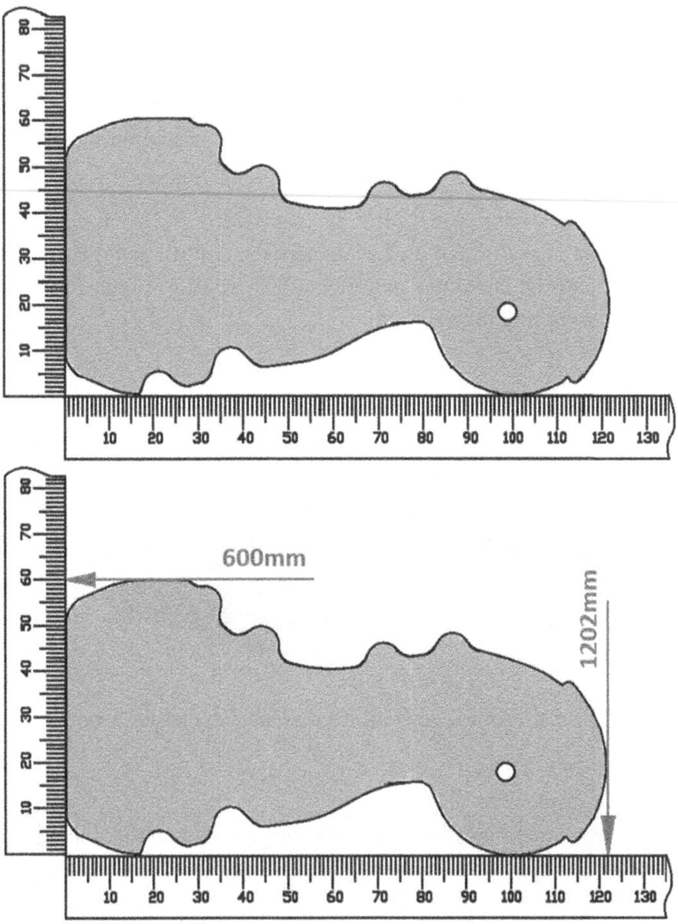

Fig. 4. The current method of measuring the length and width of an element

Thickness

Thickness measurement is performed using an electronic caliper previously zeroed and checked before measurement. The unworthy thickness of the element affects its weight. The thickness of the elements the customer requires is 12 ± 2 mm. Figure 5 shows the method of measuring the thickness of the element.

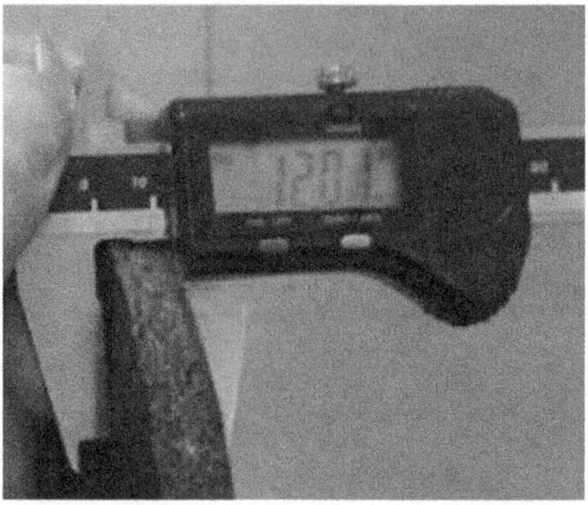

Fig. 5. Current method of the thickness measurement

Weight

Weight measurement is performed on a scale near the control and measuring table. The weight of the element the customer requires is 800 ± 80 g. Figure 6 shows the tared scale prepared for weighing the element.

Fig. 6. Current method of the measurement of element's weight

First Piece Control Card

The dimensions are entered into the computer program. Then the document is printed, which, together with the measured element, is put on the shelf for further analysis by the quality control department. Figure 7 shows the first piece of the document made during the hourly inspection.

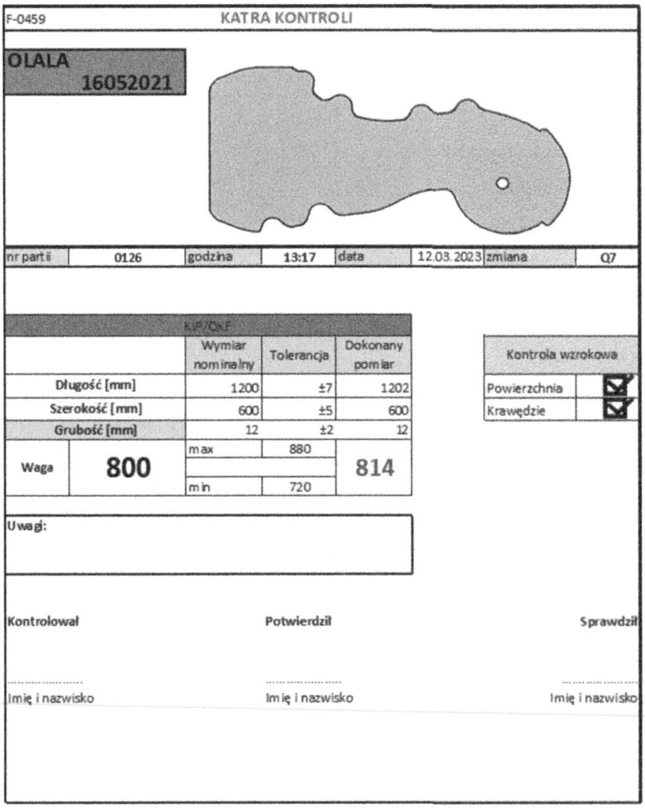
F-0459 KATRA KONTROLI

OLALA 16052021

nr partii	0126	godzina	13:17	data	12.03.2023	zmiana	Q7

KJP/OKF			
	Wymiar nominalny	Tolerancja	Dokonany pomiar
Długość [mm]	1200	±7	1202
Szerokość [mm]	600	±5	600
Grubość [mm]	12	±2	12
Waga 800	max 880 / min 720		814

Kontrola wzrokowa	
Powierzchnia	☑
Krawędzie	☑

Uwagi:

Kontrolował .... Imię i nazwisko

Potwierdził .... Imię i nazwisko

Sprawdził .... Imię i nazwisko

Fig. 7. First piece control card

The list of activities, along with their duration during the measurement of the first piece, is presented in Table 5. The time to inspect one element is 121 s and requires 21 activities.

Table 5. List of measurement activities before using Poka-Yoke

No	Activity	Number of moves	Duration of the activity [sec.]
1	Picking parts from the tape	1	5
2	Checking the batch number	1	2
3	Placing the part on the control-measurement table	1	7
4	Program opening	1	4
5	Saving the batch number in the program	1	7
6	Length measuring	1	9
7	Saving the result in program	1	7

(continued)

Table 5. (*continued*)

No	Activity	Number of moves	Duration of the activity [sec.]
8	Part offset	1	6
9	Downloading the auxiliary ruler	1	3
10	Using the auxiliary rules	1	3
11	Making a width measurement	1	9
12	Saving the result in program	1	7
13	Retrieving parts from the control and measurement table	1	6
14	Caliper download	1	3
15	Making a thickness measurement	1	9
16	Saving the result in program	1	4
17	Transporting the item on the scale	1	7
18	Weighting an item	1	6
19	Saving the result in program	1	7
20	Picking a part from the weighing scale	1	3
21	Putting the parts back on the shelf	1	7
TOTAL		**21**	**121**

The worker's path covered during the measurement activities of the first piece was presented by using the spaghetti diagram (Fig. 8).

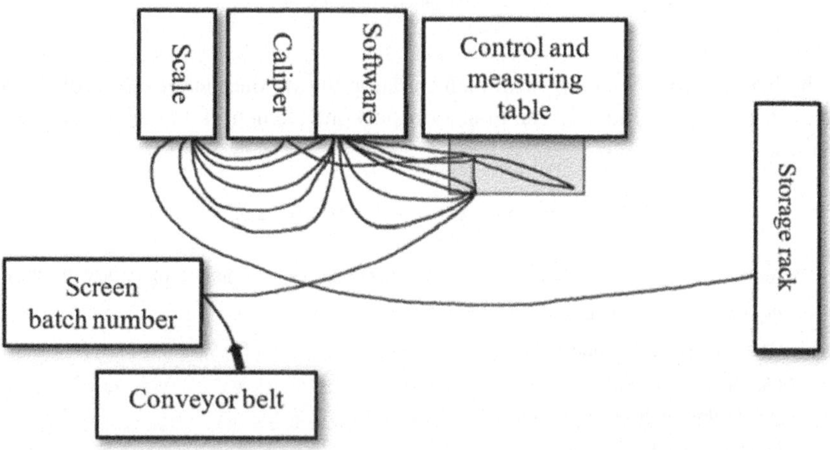

Fig. 8. Spaghetti diagram of the worker's path covered during measurement activities

4 Implementation of Poka Yoke Solutions in the Researched Production Company

Poka-Yoke1 – Design, and Execution of the Element Control Template

In order to improve the measurement process of the element in terms of determining its length and width, a template was designed and made, taking into account the tolerance ranges marked with colored lines. The outline in which the element was to be located was marked with a dashed line. The measured element was moved to the mounted limiters, which positioned it. The measurement consisted in observing the template. An important issue was the visibility of colored lines that marked the boundary of the element's dimension (Fig. 9).

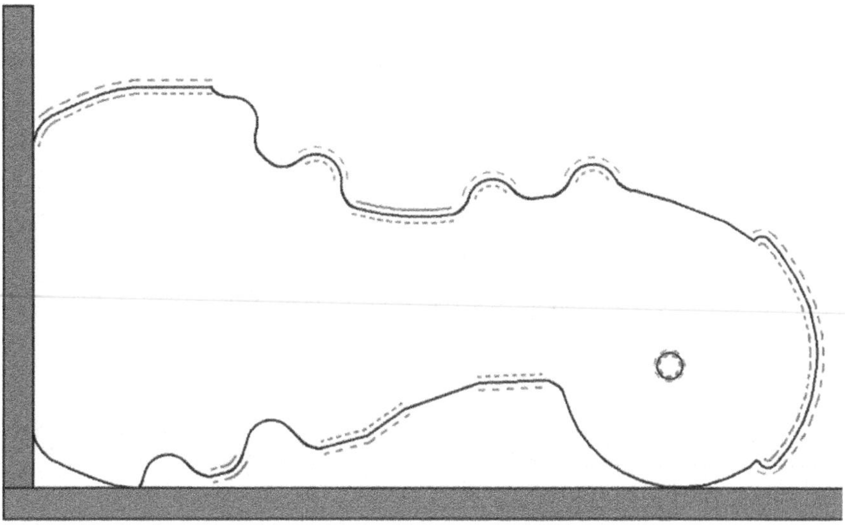

Fig. 9. Item control template

Poka-Yoke 2 – Placing the Element Control Template on the Balance

The next step was to place the control template on the balance. Taring the balance with the applied template enabled the close reading of the result (Fig. 10).

Poka-Yoke 3 - Reorganization of the Control and Measurement Station

The next step was the reorganization of the control and measurement station. The balance with the caliper was placed on the control and measuring table (Fig. 11).

Poka-Yoke 4 - Optimization of the Process of Recording Measured Elements

Recording measured parts has been optimized by downloading the batch number from another program and automatically saving it in the control card. The same solution was used for the weight of the measured element. The scale was connected to the computer, and after stabilizing the result, it was automatically sent to the card. The system for entering data regarding the length and width was replaced with a selection button, in

Fig. 10. Control template on the balance

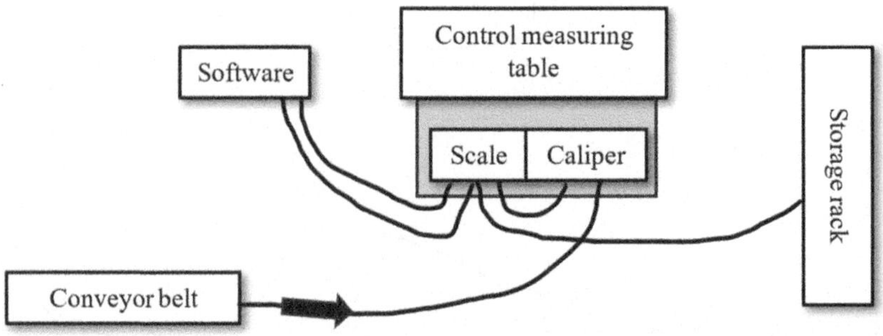

Fig. 11. Spaghetti diagram of the employee's path covered during measurement activities after the reorganization of the control and measurement station

which the employee, after analyzing the dimensions of the element with the template, made a record in the selected OK or NOK window (Fig. 12).

The list of activities, along with their duration, performed during the measurement of the first piece after applying the Poka-Yoke solutions, is presented in Table 6. The time to check one element was 54 s and required ten activities.

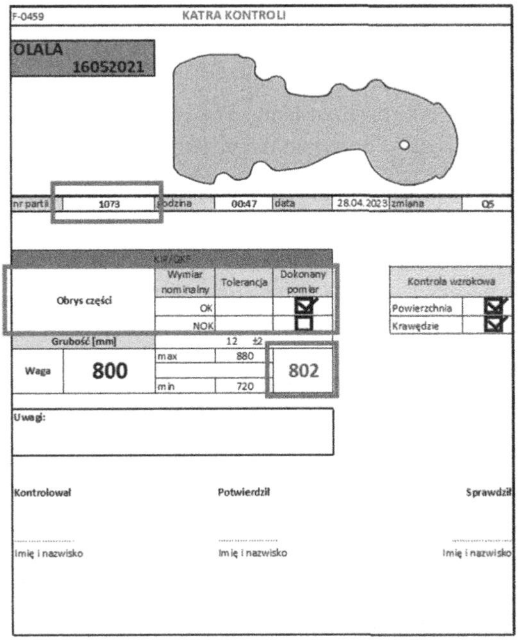

Fig. 12. First piece control card after software corrections

Table 6. List of measurement activities after using Poka-Yoke

Nr.	Activity	Number of moves	Duration of the activity [sec.]
1	Picking parts from the tape	1	5
2	Transportation of the part to the control-measurment table	1	7
3	Caliper download	1	3
4	Making a thickness measurment	1	9
5	Placing the parts on a scale with a template	1	3
6	Reading from the template	1	9
7	Program opening	1	4
8	Data saving	1	7
9	Picking a part from the weighing scale	1	3
10	Putting the parts back on the shelf	1	4
TOTAL		**10**	**54**

5 Conclusion

The article aimed to analyze the control activities performed by the employee at the measuring station and its optimization using Poka-Yoke tools to shorten the time and amount of performed activities. Based on the observations, the following were made:

- analysis of the control process of the first piece of the manufactured element for the MB order,
- diagnosis of activities performed by the employee during the control process of the first piece,
- measurement of the duration of activities performed by the employee during the inspection process of the first piece,
- analysis of the employee's route covered during measurement activities,
- reorganization of the control and measurement station,
- design and implementation of Poka-Yoke solutions concerning the control process and software.

The implemented solutions enabled the following:

- shortening the time of performing control activities from 121 s to 54 s,
- reducing the number of activities an employee performs in the control process from 21 to 10.

It should be emphasized that the implementation of the Poka Yoko tool brought many benefits. They have a different character:

- technical benefits:
  - implementation close to the source of the error,
  - simplicity of solutions,
- organizational benefits:
  - improving measurement,
  - acceleration of control activities,
  - shortening the time of performing control activities,
  - reducing the number of activities an employee performs in the control process,
  - increasing production capacity,
  - elimination of many operations related to quality control,
  - promoting activities related to work improvement,
  - immediate action when a problem occurs,
  - 100% control in the process,
  - performing quality control of the product included in the production process,
  - no significant impact on process time,
- financial benefits:
  - reducing the number of failures in processes,

  - improving the quality of products,
  - reducing the number of customer complaints
  - elimination of many operations related to quality control,
  - reduction of costs related to the time of quality control,
  - low installation and operating costs,

- personal benefits:

  - increasing work comfort,
  - less time spent on employee training.

References

1. Ahmad, A.A., Rashid, A.A., Wong, F.R., Iqbal, M.: Worker safety improvement at paper pleating production line using Poka-Yoke concept-a case study in automotive industry. J. Mech. Eng. (JMechE) **5**, 183–196 (2017)
2. Awad, M.M., Hashem, A.E., Naguib, H.M.: The impact of lean management practices on economic sustainability in services sector. Sustainability **14**(15), 9323, 1–27 (2022)
3. Bălan, E., Jantă, L.M.: Solving quality problems with the Poka-Yoke tool assistance. Case study. Ann. Acad. Roman. Sci. Ser. Eng. Sci. **11**(1), 5–16 (2019)
4. Bertagnolli, F.: Lean Management. Springer, Wiesbaden (2018). https://doi.org/10.1007/978-3-658-13124-1
5. Bhasin, S.: Lean Management Beyond Manufacturing, vol. 10. Springer, New York (2015). https://doi.org/10.1007/978-3-319-17410-5
6. Charron, R., Harrington, H.J., Voehl, F., Wiggin, H.: The Lean Management Systems Handbook, vol. 4. CRC Press, Boca Raton (2014)
7. Dąbrowska, M., Jagaczyński, D.: The reduction of machine changeover time by using the SMED method – case study. In: Antosz, K., Sa, J.C., Jasiulewicz-Kaczmarek, M., Machado, J. (eds.) Lean Thinking in Industry 4.0 and Services for Society), pp. 123–143 (2023)
8. Dąbrowska, M., Maciejewska, K.: Optimising the sealing station in production process of tubes for catalytic converters involving SMED method - case study. Technologia i Automatyzacja Montażu Assem. Tech. Technol. **2**, 15–23 (2022)
9. Dudek-Burlikowska, M., Szewieczek, D.: The Poka-Yoke method as an improving quality tool of operations in the process. J. Achiev. Mater. Manuf. Eng. **36**(1), 95–102 (2009)
10. Furmans, K.: Models of heijunka-levelled kanban-systems. In: 5th International Conference on Analysis of Manufacturing Systems-Production and Management, pp. 243–248 (2005)
11. Grout, J.: Mistake-proofing the design of health care processes. Agency for Health Care Research and Quality (2007)
12. Helmold, M.: Lean Management and Kaizen. Springer, Cham (2020). https://doi.org/10.1007/978-3-030-46981-8
13. Huang, C.C., Kusiak, A.: Overview of Kanban systems. Comput. Integr. Manuf. **9**(3), 169–189 (1996)
14. Jackson, T.L.: Implementing A Lean Management System. CRC Press, Portland (1996)
15. Krafcik, J.F.: Triumph of the lean production system. Sloan Manag. Rev. **30**(1), 41–52 (1988)
16. Lazarevic, M., Mandic, J., Sremcev, N., Vukelic, D., Debevec, M.: A systematic literature review of Poka-Yoke and novel approach to theoretical aspects. J. Mech. Eng. **65**(7–8), 454–467 (2019)

17. Liker, J.K.: The Toyota Way. McGraw-Hill, Madison (2004)
18. Liker J.K.: The Toyota Way. 14 Management Principles from the World's Greatest Manufacturer. McGraw Hill, New York (2004)
19. Liker, J.K., Meier, D.: Toyota Way Fieldbook. McGraw-Hill Education, New York (2006)
20. Lippolt, C.R., Furmans, K.: Sizing of Heijunka-controlled production systems with unreliable production processes. In: Koch, T. (ed.) Lean Business Systems and Beyond. IFIPAICT, vol. 257, pp. 11–19. Springer, Boston (2008). https://doi.org/10.1007/978-0-387-77249-3_2
21. Lizak, M.: Methods of measuring the effectiveness of lean management. Prod. Eng. Archiv. **13**(4), 31–34 (2016)
22. Mayr, A., et al.: Lean 4.0-a conceptual conjunction of lean management and Industry 4.0. Procedia CIRP **72**, 622–628 (2018)
23. Middleton, P.: Lean software development: two case studies. Softw. Qual. J. **9**, 241–252 (2001)
24. Mikhailovsky, P., Plakhin, A., Ogorodnikova, E., Kochergina, T., Guseva, T., Selezneva, M.: Lean management tools to improve the production system. Calitatea **21**(176), 65–68 (2020)
25. Miralles, C., Holt, R., Marin-Garcia, J.A., Canos-Daros, L.: Universal design of workplaces through the use of Poka-Yokes: case study and implications. J. Ind. Eng. Manag. (JIEM) **4**(3), 436–452 (2011)
26. Monden, Y.: Adaptable Kanban system maintains just-in-time production. In: Monden, Y. (ed.) Toyota Production System, pp. 15–35. Springer, Boston (1994). https://doi.org/10.1007/978-1-4615-9714-8_2
27. Naeemah, A.J., Wong, K.Y.: Selection methods of lean management tools: a review. Int. J. Product. Perform. Manag. **72**(4), 1077–1110 (2023)
28. Ohno, T.: How the Toyota production system was created. Japanese Economic Studies **10**(4), 83–101 (1982)
29. Ohno, T.: Toyota Production System: Beyond Large-Scale Production. CRC Press, Boca Raton (1988)
30. Pakdil, F., Özkök, O., Dengiz, B., Kara, I., Selvi, N., Kargi, A.: A systematic approach to reduce human and system-related errors causing customer dissatisfaction in a production environment. Total Qual. Manag. Bus. Excell. **20**(1), 129–137 (2009)
31. Parhizgar, M.M., Khabbaz, S., Foroghinia, K., Bakhshi Maleki, E.: Infallibility (Poka-Yoke) fundamentals for improving production processes, case study: an automotive parts manufacturing company. Eur. Online J. Nat. Soc. Sci. **3**(3), 475–486(2014)
32. Pettersen, J.: Defining lean production: some conceptual and practical issues. TQM J. **21**(2), 127–142 (2009)
33. Plonka, F.E.: Developing a lean and agile work force. J. Hum. Fact. Ergon. Manuf. **7**(1), 11–27 (1997)
34. Prado-Prado, J.C., García-Arca, J., Fernández-González, A.J., Mosteiro-Añón, M.: Increasing competitiveness through the implementation of lean management in healthcare. Int. J. Environ. Res. Public Health **17**(14), 4981, 1–25 (2020)
35. Radin Umar, R.Z., Tiong, J.Y., Ahmad, N., Dahalan, J.: Development of framework integrating ergonomics in Lean's Muda, Muri, and Mura concepts. Prod. Plan. Control, 1–9 (2023)
36. Sanders, A., Subramanian, K.R.K., Redlich, T., Wulfsberg, J.P.: Industry 4.0 and lean management – synergy or contradiction?. In: Lödding, H., Riedel, R., Thoben, KD., von Cieminski, G., Kiritsis, D. (eds.) APMS 2017. IFIPAICT, vol. 514, pp. 341–349. Springer, Cham (2017). https://doi.org/10.1007/978-3-319-66926-7_39
37. Saurin, T.A., Ribeiro, J.L.D., Vidor, G.: A framework for assessing Poka-Yoke devices. J. Manuf. Syst. **31**(3), 358–366 (2012)
38. Shimbun, N.: Poka-Yoke: Improving Product Quality by Preventing Defects. Productivity Press, Portland (1989)
39. Shingo, S.: Zero quality Control: Source Inspection and the Poka-Yoke System. Productivity Press (1988)

40. Solke, N.S., Sur, A.: Design and implementation of a safety Poka-Yoke for workers safety: a case study. In: Pant, P., Mishra, S.K., Mishra, P.C. (eds.) LNME, pp. 619–629. Springer, Singapore (2021). https://doi.org/10.1007/978-981-15-7779-6_58
41. Takahashi, K.: Comparing reactive Kanban systems. Int. J. Prod. Res. **41**(18), 4317–4337 (2003)
42. Tsou, J.C., Chen, W.J.: The impact of preventive activities on the economics of production systems: modelling and application. Appl. Math. Model. **32**(6), 1056–1065 (2008)
43. Wagner, T., Herrmann, C., Thiede, S.: Industry 4.0 impacts on lean production systems. Procedia CIRP **63**, 125–131 (2017)
44. Wolniak, R.: Relationships between selected lean management tools and innovations. Zeszyty Naukowe. Organizacja i Zarządzanie/Politechnika Śląska **75**, 157–166 (2014)
45. Womack, J.P.: Moving beyond the tool age (lean management). Manuf. Eng. **86**(1), 4–5 (2007)
46. Womack, J.P., Jones, D.T., Roos, D.: The Machine That Changed the World: The Story of Lean Production. Harper Collins Publishers, New York (1990)
47. Womack, J.P., Jones, D.T., Roos, D.: The Machine That Changed The Word. Free Press, New York (2007)
48. Yang, C.C., Yang, K.J.: An integrated model of the Toyota production system with total quality management and people factors. Hu. Fact. Ergon. Manuf. Serv. Ind. **23**(5), 450–461 (2013)
49. Zhang, A.: Quality improvement through Poka–Yoke: from engineering design to information system design. Int. J. Six Sigma Compet. Adv. **8**(2), 147–159 (2014)

Author Index

A

Antosz, Katarzyna 200

B

Balachandra, Kavith 228
Banaszak, Zbigniew 43
Baoyu, Dong 393
Berladir, Kristina 328
Bessa, João 300
Blanco, Danielle Barreto 515
Błażej, Ryszard 461
Bluszcz, Anna 60
Bocewicz, Grzegorz 43
Bodingerová, Michaela 159
Bryłkowski, Mateusz 533

C

Cao, Xiaofei 120
Chabior, Marcin 60
Cholewa, Mariusz 368
Chrobot, Jaroslaw 255
Cruz, João 268
Cunfu, He 393

D

Dąbrowska, Magdalena 585
De Silva, M. Mavin 228, 243
Dendera – Gruszka, Małgorzata 380
Dereń, Aldona Małgorzata 506
Dharmadasa, Nikila 243
Dimitrijevic, Sonja 28
Dopart, Dominika 406
Dostatni, Ewa 315
Dudkowiak, Anna 315
Dulina, Ľuboslav 159
Dworak, Kamil 560
Dybowski, Maciej 420

E

Esteves, João Sena 174, 268

F

Furmannová, Beáta 159

G

Giera, Julia 380
Gola, Arkadiusz 356
Golachowska, Barbara 281
Golachowski, Andrzej 281
Górski, Filip 315
Grabowik, Cezary 560
Gunaratne, Kalani 43

H

He, Cunfu 106
Helman, Joanna 368
Hewage, Harsha Chamara 43
Huzar, Jozef 585

I

Iskierka, Grzegorz 292

J

Jie, Gao 342, 393
Jokiel, Paweł 585
Jurdziak, Leszek 461

K

Kiełbowicz, Zdzisław 15
Kirjanów-Błażej, Agata 461
Kolbusz, Krzysztof 585
Kolbusz, Paula 200
Krenczyk, Damian 548
Krot, Kamil 292
Krzyżanek, Karolina 315
Kulińska, Ewa 380, 490

A. Burduk et al. (Eds.): ISPEM 2023, LNNS 790, pp. 605–607, 2023.
https://doi.org/10.1007/978-3-031-45021-1

L

Lakdinu, Manuka 228
Liang, Geng 342
Liu, Xiucheng 106
Liudmyla, Shulhina 476
Lukoszová, Xenie 490

M

Machado, José 215, 300, 515
Machado, Ricardo J. 300
Máchová, Mariana 159
Magalhães, Christophe 174
Majstorovic, Vidosav 28
Medyński, Daniel 585
Meller, Artur 356
Michulek, Tomáš 159
Miele, Lorenzo 93
Miskovic, Zarko 28
Mitrovic, Radivoje 28
Mohamed, Khaled 1
Molasy, Mateusz 368
Morais, Francisco 300
Motyka, Piotr 585

N

Nakayama, Tadachika 228, 243
Napiórkowski, Jaroslaw 75
Nielsen, Izabela Iwa 43, 243
Nielsen, Peter 228
Nowacki, Krzysztof 188
Nowicki, Tadeusz 445

O

Ottaviano, Erika 93

P

Pavlenko, Petro 146
Pereira, Ana Paula 174
Pereira, Filipe 215
Perera, H. Niles 228
Piechowski, Mariusz 356
Pihnastyi, Oleh 560
Podloch, Iwo 188
Poskart, Bartosz 292
Prządka, Przemysław 15

R

Rea, Pierluigi 93
Ribeiro, Pedro 300
Rodrigues, Marta 215
Rohman, Juni 406
Rojek, Izabela 133
Rosienkiewicz, Maria 368
Rusińska, Małgorzata 533
Rzeszowska, Aleksandra 461

S

Santos, André S. 328
Shendryk, Vira 146
Sikorsky, Igor 575
Silva, Hugo 328
Silva, Vinícius 174, 268
Simeunovic, Vladimir 28
Sirant, Marcin 434
Skonieczny, Jan 506
Smok, Anna 406
Soares, Filomena 174, 268
Soares, Nuno 300
Srikhumsuk, Phatchani 585
Stanik, Jerzy 75
Starzyńska, Beata 133
Stosic, Dragan 28
Szymańska, Elżbieta Jadwiga 490

T

Thibbotuwawa, Amila 43, 228, 243
Tomczak, Aleksander 315
Trojanowska, Justyna 146, 328
Tsesliv, Olga 575
Tubis, Agnieszka A. 406
Tunikowska, Joanna 15

U

Udoh, Salem 1

V

Varela, Leonilde R. 328, 515
Vicente, José 300, 515

W

Wang, Jinmiao 120
Wang, Yujue 106

Waszkowski, Robert 420, 434
Wicherek, Patryk 445
Woźna, Anna 533
Wu, Bin 106

X
Xavior, M. Anthony 560

Y
Yan, Lyu 342, 393

Z
Zhang, Yuejin 120
Zhou, Xinhua 106
Zhu, Ruijin 120

The manufacturer's authorised representative in the EU is Springer Nature Customer Service Centre GmbH, Europaplatz 3, 69115 Heidelberg, Germany. If you have any concerns regarding our products, please contact ProductSafety@springernature.com

Printed and bound by CPI Group (UK) Ltd, Croydon, CR0 4YY
30/09/2026
02225301-0017